Biochemistry of Nonheme Iron

Anatoly Bezkorovainy

Rush-Presbyterian-St. Luke's Medical Center
Chicago, Illinois

With a Chapter Contributed by

DORICE NARINS

PLENUM PRESS • NEW YORK AND LONDON

Library of Congress Cataloging in Publication Data

Bezkorovainy, Anatoly, 1935-
Biochemistry of nonheme iron.

(Biochemistry of the elements; v. 1)
Includes bibliographical references and index.
1. Iron metabolism. 2. Iron proteins. 3. Biological chemistry. I. Narins, Dorice. II. Title. III. Series. [DNLM: 1. Iron – Metabolism. QU130 B6144 v. 1]
QP535.F4B46 599.01'9214 80-16477
ISBN 978-1-4684-3781-2 ISI978-1-4684-3779-9 (eBook)
DOI 10.1007/978-1-4684-3779-9

Softcover reprint of the hardcover 1st edition 1980
A Division of Plenum Publishing Corporation
227 West 17th Street, New York, N.Y. 10011

Biochemistry of Nonheme Iron

BIOCHEMISTRY OF THE ELEMENTS

Series Editor: Earl Frieden
Florida State University
Tallahassee, Florida

Volume 1 BIOCHEMISTRY OF NONHEME IRON
Anatoly Bezkorovainy

A Continuation Order Plan is available for this series. A continuation order will bring delivery of each new volume immediately upon publication. Volumes are billed only upon actual shipment. For further information please contact the publisher.

**TO THE MEMORY
OF MY FATHER**

Foreword

Atomic biology has come of age. Interest in the role of chemical elements in life processes has captured the imagination of a wide spectrum of research scientists—ranging from the nutritionist to the biochemist, the inorganic chemist, and even to some biophysicists. This series, *Biochemistry of the Elements*, is a recognition of this increasing interest. When complete, the books will assemble the hard facts concerning the biochemistry of each element—singly or in a logical grouping. The series will provide a permanent reference for this active and growing field. Each volume shall represent an integrated effort by one or several authors to describe the current knowledge of an element(s) or the methods by which it is studied.

Iron is the only element to which we have devoted two volumes, because its biological and chemical role is so versatile and complex. These two volumes will treat the two main structural categories of iron: heme and nonheme iron. In this volume on nonheme iron, the first in the series, Anatoly Bezkorovainy provides what we believe is the most comprehensive treatment of this important topic.

Florida State University Earl Frieden

Preface

Professor Earl Frieden, editor of the *Biochemistry of the Elements* series (of which this book is the first volume), has charged the authors of individual volumes with the task of writing their respective works in textbook style, so that such volumes can be used as instructional tools for the various programs in the life sciences. To this author's best knowledge, no books are available that purport to present the subject of nonheme iron biochemistry in a rational and systematic fashion. In keeping with Professor Frieden's intent then, this author has endeavored to assemble the material on biologically occurring nonheme iron compounds and to present it in a concise and hopefully noncomplex language that would be understandable to advanced undergraduate and graduate students in the life sciences, to medical students, and to house staff officers. It is assumed, however, that the reader will have a command of the basic biochemical principles. This author has also made an attempt to be as definitive as possible, though many areas in the field of nonheme iron biochemistry are in such a state of flux that it is only possible to present the various arguments relative to the issues at hand.

The scope of this volume is limited. It is not a clinical manual, though certain clinical aspects of iron metabolism are prominently displayed to make certain points. It is not an exhaustive repository of information. Many areas such as the chemistry and biology of serum transferrins and the ferritins are treated in some detail; others are handled less extensively. No details on physical methodologies, especially spectroscopy, are provided. Volume 2 of this series deals with this aspect of biochemistry in a more than ample fashion. The extensive references provided with each chapter will hopefully fill any error of omission. In short, the purpose of this volume is to provide a concise overview of the field of nonheme iron biochemistry.

For additional reading, the student is directed to a number of excellent review articles and monographs that have appeared in the last decade. They are listed in alphabetical order by author or editor in the Suggested Reading section.

Last, this author would like to express his deep appreciation to Professor Earl Frieden for the opportunity to participate in the task of preparing the *Biochemistry of the Elements* series, to those colleagues and publishers who have given their permission to use their materials in this volume, to Dr. Dorice Narins for her contribution (Chapter 3), to my associates Dr. Dietmar Grohlich and Dr. James A. Hayashi for reading the manuscript, and to my wife Marilyn for typing much of this manuscript.

Rush-Presbyterian–St. Luke's Medical Center Anatoly Bezkorovainy

Contents

List of Abbreviations and Symbols xvii

1. Tissue Iron and an Overview on Nonheme Iron Biochemistry

1.1 Introduction to Iron Metabolism 1
1.2 Historical Sketch of Iron Biology 3
1.3 Some Quantitative Data on Iron Distribution in Various Life Forms 5
1.4 Iron Determination in Biological Systems 17
Summary 20
References 20

2. Ferrokinetics

2.1 Classical Approach to Ferrokinetics 25
2.2 Compartment Models 27
2.3 Approaches to Ferrokinetics Based on the Probability Theory .. 35
Summary 41
List of Additional Abbreviations Used in Chapter 2 43
References 44

3. Absorption of Nonheme Iron

3.1 Introduction 47
3.1.1 Quantitative Significance of Iron in the Body 48
3.1.2 Heme Iron 50
3.1.3 Nonheme Iron 51
3.2 Techniques of Measuring Absorption 51
3.2.1 Methods Using Stable Iron 54
3.2.2 Methods Using Radioiron 55
3.2.3 *In Vitro* Measurements 62
3.2.4 Administration of Dose 63
3.3 Mechanism of Iron Absorption 70
3.3.1 Role of the Stomach 70
3.3.2 Role of Bile 72

3.3.3 Role of the Pancreas 72
3.3.4 Role of the Intestine 73
3.4 Corporal Factors Affecting Iron Absorption 78
3.4.1 Physiological States 78
3.4.2 Iron Absorption during Disease 90
3.5 Intraluminal Factors Affecting Iron Absorption 93
3.5.1 Macronutrients 94
3.5.2 Micronutrients 98
3.5.3 Complexing Agents 101
3.5.4 Fortification Iron 103
3.6 Mucosal Factors Affecting Iron Absorption 105
3.7 Regulation of Iron Absorption 106
3.8 Groups at Risk 110
3.9 Fortification 110
Summary 112
References 113

4. Chemistry and Metabolism of the Transferrins

4.1 Introduction 127
4.2 The Levels of Transferrins in Biological Fluids 129
4.3 Isolation of the Transferrins 132
4.4 Physical Properties of the Transferrins 137
4.4.1 Hydrodynamic Parameters, Molecular Weights, and X-Ray Diffraction 137
4.4.2 Denaturation of the Transferrins 141
4.4.3 Quaternary Structure of the Transferrins 142
4.4.4 Secondary Structure of the Transferrins 143
4.5 Metal-Binding Properties of the Transferrins 144
4.5.1 Metal Ions that Are Bound by the Transferrins 144
4.5.2 The Binding of "Synergistic" Anions by the Transferrins 146
4.5.3 The Reaction of Iron with the Transferrins 148
4.5.4 Are the Two Iron-Binding Sites of the Transferrins Identical? 149
4.6 Distribution of Iron in Human Serum Transferrin *in Vivo* 158
4.7 Some Gross Differences between Iron-Free and Iron-Saturated Transferrins 159
4.8 The Iron-Binding Ligands of the Transferrins 160
4.9 Primary Structure of the Transferrins 167
4.9.1 Structure and Significance of the Carbohydrate Moiety of the Transferrins 167
4.9.2 Cyanogen Bromide Fragmentation of the Transferrins 177
4.9.3 Amino Acid Sequences of Cyanogen Bromide Fragments of Human Serotransferrin 178

4.9.4 Fragments of the Transferrins Containing a Single Iron-Binding Site 180
4.10 Microheterogeneity of the Transferrins 184
4.11 Metabolism of the Transferrins 190
Summary 193
References 194

5. Chemistry and Biology of Iron Storage

5.1 Distribution of Ferritin and Hemosiderin 207
5.2 Isolation of Ferritin 209
5.3 Physical–Chemical Properties of the Ferritins 210
5.3.1 Gross Structure of Ferritin 210
5.3.2 Molecular Weight of Ferritin and Apoferritin 213
5.3.3 Subunits of the Ferritins 215
5.3.4 Distribution of Iron among Ferritin Oligomers and Isoferritins 224
5.3.5 Composition and Primary Structure of the Ferritins 226
5.3.6 Secondary and Tertiary Structure of Ferritin Subunits 229
5.4 Metabolism of Ferritin 230
5.5 Iron Uptake and Release by Ferritin and Apoferritin 236
5.6 Ferritin and Tissue Iron Metabolism 240
5.6.1 Iron Metabolism in the Liver 244
5.6.2 Iron Metabolism in the Reticuloendothelial System 247
5.6.3 Iron Overload 248
5.6.4 Serum Ferritin 253
5.6.5 Intestinal Mucosal Cells and Iron Metabolism 258
Summary 261
References 262

6. The Interaction of Nonheme Iron with Immature Red Cells

6.1 Introduction 271
6.2 Early Investigations on Serotransferrin–Immature Red Cell Interactions 271
6.3 The Mechanism of Serotransferrin–Immature Red Cell Interaction 273
6.4 Molecular Properties of Serotransferrin and its Effect on the Interaction with Immature Red Cells 278
6.5 Transferrin Receptors in the Immature Red Cells 286
6.6 Iron Removal from Serotransferrin 292
Summary 299
References 300

7. Microbial Iron Uptake and the Antimicrobial Properties of the Transferrins

7.1 Introduction 305
7.2 Structure of Siderophores 306
7.2.1 Catechol-like Siderophores 306
7.2.2 Hydroxamate-like Siderophores 311
7.3 Chemical and Physical Properties of the Siderophores 317
7.4 Metabolism and Biological Properties of the Siderophores 319
7.4.1 Biosynthesis of Siderophores 319
7.4.2 Mode of Iron Delivery to Microbial Cells by Siderophores 319
7.4.3 Energy Requirements during Iron Uptake by Microorganisms 321
7.4.4 Control of Siderophore Biosynthesis 322
7.5 Siderophores of Mammalian Origin 323
7.6 Iron Chelators in Clinical Practice 324
7.7 Antimicrobial Properties of the Transferrins 328
7.7.1 *In Vitro* Effects of Transferrins on Microbial Growth 329
7.7.2 Iron Status and Infection in Whole Organisms 332
7.7.3 Lactoferrin and Protection of the Breast-Fed Infant against Digestive and Systemic Disease 336
Summary 337
References 338

8. The Iron–Sulfur Proteins

8.1 Introduction and Classification 343
8.2 Rubredoxins 346
8.3 Ferredoxins 351
8.3.1 Plant-Type Ferredoxins (Two-Iron Clusters) 351
8.3.2 Bacterial-Type Ferredoxins Containing Two Four-Iron Clusters 361
8.3.3 Bacterial Ferredoxins with One Four-Iron Cluster 365
8.3.4 Removal of Iron Clusters from Ferredoxins 368
8.3.5 Functions of Bacterial Ferredoxins 369
8.3.6 Evolution of Ferredoxins 373
8.4 Iron–Sulfur Proteins and Nitrogen Fixation 375
8.5 Iron–Sulfur Proteins of Mammalian Electron Transport Mechanisms 378
8.5.1 Adrenodoxin and the Biosynthesis of Steroid Hormones . 378
8.5.2 Hydroxylation of 25-Hydroxycholecalciferol 382
8.5.3 Iron–Sulfur Centers in the Oxidative Phosphorylation Pathway 382
8.6 Miscellaneous Iron–Sulfur Proteins 384
Summary 388
References 389

9. Miscellaneous Aspects of Iron Metabolism

9.1 Introduction 395
9.2 Phosvitin, an Egg Iron-Binding Phosphoprotein 395
9.3 The Oxygenases 399
9.3.1 Phenylalanine Hydroxylase 400
9.3.2 Tyrosine and Tryptophan Hydroxylases 400
9.3.3 Pyrocatechase (Catechol-1,2-dioxygenase) 404
9.3.4 Protocatechuate-3,4-dioxygenase 405
9.3.5 4-Hydroxyphenylpyruvate Dioxygenase 407
9.3.6 Proline Hydroxylase 409
9.4 Hemerythrins 410
9.5 Iron Bacteria 413
9.6 Metalloserotransferrin as a Lymphocyte Growth Promoter 415
Summary 416
References 417

SUGGESTED READING 421
INDEX 423

List of Abbreviations* and Symbols

Å ....angstrom units
Ad ....adrenodoxin
ADP .....adenosine diphosphate
AMP ....adenosine monophosphate
ATP .....adenosine triphosphate
BCG ....Calmette–Guerin bacillus
CD ....circular dichroism
CHA ....cholylhydroxamic acid
CoA .....coenzyme A
$D_{20,w}$ ....diffusion constant
dl ....deciliter (100 ml)
DNA ....deoxyribose nucleic acid
DOPA ..dihydroxyphenylalanine
DTPA ...diethylenetriaminepentaacetic acid
E ....absorbancy
E'_0 ....standard redox potential
$\Delta E'_0$ ......standard redox potential
EDHPA ethylenediamine-N,N'-bis(2-hydroxyphenylacetic acid)
EDTA ..ethylenediaminetetraacetic acid
EPR .....electron paramagnetic resonance
ESR .....electron spin resonance
e.u. ......entropy unit
FAD ....flavinadenine dinucleotide
Fd ....ferredoxin
FMN ....flavin mononucleotide
ΔG ....Gibbs free energy
Gal ......galactose
GalNAc *N*-acetylgalactosamine
GlcNAc *N*-acetylglucosamine
ΔH ....enthalpy change
Hb ....hemoglobin
HIPIP ...bacterial high-potential iron protein
INH .....isonicotinic acid hydrazide
K_m ....Michaelis constant
Man .....mannose
m.w. ....molecular weight
n ....sample number
NAD ....nicotinamide adenine dinucleotide
NADP ..nicotinamide adenine dinucleotide phosphate
NANA ..*N*-acetyl neuraminic acid (sialic acid)
nm ....nanometer
ORD ....optical rotatory dispersion
PHA ....phytohemagglutinin

* A list of abbreviations used in Chapter 2 is given on pages 43 and 44. For amino acid abbreviations, consult a standard biochemistry text.

pI ....isoelectric point
PIH ......pyridoxal isonicotin-oylhydrazone
RBC .....red blood cell
Rd ....rubredoxin
RES .....reticuloendothelial system
RNA ....ribonucleic acid
$S^{\circ}_{20,w}$ ....sedimentation constant
ΔS ....entropy change
sTf ....serum transferrin
S unit ...Svedberg unit
Tf ....serum transferrin
TI ....total iron
TIBC ....total iron-binding capacity
UIBC ...unbound iron-binding capacity
V ....volt
$\bar{V}$ ....partial specific volume
[η] ....intrinsic viscosity
μg ....microgram

Tissue Iron and an Overview on Nonheme Iron Biochemistry

1.1 Introduction to Iron Metabolism

It is difficult to decide exactly where a treatise on nonheme iron metabolism should begin, especially in view of the fact that heme and nonheme iron biology are so intricately interconnected. It is perhaps prudent to introduce some basic vocabulary and concepts in an abbreviated form and to present a brief unified overview of iron metabolism which can serve as a framework upon which the more detailed material of this and the following chapters can be constructed.

It is said that the essentiality of iron in the human diet has been known since the seventeenth century, though it was much later that the exact function of iron was completely understood. The approach termed *ferrokinetics* has shown that serum iron is rapidly taken up by the bone marrow, which incorporates the greater part of iron into the red cells and sends a minor portion to the reticuloendothelial system as "wastage," whence it returns to serum. Serum iron is also in equilibrium with the extravascular (lymphatic) iron pool and liver storage sites. By far the greatest source of serum iron is the reticuloendothelial system, which converts the iron from senescent red blood cells (0.8% of all RBCs per day) to serum iron. A minor portion of serum iron comes from the diet, extravascular pool, and liver. Since in humans the loss of body iron amounts to only 0.5–2 mg/day and since the total plasma iron turnover is some 35 mg/day, it is clear that a very efficient iron preservation mechanism exists in the body.

Iron absorption in the adult amounts to about 0.5–2 mg/day, and this occurs in the small intestine (largely duodenum), apparently in the ferrous state. In the intestinal mucosal cells, iron may combine with intracellular storage proteins called ferritin (m.w. 420,000) and mucosal transferrin (m.w. 78,000). One theory has proposed that intestinal mucosal cell ferritin is crucial in the regulation of iron absorption by the organism. Iron may be mobilized from the mucosal cells as necessary into the circulation, where it combines with the iron transport protein called transferrin.

Transferrin (m.w. 78,000) is a glycoprotein that can bind a maximum of two ferric iron atoms per molecule of protein. Normally, serum transferrin is saturated with iron to one-third or less of its maximum iron-binding capacity. The main target of transferrin-bound iron is the hemopoietic system, which incorporates iron into hemoglobin. In a normal adult male, some 22 mg of iron per day are incorporated into hemoglobin.

To accomplish this, transferrin carrying the iron must initially interact with a specific receptor located on the plasma membrane of the immature red cell. It is not exactly clear what happens thereafter in the passage of iron from the cell surface into hemoglobin. Some authorities have claimed that iron–transferrin complex is actually internalized by the cell, where iron is removed therefrom, reduced, and incorporated into hemoglobin, the iron-free transferrin then being "regurgitated" into the circulation. Others have claimed that iron is lost from the serum transferrin during its sojourn on the surface of the red cell precursor.

Another target tissue of the circulating transferrin-bound iron is the iron storage organs, principally the liver. It can take up iron that is either bound or not bound to transferrin, and it has been theorized that specific iron–transferrin receptors are present on hepatocyte membranes and that iron–transferrin is internalized by the hepatocyte cells. In the liver cells, iron is reduced and combines with either ferritin or apoferritin to be stored in the ferric state until required. Mobilization from ferritin apparently occurs via reduction with $FADH_2$ or $FMNH_2$, and upon diffusing into the circulation, iron once more is oxidized into the ferric state by a circulating ferroxidase (ceruloplasmin) to combine with transferrin.

Some authorities have claimed that the two iron-binding sites of transferrin are not identical in regard to their affinity for the various cells with which they must interact. Thus, one site may interact preferentially with intestinal mucosal and immature red cells, whereas the other may prefer liver cells. Recently developed techniques of differential iron-binding site labeling and splitting the transferrins of various animal species in half (each carrying an iron-binding site) should provide an unambiguous answer to these questions.

Transferrin as well as the other nonheme iron-binding proteins conalbumin from egg white and lactoferrin from milk exert antimicrobial properties when their iron saturation levels are low by making iron unavailable to microorganisms. Many microorganisms, however, can secrete small-molecular-weight iron chelators called siderophores which can successfully counteract the iron-binding effects of the transferrins. The laboratory synthesis of many iron chelators for use in medicine to treat iron overload is based on the known structures of microbial siderophores.

In the plants and microorganisms, iron is required for the biosynthesis of heme-containing enzymes such as the cytochromes and the iron–sulfur proteins (e.g., ferredoxins); both are involved in oxidative phosphorylation and photosynthetic mechanisms. In fact it has been claimed that 70–90% of all iron found in the green plants is localized in the chloroplasts. Iron is absorbed by the roots in the ferrous state. To create ferrous iron from ferric iron, which is usually what the soil contains, iron is reduced by a reductant secreted by the roots into the medium. Iron is transported (translocated) via the xylem as the citrate complex.

The iron–sulfur proteins are believed to be the most primitive iron-containing proteins from the evolutionary point of view. They arose at the time when earth contained a reducing atmosphere, with a very low oxygen pressure and a methane pressure of 0.01–0.001 atm. At that time iron was present in the sea sediments in the form of sulfides and spontaneously combined with the sulfhydryl groups of proteins that were formed prebiotically (Osterberg, 1974). Today, iron–sulfur proteins persist both in the animal and plant kingdoms and are concerned with electron transport mechanisms associated with respiration. Iron, in fact, was the crucial substance that permitted the primitive organisms to survive in the oxygen atmosphere. It not only provided a means for transporting electrons via the iron–sulfur proteins but, more importantly, provided a defense mechanism against toxic products created through oxygen: hydrogen peroxide and superoxide anion. Iron-containing heme proteins, catalases, and peroxidases serve to destroy hydrogen peroxide. Superoxide dismutase, usually a copper and zinc protein, destroys the superoxide anion; some superoxide dismutases may also contain iron (Fridovich, 1975). The transferrins and ferritin apparently evolved from some heme-containing peroxidases and catalases (Frieden, 1974).

1.2 Historical Sketch of Iron Biology

Iron has been associated with life since time immemorial, and especially with blood, probably because of an obvious similarity between the colors of blood and that of rust. Iron preparations were used as remedies for a variety of disorders as long ago as 2735 B.C., when iron was listed by China's Emperor Shen Nung as a cure for "anemia" (Mettler, 1947, p. 177), and was used through many centuries thereafter as an empirical remedy by the ancient Greeks, Romans, Byzantines, Arabs, and

Europeans. Among the latter, Nicolas Menardes (1493–1588) appears to have been a notable advocate of iron therapy for a number of disorders, including baldness.

The first rational use of iron therapy was made by Thomas Sydenham (1624–1689), who used iron preparations to treat chlorosis, an old term for an iron deficiency disease. He combined iron therapy with bleeding and purging (Guthrie, 1946, p. 204). Others followed him into the eighteenth and nineteenth centuries, as, for instance, did F. Magendie (1783–1855), who wrote that iron ". . . possess(es) the property of recomposing the blood" (Mettler, 1947, p. 433). It is, however, doubtful that these doctors knew why iron administration helped their chlorosis patients even in spite of the discovery of iron in ashed blood by Nicholas Lemery (1645–1715) in 1713, by V. Menghini in 1747, and by Berzelius in 1812 (Garrison, 1929; McCay, 1973).

It appears that the rationale for the use of iron in chlorosis was discovered on the one hand by F. Fodisch in 1832, who found low blood iron levels in chlorosis patients, and by Gabriel Andral (1797–1876) on the other, who in 1843 found low red cell counts in chlorotic patients using a then relatively new tool—the microscope. And finally, Gustav V. Bunge (1844–1920) introduced the concept of iron deficiency anemia in 1895 (Schmidt, 1959). This and other findings prompted various investigators to study methods for oral iron administration. Although Pierre Band had shown as early as 1832 that ferrous rather than ferric iron was effective in curing chlorosis and confirmed this in 1893 (Fairbanks *et al.*, 1971) and moreover Field Marshal Bestuzhev-Ryumin, a favorite of Catherine the Great of Russia, had concocted a ferrous sulfate tincture to bring back pink color to pale cheeks of young ladies during the eighteenth century, a controversy developed as to whether or not inorganic iron could be assimilated by the human being. The general view was that it could not, and it was not until the 1930s that it was clearly shown that parenterally administered inorganic iron was readily incorporated into hemoglobin (Fairbanks *et al.*, 1971, p. 29).

The involvement of the liver in iron metabolism was clearly demonstrated by the discovery of Prussian blue positive *granular* (1880) and *nongranular* (1894) pigments therein. These were, of course, hemosiderin and ferritin, the latter being crystallized in 1937 by Laufberger. The presence of plasma-bound iron was suggested as early as in 1899 by G. Swirski, and evidence for the existence thereof was provided by several authors in the 1920s and 1930s, most notably by G. Barkan. A timely formulation of a comprehensive iron metabolism theory was provided in a monograph by Heilmeyer and Plötner (1937). Their essential concept is illustrated in Figure 1-1. It is noteworthy that these authors correctly viewed serum

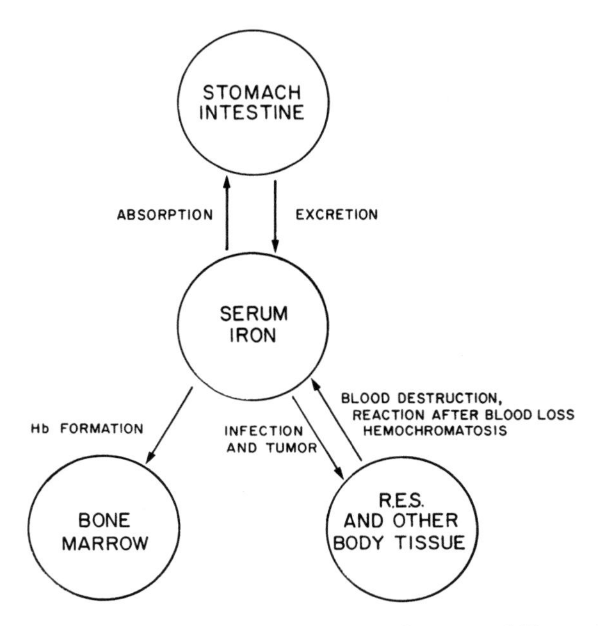

Figure 1-1. Iron metabolism as visualized by Heilmeyer and Plötner in 1937.

iron as the central figure in the overall scheme of mammalian iron metabolism.

Conalbumin, the egg white iron-binding protein, though discovered by Osborne in 1899, was not known to bind iron until 1944, when Schade and Caroline discovered its antimicrobial and metal-combining properties. The isolation of transferrin soon followed through the efforts of Schade and Caroline and of Holmberg and Laurell. When in 1959 Jandl and coworkers discovered that transferrin-bound iron was transferred to reticulocytes *in vitro,* a new era in the area of iron metabolism on the cellular and molecular levels was initiated which to this day is yielding a rich harvest of new data.

1.3. Some Quantitative Data on Iron Distribution in Various Life Forms

With a few exceptions, iron is a component of every known living organism. Data have been collected on the iron content of entire organisms, individual tissues, body fluids, etc., by many investigators such as E. M. Widdowson, P. F. Hahn, D. L. Drabkin, and others and have been tabulated in a number of handbooks, review articles, and texts. Because of the sheer quantity of primary literature on this subject, most data presented in this section were gathered from such secondary sources.

There have been serveral attempts to evaluate the whole body iron content for humans and various other animals. Clinically, such assessments may be performed on the basis of blood analyses [e.g., storage iron is taken at one-third of hemoglobin iron (Moore and Dubach, 1962] or on the basis of serum ferritin analyses (Engel and Pribor, 1978) and have a value in the assessment of the iron nutritional state of the individual or animal (the latter in the practice of veterinary medicine). Table 1-1 presents some total body iron values for a number of animals. In the case of humans, the iron status of infants and children is of importance in pediatric practice. As a rule infants are born with rather large iron stores, which are maintained if nutrition is adequate. Table 1-2 gives whole body iron contents of human children at various ages. It is seen that the human newborn has the highest amount of iron, which within 6 months rapidly declines and then stabilizes itself into adulthood. An interesting observation in this connection is the contrastingly low iron content in piglets. For this reason these animals are given iron injections after birth, which, however, may lead to complications in regard to their susceptibility to infection (see Chapter 7).

The next level of analysis for body iron is the tissue iron content. Practically all tissues contain iron, though the quantity therein, the form in which it is present, and its function in each case may be different. Table 1-3 presents iron data for various tissues of humans and some animals, and Table 1-4 presents iron contents of various iron storage organs in humans during successive phases of life. It should be noted, however, that in the short run muscle iron, which is present largely in the form of myoglobin (see Table 1-5), is metabolically inert.

Table 1-1. Total Body Iron in Various Animals (in mg/kg of fat-free weight)[a]

Animal	Adult	Newborn
Human	74	94
Pig	90	29
Cat	60	55
Rabbit	60	135
Guinea pig	—	67
Rat	60	59
Mouse	—	66

[a] Adapted from Underwood (1977b).

Table 1-2. Whole Body Iron Content During Childhood (in mg/kg of body weight)[a]

Age	Total body iron	Mean weight (kg)
Adult	48.9	70
Newborn	77.9	3.4
6 weeks	53.4	4.4
6 months	40.1	7.6
1 year	43.0	10.1
2 years	42.9	12.6
4 years	43.6	16.5
6 years	45.0	21.9
8 years	44.1	27.3
10 years	44.3	32.6
12 years	43.7	38.3
14 years	46.6	48.8
16 years	48.9	58.8

[a] Adapted from Burman (1974).

It is possible to separate body iron into specific chemical entities with which iron is associated. Surprisingly, there is a rather limited number of compounds in the mammalian organism that can account for nearly the entire iron complement of the organism. For human beings, such iron distribution is given in Table 1-5 and for some animals in Table 1-6. It should be noted that total body iron, hemoglobin iron, and storage iron contents are remarkably similar in the species studied, whereas myoglobin and cytochrome c iron shows rather wide variations. Many of the values given for human beings are estimates based on results obtained with animals. Thus, it is frequently assumed that storage iron is generally one-third that of hemoglobin iron and that there is 1.2 mg of myoglobin for each gram of muscle tissue (muscle mass is generally taken as 42% of body weight). By measuring hemoglobin iron and the person's weight, one can thus arrive at a very rough estimate of the total body iron content. For practical purposes, transferrin-bound iron and iron present in cytochrome c and catalase may be ignored in such calculations, as they account for a very small percentage of total body iron.

The various plant tissues, of course, show huge differences in iron content. Generally, leaves and shoots with their high ferredoxin levels contain the largest amounts of iron. Table 1-7 presents some representative data on iron contents of plant tissues.

The largest volume of information on iron content of animal tissues refers to biological fluids, especially the various components of blood and milk. The measurement of blood iron content has been used heavily in

Table 1-3. Iron Content of Human Tissues (in μg/g of wet tissue)[a]

Tissue	Number of samples	Iron content (with standard deviation)
Adrenals	10	38 ± 20
Aorta	93	56 ± 36
Brain	108	58 ± 17
White matter[b]	—	25.9
Gray matter[b]	—	29.9
Diaphragm	91	47 ± 23
Heart	123	55 ± 18
Rat[b]	—	72
Rabbit[b]	—	124
Intestine		
Duodenum	60	41 ± 18
Jejunum	84	38 ± 28
Ileum	78	27 ± 13
Kidney	123	76 ± 31
Liver	127	195 ± 113
Lung	120	319 ± 176
Muscle	120	42 ± 14
Ovary	16	59 ± 50
Pancreas	119	51 ± 42
Spleen	124	336 ± 210
Skin	19	15 ± 9
Testis	68	29 ± 14
Thyroid	14	62 ± 29

[a] From Tipton and Cook (1963) unless otherwise indicated.
[b] From Altman (1959).

Table 1-4. Storage Iron as a Function of Age and Sex (in μg/g of wet tissue)

Age group	Liver	Spleen	Kidney
Less than 1 month	343 ± 103	334 ± 146	43 ± 7
1 month–2 years	108 ± 91	61 ± 25	27 ± 8
3–14 years	144 ± 115	99 ± 78	36 ± 7
Adult male	233 ± 123	243 ± 152	35 ± 9
Premenopausal female	130 ± 87	140 ± 83	33 ± 9
Postmenopausal female	303 ± 217	343 ± 253	28 ± 1

[a] From Worwood (1977).

Table 1-5. Distribution of Iron among the Various Compounds in a 70-kg Normal Adult Male

	Hemoglobin	Myoglobin	Cytochrome c	Catalase	Hemosiderin	Ferritin	Transferrin	Total body iron	Reference
Total compound in body (g)	900	40	0.8	5.0	—3.0—[a]		7.5	4.2	Drabkin, 1951
	750	120	0.8	5.0	1.2	2.0	14.0	4.0	Pollycove, 1978
	800	35.3	0.8	5.0			6.2	3.85	Damm and King, 1965
Iron content (%)	0.34	0.34	0.43	0.09	—23—[a]		0.04–0.12		Drabkin, 1951
	0.34	0.345	0.43	0.09			0.05		Damm and King, 1965
Total iron in compound (g)	3.06	0.14	0.0034	0.0045	—0.69—[a]		0.003		Drabkin, 1951
	2.60	0.40	0.004	0.004	0.36	0.40	0.007		Pollycove, 1978
	2.72	0.122	0.0034	0.0045	—1.0—[a]		0.003		Damm and King, 1965
	2.0		—0.200—[b]		—1.25—[a]		0.001	3.451	Van Laucker, 1975
	2.6	0.40	0.017	0.005	—1.49—[c]		0.004		Worwood, 1977
(dog)	0.90	0.11						1.55	Moore and Dubach, 1962
Percent of total body iron	72.9	3.3	0.08	0.11	—16.4—[a]		0.07		Drabkin, 1951
	70.5	3.2	0.1	0.1	—26—[a]		0.1		Damm and King, 1965
	57.6	8.9	0.4	0.1	—33—[a]		0.1		Worwood, 1977

[a] Hemosiderin and ferritin.
[b] Myoglobin, cytochrome c, and catalase.
[c] Storage iron: hemosiderin plus ferritin distributed as follows: liver, 0.35; spleen, 0.02; muscle, 0.86; bone marrow, 0.26 g.

Table 1-6. Distribution of Iron among the Various Iron-Containing Compounds of Several Animal Species[a]

Animal	Average body weight (kg)	Total body iron (mg/kg)	Myoglobin (mg/kg)	Hemoglobin (mg/kg)	Cytochrome c (mg/kg)	Ferritin–hemosiderin (mg/kg)
Rat	0.26	60.0	1.2	44.0	0.240	14.4
Dog 1	6.35	69.4	6.4	47.2	0.093	15.8
Dog 2	9.88	66.3	2.6	47.7	0.108	15.9
Dog 3	20	77.5	5.5	45.0	—	27
Human	70	60.8	1.71	44.3	0.048	14.8
Heifer	182	61.2	5.8	41.5	0.029	13.8
Horse 1	455	63.0	10.2	39.5	0.230	13.1
Horse 2	500	66.0	12.9	39.7	0.143	13.2

[a] From Moore and Dubach (1962).

clinical medicine as a diagnostic tool. The human being takes in approximately 10–15 mg of iron per day, of which only 10% is absorbed. The breast-fed infant, however, is an exception and absorbs nearly 50% of the iron present in human milk (Engel and Pribor, 1978). An adult human male excretes 0.5–1.5 mg of iron per day (all routes considered), whereas the figure is 1–2 mg/day for the female; during pregnancy this figure becomes nearly 4 mg/day, and hence iron intake must be increased accordingly. Fecal iron excretion amounts to between 6 and 16 mg/day. This represents iron that was present in the food and had not been absorbed. Table 1-8 shows iron concentrations of various biological fluids other than blood for a variety of animal species. A look at iron and transferrin contents of human cerebrospinal fluid may be noteworthy. As reported by Bleijenberg *et al.* (1971), these are 43 μg/liter and 17.2 mg/liter, respectively. Since saturated transferrin contains 1.4 μg of iron per mg of transferrin, the value of 2.5 μg of Fe per mg of transferrin in human cerebrospinal fluid indicates that iron may also be present in a form other than bound to transferrin.

Blood and its components have been analyzed for iron most extensively both in the research laboratory and the routine clinical chemistry laboratory. In addition to heme iron, which is usually measured in terms of blood hemoglobin content or hematocrit, iron-associated measurements include the so-called total plasma or serum iron (TI) and total iron-binding capacity of serum or plasma (TIBC). The former is merely the iron concentration, most if not all of it transferrin-bound; the latter is the highest amount of iron that the plasma or serum can bind specifically. In effect, it should be a measure of total transferrin content of the serum or plasma, though calculation of plasma transferrin content from TIBC meas-

Table 1-7. Distribution of Iron among the Various Portions of Plants (in mg/kg of dry weight)[a]

Portion	American ash	American elm	Cotton	Bean	Oat	Rice	Wheat	Corn
Shoot	—	—	—	1350	—	—	290–580	312–321
Leaf	195	245–810	1754	—	—	—	—	41–810
Stem	—	—	610	—	61–860	—	60–630	400–740
Root	—	—	—	—	—	—	—	500–760
Seed	—	—	150–590	120	7–350	76–350	3–420	13–550
Bark	78	145	—	270	—	—	—	—
Twig	55	68	—	—	—	—	—	—
Heartwood	16	22	—	—	—	—	—	—
Sapwood	15	48	—	—	—	—	—	—

[a] From Altman and Dittmer (1972).

Table 1-8. Iron Content of Various Biological Fluids Other than Blood

Species	Fluid	Units	Iron concentration	Reference
Human	Amniotic fluid	mg/100 ml	0.07	Altman and Dittmer, 1961
	Bile (gallbladder)	mg/100 ml	0.031–1.680	Altman and Dittmer, 1961
	Bile (liver)	mg/100 ml	6.8 (4.8–7.8)	Altman and Dittmer, 1961
	Cerebrospinal fluid	mg/100 ml	0.035	Altman and Dittmer, 1974
			0.01–0.02	Altman and Dittmer, 1974
			0.023–0.052	Searcy, 1969
			0.0043 (0–0.0089)	Bleijenberg et al., 1971
	Ileal secretion	mg/100 ml	17 (2.2–17.7)	Altman and Dittmer, 1961
	Colostrum	mg/100 ml	0.04 (0.02–0.05)	Altman and Dittmer, 1961
	Mature milk	mg/100 ml	0.03 (0.02–0.09)	Altman and Dittmer, 1961
			0.36 (0.29–0.45)	Altman and Dittmer, 1961
	Sweat	mg/100 ml	0.027 (0.022–0.045)	Altman and Dittmer, 1968
	Urine	μg/kg/day	1.4–4.3	Altman and Dittmer, 1974
		mg/day	0.2–0.3	Underwood, 1977a
			0.045	Searcy, 1969
	Gastric juice	mg/100 ml	0.3	Searcy, 1969
	Lymph	μg/100/ml	71 (24–135)	Altman and Dittmer, 1974
Cattle	Aqueous humor	mg/100 g	0.016	Altman and Dittmer, 1961
	Vitreous humor	μg/100 g	12	Altman and Dittmer, 1961
	Bile (liver)	mg/100 ml	3–6	Altman and Dittmer, 1961
	Semen	mg/100 ml	0.6 (0.3–0.9)	Altman and Dittmer, 1961
Dog	Bile (liver)	mg/100 ml	1.8–16	Altman and Dittmer, 1961
	Bile (gallbladder)	mg/100 ml	0.09–0.18	Altman and Dittmer, 1961
Rabbit	Bile (liver)	mg/100 ml	0.13	Altman and Dittmer, 1961
Guinea pig	Bile (gallbladder)	mg/100 ml	0.09–0.18	Altman and Dittmer, 1961
Sheep	Semen	mg/100 ml	0.8	Altman and Dittmer, 1961

urements leads to somewhat higher values than those obtained by immunological methods (Haeckel *et al.*, 1973). TIBC is generally determined by adding excess iron to plasma or serum, removing the unbound iron, and then measuring the iron that remains behind. If TIBC and TI are known, it is possible to calculate the so-called unbound iron-binding capacity (UIBC), also called latent iron-binding capacity, by subtracting TI from TIBC. UIBC may also be determined directly. TI, TIBC, and UIBC determinations have been utilized heavily for diagnostic purposes in clinical biochemistry.

Human plasma contains between 200 and 400 mg of transferrin per 100 ml; since transferrin in circulation is only about one-third or less saturated with iron, then given the fact that one molecule of transferrin (m.w. 80,000) can bind a maximum of two atoms of ferric iron, TI should be between 94 and 188 μg/100 ml. TIBC should be three times that, or between 282 and 564 μg/100 ml. In practice, there are numerous reports on the so-called normal TI values. Weippl *et al.* (1973) collected such data from 608 adults and arrived at a figure of 109 ± 25 μg/100 ml of serum in men and 91 ± 27 μg/100 ml in females. More recently, Van Eijk and Kroos reported values of 111 ± 28 μg/100 ml for males and 88.3 ± 28 μg/100 ml in females. They reported transferrin levels of 264 ± 64 μg/100 ml of serum in both males and females, so that the degree of transferrin saturation with iron came out to exactly 30% (Van Eijk and Kroos, 1978). A well-known clinical chemistry text gives a normal TI range for males of 60–150 μg/100 ml and 50–130 μg/100 ml in females, with TIBC values of 270–380 μg/100 ml (Tietz, 1976). Table 1-9 provides TI and TIBC values in human beings, and Table 1-10 lists blood iron levels for a variety of animals. It is interesting to note that in laying hens TIBC is lower than TI and UIBC is 0, because laying hens have conalbumin in their circulation which does not register in the commonly used Ramsay TIBC and UIBC procedures (Planas, 1967). Conalbumin seems to differ from fowl serum transferrin in regard to carbohydrate content only.

As can be seen from the preceding discussion and Table 1-9, serum iron levels definitely vary with sex of the individual and with age, according to some investigators. In Zebu cattle, in fact, the TI and TIBC levels of aged animals are very much lower than those in young adult animals, though the degree of iron saturation remains constant (Table 1-11). On the other hand, in a Welsh human population sample, no significant age difference was observed; males had a TI of 100.7 ± 35.2 and a TIBC of 352.8 ± 51.1 μg/100 ml, whereas the respective figures for females were 90 ± 41.3 and 379.7 ± 70.1 (Marx, 1979). In addition, serum TI levels vary during the course of a day (this is called diurnal variation), being highest in the morning and decreasing toward the afternoon. Morn-

Table 1-9. Total Iron and Total-Binding Capacity Levels in Human Beings (in μg/100 ml of serum)

Age group	Total iron	Total iron-binding capacity	Reference
Newborn	160 ± 65	—	Altman and Dittmer, 1977
1–6 months	89 ± 27	—	Altman and Dittmer, 1977
4 months–5 years	85 ± 25	—	Altman and Dittmer, 1977
3–6 years	117 ± 40	402 ± 63	Altman and Dittmer, 1977
9–17 years	100 (50–200)	—	Altman and Dittmer, 1977
12–17 years	110 (60–200)	380 (330–490)	Altman and Dittmer, 1977
Adult male	95 (70–180)	385 (240–490)	Altman and Dittmer, 1977
Adult female	90 (60–180)	430 (300–520)	Altman and Dittmer, 1977
Adult male	127 (67–191)	333 (253–416)	Underwood, 1977b
Adult female	113 (63–202)	329 (250–416)	Underwood, 1977b
Iron deficiency anemia	32 (0–78)	482 (204–705)	Underwood, 1977b
Late pregnancy	94 (22–185)	532 (373–712)	Underwood, 1977b
Hemochromatosis	250 (191–290)	263 (205–330)	Underwood, 1977b
Infections	47 (30–72)	260 (182–270)	Underwood, 1977b
Young males (age 30)	140 (95–224)	347 (224–470)	Marx, 1979
Young females (age 25)	140 (101–179)	375 (319–447)	Marx, 1979
Aged males (age 73)	129 (61–218)	302 (240–375)	Marx, 1979
Aged females (age 70)	101 (78–134)	319 (257–402)	Marx, 1979

ing and afternoon samples in an individual may vary by as much as 40% (Werkman *et al.*, 1974), and it has been suggested that TI levels in patients be measured twice in the morning and twice in the afternoon. TIBC does not show such diurnal variation.

TI and TIBC data are valuable in the diagnosis of various diseases. This includes not only hematologic diseases but other ailments as well, where iron metabolism irregularities appear as secondary phenomena. Table 1-12 gives several such diseases with the concomitant TI and TIBC alterations.

Another biological fluid whose iron content has been studied extensively is milk. This comes about because of the nutritional importance of milk in infancy and early childhood as well as the age-old debate on the relative virtues of human vs. bovine milks in human infant nutrition. With respect to iron content of milks of the different animal species, there exist vast differences, which is consistent with Bunge's 1898 hypothesis that the milk contents of the different animals are uniquely suited for their respective needs (Picciano and Guthrie, 1976). The iron contents of milks of various species are given in Table 1-13, though it should be kept in mind that tremendous variations do exist from one individual to another within the same species. Milk samples from even the same individual contain decreasing amounts of iron with increasing time of lactation; thus,

Table 1-10. Total Iron Content and Total Iron-Binding Capacity of Blood Components in Various Animals (in μg/100 ml in the case of plasma and serum and in mg/100 ml in the case of whole blood and erythrocytes)[a]

Animal	Sex	Fluid	TI[a]	TIBC	UIBC	Reference
Mallard duck	F	Serum	132	—	—	Altman and Dittmer, 1974
	F[b]	Serum	1065	—	—	Altman and Dittmer, 1974
	M	Serum	159	—	—	Altman and Dittmer, 1974
Hen	F	Serum	140	250	110	Planas, 1967
	F[b]	Serum	500	370	0	Planas, 1967
Chicken	F	Serum	100	200	100	Planas, 1967
	M	Serum	100	255	155	Planas, 1967
Carp	—	Serum	25 (16–33)	—	—	Planas, 1967
Cow	F	Plasma	175	—	—	Planas, 1967
	F	Serum	97 ± 29	659 ± 140	131 ± 36	Kaneko, 1971
	F	Serum	97 ± 29	100 ± 44	—	Melby and Altman, 1974
Zebu	F	Serum	150 ± 17.2	406 ± 32	—	Tartour, 1973
Horse	—	Serum	111 ± 11	—	218 ± 21	Kaneko, 1971
Sheep	M	Plasma	169 ± 12	—	—	Altman and Dittmer, 1974
	—	Serum	193 ± 7	—	141 ± 10	Kaneko, 1971
	—	Serum	194 ± 64	437 ± 56	—	Melby and Altman, 1974
Swine	—	Serum	107 ± 15	817 ± 89	—	Melby and Altman, 1974
	—	Serum	121 ± 33	—	196 ± 39	Kaneko, 1971
Dog	M	Serum	199 ± 53	—	—	Altman and Dittmer, 1974
	—	Serum	108 (94–122)	—	200 (170–222)	Kaneko, 1971
Cat	—	Serum	140 (68–215)	—	150 (105–205)	Kaneko, 1971
	—	Serum	68–215	—	—	Melby and Altman, 1974
Rabbit	—	Serum	235 ± 17	334 ± 93	—	Melby and Altman, 1974

continued overleaf

Table 1-10. (Continued)

Animal	Sex	Fluid	TI[a]	TIBC	UIBC	Reference
	—	Serum	199 ± 37	378 ± 58	—	Melby and Altman, 1974
	—	Serum	202 ± 11	420 ± 38	—	Van Vugt *et al.*, 1975
	—	Serum	400	600	—	Jordan *et al.*, 1967
Rat	—	Plasma	230	530–610	—	Itzhaki and Belcher, 1961
Racer snake	—	Plasma	182	204	—	George and Dessauer, 1970
Giant garter snake	—	Plasma	75	155	—	George and Dessauer, 1970
Human	M	Whole blood	35–45	—	—	Altman and Dittmer, 1974
	M	Erythrocytes	96[c]	—	—	Altman and Dittmer, 1974
	M	Plasma	115 ± 42	—	—	Altman and Dittmer, 1974
Gorilla	M & F	Serum	139 ± 36	—	—	Altman and Dittmer, 1974
Rhesus	M & F	Serum	87	—	—	Altman and Dittmer, 1974
		Whole blood	29–48	—	—	Altman and Dittmer, 1974

[a] Abbreviations: M, male; F, female; TI, total iron; TIBCs, total iron-binding capacity; UIBC, unbound iron-binding capacity.
[b] Laying bird.
[c] Nonheme iron was 2.5 μg/100 ml.

Table 1-11. Variation of Total Serum Iron and Total Iron-Binding Capacity with Age in Zebu Cattle (in μg/100 ml)[a,b]

Age of cow (years)	*n*	TI	TIBC	% of saturation
0–1	15	146 ± 10 (122–167)	394 ± 25 (352–422)	37 ± 2 (35–40)
2–3	9	153 ± 16 (141–192)	357 ± 25 (371–438)	39 ± 5 (34–52)
4–7	18	150 ± 17 (136–212)	406 ± 32 (343–468)	37 ± 5 (33–55)
8–11	20	126 ± 15 (100–163)	346 ± 37 (282–411)	36 ± 4 (29–42)
12–15	18	107 ± 10 (88–127)	290 ± 17 (247–318)	36 ± 4 (31–43)
16–17	7	95 ± 16 (65–109)	263 ± 20 (230–289)	35 ± 6 (24–41)

[a] From Tartour (1973).
[b] Abbreviations: *n*, number of observations; TI, total iron; TIBC, total iron-binding capacity.

in one study at 2 weeks of lactation the medium iron concentration of human breast milk ws 0.56 mg/liter, whereas at 5 months of lactation this values was 0.3 mg/liter, remaining constant thereafter. Variations from one individual to another were greatest during the first 5 weeks of lactation, when the highest iron content found was 1.14 mg/liter and the lowest was 0.11 mg/liter (Siimes *et al.*, 1979).

1.4 Iron Determination in Biological Systems

The traditional methods for the determination of iron in biological systems have been colorimetric in nature, since physical methods such as atomic absorption do not show an advantage. An exception is the recently developed method of X-ray activation analysis, which can provide a wholly new dimension in trace metal analysis of biological systems.

Table 1-12. Variation of Serum Iron and Total Serum Iron-Binding Capacity with Disease States[a,b]

Disease	TI	TIBC	% Saturation of transferrin
Iron deficiency anemia	dd	i	dd
Infection or malignancy	d	0	Variable
Hemolytic anemia	i	0	i
Hemochromatosis	ii	0 or d	ii
Nephrosis	d	dd	0
Hepatic cirrhosis	0 or d	0 or d	i
Hepatitis	ii	i	i

[a] From Tietz (1976).
[b] Abbreviations: TI, total iron; TIBC, total iron-binding capacity; i, increased; ii, greatly increased; d, decreased; dd, greatly decreased.

Table 1-13. Iron Content of Milks of Various Animals (in μg/100 ml)

Animal	Iron content	Reference
Human	100	Blaxter, 1961
	50 (20–80)	McLaren and Burman, 1976
Cattle	45 (25–75)	McLaren and Burman, 1976
	30	Blaxter, 1961
Rabbit	120	Blaxter, 1961
	400	Jordan *et al.*, 1967
Pig	180	Blaxter, 1961
Rat	700	Blaxter, 1961
Dog	900	Blaxter, 1961

The colorimetric methods for iron determination may involve the removal of iron from iron-binding proteins by either reduction or by acidification of the fluid. This may be succeeded by deproteinization, following which an iron-sequestering reagent is added which forms a colored complex with iron. The color is then determined in a colorimeter. There are numerous variations on this principal theme, some methods requiring reduction of the iron to the ferrous state, some determining the iron in the ferric state, and some requiring deproteinization and some not. The trend is, of course, toward the use of simplified methodology with minimal manipulation. TIBC is measured by adding an excess of ferric chloride or another ferric salt to the serum, removing iron which is not bound to transferrin by either ion exchangers or a sorbent such as magnesium carbonate, and then measuring the iron remaining via a colorimetric procedure. UIBC of serum may be calculated by subtracting TI from TIBC or may be determined directly by adding to serum an excess of "saturating" iron solution containing radioactive iron and then measuring the amount of radioactive iron bound to transferrin. If TIBC and UIBC are performed directly, TI can be calculated by subtracting UIBC from TIBC.

The most venerable TI and TIBC method is the Ramsay procedure (Ramsay, 1958), which involves the reduction of transferrin-bound iron to the ferrous state with sulfite, development of color with 2,2′-dipyridyl, and deproteinization by heating and with chloroform. The determination of TIBC involves the removal of the excess iron ($FeCl_3$) added to serum with "light" magnesium carbonate. Other color reagents that may be used following the reduction of the ferric to ferrous iron in order of their sensitivity are 2,2′,2″-tripyridine, 1,10-phenanthroline, bathophenanthroline, ferrozine [3-(2-pyridyl)-5,6-bis(4-phenyl)sulfonic acid], and terosite [2,6-

bis(4-phenyl-2,2′-pyridyl)-4-phenylpyridine] (Tietz, 1976). TPTZ (tripyridyl-*s*-triazine) and PPDT [3-(4-phenyl-2-pyridyl)-5,6-diphenyl-1,2,4-triazine] have also been advocated (Ichida *et al.*, 1968; Schilt and Hoyle, 1967). The latter is said to possess a high degree of sensitivity. Thiocyanate is one of the few reagents that gives a color with ferric iron.

Iron can be determined in serum without deproteinization using the nitroso-R-salt (1-nitroso-2-napthol-3,6-disulfonic acid–disodium salt) as the ferrous iron chromogen (Ness and Dickerson, 1965). The method has been adapted to TIBC measurements (Martinek, 1973).

The International Committee for Standardization in Hematology (1978a, 1978b) has surveyed a number of laboratories and recommended a procedure for the determination of total serum–plasma iron, TIBC, and UIBC. This procedure has given results with minimal variation among the laboratories surveyed. It is based for the most part on the Ramsay procedure and uses trichloroacetic acid for protein precipitation, thioglycollic acid as the reducing agent, and bathophenanthroline as the chromogen. For TIBC measurement, $FeCl_3$ is used as the saturating agent, and light magnesium carbonate is used to absorb iron unbound by transferrin. For UIBC determination, the use of radioactive iron is recommended.

Iron can, of course, be determined by atomic absorption spectrometry, and the method has been adapted in some instances for use in the clinical biochemistry laboratory for the analysis of biological materials. It is not necessary to reduce the iron, though most methods call for deproteinization of the serum before using the instrument. Tavenier and Hellendoorn (1969) report a mean serum total iron value of 122 ± 29 μg/100 ml as determined by atomic absorption on deproteinized samples and 121 ± 31 and 107 ± 27 μg/100 ml by two colorimetric methods. Pronk *et al.* (1974) pointed out that in order to avoid errors implicit in atomic absorption spectrophotometry, it is desirable to use a double-beam instrument with an internal standard. It should also be noted that the atomic absorption methods make no distinction between heme and nonheme iron, whereas most colorimetric methods do not include heme iron.

Iron can be rapidly determined by the coulometry method, utilizing a specially constructed electrode that measures the current generated when iron is reduced from the ferric to the ferrous state. The method requires minimal amounts of serum (25 μl), and no interference from hemoglobin-bound iron is apparent. Instrumentation for the coulometric determination of iron has only recently become available (Environmental Sciences Associates, Inc., Bedford, Mass.), and rigorous correlations of serum data obtained therewith with those of the more established methods have not yet been available.

Summary

In this chapter we have established some basic vocabulary in the field of nonheme iron biochemistry. Transferrin, the circulating iron-binding protein, is essential for iron transport to the hemopoietic tissue and iron storage sites in the liver. The largest amount of circulating iron originates from the reticuloendothelial system. Ferritin is the iron storage protein present in liver and other tissues. Transferrin as well as the homologous proteins conalbumin and lactoferrin are antimicrobial in their iron-free forms. Major nonheme iron-containing proteins of plants and microorganisms are the so-called iron–sulfur proteins, which are believed to be the most primitive from the evolutionary point of view. The various aspects of iron biochemistry were discovered gradually through the centuries of human civilization, the earliest use thereof in medicine being recorded about 3000 B.C. in China. Iron has been used successfully to treat chlorosis, an iron deficiency anemia, since the seventeenth century, and iron was discovered in blood in the eighteenth century. The concept of iron deficiency anemia was delineated in 1895 by Bunge; however, the details of iron metabolism did not become clear until well into the twentieth century with the crystallization of ferritin; the delineation of the first ferrokinetic model involving plasma, bone marrow, and liver in 1937; and the discovery of transferrin and its function in the 1940s.

Iron is widely distributed in nature, and it is possible to define iron content on the basis of the whole organism, on the level of different tissues, and on the bassis of the chemical nature in which it exists. Of the approximately 4 g of iron present in the human organism, over 95% can be accounted for by the iron content of hemoglobin, myoglobin, ferritin and hemosiderin, cytochrome c, catalase, and transferrin. The greatest amount of information on iron content of biological materials is available on blood and its components, because of its diagnostic value in medical practice. Human plasma contains approximately 100 μg of iron/dl and the total iron-binding capacity of plasma is about three times that much. Determination of iron and total iron-binding capacity of plasma is generally performed by colorimetric methods, which require the reduction of the ferric iron to ferrous so that colored complexes can be formed with the appropriate chromogens.

References

Altman, P. L. (ed.), 1959. *Handbook of Circulation,* Saunders, Philadelphia, pp. 38–45.

Altman, P. L., and Dittmer, D. S. (eds.), 1961. *Blood and Other Body Fluids,* Federation of American Societies for Experimental Biology (FASEB), Washington, D.C.

Altman, P. L., and Dittmer, D. S. (eds.), 1968. *Metabolism*, FASEB, Bethesda, Md.

Altman, P. L., and Dittmer, D. S. (eds.), 1972. *Biology Data Book*, Vol. I, 2nd ed., FASEB, Bethesda, Md., pp. 407–409.

Altman, P. L., and Dittmer, D. S. (eds.), 1974. *Biology Data Book*, Vol. III, 2nd ed., FASEB, Bethesda, Md.

Altman, P. L., and Dittmer, D. S. (eds.), 1977. *Human Health and Disease*, FASEB, Bethesda, Md., p. 165.

Blaxter, K. L., 1961. Lactation and the growth of the young, in *Milk: The Mammary Gland and Its Secretion*, Vol II, S. K. Kon and A. T. Cowie (eds.), Academic Press, New York, pp. 344–347.

Bleijenberg, B. G., von Eijk, H. G., and Leijnse, B. 1971. The determination of non-heme iron and transferrin in cerebrospinal fluid, *Clin. Chim. Acta* 31:277–281.

Burman, D., 1974. Iron metabolism in infancy and childhood, in *Iron in Biochemistry and Medicine*, A. Jacobs, and M. Worwood (eds.), Academic Press, New York, p. 552.

Damm, H. C., and King, J. W. (eds.), 1965. *Handbook of Clinical Laboratory Data*, Chemical Rubber Publishing Company, Cleveland, p. 291.

Drabkin, D. L., 1951. Metabolism of the hemin chromoproteins, *Physiol. Rev*. 31:345–431.

Engel, R. H., and Pribor, H. C., 1978. Serum ferritin: A convenient measure of body iron stores, *Lab. Manage*. Oct.:31–34.

Fairbanks, V. F., Fahey, J. L., and Bentler, E. (eds.), 1971. *Clinical Disorders of Iron Metabolism*, Grune & Stratton, New York, pp. 1–41.

Fridovich, I., 1975. Superoxide dismutases, *Annu. Rev. Biochem*. 44:147–159.

Frieden, E., 1974. The evolution of metals as essential elements, in *Protein–Metal Interactions*, M. Friedman (ed.), Plenum, New York, pp. 1–32.

Garrison, F. H. (ed.), 1929. *An Introduction to the History of Medicine*, Saunders, Philadelphia.

George, D. W., and Dessauer, H. C., 1970. Immunological correspondence of transferrins and the relationships of colubrid snakes, *Comp. Biochem. Physiol*. 33:617–627.

Guthrie, D. (ed.), 1946. *A History of Medicine*, Lippincott, Philadelphia.

Haeckel, R., Haindl, H., Hultsch, E., Mariss, P., and Oellerrich, M., 1973. Comparison of 8 different colorimetric, radiochemical, and immunological procedures for the determination of iron binding capacity, *Z. Klin. Chem. Klin. Biochem*. 11:529–534.

Heilmeyer, L., and Plötner, K., 1937. *Das Serumeisen und die Eisenmangelkrankheit (Pathogenese, Symptomatologie, und Therapie*), G. Fischer, Jena, Germany.

Ichida, T., Osaka, T., and Kojima, K., 1968. A simple method for the determination of serum iron, *Clin. Chim. Acta* 22:271–275.

International Committee for Standardization in Hematology, 1978a. The measurement of total and unsaturated iron-binding capacity in serum, *Br. J. Haematol*. 38:281–294.

International Committee for Standardization in Hematology, 1978b. Recommendations for measurement of serum iron in human blood, *Br. J. Haematol*. 38:291–294.

Itzhaki, R. F., and Belcher, E. H., 1961. Studies on plasma iron in the rat. II. Plasma iron concentration and plasma iron-binding capacity. *Arch. Biochem. Biophys*. 92:74–80.

Jordan, S. M., Kaldor, I., and Morgan, E. H., 1967. Milk and serum iron and iron-binding capacity in the rabbit, *Nature (London)* 215:76–77.

Kaneko, J. J., 1971. Iron metabolism, in *Clinical Biochemistry of Domestic Animals*, J. J. Kaneko and C. E. Cornelius (eds.), Academic Press, New York, p. 388.

Martinek, R. G., 1973. Simplified determination of serum iron and binding capacity by nitroso-R-salt without deproteinization, *Clin. Chim. Acta* 43:73–80.

Marx, J. J. M., 1979. Normal iron absorption and decreased red cell iron uptake in the aged, *Blood* 53:204–211.

McCay, C. M. (ed.), 1973. *Notes on the History of Nutrition Research,* Hans Huber, Berne, pp. 156–171.

McLaren, D. S., and Burman, D. (eds.), 1976. *Textbook of Paediatric Nutrition,* Churchill Livingstone, Edinburgh, p. 59.

Melby, E. C., Jr., and Altman, N. H. (eds.), 1974. *Handbook of Laboratory Animal Science,* Vol. II, CRC Press, Inc., Cleveland, pp. 378–379.

Mettler, C. C. (ed.), 1947. *History of Medicine,* McGraw-Hill, Inc., Blakiston Division, New York.

Moore, C., and Dubach, R., 1962. Iron, in *Mineral Metabolism,* Vol. 2B, C. L. Comar and F. Bronner (eds.), Academic Press, New York, pp. 287–348.

Ness, A. T., and Dickerson, H. C., 1965. The determination of serum iron by nitroso-R-salt without deproteinization, *Clin. Chim. Acta* 12:579–588.

Osterberg, R., 1974. Origins of metal ions in biology, *Nature (London)* 249:382–384.

Picciano, M. F., and Guthrie, H. A., 1976. Copper, iron and zinc contents of mature human milk, *Am. J. Clin. Nutr.* 29:242–254.

Planas, J., 1967. Serum iron transport in the fowl and the mammal, *Nature (London)* 215:289–290.

Pollycove, M., 1978. Hemochromatosis, in *The Metabolic Basis of Inherited Disease,* J. B. Stanbury, J. B. Wyngaarden, and D. S. Fredrickson (eds.), McGraw-Hill, New York, p. 1128.

Pronk, C., Oldenziel, H., and Lequin, H. C., 1974. A method for determination of serum iron, total iron binding capacity, and iron in urine by atomic absorption spectrophotometry with manganese as internal standard, *Clin. Chim. Acta* 50:35–41.

Ramsay, W. N. M., 1958. Plasma iron, *Adv. Clin. Chem.* 1:12–19.

Schilt, A. A., and Hoyle, W. C., 1967. Improved sensitivity and selectivity in the spectrophotometric determination of iron by use of a new ferroin-type reagent, *Anal. Chem.* 39:114–117.

Schmidt, J. E. (ed.), 1959. *Medical Discoveries,* Charles Thomas, Springfield, Ill.

Searcy, R. L. (ed.), 1969. *Diagnostic Biochemistry,* McGraw-Hill, New York, p. 328.

Siimes, M. A., Vuori, E., and Kuitanen, P., 1979. Breast milk iron—a declining concentration during the course of lactation, *Acta Paediatr. Scand.* 68:29–31.

Tartour, G., 1973. The variation with age of the serum iron concentration and iron-binding capacity in Zebu cattle, *Res. Vet. Sci.* 15:389–391.

Tavenier, P., and Hellendoorn, H. B. A., 1969. A comparison of serum iron values determined by atomic absorption and by some spectrophotometric methods, *Clin. Chim. Acta* 23:47–52.

Tietz, N. W. (ed.), 1976. *Fundamentals of Clinical Chemistry,* 2nd ed., Saunders, Philadelphia, p. 929.

Tipton, I. H., and Cook, M. J., 1963. Trace elements in human tissue, Part II, Adult subjects from the United States, *Health Phys.* 9:103–143, quoted by M. Worwood, The clinical biochemistry of iron, *Semin. Hematol.* 14:3–30, 1977.

Underwood, E. J. (ed.), 1977a. *Trace Elements in Human and Animal Nutrition,* Academic Press, New York, pp. 13–55.

Underwood, E. J. (ed.), 1977b. *Trace Elements in Human and Animal Nutrition,* Academic Press, New York, p. 14.

Van Eijk, H. G., and Kroos, M. J., 1978. The iron status in healthy individuals aged from 18–25 years, *Folia Haematol.* (Leipzig) 105:93–95.

Van Laucker, J. L. (ed.), 1975. *Molecular and Cellular Mechanisms in Disease,* Spring, Berlin, p. 365.

Van Vugt, H., Van Gool, J., Ladiges, N. C. J. J., and Boers, W., 1975. Lactoferrin in rabbit bile: Its relation to iron metabolism, *Q. J. Exp. Physiol.* 60:79–88.

Weippl, G., Pantlitschko, M., Bauer, P., and Lund, S., 1973. Serumeisen-Normalwerte und statistische Verteilung der Einzelwerte bei Mann und Frau, *Blut* 27:261–269.
Werkman, H. P. T., and Trijbels, J. M. F., and Schretlen, E. D. A. M., 1974. The "short-term" iron rhythm, *Clin. Chim. Acta* 53:65–68.
Worwood, M., 1977. The clinical biochemistry of iron, *Semin. Hematol.* 14:3–30.

Ferrokinetics 2

Ferrokinetics is a term generally applied to the tracing of radioactive iron following its parenteral administration to an animal or human patient. This involves the mixing of isotopic ferrous or ferric iron in the citrate form with a serum sample to assure its binding with transferrin, injection into the organism, and the measuring of the levels of radioactive iron in plasma as a function of time. When the logarithm of plasma radioactive iron is plotted against time, a so-called iron decay curve is obtained which can yield valuable information in regard to the fate of the injected iron in the organism. In addition, various areas on the surface of the organism can be monitored for radioactive iron. This provides an estimate of radioactive iron moving into internal areas in the vicinity of such surface areas. The organs of interest are the liver, spleen, and bone marrow. In the case of the last, radioactivity present over the sacral region of the spinal column is believed to represent iron moving into the bone marrow.

2.1 Classical Approach to Ferrokinetics

It is said that Hahn *et al.* (1939) were first to use radioactive iron to study the fate of iron in a mammalian organism (Cavill and Rickets, 1974). This was followed up by numerous investigations in the post-World War II period as radioactive isotopes became commonly available. The work of Huff *et al.* (1950) is generally considered to be a classic in the area of ferrokinetics and the utilization of this method for the diagnosis of various hematologic disorders.

Upon the injection of ^{59}Fe into normal human beings, a typical straight-line decay of plasma ^{59}Fe was observed for at least 6 hr, as shown in Figure 2-1. The line can be described by the usual first-order reaction

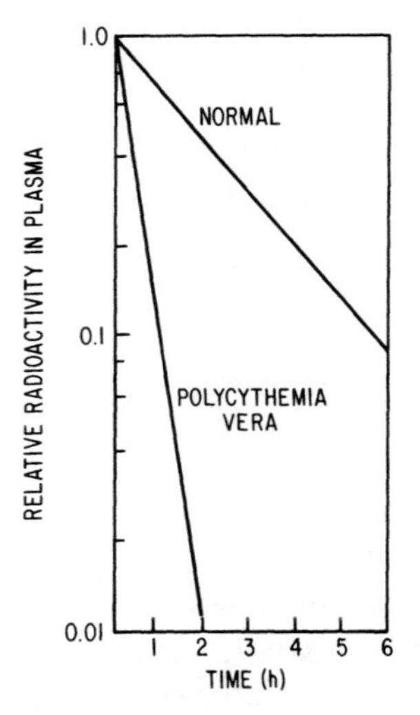

Figure 2-1. Six-hour plasma radioiron decay curve of a normal and polycythemic subject.

equation,

$$\frac{dx}{dt} = -kt \tag{2-1}$$

where t is time, x is radioactive iron, and k is a proportionality constant defined as the fraction of plasma turning over per unit time. The integrated form of the equation is

$$\ln x_t = \ln x_0 - kt \tag{2-2}$$

where x_t is plasma iron radioactivity at time t and x_0 is the initial radioactivity at 0 time. If $\ln x_t$ is plotted against t, as in Figure 2-1, the slope is given by $-k$.

The proportionality constant k may be written as the ratio of PIT, the plasma iron turnover (in mg per unit time) and total plasma iron Fe_p,

$$k = \frac{\text{PIT}}{\text{Fe}_p} = \frac{0.693}{t_{1/2}} \tag{2-3}$$

where $t_{1/2}$ is the half-life of radioactive plasma iron. Total plasma iron can be determined from iron content of plasma and total blood or plasma volume. PIT may be used to calculate erythrocyte iron turnover (EIT),

$$\text{EIT} = \text{PIT}\left(\frac{A_t}{A_0}\right) \tag{2-4}$$

where A_t is the radioactivity in the red cell mass upon equilibrium and A_0 is the total amount of radioactivity that was injected. A_t/A_0 is also called the red blood cell utilization (RBCU) and is commonly expressed in terms of a percentage. If EIT is divided by total red cell iron (Fe_c), the fraction of red cell iron incorporated per unit time (q) is obtained:

$$q = \frac{\text{EIT}}{\text{Fe}_c} \tag{2-5}$$

For normal male adults, Huff *et al.* (1950) determined the PIT to be 0.35 mg/day/kg of body weight, k was 0.41 (per hour), EIT was 0.26 mg/day/kg of body weight, and q was 0.0085/day. These parameters were also determined for a number of hematologic disorders and advocated as useful aids in diagnosis. Some of these determinations are shown in Table 2-1.

In a later paper, Huff *et al.* (1951) determined that radioactive iron was removed from plasma primarily by the bone marrow, which showed maximum radioactivity about 1 day after the iron injection. Radioactivity was then rapidly lost from the bone marrow to appear in the circulating erythrocytes, so that after about 2 weeks from the beginning of the experiment, some 80–95% of all injected radioactive iron was located there. In disease process situations where iron failed to enter the bone marrow, radioactivity was localized in the spleen and liver.

2.2 Compartment Models

The pioneering work of Huff *et al.* was further refined by Pollycove and Mortimer (1961), who pointed out that previous models had overestimated the rate of hemoglobin biosynthesis. To overcome this and other shortcomings, Pollycove and Mortimer proposed to analyze ferrokinetic data via the compartment theory, utilizing the iron clearance curve constructed from data collected over a period of at least 15 days. This type of curve (Figure 2-2) normally shows a biphasic character: a rapid iron disappearance portion as observed and analyzed by Huff *et al.* (1950) and a slowly disappearing portion that becomes manifest if observations are continued for more than 48 hr. The latter part of the clearance curve was postulated to represent a feedback or reflux of radioactive iron back into plasma from a tissue iron pool. Circulating erythrocytes were excluded as a source of this reflux iron, since their life span is over 100 days. Instead, it was proposed that there existed in the bone marrow a "labile" iron pool, which could return radioactive iron to plasma following its clearance. This labile pool was believed to be associated with erythrocyte stroma. By assuming now that most if not

Table 2-1. Ferrokinetic Data on Normal Individuals and in Some Erythropoietic and Iron Storage Disorders[a]

				D		E		F				
Category	A	B	C	J	K	L	M	N	O	G	H	I
Normals	0.35[b]	—	0.26[b]	—	—	—	—	—	—	—	—	1
	0.65		—	—	—	—	—	—	—	—	—	—
	32.5[c]	—	0.43	—	—	—	—	—	—	112.8	128	2
	0.75	82	—	10.3	0.06	7.7	0.19	0.41	0.09	85	110	3
	0.70	80	0.56	—	—	—	—	—	—	86	105	4
Hemochromatosis	51.9[c]	—	0.46	—	—	—	—	—	—	132	210	2
	1.19	78	—	10.3	0.12	4.7	0.23	0.66	0.18	107	197	3
	0.98	74	0.73	—	—	—	—	—	—	121	213	4
Polycythemia vera	1.81[b]	—	1.34[b]	—	—	—	—	—	—	—	—	1
	0.99	—	—	—	—	—	—	—	—	33.6	85.8	4
Iron deficiency	0.69	99	—	8.5	0.10	7.2	0.27	0.32	0.01	18	29	3
anemia (proliferative disorder)	0.80	94	0.97	—	—	—	—	—	—	20	21	4
Pernicious anemia	2.3[b]	—	0.38[b]	—	—	—	—	—	—	—	—	1
(maturation disorder)	2.39	28	0.62	—	—	—	—	—	—	51	159	4
Hemolytic anemia	3.2	59	—	10	0.15	15.3	1.49	1.43	0.13	—	—	3
(red cell disorder)	3.51	61	2.04	—	—	—	—	—	—	27	118.5	4

[a] Headings: A, plasma iron turnover, mg/100 ml of whole blood/day unless marked otherwise; B, red blood cell utilization, %; C, erythrocyte iron turnover, mg/100 ml of whole blood/day unless marked otherwise; D, early reflux; E, late reflux; F, fixed tissue iron turnover, mg/100 ml of whole blood/day; G, half-life, min; H, plasma iron concentration, μg/100 ml; I, Reference: (1) Huff *et al.* (1950), (2) Pollycove and Mortimer (1961), (3) Cook *et al.* (1970), and (4) Finch *et al.* (1970); J, early reflux: half-life, hr; K, early reflux: mg/100 ml of whole blood/day; L, late reflux: half-life, days; M, late reflux: mg/100 ml of whole blood/day; N, fixed tissue iron turnover: red blood cells; O, fixed tissue iron turnover: liver.

[b] mg/kg of body weight/day.

[c] mg/day.

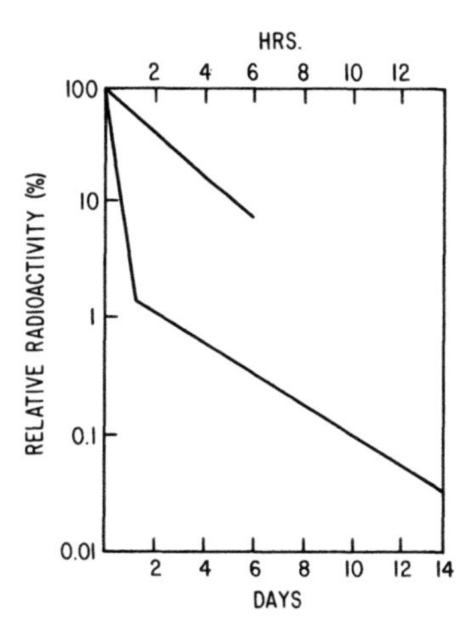

Figure 2-2. Fourteen-day plasma radioiron decay curve of a normal subject.

all radioactive iron leaves plasma to enter the bone marrow and that little if any goes into liver and other storage sites, the compartment model in Figure 2-3 can be proposed for the clearance of radioactive plasma iron in a normal individual.

On the basis of the model illustrated in Figure 2-3 and by looking at the first 2 weeks following the radioactive iron injection, the following series of differential equations can be written, where x_n is the radioactive iron in compartment n, t is time, and k_{nm} is the proportionality constant characterizing the movement of iron from compartment n to m. It is also assumed that all changes are of the first-order type:

$$\frac{dx_1}{dt} = -k_{12}x_1 + k_{21}x_2 \tag{2-6}$$

$$\frac{dx_2}{dt} = -k_{21}x_2 - k_{23}x_2 + k_{12}x_1 \tag{2-7}$$

$$\frac{dx_3}{dt} = -k_{34}x_3 + k_{23}x_2 \tag{2-8}$$

$$\frac{dx_4}{dt} = k_{34}x_3 \tag{2-9}$$

Solving these differential equations simultaneously using the Laplace transforms, for plasma we can write

$$x_{1t} = Ae^{-\lambda_1 t} + Ce^{-\lambda_3 t} \tag{2-10}$$

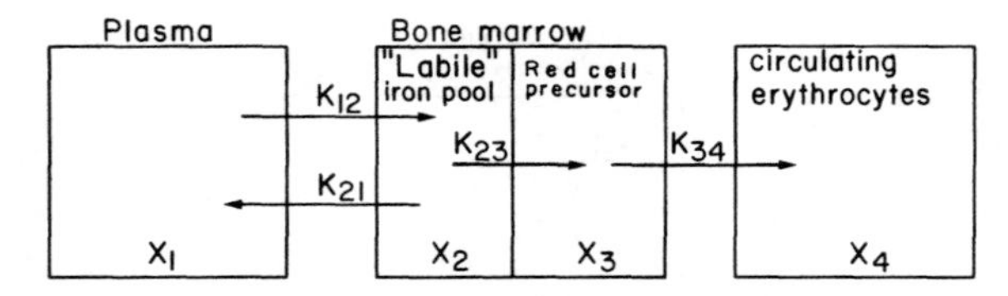

Figure 2-3. Ferrokinetic compartment model proposed for normal subjects by Pollycove and Mortimer (1961).

where A and C are constants (y-axis intercepts for each of the two portions of the curve in Figure 2-2), λ is the slope of each portion of the curve, and t is time. λ_1 characterizes the slope of the first (steepest) portion of the radioactive iron decay curve, and it was proposed that it is determined by the transfer of plasma iron into the labile iron pool of the bone marrow. λ_3, on the other hand, is determined by the rate of transfer of radioactive iron from the labile iron pool to the maturing erythrons.

In certain pathologic conditions where appreciable quantities of iron are transferred from the plasma into its liver storage sites (e.g., hemochromatosis or any condition of decreased rate of erythropoiesis), the model shown in Figure 2-3 must be amended to that shown in Figure 2-4. The equation characterizing the compartment system shown in Figure 2-4 is

$$x_{1t} = Ae^{-\lambda_1 t} + Be^{-\lambda_2 t} + Ce^{-\lambda_3 t} + E[1 - \psi(t)] \tag{2-11}$$

The terms $Ae^{-\lambda_1 t}$ and $Ce^{-\lambda_3 t}$ in equation (2-11) are identical to those of equation (2-10) whereas $Be^{-\lambda_2 t}$ describes the exchange of radioiron between plasma and a labile iron storage pool. The slope λ_2 is determined largely by the transfer of iron from the labile iron storage pool to the "fixed" iron storage pool. Ferrokinetic curves from hemochromatosis patients can usually be resolved into three portions with well-defined slopes λ_1, λ_2, and λ_3, as shown in Figure 2-5. The term $E[1 - \psi(t)]$ represents feedback from the fixed iron stores into plasma.

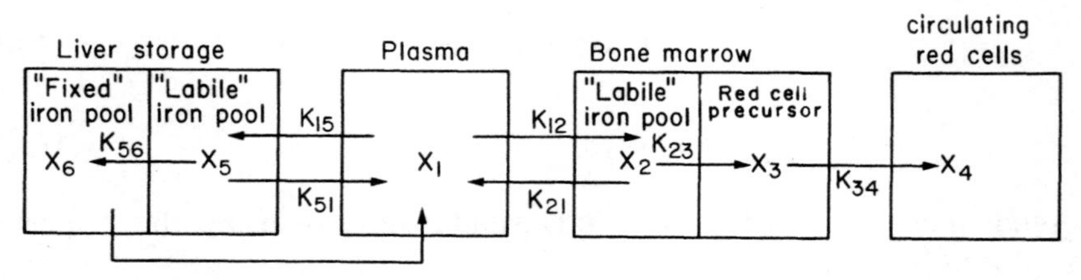

Figure 2-4. Ferrokinetic compartment model proposed for subjects that show considerable radioiron movement into the liver.

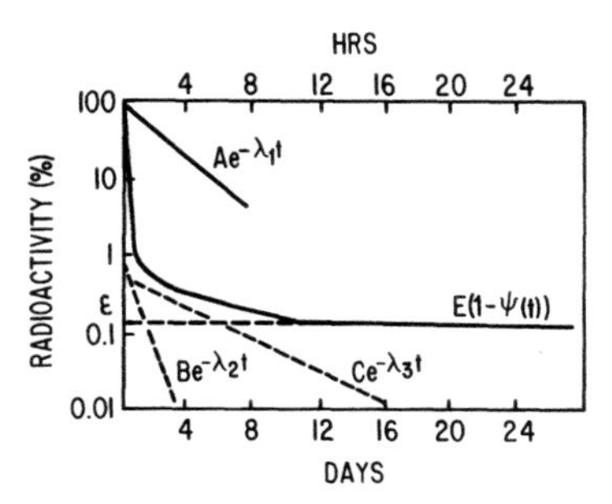

Figure 2-5. Plasma radioiron decay curve for subjects that show considerable movement of iron into the liver.

In the absence of hemolysis and if the experiment is restricted to less than 2 weeks, the ferrokinetic decay curve in a patient with considerable radioactive iron movement into the liver is approximated by

$$x_{1t} = Ae^{-\lambda_1 t} + Be^{-\lambda_2 t} + Ce^{-\lambda_3 t} \tag{2-12}$$

If the experiment is continued for over 2 weeks, then even in a normal individual what little radioiron did move into the storage sites begins to reappear in plasma, and this is especially true for patients with high initial liver radioiron uptake. The applicable ferrokinetic equation describing the clearance curve then must remain equation (2-11) where $\psi(t) = 1$ at $t = 0$, and at $\psi(t) = 0$, $t = \infty$. $\psi(t)$ is generally approximated to be equal to 0, so that E represents the final (nearly horizontal) portion of the radioiron clearance curve as indicated in Figure 2-5.

In the presence of hemolysis (e.g., hemolytic anemia of any origin), the appropriate ferrokinetic model system was modified to that shown in Figure 2-6, and the appropriate equation describing the curve was given as

$$x_{1t} = Ae^{-\lambda_1 t} + Ce^{-\lambda_3 t} + D[1 - \phi(t)] \tag{2-13}$$

where $D[1 - \phi(t)]$ has the same significance as $E[1 - \psi(t)]$. The iron clearance curve in a hemolytic patient is shown in Figure 2-7. Clearly, if both hemolysis and increased movement of radioiron into the liver are present, then the final nearly horizontal portion of the radioiron clearance curve will be the sum of $D[1 - \phi(t)]$ and $E[1 - \psi(t)]$, and the equation describing such a curve is

$$x_{1t} = Ae^{-\lambda_1 t} + Be^{-\lambda_2 t} + Ce^{-\lambda_3 t} + D[1 - \phi(t)] + E[1 - \psi(t)] \tag{2-14}$$

As was done in the previous work of Huff *et al.* (1951), Pollycove and Mortimer (1961) monitored the surface area of their patients for

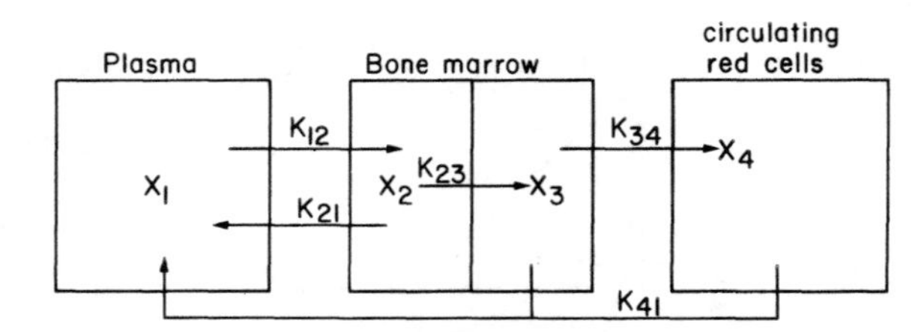

Figure 2-6. Ferrokinetic compartment model proposed for subjects with hemolytic anemia.

appearance of radioactivity. They, too, found that practically all the injected radioiron was normally localized in the circulating red cells some 7–10 days after radioiron administration. No radioiron was found over the liver and spleen (except that ascribable to blood circulating through that area). The bone marrow (as measured over the sacral area) showed maximum radioactivity 1–2 days following radioiron administration. These organ relationships are summarized in Figure 2-8.

When liver did show a slowly increasing accumulation of radioiron and subnormal incorporation of radioiron into red cells was observed, an iron storage disorder (e.g., hemochromatosis) or a condition with reduced erythropoiesis was suspected. This mandated the use of equations (2-11) or (2-12). Hemolytic conditions could be detected by an early stabilization of plasma radioiron levels and mandated the use of equation (2-13) to describe the radioiron decay curve.

The radioiron decay curve was used by Pollycove and Mortimer (1961) to estimate the half-life of plasma iron: the slopes λ_1, λ_2, and λ_3; the y-axis intercepts A, B, C, where $A + B + C = 1$; and the final "constant" radioiron levels E (in normal subjects and subjects with iron storage disorders) and D (in the case of hemolytic disorders). These experimentally determined parameters can then be utilized to calculate the various intercompartmental rate constants by the simultaneous

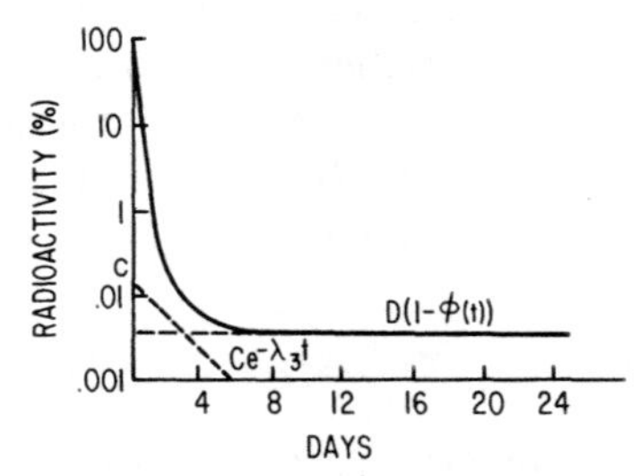

Figure 2-7. Plasma radioiron decay curve for subjects with hemolytic anemia.

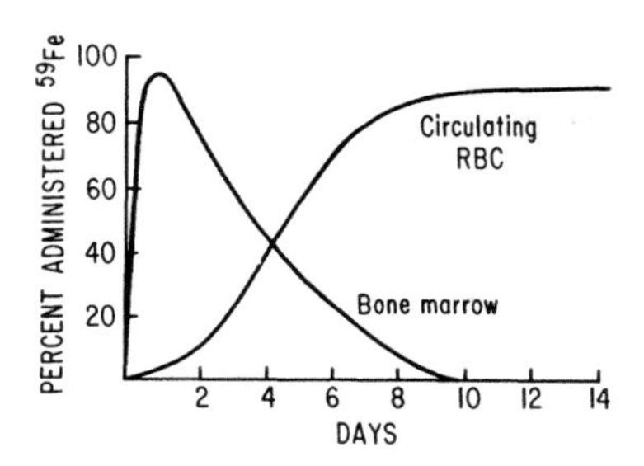

Figure 2-8. Fate of injected radioiron in a normal subject.

solution of the following equations:

$$A(\lambda_2 + \lambda_3) + B(\lambda_1 + \lambda_3) + C(\lambda_1 + \lambda_2) = k_{51} + k_{56} + k_{21} + k_{23} \tag{2-15}$$

$$A(\lambda_2\lambda_3) + B(\lambda_1\lambda_3) + C(\lambda_1\lambda_2) = (k_{51} + k_{56})(k_{21} + k_{23}) \tag{2-16}$$

$$\lambda_1 + \lambda_2 + \lambda_3 = k_{12} + k_{15} + k_{21} + k_{23} + k_{51} + k_{56} \tag{2-17}$$

$$\lambda_1\lambda_2 + \lambda_1\lambda_3 + \lambda_2\lambda_3 = k_{12}(k_{23} + k_{51} + k_{56}) + k_{15}(k_{56} + k_{21} + k_{23}) + (k_{51} + k_{56})(k_{21} + k_{23}) \tag{2-18}$$

$$\lambda_1\lambda_2\lambda_3 = k_{12}k_{23}(k_{51} + k_{56}) + k_{15}k_{56}(k_{21} + k_{23}) \tag{2-19}$$

$$f = \frac{[k_{23}/(k_{21} + k_{23})]k_{12}}{[k_{23}/(k_{21} + k_{23})]k_{12} + [k_{56}/(k_{51} + k_{56})]k_{15}} \tag{2-20}$$

where f is the fraction of injected radioiron appearing in circulating red cells at approximately 2 weeks following radioiron injection.

The various k values can then be used to calculate the various hematologic parameters as follows:

1. Amount of iron in the labile bone marrow iron pool:

$$x_2 = \frac{k_{12}\mathrm{Fe}_p}{k_{21} + k_{23}} \quad \text{(mg)} \tag{2-21}$$

2.

$$\mathrm{EIT} = k_{23}x_2 \quad \text{(mg/day)} \tag{2-22}$$

or

$$\text{daily hemoglobin synthesis} = \frac{k_{23}x_2}{3.4} \quad \text{(g)} \tag{2-23}$$

where 1 g of hemoglobin contains 3.4 mg of iron.

3. Mean erythron life span (MELS):

$$\text{MELS} = \frac{3.4(\text{total body hemoglobin})}{k_{23}x_2} \quad \text{(days)} \qquad (2\text{-}24)$$

where total body hemoglobin may be estimated from blood hemoglobin content and blood volume.

4. The mean effective erythron hemoglobinization time (MEEHT) is the time interval between the appearance of a certain amount of radioiron in the bone marrow and an identical amount of radioiron in the circulating red cells. This is illustrated graphically in Figure 2-9.

$$\text{MEEHT} = t_{1/2}N - t_{1/2}F \quad \text{or} \quad t_{1/5}N - t_{1/5}F \qquad (2\text{-}25)$$

whichever is greater (days or hours). N is here defined as the amount of radioiron present in ciculating red cells, whereas F is the radioiron level in maturing erythrons.

5. Amount of iron in the labile iron storage pool, x_5:

$$x_5 = \frac{k_{15}x_1}{k_{51} + k_{56}} \quad \text{(mg)} \qquad (2\text{-}26)$$

and

$$\text{iron stored per day} = k_{56}x_5 \qquad (2\text{-}27)$$

6. Iron leaving plasma per day, PIT:

$$\text{PIT} = k_{12}x_1 \qquad (2\text{-}28)$$

Data obtained by Pollycove and Mortimer in normal subjects and patients with endogenous hemochromatosis are summarized in Table 2-2.

Many modifications of the compartment approach (also termed *deterministic*) to ferrokinetics have been proposed at one time or another,

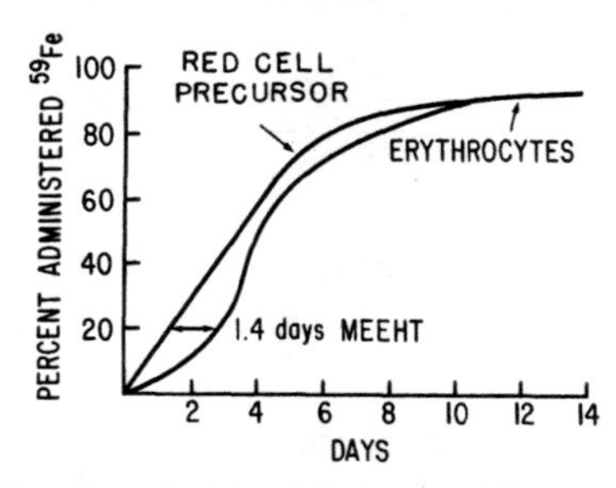

Figure 2-9. Curves illustrating the principle of the mean effective erythron hemoglobinization time (MEEHT).

Table 2-2. Hematologic Parameters Reported by Pollycove and Mortimer (1961)

Parameter[a]	Normal males	Endogenous hemochromatosis
$T_{1/2}$ (hr)	1.88 ± 0.38	2.20
Fe_p (mg)	3.6 ± 0.7	7 ± 1.65
λ_1	9.3 ± 2.4	7.70
λ_2	—	1.483
λ_3	0.254 ± 0.039	0.241
A	0.9835	0.981
B	—	0.0142
C	0.0165 ± 0.0054	0.0053
PIT (mg/day)	32.5 ± 5.7	51.9 ± 10.1
EIT (mg/day)	21.4 ± 3.2	24.7 ± 7.6
"Labile" Fe pool (mg)	84.9 ± 19.9	109.1 ± 18.3
MELS (days)	117 ± 7.5	108 ± 5
MEEHT (days)	1.4 ± 0.2	1.6 ± 0.5

[a] For abbreviations, see the list at the end of this chapter.

and several are reviewed by Cavill and Ricketts (1974). The theory of Pollycove and Mortimer (1961), however, has remained the staple of many hematology and clinical chemistry texts and as a diagnostic tool in some hematologic laboratories (see, e.g., Pollycove, 1978) even in view of more recent theoretical and methodological developments in the area of ferrokinetics.

2.3 Approaches to Ferrokinetics Based on the Probability Theory

A more modern approach toward the analysis of the basic 14-day plasma radioiron decay curve is to view plasma iron levels in terms of a percolating system, where radioiron levels would be the sum of that which is yet to be cleared and that which has refluxed back into the plasma from whatever source. This is the so-called stochastic model of iron turnover, and it utilizes the probability theory to calculate the sojourn time for radioiron from each reflux source. The stochastic approach makes no assumptions in regard to the existence of iron compartments in the organism. On the contrary, the stochastic analysis of the plasma radioiron decay curve should provide a basis for the construction of such models.

The basis for the stochastic approach to multicompartment systems in general was provided by Marsaglia (1963), who proposed that in a system with multiple refluxes, the decay curve can be described by a sum of exponential terms. The number of such terms is one more than the number of refluxes present. This proposal was applied to ferrokinetic data by Hosain *et al.* (1967), who pointed out that the previous compartment models had overestimated red blood cell turnover rates and that the existence of the labile bone marrow iron pool proposed by Pollycove and Mortimer (1961) could not be substantiated experimentally (Noyes *et al.*, 1964). Moreover, previous workers had not taken into account diurnal plasma iron fluctuations, where plasma samples taken during a 24-hr period can differ in iron content by as much as 40%. Taking such fluctuations into consideration allows for a much more precise curve fitting. The equation derived to represent the disappearance of plasma radioactive iron [$f(t)$ is the amount of radioiron in plasma at time t] was

$$\frac{df(t)}{dt} = -hf(t) + h\int_0^t g(y)f(t-y)\,dy \tag{2-29}$$

where h is the slope of the initial leg of the plasma radioiron disappearance curve, and the integral term represents the sum total of all radioactive feedback (refluxes) into plasma, where $g(y)$ is the reflux density function. This equation thus represents a situation where there is a sustained and constant disappearance of radioactive iron from plasma [$-hf(t)$] which is affected by an indeterminate number of refluxes.

Hosain *et al.* (1967) studied a number of normal radioiron clearance curves, making appropriate corrections for diurnal plasma iron variations, and determined that such curves could be described by the three-term exponential equation

$$f(t) = Ae^{-\lambda_1 t} + Be^{-\lambda_2 t} + Ce^{-\lambda_3 t} \tag{2-30}$$

indicating, from Marsaglia's theory, that there were two refluxes operating. Note that equation (2-30) is identical to equation (2-12) as derived by Pollycove and Mortimer for special cases where substantial amounts of radioiron were taken up by the liver. Equation (2-30) was then used to solve the more general equation (2-29), from which it was estimated that some 35% of radioiron leaving plasma is refluxed back. One reflux comprising 5–10% of total plasma iron has a half-life of about 8 hr and is called the early reflux. The other reflux comprising about 25% of plasma iron has a half-life of about 8 days and is called the late reflux. Early reflux represents the exchange of transferrin-bound iron between plasma and the extravascular space (lymph), whereas the late reflux is representative of erythropoietic "wastage" iron, i.e., iron which is incorporated into the bone marrow but does not find its way into circulating red cells. It returns to plasma via the reticuloendothelial system.

Another and easier way of estimating the degree of radioactive iron refluxing into plasma is by looking at the area(s) subtended by the radioactive iron decay curve (Cook *et al.*, 1970). If the initial rapid decay portion of the curve is extended, the area subtended by it represents the radioactive iron fixed by the tissues (FTT), including circulating red cells and the liver. The area subtended by the entire curve represents both the fixed and refluxed radioactivity. Refluxed radioactivity is thus (in percentage of total radioactivity) equal to 100 − FTT. The ferrokinetic parameters, in addition to FTT, calculated by Finch's group (Hosain *et al.*, 1967; Cook *et al.*, 1970; Finch *et al.*, 1970) were the following:

1. $$\text{PIT} = \frac{\text{PI}(0.693)(1440)}{t_{1/2}\,(\text{min})}\,\frac{100 - 0.9\text{Hct}}{100} \qquad (2\text{-}31)$$

   where PIT is in mg/100 ml of whole blood/day, PI is plasma iron concentration, and Hct is hematocrit.

2. $$\text{RBCU} = \left(\frac{A_{14}}{A_0}\right)(0.9)(\text{Hct})(\text{blood volume}) \quad (\%) \qquad (2\text{-}32)$$

3. $$\text{EIT} = \text{PIT}(\text{RBCU}) \quad (\text{mg/100 ml of whole blood/day}) \qquad (2\text{-}33)$$

4. Fixed erythrocyte turnover:

   $$\text{FET} = \text{FTT}(\text{RBCU}) \qquad (2\text{-}34)$$

5. Fixed parenchymal (liver) iron turnover:

   $$\text{FPT} = \text{FTT} - \text{FET} \qquad (2\text{-}35)$$

6. Erythron iron turnover:

   $$\text{EIT} = \text{FET} + \text{late reflux} \qquad (2\text{-}36)$$

7. $$\text{nonerythron iron turnover} = \text{FPT} + \text{early reflux} \qquad (2\text{-}37)$$

A physiological model has been proposed for iron metabolism on the basis of the stochastic approach to ferrokinetics and is represented in Figure 2-10. This model has so far withstood the test of time.

Information derived from ferrokinetic measurements has been used as an aid in the diagnosis of a variety of hematologic diseases. In fact, it has been pointed out that these measurements have been used as the basis for classifying various anemias into disorders of proliferation (decreased levels of erythropoietin, renal failure, hypothyroidism, iron deficiency anemia, infection, rheumatoid arthritis, nonhematologic malignancy, Hodgkin's disease, primary hypoplasia, multiple myeloma, several types of leukemia), disorders of maturation (megaloblastic or macrocytic anemias, sprue, the thalassemias, refractory anemia), and

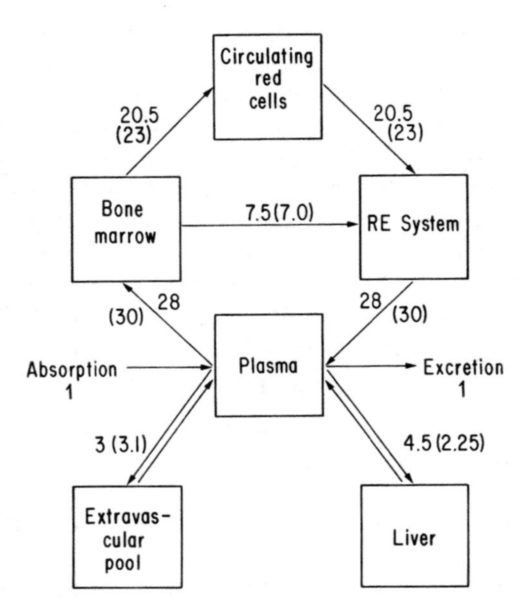

Figure 2-10. Currently accepted model of iron metabolism. Numbers indicate the approximate daily amount (in mg) of iron turned over; numbers without parentheses were obtained by Finch's group and numbers in parentheses by the British group (see the text).

disorders of red cell physiology or structure (spherocytosis, sickle cell anemia, various hemolytic anemias) (Finch *et al.*, 1970). Finch's group, like its predecessors, has provided ferrokinetic data on a number of hematologic conditions, some of which are summarized in Table 2-1.

It is interesting to note that the ineffective erythropoietic pathway ("wastage") is considerably greater in aged individuals than it is in normal adults (Marx, 1979). Careful ferrokinetic studies on aged individuals have not yet been done and should be able to provide some very interesting and valuable information on the relationship between the aging process and iron metabolism.

A refinement of the iron reflux model proposed by Finch's group was provided by Ricketts *et al.* (1975). It was claimed that their approach would provide a better estimate of "total erythroid activity, effective and ineffective erythropoiesis, and the red cell life span." It was pointed out that radioactive iron channeled daily to the red cells may have been refluxed through plasma several times. If, as before, we define $f(t)$ as the amount of radioactive iron remaining in plasma at any time, h as the plasma iron clearance rate [$h = df(t)/dt$], $u(t_1)$ as the fraction of the administered dose of radioiron found in red cells after some time interval (e.g., 14 days), and z_1 as the constant proportion of plasma iron incorporated into red cells, then we can write

$$u(t_1) = z_1 h \int_0^{t_1} f(t)\, dt \tag{2-38}$$

$$z_1 = u(t_1) \Big/ h \int_0^{t_1} f(t)\, dt \tag{2-39}$$

This equation simply states that total iron incorporated into red cells after, say, 14 days is the product of z_1 and all the radioactive iron that has passed through plasma up to 14 days. It then follows that

$$\text{red cell iron turnover (RCIT)} = z_1(\text{PIT}) \tag{2-40}$$

The term z_1 can be evaluated from the plasma radioiron decay curve utilizing a computer program.

Additionally, another constant z_2 can be defined as

$$z_2 = \frac{C_2}{b_2} \tag{2-41}$$

where C_2 and b_2 are terms in the reflux density expression $g(y)$,

$$g(y) = C_1 e^{-b_1 y} + C_2 e^{-b_2 y} \tag{2-42}$$

Equation (2-42) describes the late reflux of Cook *et al.* (1970). The constant z_2 can be used to arrive at the so-called ineffective red cell iron turnover (IIT) by

$$\text{IIT} = z_2(\text{PIT}) \tag{2-43}$$

Total marrow iron turnover (MIT) then becomes

$$\text{MIT} = \text{RCIT} + \text{IIT} \tag{2-44}$$

Similarly, if the reflux density term representing the early reflux in equation (2-42) is considered, we can define another constant z_3 as

$$z_3 = \frac{C_1}{b_1} \tag{2-45}$$

The constant z_3 is in turn used to calculate extravascular iron turnover (XIT):

$$\text{XIT} = z_3(\text{PIT}) \tag{2-46}$$

Tissue iron turnover (TIT) then becomes

$$\text{TIT} = \text{PIT} - \text{MIT} - \text{XIT} \tag{2-47}$$

Table 2-3. Ferrokinetic Data Compiled from Ricketts *et al.* (1975), Ricketts *et al.* (1977), and Cavill *et al.* (1977)[a]

Category	A	B	C	D	E	F	G	H	I	J	K	L
Normals	106.4	86	84	105	0.72	0.60	0.60	0.46	0.14	23	0.062	0.045
Megaloblasic anemia	128.8	37	48	56	2.56	1.05	2.24	0.58	1.66	74	0.031	0.017
Iron definiency anemia	22.4	26	85	78	0.65	0.55	0.54	0.36	0.18	33	—	—
Hemolytic anemia	89.6	32	66	—	2.35	1.33	1.80	1.03	0.77	43	—	—
Hemochromatosis	207.2	118	66	—	1.06	0.71	0.77	0.49	0.28	36	—	—

[a] Headings: A, serum iron, μg/100 ml; B, half-life, min; C, red blood cell utilization, %; D, mean red cell life span, days; E, plasma iron turnover, mg/100 ml of whole blood/day; F, erythrocyte iron turnover, units as in E; G, marrow iron turnover, units as in E; H, red cell iron turnover, units as in E; I, ineffective red cell iron turnover, units as in E; J, percent ineffective iron turnover; K, extravascular space iron turnover, mg/100 ml of whole blood/day; L, tissue iron turnover, units as in K.

Red cell life spane was defined as

$$\text{MELS} = \frac{\text{total red cell iron}}{\text{RCIT}} \tag{2-48}$$

Ferrokinetic data obtained through the use of the preceding equations are summarized in Table 2-3 and Figure 2-10.

It has been pointed out by several authors that in spite of many warnings, routine hematological laboratories have used $t_{1/2}$, RBCU, and PIT as the sole criteria for assessing erythropoietic situations in human patients (e.g., Cavill and Ricketts, 1978). As it became clear that these simple determinations were valid diagnostic tools only in certain limited instances, ferrokinetic measurements have become less popular in the routine laboratory. To arrive at a meaningful diagnosis through the use of ferrokinetic information, it is necessary to handle the data rigorously by the methods of Pollycove and Mortimer (1961), Finch *et al.*, (1970), or Cavill and Ricketts (1978). This requires specialized computer facilities that are available only in large medical centers. More importantly, however, the ferrokinetic approach remains a powerful tool in the hands of the biomedical researcher who is interested in studying the fate of iron in various pathological conditions (as opposed to routine diagnosis) or the metabolism of iron in lower animal forms (e.g., Barosi *et al.*, 1978; Cavill and Jacobs, 1971; Najean *et al.*, 1978; Douglas *et al.*, 1971; and Balasch and Planas, 1972).

Summary

The so-called ferrokinetic studies involve the parenteral administration of trace doses of radioactive iron to an animal, followed by the determination of the remaining radioactivity in plasma and red cells for 14 or more days. Normally the decay curve shows two phases: a rapidly descending leg representing the constant removal of iron from the bloodstream and a flatter portion, which represents the reflux of iron into plasma from various tissues. After 14 days, most if not all of the administered iron is found in the erythrocytes.

The most readily determinable parameters are the half-life, which is about 90 min in the normal adult; the plasma iron turnover (PIT), which is near 35 mg/day; red blood cell utilization (RBCU), which is the percentage of administered radioactivity found in circulating red cells after 14 days; and erythrocyte iron turnover (EIT), which is PIT(RBCU). Although these parameters do provide a good view of iron metabolism in the normal individual, for most hematologic disorders they alone are not sufficient to provide a reliable diagnostic tool.

Several attempts have been made to provide a mathematical model for the iron clearance curve constructed over a sampling period of 14 days or longer. The compartment model of Pollycove and Mortimer proposed the existence of a labile erythron iron pool and resulted in the derivation of a two-term exponential equation for the normal subject, where radioiron leaves plasma almost exclusively for the red blood cells:

$$x_1 = Ae^{-\lambda_1 t} + Ce^{-\lambda_3 t}$$

Where significant amounts of iron are also transported into the liver iron stores, a three-term equation best describes the radioiron clearance curve:

$$x_1 = Ae^{-\lambda_1 t} + Be^{-\lambda_2 t} + Ce^{-\lambda_3 t}$$

where each term describes the appropriate phases of the radioiron clearance curve.

A mathematical model may also be constructed by assuming that radioactivity in plasma is the sum of the as yet uncleared radioiron and iron that has refluxed back into the plasma from various tissues or spaces. Not suprisingly, the three-term equation best describes the radioiron clearance curve. It follows that there must be two significant iron refluxes present in plasma. These are extravascular transferrrin-bound iron and "wastage" iron incorporated into bone marrow but not used for circulating red cell biosynthesis. These are termed the early and late refluxes, respectively. Finch's group has provided methods for the calculation of the half-lives of the two refluxes and fixed erythrocyte and liver iron turnovers. Cavill and Ricketts have refined the reflux model of Finch *et al.* and have defined such terms as effective and ineffective erythropoiesis, tissue and extravascular iron turnover, and marrow iron turnover. The calculation of the various parameters defined by Pollycove and Mortimer, Finch *et al.*, and Cavill and Ricketts through computer-assisted curve fitting and solution of the three-term exponential equation can provide significant information for the diagnosis of various hematologic disorders.

The model that best describes iron metabolism, with the daily amounts of iron turned over, may be described as follows: Plasma loses 28–30 mg of iron to the bone marrow, of which 20–23 mg end up in circulating red cells and the rest ("wastage") in the reticuloendothelial system. The latter returns to plasma 28–30 mg of iron, 20–23 mg of which comes from senescent red cells and the rest from wastage. Plasma also exchanges about 3 mg of iron with extravascular space and 2.5–4.5 mg with the liver storage sites. Some 1 mg is absorbed and excreted.

List of Additional Abbreviations Used in Chapter 2

A to *E* ....*y*-axis intercepts of the various portions of the plasma radioiron decay curve
A_0 ....total radioiron administered to animal or patient
A_t ....total radioiron present in circulating red cells at time *t*
EIT ....erythrocyte iron turnover
F ....radioactivity in maturing erythrons, exclusive of labile iron pool
Fe_c ....total red cell iron
Fe_p ....total plasma iron
FET ....fixed erythrocyte iron turnover
FPT ....fixed parenchymal (liver) iron turnover
FTT ....fixed tissue iron turnover
f ....fraction of administered radioiron appearing in red cells in 14 days
$f(t)$ ....amount of radioiron in plasma at time *t*
$g(y)$ ....reflux density function
h ....initial slope of plasma radioiron decay curve
Hct ....hematocrit
IIT ....ineffective red cell iron turnover
k_{nm} ....proportionality constant characterizing movement of radioiron from compartment *n* to *m*
MEEHT .mean effective erythron hemoglobinization time
MELS ....mean erythron life span
MIT ....marrow iron turnover
N ....net incorporation of iron into circulating erythrocytes
PI ....plasma iron concentration
PIT ....plasma iron turnover
q ....fraction of total red cell iron turned over per unit time
RBCU ....red blood cell utilization
RCIT .....red cell iron turnover
TIT ....tissue iron turnover
t ....time
$t_{1/2}$ ....half-life
$u(t_1)$ ....fraction of injected radioiron in red cells at time *t*
XIT ....extravascular space iron turnover
x_n ....amount of radioiron in compartment *n*
z_1 ....proportion of plasma iron which is incorporated into red cell mass
z_2 ....fraction of PIT which returns to plasma from the late reflux

z_3 ....fraction of PIT which returns to plasma from the early reflux
λ_n ....slope of the n portion of the plasma radioiron decay curve

References

Balasch, J., and Planas, J., 1972. Iron metabolism in duck and turkey, *Rev. Esp. Fisiol.* 28:125–128.

Barosi, G., Cazzola, M., Morandi, S., Stefanelli, M., and Perugini, S., 1978. Estimation of ferrokinetic parameters by a mathematical model in patients with primary acquired sideroblastic anemia, *Br. J. Haematol.* 39:409–423.

Cavill, I., and Jacobs, A., 1971. Iron kinetics in the skin, *Br. J. Haematol.* 20:145–153.

Cavill, I., and Ricketts, C., 1974. The kinetics of iron metabolism, in *Iron in Biochemistry and Medicine,* A. Jacobs and M. Worwood (eds.), Academic Press, New York, pp. 613–647.

Cavill, I., and Ricketts, C., 1978. Erythropoiesis and iron kinetics, *Br. J. Haematol.* 38:433–437.

Cavill, I., Ricketts, C., Napier, J. A. F., and Jacobs, A., 1977. Ferrokinetics and erythropoiesis in man: red cell production and destruction in normal and anemic subjects, *Br. J. Haematol.* 35:33–40.

Cook, J. D., Marsaglia, G., Eschbach, J. W., Funk, D. D., and Finch, C. A., 1970. Ferrokinetics: A biologic model for plasma iron exchange in man, *J. Clin. Invest.* 49:197–205.

Douglas, T. A., Renton, J. P., and Watts, C., 1971. Placental transfer of iron in the rabbit, *Br. J. Haematol.* 20:185–194.

Finch, C. A., Denbelbeiss, K., Cook, J. D., Eschbach, J. W., Harker, L. A., Funk, D. D., Marsaglia, G., Hillman, R. S., Slichter, S., Adamson, J. W., Ganzoni, A., and Giblett, E. R., 1970. Ferrokinetics in man, *J. Clin Invest.* 49:17–53.

Hahn, P. F., Bale, W. F., Lawrence, E. O., and Whipple, G. H., 1939. Radioactive iron and its metabolism in anemia; its absorption, transportation and utilization, *J. Exp. Med.* 69:739–753.

Hosain, F., Marsaglia, G., and Finch, C. A., 1967. Blood ferrokinetics in normal man, *J. Clin. Invest.* 46:1–9.

Huff, R. L., Hennessy, T. G., Austin, R. E., Garcia, J. F., Roberts, B. M., and Lawrence, J. H., 1950. Plasma and red cell iron turnover in normal subjects and in patients having various hematologic disorders, *J. Clin. Invest.* 29:1941–1052.

Huff, R. L., Elmlinger, P. J., Garcia, J. F., Oda, J. M., Cockrell, M. C., and Lawrence, J. H., 1951. Ferrokinetics in normal persons and in patients having various erythropoietic disorders, *J. Clin. Invest.* 30:1512–1526.

Marsaglia, G., 1963. Stochastic analysis of multicompartment systems, *Math. Note 313, Di-82-0280,* Boeing Scientific Research Laboratories, Seattle, quoted by Hosain *et al.*, 1967, and by Finch *et al.*, 1970.

Marx, J. J. M., 1979. Normal iron absorption and decreased red cell iron uptake in the aged, *Blood* 53:204–211.

Najean, Y., Cacchione, R., Castro-Malaspina, H., and Dresch, C., 1978. Erythrokinetic studies in myelofibrosis: Their significance for prognosis, *Br. J. Haematol.* 40:205–217.

Noyes, W. D., Hosain, F., and Finch, C. A., 1964. Incorporation of radioiron into marrow heme, *J. Lab. Clin. Med.* 64:574–580.

Pollycove, M., 1978. Hemochromatosis, in *The Metabolic Basis of Inherited Disease,* J. B. Stanbury, J. B. Wyngaarden, and D. S. Fredrickson (eds.), McGraw-Hill, New York, pp. 1127–1164.

Pollycove, M., and Mortimer, R., 1961. The quantitative determinants of iron kinetics and hemoglobin synthesis in human subjects, *J. Clin. Invest.* 40:753–782.

Ricketts, C., Jacobs, A., and Cavill, I., 1975. Ferrokinetics and erythropoiesis in man: The measurement of effective erythropoiesis, ineffective erythropoiesis, and red cell life span using ^{59}Fe, *Br. J. Haematol.* 31:65–75.

Ricketts, C., Cavill, I., Napier, J. A. F., and Jacobs, A., 1977. Ferrokinetics and erythropoiesis in man: An evaluation of ferrokinetic measurements, *Br. J. Haematol.* 35:41–47.

Absorption of Nonheme Iron*

3.1 Introduction

McCance and Widdowson (1937) proposed the theory that the amount of iron in the body must be regulated by controlled absorption with excretion by the bowel practically nil and with very little excretion via the kidney. Subsequent data reinforced this view, and research focused on absorption. Despite the volume of research conducted since that time, there are still considerable gaps in our understanding of the mechanism of iron absorption and its control. Discrepancies in the data on which present concepts are based complicate interpretation.

Historically the study of iron absorption has gone through a number of phases coinciding with changes in technology. Before 1930, methods were poor, and although there were a few balance studies, the results obtained are questionable. Between 1930 and 1940 the methods were sufficiently developed to permit better clinical studies using the chemical balance technique. After 1940 there was more widespread use of radioisotopes, and the total number of studies and the number of subjects per study increased. With changes in technique, studies concentrated not only on absorption in whole animals but also on intestinal absorption by intestinal segments, isolated mucosa, and fractions of the mucosal cells. Studies of iron absorption can be categorized as those investigating the mechanism and regulation of absorption, those investigating absorption from food and interactions between dietary components, and those comparing different iron salts for use in fortification. Practical applications are the diagnosis of prelatent iron deficiency and prevention of iron deficiency by appropriate fortification.

* By Dorice Narins, Ph.D., Rush-Presbyterian–St. Luke's Medical Center.

Absorption of minerals, such as iron, appears to involve several steps: reduction and release from conjugation, uptake by the epithelial cells of the intestine, and serosal transfer. Iron is sequestered by the epithelial cell in variable amounts and may or may not reach the plasma. There appears to be a separate control mechanism for the uptake by the cell and by the plasma. Epithelial cells monitor the body's requirements by a mechanism not yet completely understood.

In this chapter, the amount of iron absorbed means the amount which passes from the lumen of the intestine into the bloodstream. The process of absorption includes mucosal uptake, mucosal transfer, and serosal transfer of a proportion of the iron present in the cell to the plasma. Important factors in total food iron absorption are the form of the iron, heme, or nonheme; total iron in the diet; conditions within the gastrointestinal tract; amounts and proportions of various other components in the diet; and the physiological state of the individual.

Partial and complete reviews of iron absorption have appeared with regularity (Josephs, 1958; Crosby, 1968; Pinkerton, 1969; Bothwell and Charlton, 1970; Conrad, 1970; Heinrich, 1970; Fairbanks *et al.,* 1971; Turnbull, 1974; Forth and Rummel, 1973; Jacobs, 1973; Cook, 1977; and Cook and Lipschitz, 1977). In this chapter we attempt to cover recent work without neglecting the historical development of studies of iron absorption.

3.1.1 Quantitative Significance of Iron in the Body

Sixty to seventy percent of the body iron is functional iron incorporated in hemoglobin, myoglobin, and certain respiratory enzymes. The remainder is storage iron. In the healthy adult there are 3–5 g of iron, the male adult has 40–50 mg of iron per kg of body weight, and the female 35–50 mg/kg of body weight. In the newborn there is approximately 250 mg total or 70 mg/kg of body weight. Premature infants are born with lower iron stores. Growing children must absorb approximately 0.5 mg in excess of body losses to meet the needs of growth and build up storage. Adult males lose 1 mg of iron daily, two-thirds as desquamated epithelium and secretions from the gut and most of the remainder from the skin and urine. Losses are reduced to half normal in patients with iron deficiency and increased in patients with iron overload (Bothwell and Charlton, 1970). Stores in the male are approximately 900–1000 mg of iron. During the childbearing years adult females lose, on the average, an additional 30 mg/month (ranges, 3–80 mg) from menstrual bleeding and about 800 mg with each pregnancy. In the adult female, approximately 300 mg are stored. The lower storage by the female is due to the lower intake and greater losses. Various surveys have shown that 10–30% of adult women have no detectable iron reserves even in countries with iron fortification programs. Iron deficiency is also common among infants under 2 years of age and adolescent females.

Normal diet provides a sufficient total amount of iron to more than adequately replace losses. However, much of that iron is not available. Since the average diet contains 6 mg of iron per 1000 kcal, adult males can maintain a positive iron balance with an absorption rate of only 5% of their daily intake. With an increase in absorption to 20% in iron deficiency, men can tolerate losses of approximately 3.5 mg daily without becoming progressively anemic. In women, the balance is more difficult to maintain. With an intake of 10–12 mg of iron daily, women average 12–15% absorption and with iron deficiency can only reach a maximum of 2 mg. Because of their increased iron requirement for growth and lower intake, children also have a difficult time maintaining a positive iron balance.

The level of absorption is sensitive to the iron status of the body. As stores are reduced, absorption is greatly increased. When reduced intake continues for a period of time, there are several stages in the development of iron deficiency and finally iron deficiency anemia. Figure 3-1 shows the developmental pattern of iron depletion, showing the biochemical parameters indicative of the approximate iron status at each stage of de-

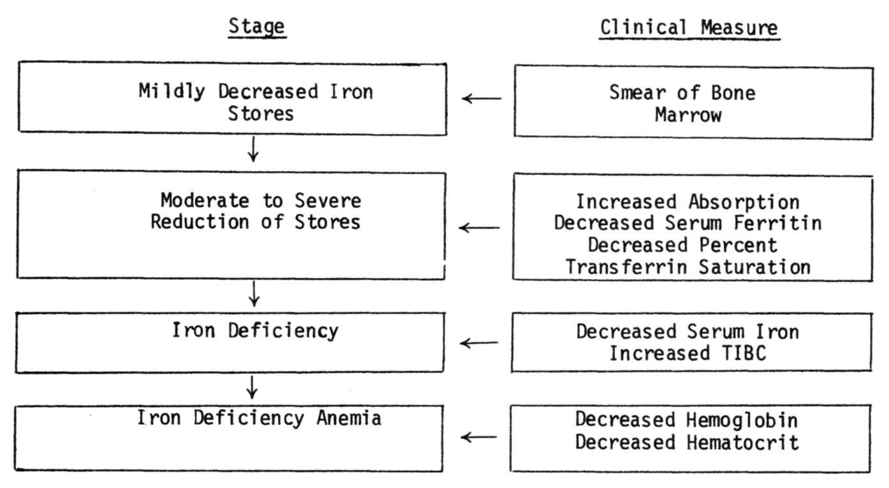

Figure 3-1. Stages in the development of iron deficiency and iron deficiency anemia together with the biochemical measurements used to identify the various stages.

velopment. There is disagreement regarding the relative sensitivity of some of the measures, particularly in the early stages of the development of the deficiency. There are also some person-to-person variations which make it difficult, at the present state of our knowledge, to rank the various measures with complete accuracy. According to some data (Heinrich, 1970), patients with reduced iron stores have greatly increased iron absorption in the presence of normal hemoglobin and serum iron concentrations. If iron stores are estimated by measuring serum ferritin, values indicative of reduction of stores occur concurrently with changes in percent transferrin saturation. Elevation of total iron binding capacity occurs before the decrease in serum iron (Ballas, 1979). Absorption of iron is not a clinical tool at the present time; therefore, other measures must be used to determine iron status. The study of iron absorption does, however, provide us with information on the factors which play a role in the pathogenesis of clinical disorders of iron balance and on which of the forms of iron used for supplementation will be most effective in the prevention of iron deficiency anemia.

3.1.2 Heme Iron

Absorption of heme iron is covered in another volume of this series; however, for the sake of completeness and because many investigations cover absorption of both heme and nonheme iron, a few brief comments are in order. Heme iron released from foods in the stomach enters the intestinal epithelial cell as the heme moiety with the porphyrin ring intact. Within the mucosal cell, heme is catabolized by heme oxygenase and enters the same pathways to either storage or transport as does nonheme iron.

Absorption of heme iron is less affected by the amount of iron administered or by the iron status of the individual in the usual physiological range (Wheby *et al.*, 1970), but at high doses there is a strong correlation with the subject's iron status (Hallberg *et al.*, 1979). Absorption of nonheme iron is affected by blocking agents in food, reducing agents, ascorbic and hydrochloric acids and synthetic chelating agents which bind ionic iron (Callender *et al.*, 1957; Turnbull *et al.*, 1962; Conrad *et al.*, 1966; Wheby *et al.*, 1970; Callender *et al.*, 1970). Heme iron absorption is not affected by ascorbic acid and is only slightly reduced by desferrioxamine and vegetable foods (Martinez-Torres and Layrisse, 1971). Meat enhances the absorption of heme iron possibly by stimulating the digestion of food so that either form of iron is more efficiently released and made available (Hallberg *et al.*, 1979).

One-third of the iron in animal tissue, beef, or poultry, is heme iron. Although only 5–10% of dietary iron is heme, absorption of the heme is five- to tenfold higher than nonheme iron; therefore, heme iron may supply as much as one-third of the daily iron requirement. Having animal protein in the diet facilitates the absorption of nonheme iron.

3.1.3 Nonheme Iron

The world's supply of food iron is predominantly in the form of nonheme iron. In most countries, cereals are the largest single source of both calories and iron. In Western countries, cereals and cereal products contribute one-fourth to one-third of the total calories and total iron in the diet, and the greater part of these cereals is eaten as white bread or other products of flour. Further, when the diet is inadequate either in terms of its total iron content or its quality, then a negative iron balance occurs. Females and children, because of their more precarious iron balance, are more sensitive to diets of low iron quality. It is therefore urgent to understand the factors that affect the absorption of nonheme iron. Older methods of measuring iron absorption were tedious and prone to error by contamination. Use of radioiron and the finding that absorption of intrinsic and extrinsic labels was identical when both are mixed in a meal have greatly facilitated the study of iron absorption. For the remainder of this chapter, the term *iron* will be used for nonheme iron (heme iron will be noted as heme iron).

3.2 Techniques of Measuring Absorption

A fast, sensitive, and reliable method to estimate iron absorption is of both research and diagnostic interest. Methods most commonly used to study iron absorption have been summarized in Table 3-1. None of these methods is altogether satisfactory, and they do not all give the same information, although the results of all methods are frequently labeled "absorption." In addition, there are variations in how the dose is administered (i.e., as the iron salt, intrinsic label, or extrinsic label) and how it is given (alone or with a meal).

Even with precise measurements, there are still large differences in iron absorption among normal subjects (biological variability). In addition, there are variations in absorption of iron using the same subject several times (physiological variability). Data suggest that the absorption

Table 3-1. Survey of Methods Commonly Used to Study Iron Absorption in Humans and Animals *in vivo* and *in vitro*

Experimental set	Measurements	Duration of experiment	Comments
In vivo			
Oral ingestion	Chemical balance	Days, weeks	Laborious and time-consuming; unabsorbed Fe indistinguishable from mucosal Fe
	Serum iron curves	Hours	Dependent on absorption rate and plasma clearance rate; only qualitative
	Plasma tolerance curves	Hours	Correlates with RBC utilization but not with absorption
	RBC incorporation	Days	Semiquantitative correlation; does not measure true absorption
	Unabsorbed radioiron	Days, weeks	Difficult to collect; errors in sample preparation
	Whole-body retention of radioiron	Days, weeks	Instrument expensive; most accurate
	Hemoglobin response	Days, weeks	Indirect method
	Radioiron distribution in subcellular fractions of mucosal tissue	Hours	Animal studies only
Intestinal tubes	Concentration of radioiron in lumen	Hours	Animal studies only
	Radioiron in body	Days, weeks	Humans
Injection into stomach, duodenum, or tied-off intestinal segments (single dose or perfusion)	Radioiron in carcass	Days	Animal studies only
	Radioiron in mucosal tissue	Minutes, hours	Animal studies only
	Radioiron distribution in subcellular fractions of mucosal tissue	Minutes, hours	Animal studies only
In vitro			
Isolated intestinal segments without blood supply	Concentration of radioiron or iron at serosal side	Minutes, hours	Animal studies only
	Radioiron or iron in mucosal tissue	Minutes, hours	Animal studies only
Chemical methods	α, α′-Dipyridyl	—	Does not measure interactions
	Tripyridyl triazine	—	

of iron is not haphazard but is rather a highly regulated process not yet well understood.

About half the variation in iron absorption is due to biological variability. When 300 apparently healthy volunteers were given a test dose of elemental iron as a solution of ferrous sulfate plus ascorbic acid, absorption measured by red cell incorporation was quite variable (see Figure 3-2). Average absorption was 30% with a range from 0 to 100%, making interpretation of an individual value difficult. Differences between populations presumably reflect the type and level of dietary iron supply. Within a population group, however, the variations must be due to the behavior of the intestinal mucosa and reflect changes in iron stores. Absorption of iron from a standard dose of inorganic iron is a sensitive index of iron stores of a population (Heinrich, 1970). Absorption of ferrous iron is reported to be higher in England than values reported for the United States.

Cook *et al.* (1969) compared published data obtained on Swedish and Venezuelan populations. The frequency distribution curves of the populations were quite different, the Venezuelan population apparently having a higher proportion of subjects with iron deficiency anemia. However, both curves were highly skewed in a positive direction, and neither resembled the normal distribution curve. Intrasubject differences become very important when trying to determine if a group of subjects is suspected to have an abnormal iron absorption compared to a control group. If the difference in absorption is 30% and one wanted an 80% chance of dem-

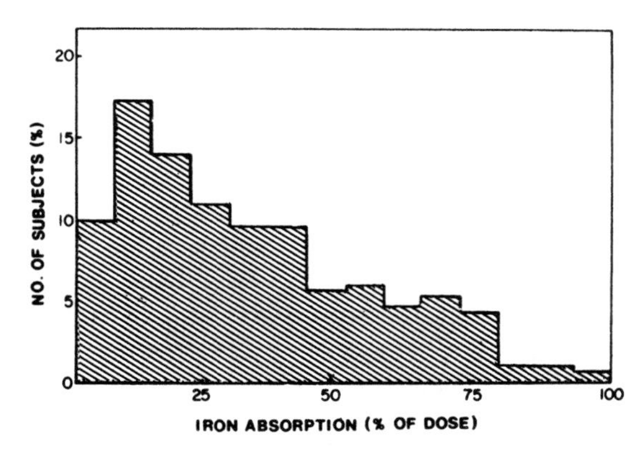

Figure 3-2. Variations in absorption as measured by incorporation of iron into red blood cells. Three hundred volunteers were given a test dose of elemental iron. From Cook and Lipschitz (1977).

onstrating a difference at the 5% level, the group size would have to be 156 if the change represented a decrease in absorption and 256 if the change represented an increase in absorption. For a 60% difference in absorption the numbers of subjects would have to be 27 and 90, respectively.

Intersubject variations can be reduced by the selection of individuals with a more uniform iron requirement, by applying appropriate statistical methods to the skewed data, and by studying sufficient subjects to provide statistical validity to the results. Brise and Hallberg (1962) eliminated biological variability by comparing the absorption of a given form of therapeutic iron for each subject with the absorption of ferrous sulfate using two radioiron tags. Calculating absorption ratios made it possible to detect small differences in availability with a small number of test subjects and makes it possible to assess absorption in different populations and groups of subjects who may differ in iron status.

Physiological variability can be determined by giving the same test meal labeled with either ^{55}Fe or ^{59}Fe on 2 successive days in the same subject. Cook *et al.* (as quoted in Cook and Lipschitz, 1977) state that physiological variability accounts for one-half of the overall variation. Variation is somewhat less when food substances are studied than when the test dose is inorganic iron. This source variability can be reduced by giving the radioiron as multiple test doses over several days (Brise and Hallberg, 1962).

3.2.1 Methods Using Stable Iron

Chemical Balance. The oldest technique is chemical balance. By this method absorption equals the difference between the amount of dietary iron consumed and the fecal excretion of iron: amount excreted in the feces × 100 divided by the total intake. Four studies of total iron absorption using this technique are summarized by Moore (1968). These studies lasted 28–70 days. In 21 healthy female subjects, absorption calculated on the basis of positive balance varied from 10.3 to 21%. In one of the studies (McMillan and Johnston, 1951) 3 mg of iron as spinach were added to the basal diet after 28 days of observation on the basal diet. The authors assumed that the fecal excretory loss should have been approximately the same after the addition and estimated that 13% (0–27%) of the iron in spinach was absorbed. Compared to data obtained by other techniques, these values are very high. Bothwell *et al.* (1955) measured absorption using both stool recovery and the double isotope technique in four subjects. Absorption averaged 12% by the stool method and 9%

by the double isotope techniques. The method has been used in more recent studies (Senchak *et al.*, 1973; King *et al.*, 1978). With this technique it is not possible to differentiate between absorption of heme and nonheme iron or among unabsorbed iron, iron in the sloughed mucosal cells, and iron secreted into the gastrointestinal tract. Small errors in fecal collection are reflected by marked changes in estimated absorption. There are potential analytical errors which may invalidate results if not carefully avoided. Losses in iron can occur during ashing, during removal of interfering phosphate, or during colorimetry or titration due to a number of factors. Iron balance studies have the decided advantage of measuring absorption over a period of several days rather than at one to three points in time. Meticulously done studies still merit attention, and this method should perhaps be reinvestigated for use in infants and children and in pregnant women, particularly in studies which compared absorption from different combinations of foods or iron sources. Negative values and unrealistically high values have given the method a bad name. Data obtained before 1935 are invalid because of inaccuracies of methods. Data from 1935 on must be carefully evaluated.

King *et al.* (1978) modified the basic method by feeding their subjects a stable isotope of iron. The isotopic iron in the feces was activated by neutron activation analysis. Mean absorption for 14 women, 19–22 years of age, was 13.5 ± 12.3%.

Postabsorption Serum Iron Curves. During the period 1943–1953 another historical method was used to calculate iron absorption. An unphysiologically high dose of oral iron (100–250 mg) was given to the subject and blood samples taken for determination of changes in serum iron during the first 2–4 hr after the load. Changes in serum iron depend not only on the amount of iron absorbed but also on gastric emptying, rate of iron absorption, and the iron plasma clearance rate. When compared with intestinal ^{59}Fe absorption, there was no correlation between true absorption as determined by fecal excretion and the postabsorption ^{59}Fe plasma tolerance curve.

3.2.2 Methods Using Radioiron

When radioisotopes became available, it was much easier to study iron absorption. ^{59}Fe, which first became available in the late 1930s, has a relatively short half-life, 45 days, and emits high-energy gamma rays (1.1 and 1.3 meV) which can be measured by a variety of radiation detectors as well as beta. ^{59}Fe is very suitable for measurements of whole body radioactivity. In the early 1950s, ^{55}Fe became more available. ^{55}Fe

has a half-life of approximately 3 years. The energy of emission is much lower (5.9 keV) and can be measured only by highly sensitive counting techniques. ^{55}Fe disintegrates exclusively by electron capture, emitting only Mn X rays. When counting samples containing both ^{55}Fe and ^{59}Fe, the beta particles of iron-59 often disturb the counting of the Mn X rays. Measurements using scintillation counters have not been altogether satisfactory because of quenching by the iron-rich scintillation liquid. Erroneous results can be obtained if the sample is counted using a proportional counter if the patient has had other radionucleides. Kumpulainen and Saukkonen (1979) suggest a method using a semiconductor detector connected to a multichannel analyzer.

Plasma Tolerance Curves. Postabsorption ^{59}Fe (^{55}Fe) plasma tolerance curves are similar to postabsorption serum iron values but have the advantage of using physiological oral radioiron doses. Comparative studies of ^{59}Fe-absorptive fecal excretion and the tolerance test have shown that even with a correction for different plasma radioiron clearance, there is very poor agreement in the results obtained in at least 60% of the subjects. Therefore, it is not a reliable quantitative measure of iron absorption (Bothwell *et al.*, 1955).

Incorporation of Radioiron into Erythrocytes. Based on data obtained in several studies of iron absorption in anemic animals and subjects, Hahn *et al.* (1940) suggested that the ^{59}Fe activity in the circulating erythrocytes after 10–14 days should be used as a quantitative measure of total ^{59}Fe absorption. At the present time, this is the most common method used to determine iron absorption in human subjects. With this method, errors are due to variations in the red blood cell incorporation and estimating total blood volume from height and weight. In subjects with normal iron stores, about 90% of the absorbed iron is assumed to be incorporated into the red blood cells. Studies in which incorporation of radioiron into red blood cells is compared to absorption as indicated by whole body counting verify that this is generally true (Table 3-2). However, in the elderly subjects studied by Marx (1979), there was a highly significant reduction in red cell uptake. Considering the wide range obtained by Lunn *et al.* (1967) in their study of three different methods, it is possible that some of the subjects included in the study were elderly. There are some data on absorption of subjects aged 40–50; however, only one method of absorption was used, so there is no information regarding the age at which the decreased red cell uptake begins. Using the 90% correction for all ages could lead to substantial errors.

In patients with iron deficiency anemia, approximately 100% of the absorbed ^{59}Fe is incorporated into erythrocytes. Patients with untreated pernicious anemia, hypoplastic anemia, hemolytic anemia, Hodgkin's

Table 3-2. Comparison of Percent of Iron Absorbed by Two Techniques: Whole Body Counting and Erythrocyte Incorporation

Reference	Number[a]	Age	% Absorption by Whole Body Count / % Absorption by RBC Incorporation
Lunn *et al.*, 1967	17	Not specified	89 (42–100)
Larsen and Milman, 1975	8M	[b]	96[c]
	19F		91[c]
Marx, 1979	12	19–33	91 .2 (63–110)
	10M	65–71	66 .2 (40–84)
	11M	72–83	66 .3 (40–92)
	6F	65–69	66 .8 (43–88)
	6M	70–77	64 .5 (40–74)

[a] Abbreviations: M, male; F, female.
[b] Eleven were medical students or members of the hospital staff; 16 were patients suffering from minor disorders which would not influence iron absorption.
[c] Calculated from mean values given in article. Authors give an overall average of 92.9%.

disease, hemochromatosis, or fever incorporated less iron into the red blood cells. Sayer and Finch (1953) suggested a correlation technique using oral ^{59}Fe and simultaneous intravenous injection of ^{55}Fe or vice versa. After an interval, the ratio of the two isotopes in a blood sample gives a measure of absorption. This method assumes that the oral and parenterally administered irons are handled the same.

The error in estimated blood volume is approximately $\pm 5\%$. Precise measurements using ^{51}Cr-tagged red cells to calculate blood volume are usually not justified. Using the same subject for multiple tests and expressing the absorption of one test relative to another, one eliminates the error in estimated blood volume and reduces the error of assuming constant red cell incorporation. Four absorption measurements can be done on the same person by administering test meals labeled with ^{59}Fe and ^{55}Fe on 2 successive days and determining blood radioactivity 14 days later. An additional two tests can be done by repeating the procedure and measuring the rise in radioactivity between day 14 and day 28. This multiple testing enables one to include absorption from a standard, thus further improving the validity and enabling an accurate comparison to take place. Methodology for use of stable isotopes of iron as tracers has been developed as an altenative to radioisotope tracer techniques (Miller and van Campen, 1979). Iron was chelated with 2,4-pentadione before analysis in the mass spectrometer. With the 3 stable isotopes of iron it is possible to double or triple label without additional sample preparation. This technique may find its greatest use in studies involving infants and pregnant women, in whom use of radioactive isotopes tracers is not always considered acceptable. Present sensitivity levels do not permit labeling hemoglobin iron with a single oral dose. Feeding the isotope with several meals would permit use of the hemoglobin iron technique and reduce the problem of day-to-day intrasubject variations.

Hemoglobin Response. There are two variations to this indirect technique of assessing iron absorption, a curative or preventive method. In using the curative method, the subject is depleted of hemoglobin, and hemoglobin regeneration is monitored. Iron deficiency can be produced in both humans and animals by bleeding and only in animals by feeding a low-iron diet or by producing intravascular hemolysis by injection of phenylhydrazine. The main objection to this technique is that iron-deficient animals absorb iron at a higher rate than iron-replete animals and do not differentiate among various foodstuffs. In using the prophylactic response technique, the prevention of hemoglobin loss and the improvement of iron status is observed.

Protocols which are based on hemoglobin response are also limited by certain components of control and experimental diets, for example, the phosphate content, which may in themselves alter iron absorption. Both the slope ratio assay or the parallel level assay for testing various forms of fortification iron require that the animals be given graded doses of iron. Errors in both dilution due to body weight differences and of quality due to differences of dietary iron content should be considered in assays using hemoglobin response as a measure of bioavailability. The efficiency of conversion of food iron to hemoglobin is minimally affected by differences in dietary iron concentration and food intake under ad libitum (Mahoney *et al.*, 1974) and constant intake (Mahoney and Hendricks, 1976) conditions.

Another disadvantage of this method is the need to estimate "diffuse storage iron." This would allow the identification of prelatent iron deficiency, but smears of bone marrow cells are not done routinely, and there is still a lack of correlation between other estimates of iron stores and iron absorption.

Fecal Excretion Test. Absorption of the compound in question is calculated as the unrecovered amount, i.e., the difference between known oral intake and the amount recovered in the feces. This technique requires the cooperation of the patient to collect samples, may be subject to errors in preparing material for counting, and requires a greater number of samples per test.

The most common error is incomplete collection of stools during the week or 10-day period necessary. As reported by Callender *et al.* (1966), of the 54 patients where feces collection was attempted, 16 were incomplete. The completeness of sample collection can be improved by using a second tracer, $^{51}CR^{3+}$, ^{140}La, or ^{133}Ba. Unfortunately, use of the second tracer does not allow calculation of a correction factor since these tracers are excreted much faster than the unabsorbed iron. Incomplete feces collection would result in an apparently higher iron absorption.

If a small volume NaI well detector is used and if the whole sample is not counted, there could be an error in taking an aliquot. Poor homogenization could result in the aliquot containing more or less per gram than the whole sample. Counting the whole sample within a plastic scintillation counter or well detector eliminates the possible sampling error. However, the radioiron may be unevenly distributed and introduce other types of geometry errors. An additional disadvantage to this test is that 7–14 samples have to be prepared and measured. An average laboratory probably could handle only about 10 tests/day.

Whole Body Retention. Whole body retention of radioiron is the most reliable, sensitive, and sole quantitative method for measuring absorption in humans. This technique can be used both clinically and experimentally. The general procedure is to count the natural background radioactivity in the fasting subject, give an oral dose of 1–10 μc of ^{59}Fe either alone or with food, and count 4 hr and 14–20 days later (Price *et al.*, 1962). The 4-hr count is the 100% value. During the initial period, the distribution of the radioiron is affected by the distribution within the body and by the size of the subject. Tracer radioiron can be found in the stomach, upper small intestine, portal circulation, and liver shortly after ingestion. As the iron is absorbed and distributed throughout the body, the total number of counts fluctuates from 10 to 30%. Differences between counts following oral and intravenous injection of radioiron in the same subject varied as much as 25% for 4π liquid counters and 35% for single-crystal detectors. When the second count is made at 2 weeks, the radioiron is distributed among the red cell mass, liver, spleen, bone marrow, and other storage areas. Counting earlier than 4 hr to obtain a 100% value is not as accurate as a count 4–5 hr after ingestion of the oral dose (Dymock *et al.*, 1971). The clinical usefulness of the 100% value has been investigated further by giving an intravenous dose to calibrate each patient (Lunn *et al.*, 1967). Although this dose may not exactly duplicate the ultimate distribution of the ingested iron in patients with large stores of iron, it is close for the patient with normal or low stores (Schiffer *et al.*, 1964).

A variety of counters has been used in the past; the basic counter consists of a chamber made of old steel, to contain as little intrinsic radioactivity as possible, and a thalium-activated sodium iodide crystal detector. Very sensitive and elaborate counters using 4π geometry allow use of lower amounts of radioactivity (Figure 3-3). Physical characteristics, reproducibility of counting rate, variation in counting rate due to patient position, and efficiency for ^{59}Fe for 4π counters have been de-

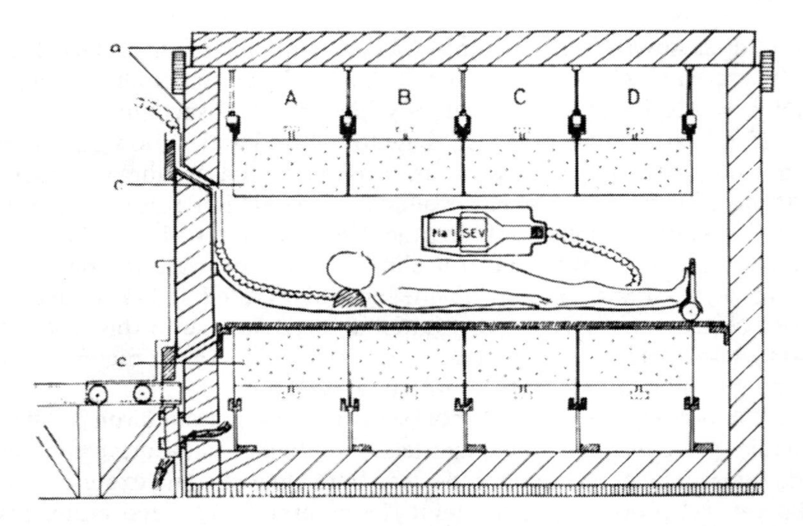

Figure 3-3. Two types of whole body counters. (Left) Longitudinal view of 4π whole body counter—very sensitive but expensive. a—Steel plates, c—liquid scintillator in tanks; NaI–sodium iodide crystal. Reprinted from Heinrich, 1970, with permission. (Right) Two-detector system utilizing shadow shield principle and longitudinal scan—quicker and less expensive. Reprinted from Callender et al, 1966, with permission.

scribed by Price *et al.*, (1962). A full steel room is expensive, and subjects can become uncomfortable.

A two-detector system which makes use of the *shadow shield* principle has been designed (Warner and Oliver, 1966). The use of two detectors, one above and one below the subject, reduces the dependence of count on the position of the subject (Figure 3-3). The subject is placed on a motor-driven couch and a 15-min longitudinal scan is made (Callender *et al.*, 1966). Some newer instruments use four sodium iodide detectors. Accumulating the counts from all crystals simultaneously allows the count to be completed in a short period of time, cost is reduced with the reduced amount of shielding, and traverse counts are less affected by the distribution of radioactivity within the body. To improve the counting efficiency still further, plastic pellets, Delrin, having the same density as the human body, are layered over the subject until they are level with the top of the carrier. These provide a uniform mass of scattering material (Palmer *et al.*, 1970). With this technique, the standard deviation of measurements following oral ^{59}Fe administration was 1%, and the maximum difference between oral and intravenous radioiron injections was 1.2%.

Figure 3-4 shows the change in retention with time for 10 female and 10 male volunteers (Celada *et al.*, 1978). All volunteers had serum values

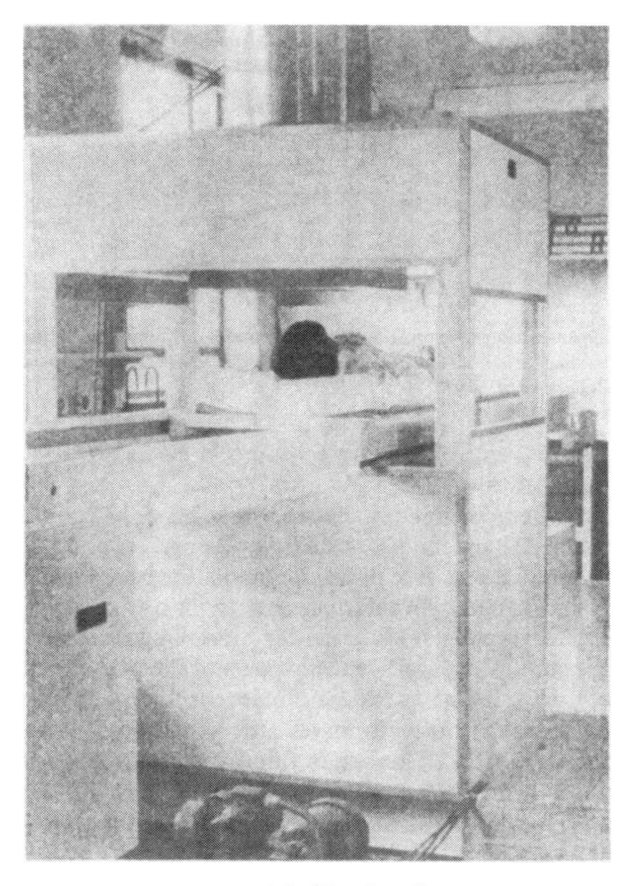

Figure 3-3. (Continued)

above 80μg/100 ml and percent transferrin saturation values above 28. It appears that there are two phases in the elimination of the radioactivity (early, i.e., first week, and late, 7+ days). The initial decrease, first 4 days, is probably the amount excreted in the stools; for some adults, this may take up to 8 days. Between 4 and 7 days, the somewhat slower decrease is probably related to the loss of iron in the mucosal cells which has not been transported across the serosal side. From 7 to 14 days, the rate of decrease is slow, reflecting obligatory physiological loss. In men the slow rate continues. In women the rate of loss is greater after 90 days, probably due to menstrual losses. Therefore, measurement of iron absorption less than 2 weeks after ingestion is more valid, although obsti-

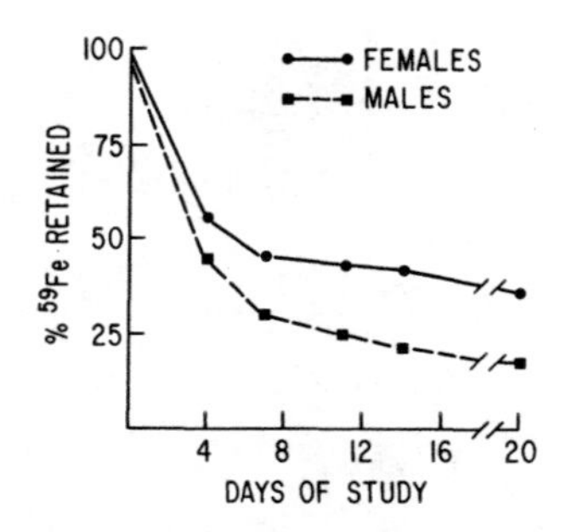

Figure 3-4. Changes in iron retention with time in male and female volunteers. Initial decrease is due to loss in stools; from 7 days, decrease due to obligatory physiological losses. Adapted from Celada *et al.* (1978).

pation, bling loops, and blood loss during this period would still suggest a higher absorption rate.

Use of Intestinal Segments, Tubes, and Cell Fractions. Much information regarding the mechanisms of iron absorption and its regulations has been obtained using intestinal segments, tubes, or cell fractions. Wheby (1966) confirmed by intubating directly into various intestinal sites that humans, like animals studied earlier, absorbed iron most efficiently from the duodenum. Most studies using parts of the intestine have utilized animal models. Everted gut sacs can be prepared from a segment of the duodenum just distal to the pylorus (Brown and Justus, 1958; Manis and Schachter, 1962a, 1962b). The sac is filled with an appropriate medium and incubated in a medium containing radioactivity. Following incubation, the sac is drained, and the contents measured and/or the mucosal layer scraped and processed further.

Closed loops of intestine can be prepared by applying a proximal and a distal ligature around the serosal surface of the segment to be studied (Wheby *et al.*, 1964). Care must be taken not to occlude the observable blood vessels. Suspensions of mucosal cells can also be prepared (Halliday and Powell, 1973). Scrapes may include subepithelial cells and result in confusing results. Studies of cell fractions, obtained by commonly accepted techniques, either alone or with electron micrographs, contribute to our knowledge of iron absorption by allowing for the measurement of iron in the various parts of the cell.

3.2.3 *In Vitro* Measurements

Some research groups have suggested *in vitro* methods for the determination of availability of nonheme iron from foods and diet using α,α'-

dipyridyl or tripyridyl triazine. Studies using tripyridyl triazine showed that more iron was in an ionizable form from enriched white bread than from whole wheat bread (Hart, 1971). Jacobs and Greenman (1969), evaluating 25 common cooked foods, obtained values which ranged from 17.8% for eggs to 70% for potatoes. Most of the values fell between 30 and 50%. In the α,α'-dipyridyl method food was extracted with pepsin–HCl at pH 1.35. The pH was adjusted to 7.5, the solution filtered, and the ionizable iron determined in the filtrate. In a recent study by Rao and Prahavathi (1978) a direct correlation between the percent ionizable iron at pH 7.5 from six meals and the percent of iron absorbed by adult males from the same meals was found. The correlation coefficient was 0.94. A prediction equation (percent iron absorption in adult men $= 0.4827 + 0.4707 \times$ percent ionizable iron at pH 7.5) was developed. This equation predicted absorption with a fair degree of accuracy. Neither method has been tested for women, over a wider range of iron absorption, or with more varied diets.

3.2.4 Administration of Dose

In addition to variations in the technique of measuring the dose administered, there are also variations in how the oral dose of radioiron is fed. Some investigators gave iron salts, such as ferrous or ferrous ascorbate, to fasting subjects. Other investigators gave the test dose as a drink with a standard meal. Sharpe *et al.* (1950) found that adding food to the test dose of radioiron produced a marked inhibition of absorption. They suggested that the degree of inhibition was related to the bulk of the food ingested. This method of administration led to an overestimation of the real absorption. Absorption was reduced when small amounts of iron were mixed with various types of bread, before baking, that were later eaten in a meal (Elwood *et al.*, 1968; Callender and Warner, 1968). The standard meal technique was widely used, but the implication that the results provided a valid measure of iron absorption from the meal was carefully avoided. Table 3-3 lists the composition of the meals in the four studies and the average percent absorption by controls and iron-deficient subjects. The meals contained nonlabeled heme iron in different amounts and other dietary components. At the present time, absorption of iron salts is used as a reference standard in order to compare data from populations of different iron status or in studies of the mechanism or regulation of iron absorption or to determine the best salt for fortification. Absorption of food iron from single foods or from meals has been studied most thoroughly using either intrinsic or extrinsic tag techniques.

Table 3-3. Absorption of Radioiron from Standard Meals

Reference	Control (av%)	Fe deficient (av%)	Composition of Meal
Sharpe *et al.*, 1950	6 8		Milk, 200 g; cooked rolled oats, 285 g; white bread, 34 or 56 g; tomato juice, 150 g
Turnbull, 1965	3	25[a] 46	Canned stewing steak, 50 g; potato powder, 15 g; frozen peas, 50 g; butter, 10 g; orange, 100 g; cream, 20 g
Pirzio-Biroli *et al.*, 1958	6 5	22	Canned corn-beef hash, 200 g; apple sauce, 200 g; tomato juice, 200 g; soda crackers, 13 g; cocoa, 3 g
Goldberg *et al.*, 1963		58	Corned beef, 100 g; apple sauce, 100 g; tomato juice, 200 g; cream crackers, 13 g; cocoa, 3 g

[a] Fe deficiency without anemia.

Intrinsic Label. Biosynthetic labeling (intrinsic labeling) was first introduced by Moore and Dubach (1951) and is considered by some to be the most valid approach to measuring absorption of food iron. Vegetable foods were labeled by growing them in a hydroponic medium containing radioiron. Animals or fish were labeled by being fed high-specific-activity radioiron in the months before sacrifice. Test meals containing the labeled food were prepared and given to fasting subjects in amounts that would provide about 25% of the total daily intake.

Moore (1968) summarized the studies done to that time, 219 determinations on adults with apparently adequate stores. The average absorption from individual foods which included liver, hemoglobin, and muscle as well as eggs, wheat corn, etc., was 8.5%. Average values for individual foods ranged from 2 to 18%. With the exception of soybeans, with an average absorption of about 18%, nonheme iron from vegetable sources was less well absorbed. In these studies, absorption was determined by the amount of unabsorbed radioactivity recovered in the feces or the amount utilized for hemoglobin synthesis or both. In a few cases absorption was measured by whole body retention.

The most extensive single study, a collaborative project by workers in Seattle and Caracas, dealt with absorption of 12 different foods in a total of 520 subjects. Data were summarized by Martinez-Torres and Layrisse (1974). To correct for differences in iron status, all studies included a measurement of absorption of ferrous sulfate. Absorption was measured by incorporation of radioiron into red blood cells. Absorption from single foods was lowest from rice and spinach (1%); intermediate from black beans, maize, lettuce, and wheat (3, 3, 4, and 5%, respectively); and highest from soybeans (6%). Mean absorption of iron was greater than 10% from veal liver, fish, and veal muscle. In some cases absorption exceeded that from the reference standard.

Hussain *et al.* (1965) studied intrinsically labeled wheat and also found absorption to be about 4.5% in subjects with good stores and 7.8% in patients with iron deficiency anemia. Absorption of ferritin iron was 7.6 and 12.5%, respectively. These values are in good agreement with those of Martinez-Torres and Layrisse (1974). Although these data are only for single foods, they provide information which is valuable for some populations because these foods are main staples and may account for 95% of the dietary iron.

The values reported for absorption of iron from soybeans (6 and 18%) may be at variance due to small sample size (7 subjects in one case and 63 in the other). Also, the evaluation of iron status has improved with time, and if only 1 or 2 of the 7 subjects with very high iron absorption from soybeans had iron deficiency without anemia, their values would have a large effect on the data.

For other populations, it is not sufficient to measure absorption from single foods; it is necessary to evaluate absorption from two foods and ultimately from whole meals. Studies using intrinsic label were limited to two foods because of the availability of only two radioisotopes of iron. Theoretically a third food could be labeled by a stable isotope which could later be activated by neutron activation analysis.

Iron absorption studies using two foods revealed major interactions. Table 3-4 shows the interactions of black beans, maize, fish, and veal muscle in individuals with adequate stores and those with iron deficiency (Layrisse *et al.*, 1968). Mixing the vegetable and animal products resulted in increased absorption from the vegetable food and decreased absorption from the animal food. One series of studies was also done to determine the effect of free amino acids. The amino acids were in the same number

Table 3-4. Absorption of Iron from Single Foods or Mixed Foods

	Alone	Beans–Veal	Maize–Veal	Maize–Fish	Beans–Fish
Black beans	5.9	13.4	—	—	—
	1.0				1.7[a]
Veal	22.1	18.4	—	—	—
	24.9		18.5	—	—
Maize	5.9	—	8.2		
	4.2	—	—	7.5[a]	
Fish	9.9	—	—	10.9[a]	

[a] Not significant.

and proportion as are present in 100 g of fish. Iron absorption from black beans was about three times higher than black beans alone or black beans with fish muscle.

Biosynthetically labeled foods are expensive and difficult to prepare. They can provide information on absorption from a complex diet. This method, however, is the only one available to study insoluble forms of iron.

Extrinsic Label. Validation of a model to measure iron absorption using an extrinsic tag has been the most important advance in studies of iron absorption since the introduction of radioactive iron. Several investigators had earlier suggested that an extrinsic label might be suitable. Pirzio-Biroli *et al.* (1958) gave tracer doses halfway through a meal and suggested this method for the study of dietary iron but stated that the method did not provide information on the absorption of the native iron present in food. The method was used by others; however, it was not until absorption from intrinsic and extrinsic labels was systematically compared that the extrinsic tag method was demonstrated to be a valid technique (Layrisse and Martinez-Torres, 1972; Cook *et al.*, 1972; Bjorn-Rasmussen *et al.*, 1972).

Absorption of intrinsic and extrinsic iron from 96 females and 84 male volunteers from Venezuela and Seattle was studied under a variety of conditions. A consistent extrinsic–intrinsic absorption ratio, averaging 1.10, was observed in studies with maize, black beans, and wheat and with either ferric or ferrous iron in doses ranging from 0.001 to 0.5 mg of iron in a test meal containing 2–4 mg of food iron. Addition of ascorbic acid or meat increased the rate of absorption and desferrioxamine decreased the rate of absorption but not the $E:I$ ratio.

Additional studies by Bjorn-Rasmussen *et al.* (1972) using maize, wheat, and eggs confirmed the concept of complete exchange between the biological label and added tracer. The effect of egg was of particular interest since absorption is very low and egg inhibits the absorption of inorganic iron. However, another study of the rate of isotopic exchange in the nonheme pool and the isotopic exchange between two biosynthetically labeled foods which differed in absorbability provided different results (Bjorn-Rasmussen *et al.*, (1973). Identical plasma activity curves were noted when the subjects consumed a meal prepared from intrinsically labeled soybeans or wheat or added tracer. When using rice flour the exchange was complete and absorption the same but not when rice grains were used, suggesting that the rate of diffusion of iron was too slow. These data indicate that mechanical factors, mixing, and milling have to be considered when using the extrinsic tag method to measure the absorption of nonheme iron from the diet.

Extrinsic iron tag is consistently absorbed at a slightly higher rate than intrinsic label. This difference may reflect the time needed to form the common nonheme pool. Additional studies (Cook *et al.*, 1973) confirmed that certain insoluble iron salts such as sodium iron pyrophosphate or ferric orthophosphate may have a lower absorption than the radioiron fed simultaneously.

The ratio of absorption of extrinsic to intrinsic iron averaged 1.14 in 16 studies of 113 rats in which the absorption was assessed under diverse conditions including changes in the quantity of test food, the inclusion of other foods, the addition of chelates, and the use of normal and iron-deficient animals (Monsen, 1974). These data provide further evidence that extrinsic tag can be used to assess the absorption of soluble nonheme iron over a wide range of absorption.

Layrisse *et al.* (1973) studied iron absorption of 228 subjects using the extrinsic tag method. Absorption was standardized by feeding ferric and/or ferrous salts with a complete meal and also by comparison to foods with intrinsic tag. Once again the ratio of extrinsic to intrinsic tag absorption was close to unity over a wide range of absorption and independent of the dose administered when the test meal of labeled maize was given alone or with unlabeled meat. Addition of increasing doses of $^{59}FeCl_3$ added to a test meal of labeled animal meat produced a progressive fall in the ratio of extrinsic to intrinsic tag absorption. These data indicated that assimilation of dietary iron occurred from two pools, nonheme and heme.

Fortification iron added to a vegetable food was absorbed only in a limited amount, 0.3 μg with an intake of 60 μg. When 5 μg of fortification iron were consumed with veal muscle, 0.85 μg of iron was absorbed. When mixtures of animal and vegetable foods were used, intermediate values were obtained. The absorption of meat iron was not affected by the dose of nonheme iron either as food or an iron salt.

Studies of iron absorption were further improved by the realization that absorption of both heme and nonheme iron could be measured using two different isotopes. Therefore, for the first time dietary absorption could be measured from a complete meal (Layrisse and Martinez-Torres, 1972). Labeled hemoglobin was obtained by injecting a rabbit with ^{59}Fe in the form of ferrous citrate. Two weeks later blood was withdrawn and used in the preparation of hemoglobin. Absorption of the labeled hemoglobin was determined after mixing it with raw ground veal muscle and comparing the values obtained to absorption when each was administered alone. There were no differences in the absorption of the iron over a wide range of absorption. Absorption of the hemoglobin iron was not altered by the presence of other foods in the diet. Finally the absorption of heme and nonheme iron was measured in a complete meal consisting of three vegetable foods and meat. The amount of meat, approximately one-third of the cooked meal, provided 80% of the iron either directly or indirectly, i.e., from the meat or by increasing the amount of iron absorbed from the vegetables. In normal subjects this amounted to 0.40 of the 0.46 mg absorbed, and in iron deficiency subjects the amount was 0.73 of 0.93 mg.

Additional studies were of iron absorption from meals containing five to eight items from three different parts of Venezuela (Layrisse *et al.*, 1974). Breakfast tagged with ^{59}Fe was fed in the morning of the first day after an overnight fast. Durig the 3 hr after the meal no food or drink was allowed. The next day the subjects were given lunch tagged with ^{55}Fe. Fifteen days later blood samples were taken and the subjects fed a dinner meal tagged with ^{59}Fe. On the sixteenth day a standard dose of ferrous ascorbate tagged with ^{55}Fe was given. Two weeks after this dose a blood sample was taken. Blood samples were used to determine hematological characteristics and radioactivity.

Nutrient patterns of the diets from the three areas were somewhat different. Table 3-5 shows the percent of the total intake of several nutrients consumed at each meal. From diet of the central areas the nonheme iron absorption was low from each meal but absorption from the lunch meal and supper significantly higher than from breakfast. Absorption was highest from lunch, which contained the most ascorbic acid. The im-

Table 3-5. Nutrient Distribution in Various Venezuelan Meals as Percent of Total Amount Consumed Per Day

	Breakfast	Lunch	Supper
Central areas			
Energy	34	36	30
Protein	30	31	39
Fat	42	23	36
Ascorbic acid	5	95	0
Andes			
Energy	23	38	39
Protein	27	43	30
Fat	43	34	22
Ascorbic acid	0	56	44
Coast			
Energy	28	34	39
Protein	35	30	35
Fat	40	33	27
Ascorbic acid	0	28	72

provement of absorption from the supper meal may have resulted from the increased amount of protein in this meal.

From the diet of the Andean region absorption was also low for breakfast and higher at lunch. Absorption from the dinner meal was higher than from breakfast but not as high as from lunch, which contained the majority of ascorbic acid and protein.

Absorption of nonheme iron from the diets of the coastal region was 5% at breakfast, 7% at lunch, and 13% at supper in the first study and 7% for breakfast, 11% at lunch, and 17% at supper in the second study. This diet is different from the others. It contains fish high in protein at each meal and a large portion of papaya high in ascorbic acid at supper both of which could contribute to the excellent rate of absorption. More detailed studies of the diet revealed that the ascorbic acid was the important factor. This study provided the first information which showed that calculating the total dietary intake of iron is irrelevant and that one must take into consideration the ingredients of the meal, such as beef, fish, and fruits, in order to obtain a reasonable estimate of the utilization of iron at breakfast.

Hallberg and Bjorn-Rasmussen (1972) suggested the use of the extrinsic label technique to measure absorption of heme and nonheme iron as a test of the two-pool model for iron absorption. Absorption was determined by measuring ^{55}Fe and ^{59}Fe activity in a blood sample drawn 2 weeks after ingesting the dose plus a count of whole body retention. The data provided the first evidence of the two-pool model and showed that in recently bled donors the absorption of nonheme iron was six to seven times higher and that for heme iron it was twice as high. Other studies (Bjorn-Rasmussen, 1974) revealed a significant correlation between the absorption of the two kinds of iron, although a much greater fraction of the heme iron was absorbed (37% compared to 5% of the nonheme iron).

There are several problems which must be resolved in using this technique. Bjorn-Rasmussen *et al.* (1976) evaluated absorption after several different mixing procedures. Mixing the label homogeneously into the meal, adding it dropwise, or mixing it into the most bulky component resulted in equal or almost equal absorption. Surprisingly, there was higher absorption when the components of the meal were served separately than when the meals were mixed before serving. Finally, the validity of using a reference dose for comparison in different groups was higher when serving the reference dose and meals on consecutive days rather than 2 weeks apart. The most recent study published by this group (Hallberg *et al.*, 1979) was the first step toward developing a method to measure food iron absorption from realistic common meals prepared and consumed by the subjects themselves in their own homes. Future studies will require a sufficiently long run-in period and careful instruction of the subjects, but the basic concept is feasible.

Other investigators (Cook and Monsen, 1975), examining the composite effect of substances that either block or facilitate absorption from a common pool of iron, have studied absorption from a standard meal or a semisynthetic meal having the same total chemical composition. Absorption from the standard meal in 32 healthy women averaged 10.0% as compared with a meal absorption of 1.8% from a semisynthetic meal. Most but not all of the difference was said to be due to the enhancing effect of meat.

Bing (1972) suggested that the availability from an iron salt or food is not an inherent characteristic of the substance under study but an experimentally obtained value indicative of absorption under a particular set of test conditions. Data published since that time tend to confirm the idea. The choice of how absorption is to be measured depends in part on the question to be answered. A study to determine the best way to fortify the diet of a target population will be designed quite differently from a study of the effect of the lipid content of the diet on iron absorption. One positive effect of the studies of iron absorption is that the diet of an individual is now not only evaluated in terms of the total amount of iron consumed, but estimates take into account the amount of ascorbic acid and meat in the diet as well as the iron stores.

3.3 Mechanism of Iron Absorption

The mechanism and control of iron absorption have been pursued for decades. In this section we shall deal with the events which prepare the iron for absorption and those occurring from the brush border of the mucosal cell to the transport across the serosal surface. Regulation of iron absorption will be covered later.

3.3.1 Role of the Stomach

Gastric juice and other digestive secretions are responsible for the reduction and release of iron from conjugation. Less than half the total iron in food is released by hydrochloric acid and peptic digestion, and less than a third is ionized. Patients with iron deficiency anemia and normal gastic secretion absorb more of a test dose of ferric iron then those with histamine–fast achlorhydria (Goldberg *et al.*, 1963). Patients with complete achlorhydria absorb more iron when the dose is given with 0.05 *N* hydrochloric acid (Jacobs *et al.*, 1964). The effect is greater for ferric than for ferrous iron.

Availability of the iron released depends on its chemical reactions. Ferric ions polymerize as the pH approaches neutrality and finally form a precipitate of ferric hydroxide. Ferric ions do not polymerize, are more soluble at any given pH, and are therefore more available. The unpolymerized ions of both forms are chemically reactive at pH 1–2.

The only factor in gastric juice of importance in modifying the absorption of nonheme iron is the pH. When gastric secretion is restricted or entirely interrupted, absorption of iron is generally reduced in humans. Recent *in vivo* studies of iron-adequate subjects, patients with iron deficiency, and patients with hemochromatosis revealed a relationship between solubilization and pH which was independent of diagnosis (Bezwoda *et al.*, 1978). The same relationship was found when either HCl solutions or gastric juice was incubated with bread containing radioiron. It was further noted that 8 of the 18 iron-deficient individuals in this study had pH values of the gastric juice which were higher than 2. This could explain the malabsorption of nonheme iron by some individuals. On the other side of the spectrum, the patients with idiopathic hemochromatosis had gastric juice with a low pH. There was no evidence of a promoting factor other than the low pH.

Following gastrectomy, iron deficiency was found in one series of patients and administration of hydrochloric acid to achylous patients restored the absorption of iron to normal (Baird *et al.*, 1957). In another series of 177 patients, absorption was the same as for the controls (Magnusson, 1976). Level of gastric acidity tends to fall with age and may contribute in part to the increased incidence of iron deficiency in the elderly.

Over the years a great deal of attention has been focused on the effects of other gastrointestinal secretions on iron absorption. Some investigators have suggested that gastric secretions contain factors which promote iron absorption. Other investigators have suggested factors to inhibit iron absorption. It was thought that variations in the amount of an intrinsic factor controlled iron absorption in persons who were iron deficient or had an iron overload (Jacobs, 1970). Davis *et al.* (1966) postulated the presence of a high-molecular-weight iron-binding protein which they named gastroferrin and suggested this was the physiological inhibitor of iron absorption. They further suggested that patients with iron deficiency anemia secreted less gastroferrin and therefore absorbed more dietary iron. Other investigators showed that the amount does not vary with iron status. While gastroferrin binds *in vitro*, it does not regulate iron absorption *in vivo*. The iron-binding fraction of human gastric juice is a glycoprotein, 85% carbohydrate and 15% polypeptide, of 2.6×10^5 and 180–190 iron atoms/molecule (Webb *et al.*, 1973).

Under physiological conditions iron ionizable from food may be either reduced or chelated or become unavailable during neutralization. Ascorbic acid enhances iron absorption both by reducing iron to the more soluble ferrous form and by the formation of soluble iron ascorbate chelates. Ferric iron becomes less soluble at higher pH and therefore less available.

3.3.2 Role of Bile

Bile from normal or iron-deficient dogs facilitated absorption of ferrous iron by an animal with iron deficiency (Wheby *et al.*, 1962). Ligation of the bile duct in rats resulted in a variable response. In one study there was increased absorption of ferrous sulfate and a decrease in ferrous phosphate (Webling and Holdsworth, 1966). In other studies there was reduced absorption of ferrous and ferric salts in normal rats and only ferric in iron-deficient animals (Conrad and Schade, 1968). Bile may play a small role through the formation of iron complexes. It has not been confirmed that ascorbic acid is the active agent.

3.3.3 Role of the Pancreas

In the 1960s, the pancreas and pancreatic juice were suggested as playing a role in the absorption of iron. Crosby (1968), after review and analysis of published work, came to the conclusion that the pancreas plays litle or no role in the physiological regulation of iron absorption. Bicarbonate is known to promote complexes when free iron is present; however, experimentally there is no consistent effect on iron absorption. Some patients with chronic pancreatitis or portal cirrhosis absorb iron excessively, usually because of reduced iron stores. Surgical removal of the pancreas diminished iron absorption rather than increasing it. Pancreatic extract increased absorption in these patients. Pancreatic secretions may play an indirect role in the iron absorption by releasing amino acids, which in turn detach iron from its mucopolysaccharide carrier and make the iron available to form low-molecular-weight complexes.

3.3.4 Role of the Intestine

In humans, the duodenum and jejunum have the highest capacity for iron absorption (Wheby *et al.*, 1964, 1970). An extremely limited amount of iron is absorbed in the stomach (Dagg *et al.*, 1967). The ileum has a smaller capacity for absorption than the jejunum; however, since transit time is slower, there may be more absorption than expected. The mucosal cells of the duodenum have the greatest capacity for iron transfer. In patients and animals with iron deficiency more iron appears to be absorbed lower down in the ileum, but even so the majority of the iron is absorbed in the duodenum and jejunum (Chirasiri and Izak, 1966). Absorption in suckling rats is very high and resembles the processes which occur in adult iron-deficient animals (Loh and Kaldor, 1971; Gallagher *et al.*, 1973).

The absorptive process can be divided into three components: mucosal uptake, transfer through the cell, and serosal release. Mucosal uptake starts within seconds of the iron reaching the brush border and takes a maximum of 30 to 60 min (Hallberg and Solvell, 1960). There is virtually no time lag when radioiron is injected directly into a loop of the small intestine. There are two steps in the transfer from the mucosa cell to the plasma: a rapid transport mechanism which within 2 hr of administration of the dose transfers up to 80% of the iron ultimately transferred to plasma and a slower process which transfers the remainder in the next 12–20 hr (Wheby and Crosby, 1963; Brown and Rother, 1963).

Mucosal Uptake. Depending on the dose of iron presented to the intestinal tract, there appear to be two systems. For the lower-dose range there is some rate-limiting process, for example, a carrier system with limited capacity, and at the higher-dose range there is a passive diffusion process with apparentlyunlimited capacity. In patients with iron deficiency anemia the second process becomes more important.

Manis and Schachter (1962b) using everted gut sac prepared from the proximal duodenum of the rat, mouse, and hamster, obtained data which indicated that (1) iron absorption involved at least two steps, (2) oxidative metabolism was apparently required at each of the steps, and (3) net transfer to the serosal surface is slower, more readily rate-limited, and more sharply localized to the proximal duodenum than is the mucosal uptake. Later studies by Bannerman *et al.* (1962), Wheby *et al.* (1964), and Howard and Jacobs (1972) provided additional data on absorption of iron over the range of concentrations which suggested two processes for each step, a saturable process suggestive of a carrier mechanism and simple diffusion to explain the direct linear relationship.

Loading rats with iron produced no change in uptake by the epithelial cells, but in iron-deficient animals there was an increased rate of absorption (Howard and Jacobs, 1972). Using rabbit duodenum, Acheson and Schultz (1972) obtained similar data. Intestinal hemogenates of normal and iron-loaded animals contain similar amounts of iron, but in iron-deficient animals there was a 66% decrease (Jacobs, 1975).

Specific receptors for iron on the mucosal surface were suggested by the finding that isolated brush borders from jejunal cells of rats bind more than those from ileal cells (Greenberger *et al.*, 1969). Receptor populations from the proximal and distal small intestine of iron-deficient guinea pigs increased and from iron-loaded animals decreased (Kimber *et al.*, 1973), confirming the trend seen by Greenberger *et al.* These changes did not occur immediately in the proximal portion of the intestine but were evident at 1 day, the first point measured, in the distal portion for the animals who were iron deficient. Data for the third to seventh days showed greater changes in the same directions in both portions of the intestine.

Ultrastructural cytochemical studies have added to our understanding of iron absorption. Parmley *et al.* (1978) used the Prussian blue reaction to localize nonheme iron in subcellular sites with a high degree of resolution and to differentiate between ferrous and ferric iron. After administration of iron there were numerous small stain deposits along the inner membrane of the microvilli. Similar precipitates were seen in the apical nonmembrane-based cytoplasm. The same pattern was seen in iron-deficient or iron-loaded animals. Ferrous iron appeared to be converted to ferric iron at the microvillus membrane.

Data from studies with neonatal rats and piglets suggest that the high rate of absorption in these animals might be due to an endocytotic mechanism which is not seen in the adult animal (Loh and Kaldor, 1971; Gallagher *et al.*, 1973; Furugouri, 1977). There are no ultrastructural studies of absorption in the suckling animal to confirm the postulated endocytotic mechanism.

Mucosal Transfer and Storage. Just how iron is transported across the mucosal cell is the greatest controversy in the study of iron absorption. There are extensive data, some of which are in conflict with other data. Each discussion of the process is based on a judgmental evaluation of the information available.

Once in the cell, iron is probably present as metabolically active cell components or bound to intracellular ligands. Richmond *et al.* (1972) measured the iron concentration of various cell fractions and found that the mitochondrial fraction contained the highest concentration in normal, iron-loaded, and iron-deficient animals. In the normal and iron-loaded animals, there was also a high amount in the soluble fraction. In iron-deficient animals the mitochondria still represented the major site, although the absolute amount of iron was reduced. Figure 3-5 is a composite of the data of Linder *et al.* (1975) and Worwood and Jacobs (1971, 1972) showing the distribution of iron in cell fractions for times up to 6 hr after intragastric intubation. Within 30 min, half of the total uptake was present in the cytosol. The mitochondrial uptake was slower, rising with time to become a significant part of the total activity. A more recent study (Humphreys *et al.*, 1977) examined the distribution at 1 and 18 hr after administration of ^{55}Fe by stomach tube. After 1 hr the relative specific activity of the soluble fraction was 1.8, and of the two mitochondrial fractions it was 0.82 and 0.6. The subcellular distribution of grains found on autoradiography was quite low in the mitochondrial fraction. After 18 hr the patterns had changed. The relative specific activity of the soluble fraction had decreased to about 1.0, and that of the mitochondrial fractions had increased to 2.6 and 1.4, respectively. At this time the subcellular distribution of grains by autoradiography was up in the mitochondria (from 6 to 77). Since iron in the epithelial cells is derived either from the body or from the lumen of the intestine and iron from an intravenous injection of ^{59}Fe–transferrin concentrated in the mitochondria, it can be suggested that the iron which slowly accumulates in the mitochondria has passed through the mucosal cell circulated once and returned from the serosal side, possibly to developing epithelial cells. Analysis of the grain counts over cell organelles (Table 3-6) revealed that the mitochondria of the lower

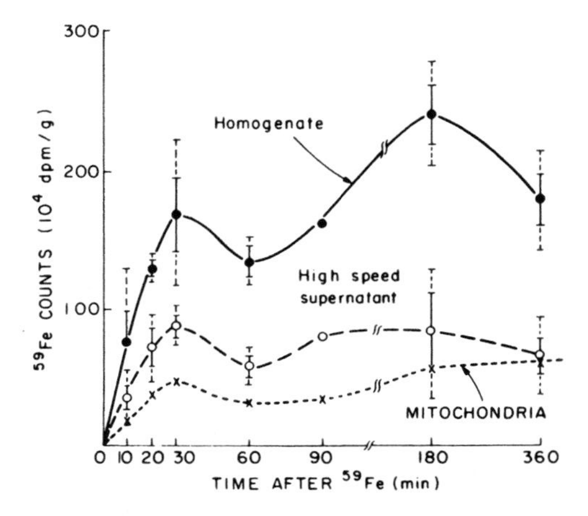

Figure 3-5. Distribution of iron in cytosol and mitochondria of small intestine at various times after intragastric intubation. From Linder and Munro (1977).

Table 3-6. Grain Counts of Cell Organelles Were Carried out over a Total Area of 1800 μm² for Each Region and Time Interval[a]

	Mitochondria		Cytosol	
	1 hr	18hr	1 hr	18 hr
10 cm	15	9	73	17
20 cm	0	53	36	35
30 cm	8	11	46	37

[a] Adapted from Humphreys *et al.* (1977).

intestinal segments retained more radioactivity. This increased retention of radioiron by cells lower down the intestine was also seen by Manis and Schachter (1962b) using everted sac preparations. Mitochondria appear to be quantitatively more important in iron metabolism within the cell but do not appear to be involved in iron transport during absorption.

Parmley *et al.* (1978) found that the transport of iron through endothelial cells into the intravascular space was associated with the localization of both large and small particles of ferric iron. In animals dosed with iron, with normal and deficient iron repletion, a gradient of large, small, and very small stain deposits was associated with the endothelial cells. Bedard *et al.* (1971, 1973, 1976), using large doses of iron and examining only the duodenum, found that after the initial uptake by the brush border much of the radioactivity in the cell was located over areas of cytoplasm containing the rough endoplasmic reticulum and free ribosomes. Neither biochemical nor more recent ultrastructural studies have confirmed these observations. The high dose of iron given may be producing an artificial situation.

Many studies have provided data to support the concept that iron in the mucosal cell is bound to one or more specific carriers which regulate its passage across the cell. Most of the iron in the soluble fraction is nondialyzable, and it is not clear how many of the various iron-binding proteins detected by different workers are identical. Most identify the iron-binding proteins as transferrin, a transferrin-like protein, ferritin, and a low-molecular-weight chelate of an amino acid. The transferrin-like protein has a more rapid turnover than ferritin, and the iron turnover correlates with iron absorption (Pollack *et al.*, 1972). According to some, most of the absorbed iron in the cytosol is present as ferritin and the low-molecular-weight form (Halliday *et al.*, 1975; Sheehan and Frenkel, 1972). According to others, 95% of the iron present is bound to the transferrin-like protein and ferritin (Huebers *et al.*, 1971). Inhibition of protein synthesis produced a decrease in iron absorption, while stimulants of protein synthesis cause an increase, adding evidence to the concept that at least one protein carrier is involved. The uptake of iron by the transferrin-like protein does not occur if the cytosol is separated from the remainder of the mucosal homogenate, indicating that an additional factor or factors essential for transport may be present (Huebers, 1975).

The transferrin-like protein is present in increased amounts in the jejunal mucosa of animals with iron deficiency. It appears to participate in the rapid phase of absorption. At least some of the transferrin present in mucosal preparations comes from contamination of plasma during the preparation of the cell suspension (Halliday *et al.*, 1975). Levine *et al.* (1972) proposed that the unsaturated transferrin binds to the epithelial cell membrane and enhances the release of iron from the cell. Saturated with transferrin and having a low affinity for the cell, the transferrin is released into the plasma. The plasma transferrin transports iron to the erythrocytes, and the transferrin molecule, now unsaturated, returns to the intestinal cell. Transferrin has been reported to enter the developing erythrocyte in the process of delivering iron to it (Morgan, 1974). Therefore, it is theoretically possible that transferrin could move into the cell to pick up iron and move back into the plasma to deliver it rather than facilitate release from the cell.

Mucosal ferritin does not appear to be involved in the rapid transfer of iron across the epithelial layer in animals with iron deficiency. It does, however, appear to participate in the slow process of absorption in normal animals. The interpretation of available data presented by Turnbull (1974) is that iron in excess of that bound by the carrier system is incorporated into ferritin. When the cell is exfoliated, the iron is lost into the lumen of the bowel, thus accounting for the decline in radioiron bound to ferritin from mucosa of iron-replete rats during the 4–18-hr period after the oral dose. Thus ferritin acts as a storage compound as it does in other cells. Unfortunately or fortunately, depending on one's viewpoint, the verdict cannot be handed down. Halliday *et al.* (1978) examined serum ferritin and duodenal ferritin in patients with iron deficiency, secondary iron overload, and idiopathic hemochromosatosis as well as normal subjects. In patients with iron deficiency who had been treated with oral iron therapy, there was an increase in mucosal ferritin concentration in spite of low serum values. This increased ferritin in the mucosa might bind iron and prevent a large influx of iron into the body, a finding consistent with the storage role for ferritin. However, the data on isoferritin profiles in the various subjects were interpreted by the authors to suggest a possible regulatory role for the more basic isoferritins, although they recognize the need for additonal research to clarify this point.

Serosal Release. Approximately 95% of the iron released from the mucosal cells enters portal blood; less than 5% enters the lymphocytes. Receptors for serosal transfer must exist because a genetic defect of the mechanism has been identified and because administration of parenteral iron quickly closes the serosal barrier. Rate of release is increased by plasma with low transferrin saturation and by pure iron-free transferrin.

There are several differences between mucosal uptake and serosal release. Uptake at the mucosal surface is more rapid than transfer at the serosal surface. With increasing iron concentration, mucosal uptake increases, while serosal transfer remains constant, and while mucosal uptake *in vitro* is similar in segments from various regions of the small intestine, serosal transfer is maximum in the proximal duodenum. Iron-loaded rats have a lower rate of absorption, and there is no difference in the uptake on the mucosal side. Several conditions, including low-iron diet, late pregnancy (Manis and Schachter, 1962a) and certain drugs (Manis and Kim, 1979) stimulate iron absorption by selectively enhancing the serosal transfer in rats. The mouse with sex-linked anemia (SLA) has a specific genetic defect in the serosal transfer of iron (Manis, 1971).

3.4 Corporal Factors Affecting Iron Absorption

Rate of erythropoiesis and decrease in tissue iron stores are usually cited as important stimuli to iron absorption. Changes in one or both parameters occur in both physiological and pathological states. Pregnancy and growth are physiological states which increase the iron requirement by increasing the rate to erythropoiesis and the need for nonerythroid iron. Failure to meet these demands either by increased absorption from food and/or by mobilization of iron stores results in the development of deficiency, and indeed iron deficiency is the most widespread form of nutritional deficiency in the world. Patients with disease states leading to changes in rate of erythropoiesis also show changes in iron absorption.

3.4.1 Physiological States

Most data on absorption of iron are for individuals 18 to 35 years of age. Data for younger and older individuals vary from limited to almost nil. Some of the available studies are difficult to evaluate because of an incomplete description of the control group; others are difficult to evaluate because of the limited number of subjects and still others because of the state of technology at the time the study was conducted. There is a definite need for more information on absorption by persons of all ages but particularly for those groups where almost nothing exists.

Pregnancy. One-third of young, nonpregnant American college women lack or have only small amounts of storage iron (Scott and Pritchard, 1967). In the Health and Nutrition Examination Study (HANES) survey, the prevalence of inadequate iron status as reflected by low transferrin saturation values is 7 and 13% in white and black females aged 12–17 years for income below the poverty level and approximately 6 and 7% for women aged 18–44 years, again for income below the poverty level (DHEW-USPHS Health Administration 1974). The percentage having low values was somewhat less for the same age groups for income above the poverty level. Similar figures were found in Sweden (Hallberg *et al.*, 1968). In many developing countries, and possibly in high-risk areas in developed countries, the prevalence of iron deficiency reaches 30–50%. During normal pregnancy, there is an increase in the demand for basal iron losses, expansion of red cell mass, fetal iron, and blood loss at delivery. The daily requirement increases from approximately 0.8 mg in early pregnancy to around 6–8 mg daily in the last month of gestation.

Studies of absorption of iron during pregnancy are very limited, and differences in protocol make comparisons difficult. The largest series (466) was reported by Hahn *et al.* (1951). All the women had uncomplicated pregnancy, were nonanemic, and were considered to be healthy. Iron was given as ferrous ascorbate at dosage levels supplying from 1.8 to 120.0 mg of elemental iron. Iron absorption was evaluated by determining the amount of incorporation into red blood cells 2 weeks after the administration of the tagged iron. Figure 3-6 shows the median percentage absorption by pregnant women at various stages during gestation. Absorption was the lowest in women tested at less than 14 weeks of gestation. Unfortunately, iron absorption was not measured in a nonpregnant control group; therefore it is not possible to state that under the conditions of this

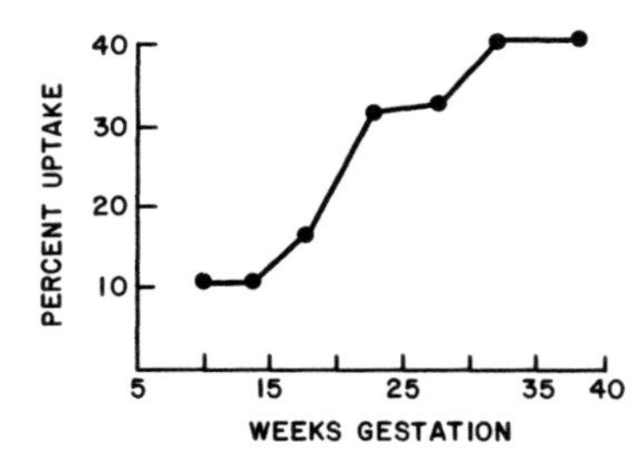

Figure 3-6. Absorption of iron (median values) by pregnant women at various stages of gestation. Adapted from Hahn *et al.* (1951).

study iron absorption was less than or equal to the prepregnancy absorption. From 15 weeks of gestation to the end of pregnancy, the percentage of absorption increased. This pattern was the same regardless of the dosage or iron; however, the percent of the higher dose which was actually absorbed is lowered.

Table 3-7 shows the data when the subjects were classified according to parity. There is an apparent increase in iron absorption with increasing parity under 20 weeks of gestation. Unfortunately, there was no statistical analysis of these data, so it is not possible to state that the findings are significant.

Using the chemical balance technique, Apte and Iyengar (1970) studied iron absorption in a total of 12 Indian women consuming their habitual diets. Not all the women were followed longitudinally during the pregnancy. Nonpregnant controls were included. The diet was vegetarian, containing 46–48 g of protein and 22 mg of iron. Considering that there is a great deal of variation from person to person in iron absorption and considering that the total number of patients was quite small, separating the values of subjects who maintained adequate iron values and those who became iron deficient does not seem valuable. The mean absorption during the period 8–16 weeks was 8.14%, 28–32 weeks it was 31%, and 36–39 weeks it was 32.8%. The values are quite high compared to those obtained in nonpregnant subjects and would suggest a strong possibility of incomplete sample collection.

In 1974, Svanberg and his group published data obtained on 27 women in which iron absorption was measured during pregnancy and once in the post-partum period and on 67 women who were studied only once during the pregnancy. Absorption was measured by consumption of radioiron-labeled test meals and by retention of radioiron by counting in a whole body counter. Sternal bone marrow smears were analyzed at about the twelfth or thirteenth week of gestation and 2 months after delivery or abortion. The average caloric intake of the women was 2360 kcal, ap-

Table 3-7. Median Percent Uptake of Radioiron among Pregnant Women According to Period of Gestation and Parity; Dosage: 9 mg and under[a]

	Parity		
Weeks of gestation	0	1.3	3+
Under 20	10.5	13.5	19.5
20 and over	35.0	34.5	36.0

[a] Adapted from Hahn *et al.* (1951).

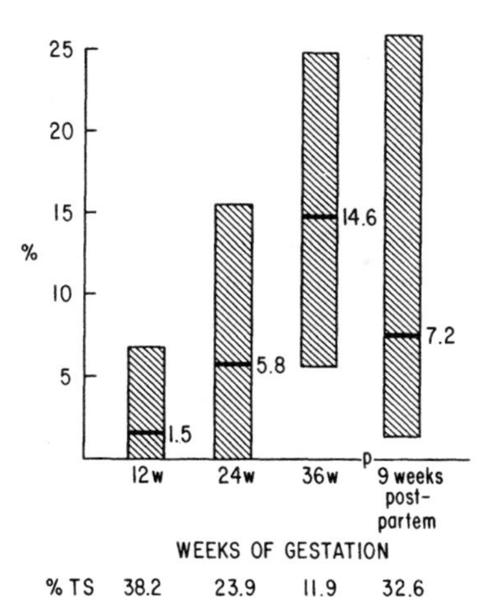

Figure 3-7. Absorption of nonheme iron and percent transferrin saturation in women during pregnancy and following delivery. Mean values are shown by dark lines, ranges by shaded areas. Adapted from Svanberg (1974).

proximately 10–20% higher than in previous Swedish studies. The authors suggest that this difference was probably due to the difference in the methods used to determine the intake. Iron consumption was 17.0 mg daily or 7.2 mg/100 kcal, again somewhat higher than previous reports. One reason for this higher intake of iron was the increased enrichment of food. One to two milligrams of the intake were heme iron, about 10 mg were native nonheme iron, and 5–6 mg were fortification iron.

Quite unexpectedly, it was found that in early pregnancy iron absorption was low in women with hemosiderin in their bone marrow smears, i.e., lower for nonpregnant women served the same test meal. In women without stainable iron in the bone marrow smear, absorption was higher. After delivery and after abortion, this difference was not observed. Figure 3-7 shows the absorption of nonheme iron during pregnancy and after delivery in women not treated with supplemental iron plus the mean percent transferrin saturation values. It can be seen that the absorption increases as the percent transferrin saturation decreases throughout the pregnancy. Concurrently the amount of iron in the bone marrow decreased until during the third trimester none of the women had stainable iron in their bone marrow smears. Eight to ten weeks after delivery the absorption value was higher than in the twelfth week of pregnancy in all but one woman. The investigators observed that absorption from a mixed pudding meal was lower than from an identical meal when the foods were served separately on a plate as an ordinary meal.

On the basis of these data, it appears that the daily absorption of nonheme iron increases from about 0.2 mg in early pregnancy to almost 1 mg in midpregnancy to about 2.2 mg in late pregnancy and that the total iron absorption increases from 0.4 mg in early pregnancy to about 3 mg in late pregnancy. After delivery, the increased absorption of iron was not related to the hemosiderin grade of the bone marrow smear. The relationship between hemosiderin grading and iron absorption may hold true for steady-state conditions, and these may not have been reached by 2 months post partum. Under non-steady-state conditions, the relationship between storage iron and absorption may be misleading as a measure of size of stores. Erythropoiesis may be slightly reduced in early pregnancy compared to menstruating women who have to compensate for menstrual blood loss, or it may be that erythropoiesis is slightly reduced for another reason. Absorption of ferrous iron was reduced to the same extent as the absorption of food iron. Endocrine, gastrointestinal, or unknown factors may be involved.

Hahn *et al.* (1951) were the only investigators to determine the amount of radioactivity in the infants' red cells at birth. From 1.8 to 3.0% of the dose given the mother was found in the infants' red cells when the dosage given the mother was 9.0 mg or less and 0.9% when the dose was greater than 18 mg. It could not be determined if there was a relationship between the percentage of iron in the red blood cells and the period of gestation since the number of cases was limited.

Manis and Schachter (1962a) studied iron absorption of rats during pregnancy both *in vivo* and using everted gut sacs of rats. Pregnancy increased the net transport of iron mainly by increasing transport to the serosal surface 113%. Mucosal uptake increases only about 6%. *In vivo* the transfer of iron from intestine to bloodstream was also increased 137%, while the mucosal uptake was only increased by 35%.

Infants. In a study of 5000 preschool children, Owen *et al.* (1970) found that approximately 7% of the children between 1 and 6 years of age were anemic and that some 45% of the children had iron deficiency on the basis of percent transferrin saturation. If the same criteria were applied to poverty-area preschool children, the authors suggest that a 20% incidence of anemia and an even higher incidence of iron deficiency would be found if plasma iron determinations were made. Karp *et al.* (1974) showed that when the school-age child was iron deficient, it was likely that the mother and siblings were also iron deficient. In view of these statistics it becomes apparent that we need to know more about the absorption of iron in the infant and child and more about the factors which regulate that rate of absorption.

Over the years it was a common observation by pediatricians that breast-fed babies in general do not become anemic. In the past this was attributed to either a higher iron content or to better rate of absorption (H. Albers, reported in Josephs, 1958). Further, it was reported that an artificially fed baby, even when more iron was received and retained than by a breast-fed baby, was still more likely to become anemic. Mackey (1928) found that breast-fed infants had persistently higher hemoglobin values than the artificially fed group. More recently, Coulson *et al.* (1977) looked at hematocrit levels of 66 clinically healthy breast-fed infants from 3 months to 2 years of age. Nine of the infants had hematocrit values below 32% and 4 below 31%.

In discussing iron absorption at different stages of development, it is important to be aware of changes in the needs of the infants from birth to approximately 3 years of age. The average-term infant gains enough iron to satisfy needs for the first 6 months from the breakdown of excess red blood cells at birth. At approximately 5 months of age the rate of erythropoiesis increases to meet the requirement during rapid growth, and whatever stores the infant had become depleted rapidly. During the period from 6 months to approximately 2 years the needs for iron must be met by consumption of an optimum diet. Sometime between ages 2 and 3 the child theoretically begins to have a surplus depending on intake. Data on tissue iron suggest that storage is at its lowest point between 12 and 24 months of age.

McMillan *et al.* (1976) and McMillan and Oski (1978) evaluated the iron status of 14 infants exclusively breast-fed for periods from 6 to 18 months. Thirteen infants had normal hemoglobin, serum iron, transferrin saturation, and free erythrocyte protoporphyrin values. One child had a hemoglobin value of 10.6 g/dl; serum iron and TIBC were not done. They then investigated the absorption of iron by adults from human milk and cow's milk by the technique of incorporation into red blood cells. Absorption from human milk was significantly higher (20.8% vs. 13.6% for cow's milk). It was suggested that the lower protein content of human milk or the higher lactose content or the lower phosphorus content or the higher vitamin C content was responsible for the difference in absorption.

In a second study (McMillan *et al.*, 1977) absorption by eight additional adults was studied. Mean absorption was 15.4%. Pooling the data from both studies would give an average of 18%. Absorption of simulated human milk was lower, 9%, and even more depressed by the addition of lactoferrin, 4.7%. Boiling the human milk failed to alter absorption (range: 5.1–48.2% and 6.5–49.0%).

The absorption of iron was studied in 45 term infants by the extrinsic tag technique combined with whole body counting of retained iron. Laboratory assessment of the iron status of the infants was based on serum ferritin, hemoglobin, mean corpuscular volume, and transferrin saturation. Infants who were breast-fed during the first 6 or 7 months of life attained greater iron stores than did the infants who received a cow milk formula. Breast-fed infants absorbed an average of 49% of the trace dose administered during breast feeding and 38.1% when iron was given after a fast in contrast to the formula-fed infants, who absorbed only about 19.5% (Saarinen *et al.*, 1977). The values are very close to the extrapolations suggested by McMillan *et al.* (1976) based on the differential absorption by adults.

Heinrich *et al.* (1969) found that infants absorbed approximately 18% of a 5-mg dose of iron when it was administered alone but only 4% when it was given with cow's milk. Schulz and Smith (1958a) found that 12–15% of a 30-mg dose of ferrous sulfate was absorbed by normal children aged 4–52 months. Giving either milk or orange juice with the iron salt resulted in less absorption of iron than when the ferrous sulfate was given alone. Iron-deficient infants absorbed more ferrous sulfate than did infants with good stores. Adding labelled iron to milk given to children less than 3 months old revealed a higher percentage absorption than in older children (Garby and Sjolin, 1959). Even within the group of 12 infants, the younger infants absorbed somewhat more than the older infants. Absorption of the tracer also decreased with the amount of iron in the dose.

The highest absorption by infants was reported by Gotze *et al.* (in Dallman *et al.*, 1980). Absorption, as measured by whole body counting after administration to fasted infants, increased from 18% at approximately 2 months of age to slightly over 70% at 18 months of age. The increased absorption appears to parallel the decrease in iron concentration during the course of lactation (Siimes *et al.*, 1979). Iron content falls from about 0.5 mg/liter during the first months to 0.3 mg/liter at 4–6 months. Absorption by preterm infants increased more rapidly, reaching peak absorption (about 65%) at 9 months of age. As expected, absorption was highest in infants with the least stainable iron in the bone marrow.

Using both intrinsic and extrinsic tag, Schulz and Smith (1958b) found that 3–13% of the iron from cow's milk was absorbed by infants 4–52 months of age. When adding the ferrous sulfate to the milk, the range was 2–17%. The absorption of milk iron by four iron-deficient children was 11–24% with a mean of 16%. Adult men absorbed approximately 2.8% of the iron from milk. Of the five normal children in whom incorporation into red blood cells was measured, 12% of the iron was retained and 10.6% utilized in hemoglobin formation (88% of the amount retained). For the six adults, 2.8% was retained and 2.1% utilized (77% utilization). Absorption of iron from enriched cereal (mixed, oatmeal, or rice) ranged from 0 to 25% with a mean of 12.3%. The iron content of the cereal was approximately 30 times that of the milk, yet the percentage of iron absorbed was approximately the same. The total amount of iron absorbed from the supplemented cereal was 0.37 mg compared to the 0.009 mg absorbed from milk. Five children 8–45 months of age were studied to determine the absorption from egg. An average of 11% of the iron was absorbed. In another group of 52 children 1–15 years of age, the absorption of egg iron was higher, 0–27% with a mean of 8.6%. Children less than 3 years of age absorbed more egg iron than did children 3–10 years of age. The differences between the two age groups was statistically significant.

To determine the adequacy of iron fortification of cereals and formulas, an extensive study of infants using these two sources of iron was conducted (Rios *et al.*, 1975). The iron salts were all added to the same mixed cereal. Absorption of ferric orthophosphate was the least, 0.7% in the 4 infants studied. In 9 infants, the absorption of sodium iron pyrophosphate was 1.0%, whereas in 25 infants the average absorption of ferrous sulfate was 2.7%. The best absorption was obtained when reduced iron of very small particle size was used as the fortification material. In the latter case, the mean absorption of 12 infants was 4.0%. Absorption was determined by total body count 10 days after the last of five feedings. The absorption of ferrous sulfate from three formulas was not statistically significantly different. For 42 infants, the mean absorption was 4.2%. Absorption of iron by 7 adults from milk-based formulas was 3.08 and 2.96% (McMillan *et al.*, 1977). Absorption of formulas containing 0.8, 6.8, and 12.8 mg of iron per liter of ferrous sulfate was also studied in 30 healthy infants 11–13 months of age. Average absorption was 12, 9, and 7%, respectively. Differences were not significant. Unfortunately, the group with the highest absorption had the lowest mean transferrin saturation and lowest range of values for transferrin saturation (Saarinen and Siimes, 1977).

Saarinen and Siimes (1979) proposed another technique to measure the absorption of iron without the introduction of radioisotopes. They measured iron absorption by using changes in the calculated total body iron (TBI) of infants. The TBI was determined as the sum of the hemoglobin iron (HbI) and the body storage iron (BSI). There was a close correlation between the concentration of serum ferritin (SF) expressed as the logarithm and the BSI expressed as milligrams per unit of body weight. Accordingly, they calculated the BSI from SF and body weight. Iron absorption was estimated from the increment of TBI and the estimated iron intake at different time intervals. The authors assume that the errors in calculations of the BSI, HbI, and amount of iron ingested would not affect the position of the regression line used in calculating BSI. This might be questioned particularly since the results obtained are much higher than values obtained in most studies except Gotze *et al.* (in Dallman *et al.*, 1980): 70% vs. 20–49% for human milk, 30% vs. 10% for cow's milk, and 10% vs. 4.7% for formula. The authors suggest that these figures are higher because of the continuous supplementation with ascorbic acid rather than because of a problem with the calculations. The most interesting observation obtained in the study was that there was a marked drop in the monthly increments of TBI in both breast-fed and formula-fed infants after the introduction of solid vegetable foods into the diet.

Animal studies confirm that suckling animals absorb more ingested iron than adults (Ezekiel, 1967; Loh and Kaldor, 1971). Mean absorption of inorganic iron salts and of iron from milk was over 85% for all suckling rats 20 days of age or less. Treating the animals with cortisone to "mature" the intestine or weaning them early resulted in a decrease in iron absorption. Gallagher *et al.* (1973) found that the ileum was an important site of iron absorption in the neonatal animal, especially when the dose exceeded the absorptive capacity of the proximal gut. Pinocytosis appears to be the major route of entry. The duodenum and jejunum are also important in the neonate. Iron-binding proteins have been isolated from the duodenal mucosa, indicating an active transfer system as seen in adults (Furugouri, 1977).

There have been a few studies of absorption in the premature or small for gestation age (SGA) infants. Early studies fail to distinguish the two groups of neonates. Ten premature infants, under seven days of age, studied by Oettinger *et al.* (1954), incorporated from 0.29 to 6.8% of a test dose of ferrous chloride into their red blood cells. Values for full-term infants, 2–4 days old, were 0.4–8.2%. Gorton *et al.* (1963) showed that low-birth-weight infants 1–10 weeks of age absorbed an average of 31.5% of iron. Heinrich *et al.* (1969) reported that preterm infants 2–80 days of age absorbed 20% of the test dose. Using the technique of whole body counting, Watanabe (1974) showed that absorption of a test dose of ^{59}Fe in preterm infants increased steadily from 4.8% at 3 weeks of age to 25% at 10 weeks of age. Dauncey *et al.* (1978) studied premature and SGA infants using the balance technique. Premature infants were in negative iron balance during the first 30 days of life. In contrast, SGA infants were in positive iron balance. Premature infants lost 12–24% of the iron present in a 1.0-kg infant at birth. SGA infants absorbed increasing amounts of iron from human milk during the period 12–30 postnatal days (from 37 to 67%). In the premature infants, it appears that there was no control over the amount of iron absorbed, the amount being determined by the concentration of iron in the diet. It is difficult to make inferences from the small numbers of children involved in the study. However, it would appear that too little iron will result in iron deficiency and too much in iron overload. It is extremely important to define the optimal range of intake for these infants to prevent both problems and also because under certain conditions modest amounts of iron in the diet may initiate lipid peroxidation of red cell membranes and cause hemolytic anemia. In infants who did not receive blood transfusions, the absorption of supplementary iron was a linear function of intake and was close to 34%. Blood transfusion produced a reduction in iron absorption.

There is some suggestion that ingestion of human milk, by itself, conditions the intestinal mucosa so that iron absorption is facilitated even when human milk is not consumed with the source of iron (Saarinen *et al.*, 1977). Factors, such as the presence of *Lactobacillus bifidus* which keeps the pH of the intestine lower, may facilitate iron absorption. The identity of these factors is not known.

Childhood and Adolescence. There is very little information about absorption by humans age 1–18 years. Darby *et al.* (1947) studied the absorption of ^{59}Fe–ferrous chloride given as a drink of lemonade with a slight excess of ascorbic acid. Absorption was measured by incorporation into red blood cells. One hundred and seventy-six children participated in the study. The mean uptake for each age group is presented in Table 3-8. The range of values for absorption was extensive (1–43%), although relatively few children had absorption values exceeding 20%. Iron status of the children was only estimated by the hemoglobin, a common practice at the time the study was done. It is possible that some of the children were iron deficient at the time of the study since hemoglobin values will not detect iron deficiency. There is a substantial increase in the percentage absorption of iron by girls at 8 years of age and by boys at 9 years. Enhancement of iron could be due to an increase in the growth rate of the children at these times.

Table 3-8. Absorption of Radioiron by Children in Two Schools in Nashville, Tenn. (absorption measured by RBC incorporation)

	% absorbed, mean	
Age	Male	Female
7	9.26	7.75
8	10.38	15.84
9	16.14	16.89
10	16.68	14.50

Using the balance technique, Abernathy *et al.* (1965) studied iron absorption in 36 healthy girls 7–9 years of age. Iron absorption ranged from 2.5 to 25.0% and was not significantly affected by the protein, riboflavin, or niacin content of the diet. However, in one study iron absorption was significantly increased when the daily protein intake was reduced by the removal of all milk from the diet, a finding seen by other investigators using different experimental protocols.

Adults and Elderly. Adults 18–50 years of age are typically used as controls, normals, or contrast groups in most studies of iron absorption; for example, "64 women and 52 men between 19 and 60 years of age volunteered for the present study" (Bjorn-Rasmussen *et al.*, 1976). Of the seven series, nine groups, the mean age of eight groups was 22, 22, 21, 25, 23, 23, 23, and 23, while only one group had a mean age of 38. These data indicate the inclusion of very few persons over 30. It would be of considerable interest if those groups of investigators having data on absorption of a large number of subjects would evaluate those data in relation to age. There is considerable intersubject variation in absorption due mainly to differences in iron stores. A man with adequate iron stores absorbs 7% or less of a dose of iron ascorbate; less than 1% of the iron from rice, egg, maize, spinach, and black beans; between 1 and 2% of the iron from lettuce, wheat, and soybeans; and between 2 and 6% of the iron from hemoglobin, fish, liver, and veal muscle (Martinez-Torres and Layrisse, 1973). Individuals with low stores would absorb 50% from ascorbate and increased amounts from vegetable and animal sources in the diet.

Many elderly people have iron deficiency anemia, which may be the result of improper or insufficient nutrition. The mean intake of 70 elderly women in private or nursing homes was found to be 9.4 and 8.3 mg/day, respectively (Jansen and Harrill, 1977). If an estimate of 15% iron absorption (indicative of some degree of iron deficiency) were used, the amount of iron absorbed by these subjects would be marginal with regard to meeting the average of 1-mg loss usually estimated for the healthy adult.

Freiman *et al.* (1963) measured absorption of ferrous sulfate in 45 patients, 25 women between 70 and 79 years of age, 11 women more than 80, and 9 men 69–85 years of age. For the women 70–79 years of age the mean absorption was 57.4%, and for the older women it was 60.9%. The mean value for the elderly men was 49.1%, and for 16 normal controls 27–60 years of age the mean value was 70.8%.

Jacobs and Owen (1969) studied a group 21–78 years of age by whole body counting. Each subject was given a test meal containing ^{59}Fe-labeled ferric citrate in cream of chicken soup. Twenty-five of the same subjects were also tested with ^{59}Fe-labeled hemoglobin given in cream of chicken soup. The absorption of heme iron did not change with age, while the absorption of nonheme iron appears to decrease with age. The mean absorption of subjects less than 30 years of age was 26.0%, while mean absorption of subjects 50 years of age was 13.1%.

Data from both studies suggested that there was a tendency toward decreased absorption with age. However, although the elderly patients were well characterized regarding iron status (i.e., hemoglobin, hematocrit, serum iron, percent transferrin saturation), the contrast group, younger persons, were not. Since premenopausal women are more prone to iron deficiency, it was important for interpretation of the data to have known more about the "control" groups. The same-year examination of bone marrow specimens obtained at 300 consecutive necropsies showed a decided tendency toward high storage in the oldest group (17+) (Benzie, 1963). The presence of disease was not significantly reflected in the bone marrow. Decreased absorption would be consistent with high iron stores; however, high iron stores are not consistent with high prevalence of iron deficiency anemia. Further, the lack of characterization of controls still left unresolved the question of changes in absorption of iron with age.

In more recent studies (Charlton *et al.*, 1970 and Gautier du Defaix *et al.*, 1980) increased storage iron was found in the liver of subjects, aged 55 to 74, and 61 to 70 years, respectively. The subjects had died of acute traumatic injuries. There is no explanation for this increased storage at this time.

A recently published study has contributed greatly to our understanding of iron metabolism in the elderly (Marx, 1979). Absorption was studied by retention of iron as measured by whole body counts at 14 days. Mucosal uptake, mucosal transfer, and red blood cell uptake were also measured. The subjects included a contrast group of 15 males and 10 females, all of whom were in good iron status as measured by hemoglobin, serum iron, and total iron-binding capacity. Forty elderly subjects were included in the study after it was determined that they were in good iron status. Twenty patients of all ages with uncomplicated iron deficiency were also included. Iron absorption was not decreased in the aged subjects compared to the young adults. Mucosal uptake, mucosal transfer, and retention were increased in all subjects with iron deficiency. A number of the healthy elderly had a higher amount of storage iron than normal, in agreement with earlier studies. Red cell iron uptake, calculated as the percentage of radioiron retained from the oral test dose and recovered in the erythrocytes, was markedly decreased in the elderly group (mean 66%) compared to the young adults (91%). The value for young adults is in good agreement with values reported by others, and 90% utilization is the factor assumed by most investigators using this method. This study has raised several questions, and more information is needed to determine if symptomatic iron therapy would be useful for the elderly.

3.4.2 Iron Absorption during Disease

Diseases which cause increased motility through the small intestine, anemia, malabsorption, or chronic bleeding produce secondary iron deficiency and alterations in absorption. There have been several studies of iron absorption by patients having various diseases. In many cases, numbers are quite small by necessity.

Pyrexia. While investigating the absorption of iron in two infants, Ashworth *et al.* observed that the iron absorption was reduced during a febrile illness (unpublished observation as quoted in Beresford *et al.*, 1971). In a more detailed study of malnourished children recovering from kwashiorkor and marasmus, Beresford *et al.* (1971) studied the absorption of ^{59}Fe-labeled ferrous ascorbate by whole body counting in 19 children. Five patients each had three absorption tests, one when afebrile and two when febrile or vice versa. During the fever absorption was decreased to 77% of the mean value in the afebrile state. For the whole series, the mean level of absorption in the afebrile state was 41.2% of the adminisered dose and in the febrile state 15.1% of the administered dose. This represented a decrease of 63.4% of the original value. Absorption of iron by children with marasmus tended to be lower than children with kwashiorkor whether they were febrile or not. One explanation for the difference might be that iron salts pass more rapidly through the duodenum during a febrile episode and are therefore not as well absorbed.

Hemolytic Syndromes. Prolonged survival of patients with thalassemia syndromes by use of repeated blood transfusions has raised the possibility of the complication of iron overload. Absorption of food iron might represent an additional possibility of iron overloading since autopsy findings reported more iron than could be accounted for by the transfusion alone. Erlandson *et al.* (1962) studied iron absorption in children with various congenital hemolytic syndromes, the majority of whom had thalassemia. They concluded that the studies of absorption of ferrous iron in children with heterozygous thalassemia were similar to those of control children, although there were only four children in each group. One of the children with thalassemia had a serum iron value of 30; therefore, the iron status of this child might be questioned. Interestingly enough, the child did not respond to the low serum iron by increasing absorption of iron. It is possible that the child had an infection which resulted in the shift of iron out of the bloodstream and not iron deficiency anemia. Four of the 10 children with thalassemia major in whom there was active erythropoiesis absorbed abnormally large amounts of iron, i.e., 42–80%. Absorption by other children was within normal limits.

Shadid and Haydar (1967) found that 11 children with thalassemia minor absorbed iron within the normal range or slightly above (2 children had had slightly high values, 34 and 36%. Four children with thalassemia major had modestly elevated iron absorption (38–40%).

Bannerman *et al.* (1964) studied iron absorption in 11 children with heterozygous thalassemia and found the absorption to be within normal limits. However, their "control" group included individuals in whom absorption exceeded 40% (approximately one-quarter of the total number examined). Therefore, the mean absorption for the control group was 30%. The paper details the iron status of the patients with thalassemia accurately. Unfortunately, the controls are not well characterized, so one is not able to determine the iron status of the controls. It is stated that there were no hematological abnormalities present, but there must have been a reason for the abnormally high absorption in some of the patients.

Grouping the data from the preceding studies (the method for measuring absorption was the same in all cases, i.e., ingestion of a sample of radioiron and collection of feces for counting), one finds that 27 of the 46 cases had iron absorption within normal limits. An additional 8 patients demonstrated a mildly elevated absorption (30–45%). Eleven, approximately 25%, of the patients had values from 46 to 80%. This group might be at greater risk for iron overload than the others. Since our knowledge of what factors regulate iron absorption is limited, there might be subtle differences in these patients when compared to others. It would certainly be beneficial to study additional patients to determine how many are at risk for iron overload.

Patients with sickle cell–hemoglobin C disease and hereditary spherocytosis who have slight anemia in the presence of increased erythropoiesis absorb an average amount of iron.

Diseases of the Gastrointestinal Tract. In a study of 10 patients with gastric achylia, Celada *et al.* (1978) found that iron absorption, as measured by whole body counter, was significantly lower than in the controls (Figure 3-8). Of the 10 patients, 1 had iron deficiency and 4 had anemia. Patients who were chronic alcoholics also absorbed less iron. There is a modest defect of food iron absorption in patients with achlorhydria (Cook *et al.*, 1964). Following partial gastrectomy, patients commonly have iron deficiency which is in part related to iron malabsorption. There is no decrease in the absorption of inorganic iron (Smith and Mallet, 1957), while the absorption of food iron is markedly impaired (Pirzio-Biroli *et*

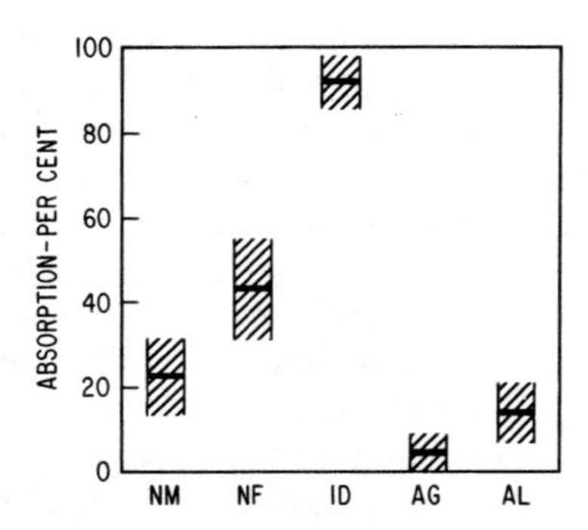

Figure 3-8. Iron absorption in normal males (NM), normal females (NF) and patients with iron deficiency (ID), gastric achylia (AG) and alcoholism (AL). Adapted from Celada *et al.* (1978).

al., 1958). In patients with partial gastrectomy, there was no difference in the absorption of ferrous sulfate and a brand of slow release iron (Baird *et al.*, 1974).

Iron absorption in patients with celiac disease is low (Callender, 1974). In 20 adult patients with celiac disease and iron deficiency, the absorption of 5 mg of iron as ferrous ascorbate averaged 14.8% [standard deviation (s.d.) 13.6] compared to 45.3% (s.d. 17.8) in 64 patients with uncomplicated iron deficiency. Sixteen of the patients with celiac disease absorbed less than 20% of the dose. Of 4 nonanemic patients with celiac disease, 3 absorbed a therapeutic dose comparably to the controls, while the fourth absorbed only 1.7%.

In a study of 31 patients with Crohn's disease and 13 patients with ulcerative colitis, Bartels *et al.* (1978) found that the ability to absorb iron was well preserved. Five of the patients in the study suffered from severe bleeding. Others had a small constant iron loss due to intestinal protein loss. There was no correlation among disease activity, site of lesion, intestinal resection, serum iron, transferrin, albumin, or hemoglobin. There was a significant inverse correlation between iron absorption and serum ferritin. There was also a significant positive correlation between serum ferritin and iron content of the bone marrow.

Intestinal motility affects iron absorption by changing the time interval iron is exposed to the absorptive surface of the gut. Patients with chronic diarrhea for prolonged intervals develop iron deficiency, and drugs which slow transit increase iron absorption.

Diseases of Iron Metabolism. Iron absorption in patients with fully developed idiopathic hemochromatosis is usually normal or slightly decreased because of the inhibitory effects of increased iron stores (Powell *et al.*, 1970). During or following therapeutic phlebotomy there is a marked elevation of absorption. Absorption of food iron was increased in 16 of 29 close relatives of patients with proven disease and may be important in screening (Williams *et al.*, 1965).

Iron absorption is enhanced by iron deficiency and increased rate of erythropoiesis in human subjects and in experimental animals. Switching animals from a low- to a high-iron diet reduces absorption, and from a high- to a low-iron diet increases absorption (Bannerman *et al.*, 1962). These changes occur too rapidly for body iron stores alone to be the most important factor in governing iron absorption. However, it is generally agreed that the body iron stores are an important factor. Since serum ferritin reflects the quantity of iron in the body, several investigators have examined the relationship between serum ferritin concentration and the rate of absorption. While there is a high degree of statistical significance the correlation coefficients have not been as high as might be expected (Charlton *et al.*, 1977; Cook *et al.*, 1974). Injecting ferritin into experimental animals does not alter the rate of absorption (Greenman and Jacobs, 1975). Serum ferritin may mirror the body stores but is apparently not the messenger which alters mucosal response; therefore, the correlation between serum ferritin levels and the rate of absorption is reasonably accurate but not perfect. Ten volunteers made slightly anemic by phlebotomy differed markedly in their ability to absorb iron from food (Olszon *et al.*, 1978). The mean value was 21% with a range of 9 to 34% in spite of the fact that none of the subjects had iron stores, again pointing to our lack of understanding of factors regulating absorption.

3.5 Intraluminal Factors Affecting Iron Absorption

The amount of iron potentially available from food depends on the adequacy of secretions of the gastrointestinal tract, the amount of iron supplied, the nature of that iron, and the composition of the meal with which the iron was consumed. Simple ferrous salts are better absorbed than more complex salts and ferric salts in humans (Brise and Hallberg, 1962) and in the monkey (Rao *et al.*, 1977). Ferric and ferrous iron are equally well absorbed by rats (Forth and Rummel, 1973), chicks (Pla and Fritz, 1970), dogs (Moore *et al.*, 1944), and guinea pigs (Rao *et al.*, 1977). Various dietary factors affect iron absorption differently in rats and humans (Cowan *et al.*, 1966).

Generally as the level of iron in the diet increases, the proportion of iron that is absorbed decreases; however, the absolute amount increases. Absorption from food varies widely. Animal sources, because of their high heme content and possibly other factors, are generally, but not always, better utilized than plant sources. If plant and animal sources are fed together, one source tends to modify the availability of the other. The two-pool model proposed by Hallberg (1974) and Bjorn-Rasmussen *et al.* (1974) and the use of extrinsic tag to label each pool differentially have contributed greatly to our understanding of the absorption of dietary iron, but there is still much to learn.

3.5.1 Macronutrients

Data on the effect of protein, carbohydrates, and fat on iron absorption are fragmentary and often contradictory. Most studies use laboratory diets that maintain constant caloric intake while varying the proportions of the constituents. It is therefore not always possible to determine if the changes in absorption are due to reduction of one constituent or increase of another.

Carbohydrates. Studies of the role of carbohydrate on iron absorption have produced conflicting data. Garretson and Conrad (1967) found that neither starch nor sucrose affected iron absorption. Amine and Hegsted (1971) found that, in general, simple sugars improved absorption, while starch inhibited it. The effect of lactose was greater than sucrose. Most important, however, the effect of carbohydrate was not uniform when iron sources of different availability were tested. Miller and Landes (1976a) found that iron stores in liver and spleen were reduced when glucose was the source of dietary carbohydrate compared to when starch was the only source of carbohydrate. Serum and liver iron concentrations were inversely related to the levels of dietary iron and were higher in rats fed diets containing starch than those given glucose diets. These data were suggestive of decreased absorption when glucose was the source of carbohydrate. In a second report using anemic rats, Miller and Landes (1976b) found that more hemoglobin was regenerated by the rats per milligram of iron consumed when starch was the source of dietary carbohydrate than sucrose or glucose. Using glucose as a source was less effective than sucrose.

Fructose enhances iron absorption. According to Pollack *et al.* (1964), metabolic products of fructose metabolism, lactic and pyruvic acids, complex with the iron and facilitate transcellular passage of the iron. When given orally with iron, lactic and pyruvic acids have been reported to increase absorption also. If this indeed is the mechanism, lactic and pyruvic acids must appear in the intestinal lumen in sufficiently high concentrations. The high sugar intake especially of fructose has been suggested as the cause of the excessive absorption of the Bantus which led to alimentary hypersiderosis (Saltman, 1965; Bothwell, 1969). Sorbitol also appears to increase iron absorption (Herndon *et al.*, 1958; Brown and Rother, 1963). Different results can be explained in part by differences in methodology.

Studies in humans follow somewhat different protocols than do the animal studies. In a pair of reports, Layrisse *et al.* (1976a, 1976b) report the results of a series of studies using sugar as a vehicle for iron fortification. Mixing iron-fortified sugar with vegetals resulted in the same absorption as from the native vegetals alone. When the fortified sugar was administered with the beverages, there was a further increase in absorption. When given with orange juice, Coca-Cola and Pepsi-Cola, the mean absorption ratio of fortified sugar was between 0.45 and 0.66, more than three times the absorption of the iron fortification mixed with vegetals. When sugar was added to coffee, the mean absorption ratio was 0.30, and if milk was added to coffee, it was 0.15. When different forms of iron were added to sugar and compared, it was found that Fe(III)–EDTA complex and ferrous sulfate were absorbed the best and that ferric ammonium citrate was the least well absorbed. However, Fe(III)–EDTA caused a discoloration in tea. If the soft drink is fortified with 3 mg of iron as ferrous sulfate, an absorption of between 0.25 and 0.80 mg of iron per soft drink can be expected on the basis of their results. Calibrated iron absorption from seven beverages ranged from 12.5 to 21.9%. If these data can be projected, two soft drinks per day between meals would provide sufficient iron to meet the needs of 95% of menstruating women even with a daily iron absorption from the diet of 0.8–1.0 mg of iron.

Lipids. Little is known about the effect of lipid on absorption of metals. Brodan *et al.* (1968) reported a decrease in absorption of iron in animals on a high-fat diet. On the other hand, Amine and Hegsted (1971) reported an enhancing effect on iron absorption by rats when the fat content of the diet was increased. Both polyunsaturated and saturated fats produced the same effect.

More recently, Bowering *et al.* (1977) determined the effect of changing the level and type of dietary fat on iron absorption in iron-depleted rats. Changes in the fat content of the diet included increasing the amount of fat from 5 to 20% and also exchanging lard for corn oil. Increasing the total amount of fat in the diet and changing the source to a more saturated fat increased the absorption by a small but significant amount. The effect was observed at suboptimal, average, and excessive levels of iron intake. With the lowest level of iron, enhancement was seen in the three criteria used to assess absorption, i.e., whole body counting, regeneration of hemoglobin, and liver iron accumulation. With the intermediate level of iron supplementation the final hemoglobin level did not show an enhancing effect of the fat. At the highest level of iron supplementation, only the liver iron accumulation was significantly enhanced by the increase in dietary fat. Monsen and Cook (1979), varying the components of a semisynthetic meal which utilized corn oil as the fat source, found that lipid had no influence on absorption.

Protein. Of the three main dietary components, proteins have received the most attention. In general protein enhances iron absorption, and protein deficiency produces diminished iron absorption. Animal proteins, such as beef, pork, lamb, chicken, and fish, have a pronounced effect on enhancing iron absorption (Johnston *et al.*, 1948; Layrisse *et al.*, 1968), whereas dairy products, such as milk, cheese, and eggs, inhibit iron absorption (Cook and Monsen, 1976).

To determine the mechanism by which absorption was enhanced by animal tissue protein, several investigators studied the effects of single amino acids on absorption. Histidine, lysine, and cysteine increase the absorption of ferric iron in isolated intestinal segments of rats (Van Campen and Gross, 1969; Van Campen, 1972). In healthy children, valine and histidine supplements promoted iron absorption, valine less than histidine. Cysteine, glutamic acid, and cystine supplementation had no effect. In children with iron deficiency there was a slight increase but to a lesser extent than in the controls (El-Hawary *et al.*, 1975). When a mixture of amino acids, which corresponded to meat, was given, there was a slight but significant effect on absorption of iron from maize in adults (Bjorn-Rasmussen and Hallberg, 1979). The effect was less than with meat. Again, no effect was seen with cysteine. The enhancing effect of meat may be due to a component not related to the protein content at all. Absorption studies with rats showed that iron in a digested muscle dialysate was more readily absorbed than that from an aqueous muscle extract. Hazell *et al.* (1978) therefore suggested that the degradation products of the hemoproteins formed by digestion might be responsible for the high availability of iron in meat. Bjorn-Rasmussen and Hallberg (1979) also found that the water extract of beef had no measurable effect on iron absorption.

The inhibitory effect of milk on iron absorption was noted by several investigators. The relationship between milk protein and iron content on hematologic values was studied in 144 infants using four formulas. Infants fed the lower-protein formulas had the highest hemoglobin and iron values irrespective of the iron content. Conversely, the infants fed the higher-protein, trace iron formulas had the lowest hemoglobin and serum iron levels (Gross, 1968). Milk also reduced the absorption of supplemental iron in adolescent males by one-third (Sharpe *et al.*, 1950). Absorption did not change significantly when dairy products, such as eggs, milk, and cheese, were substituted for egg albumin in the semisynthetic meal, but when substituted by beef in the standard meal, absorption decreased by 60–80% (Cook and Monsen, 1976). Nonfat cow's milk did not inhibit iron absorption in mice and chicks (Carmichael *et al.*, 1975), nor did dietary casein affect efficiency or iron use for total hemoglobin regeneration in rats (Conley and Hathcock, 1978).

The iron of egg yolk was approximately one-third as well utilized for hemoglobin regeneration as ferrous sulfate by anemic chicks and rats (Fritz *et al.*, 1970). Egg yolk iron is also poorly utilized by humans. The availability of iron was improved by ascorbic acid in humans (Callender *et al.*, 1970) and rats (Morris and Greene, 1972). Egg albumin also does not enhance iron absorption; in fact, it may actually inhibit it (Monsen and Cook, 1979; Bjorn-Rasmussen and Hallberg, 1979). Phosvitin, a phosphoprotein of egg yolk, binds iron and prevents its utilization. At present there is no explanation for the effect of egg albumin.

Absorption of biosynthetically labeled hemosiderin and ferritin has been studied. Mean absorption by subjects with good stores was 19% for hemosiderin and 0.9% for ferritin; for subjects with moderate iron deficiency, the values were 4.7 and 2.5%, respectively. Feeding either with wheat in a meal showed that its iron absorption was distinctly lower. Data from these studies support the possibility that hemosiderin and ferritin form a third pool different from the nonheme pool formed by vegetal iron, egg iron, and ferric and ferrous salts (Layrisse *et al.*, 1975; Martinez-Torres *et al.*, 1976).

Of the vegetable sources, the best absorption is from soybeans which are high in protein, 40%. Iron added to diets containing isolated soybean protein was absorbed by rats as well as the iron present in soybeans (Steinke and Hopkins, 1978). Sodium iron pyrophosphate added to commercially available soy isolate infant formula had a relative availability of 77 as compared to ferrous sulfate (Theuer *et al.*, 1971). Processing resulted in enhancement of absorption, possibly due to one of the other ingredients of the formula.

The importance of dietary protein in iron absorption is confirmed by several reports of defects in iron absorption in protein–calorie-deficient children (Klavins *et al.*, 1962) and experimental animals (Enwonwu *et al.*, 1972). Intestinal mucosal abnormal abnormalities did not play a significant role in malabsorption by 16 infants suffering from protein–calorie malnutrition since ferrous salts and hemoglobin iron were well absorbed, while ferric iron was not (Lynch *et al.*, 1970). Offspring of malnourished rats had a lowered iron absorption at weaning compared to progeny of adequately fed brothers (Enwonwu *et al.*, 1972). Refeeding a diet high in protein resulted in increased body weight and concomitant enhancement of retained iron. These animal data suggest that availability of iron from inadequate weaning diets should be examined to determine if low absorption of the small amount of iron present is a factor in the development of iron deficiency.

3.5.2. Micronutrients

The interrelationship of a few micronutrients has been well investigated. However, like all other aspects of iron absorption, more information is needed.

Vitamins. That ascorbic acid enhances the absorption of iron has been known for a long time (Moore *et al.*, 1939). Ascorbic acid is one of two enhancing substances felt to be sufficiently well defined on a quantitative basis to be considered in making estimates of iron intake (Monsen *et al.*, 1978). The amount of ascorbic acid in a given meal, as eaten, is a determinent of low, medium, or high availability.

Ascorbic acid increases the absorption of ferrous and ferric iron (Moore *et al.*, 1939; Hoglund and Reizenstein, 1969), of iron in vegetables and eggs (Sayers *et al.*, 1973), and of ferritin (Kuhn *et al.*, 1968). Ascorbic acid enhances iron absorption by its capability to reduce ferric iron and by the formation of a soluble iron–ascorbate chelate (Conrad and Schade, 1968). The effect of vitamin C is quite pronounced; adding 60 mg to a meal of rice more than tripled absorption, and an equivalent amount as papaya enhanced iron absorption from maize fivefold (Layrisse *et al.*, 1974). The binding of iron by phosphoprotein is decreased by ascorbic acid (Peters *et al.*, 1971). Orange and other fruit juices, most likely due to their ascorbic acid, citric acid, and fructose content, also improve the absorption of iron.

Vitamin A deficiency results in an anemia which has some of the characteristics of iron deficiency anemia. However, the anemia does not respond to medicinal iron and may occur in spite of an adequate intake of iron (Mejia *et al.*, 1977) in contradiction to animal data which suggested that iron absorption was increased in vitamin-A-deficient rats (Amine *et al.*, 1970). More recent data showed there was no significant differences in absorption of iron by vitamin-A-deficient rats (Mejia *et al.*, 1979). Pair-fed control rats, whose food intake was restricted, absorbed and retained less iron than the vitamin-A-deficient animals. Either the meal pattern or the reduced intake may have resulted in increased desquamation of the intestinal epithelium. In addition to decreased absorption, incorporation of the absorbed iron into red blood cells was significantly reduced, and there was a greater accumulation of isotope in the liver and spleen of the vitamin-A-deficient animals. Goldberg and Smith (1960) also found that vitamin-A-deficient rats had an extensive deposition of iron in the liver when given large oral doses of iron. The apparent contradiction between the data of Amine *et al.* (1970) and that of Mejia *et al.* (1979) illustrates the hazards of studies of iron absorption. The earlier study determined iron absorption by whole body counting, which does not detect shifts within the body itself, and the apparent increase in hematopoietic activity may have been the result of hemoconcentration. Mejia *et al.* (1979) found a significant reduction in plasma and blood volumes in the vitamin-A-deficient rats. The authors suggest that in animals with vitamin A deficiency either plasma iron is trapped in the liver or spleen and not effectively released into circulation for utilization or it accumulates in the liver because it is not properly utilized in the bone marrow. Only further studies will clarify which mechanism is true.

Minerals. Evaluating data on the effects of calcium and phosphate on iron absorption is complicated by the frequent use of unphysiological levels of the salts. Phosphoprotein, such as vitellin in egg yolk and casein in milk, or insoluble calcium phosphate bind iron more effectively than soluble phosphate salts and cause the greatest decrease in iron absorption in humans. In the presence of phosphoprotein or insoluble calcium phosphate, iron was not dissolved even at pH 1.3 unless ascorbic acid was present (Peters *et al.*, 1971). Phosphates appear to act by accelerating oxidation to the ferric state and sequestering iron as ferric phosphate.

Monsen and Cook (1976) investigated the effects of calcium and phosphate on absorption of iron using healthy men and woman with adequate iron status. With a semisynthetic meal, iron absorption was reduced from a mean value of 1.7% to 0.8% with the addition of calcium and phosphorus in the form of $CaHPO_4$. A subset of the same group was used to evaluate the effects of the salts when the meal contained some beef as the protein source rather than the albumin used in the first series. With beef in the diet, absorption of iron was 8.8%, but addition of calcium and phosphorus salts again resulted in a reduction. The absorption ratio with/without CaP was essentially the same in both studies. When calcium or phosphate was added individually, there was no significant effect on nonheme iron absorption. Monsen and Cook suggest that the effects are best explained by the formation of a complex of calcium, phosphate, and iron phosphate which is more effeciently precipitated by iron in pure solutions in the presence of calcium salts, particularly when the pH level is above neutrality.

Rats fed a diet low in phosphorus and high in ferric citrate absorbed excessive amounts of iron and deposited some in the liver. The amount deposited was inversely related to the phosphate content of the diet. Absolute amounts of phosphate and/or iron salts are as important as the iron–phosphorus ratio (Hegsted *et al.*, 1949).

Removal of either phosphate of calcium from the salt mixture resulted in inhibition of iron absorption. Increasing the total concentration of salts in the diet of iron-deficient rats also decreased the absorption of iron. With no added salt, 78.7% of the iron dose was retained. With a 2% salt mix, absorption was reduced to 57.5%, and when the salt content was 4%, the mean absorption was 41.0%. When the diet contained 2% of either calcium or phosphate-free salt mix, the retention was decreased equally. Retention of iron from the diet with 4% phosphate-free salt mix was not affected, while retention from the 4% calcium-free salt mix was significantly lower from either: the complete 4% (Amine and Hegsted, 1971). The design of the study does not permit one to distinguish between the effects of the material removed from those of the material added.

Buttner and Muhler (1959) also found that phosphate produced a significant reduction in the liver iron content as well as a decrease in the calcium and phosphorus content of newly forming incisor dentine. These data suggest a significant reduction in blood phosphorus concentration.

Dams and their weanlings were severely anamic when fed a diet containing 2% calcium carbonate. Animals maintained on 0–0.5% added calcium carbonate had normal values (Greig, 1952). Addition of 10 ppm of iron as ferric citrate prevented the anemia. Chapman and Campbell (1957) studied the effects of adding bone meal, calcium carbonate, calcium lactate, calcium chloride, disodium phosphate, or commercial sodium hexametaphosphate. The calcium salts interfered with iron utilization as reflected by reduced liver iron stores, hemoglobin regeneration, and increase in heart weight.

It would appear that the addition of either calcium or phosphate salts may improve absorption under certain circumstances, whereas excessive amounts of either are detrimental to the absorption by rats. Robertson and Worwood (1978) studied iron and lead absorption in iron-deficient and iron-replete rats. Absorption was defined as the percentage of the dose remaining in the carcass after removal of the whole gastrointestinal tract. Iron absorption, as expected, was increased in iron-deficient rats, but lead absorption was not consistently increased in either long- or short-term depletion, and there appeared to be no direct relationship between transfer of iron and lead across the intestinal mucosal.

Earlier studies found that the rats consuming an iron-deficient diet might accumulate more lead in their tissues (Six and Goyer, 1972).

3.5.3. Complexing Agents

Desferrioxamine forms a stable complex and reduces absorption of inorganic iron (Hwang and Brown, 1965), as well as iron from wheat, corn, soybeans, and ferritin (Kuhn *et al.*, 1968). It appears to bind iron released from the ferritin in veal muscle (Martinez-Torres and Layrisse, 1971).

Cholesytramine binds ionic iron, reducing its absorption (Greenberger, 1973). Tetracycline chelates iron, but its effect on absorption has not been demonstrated (Greenberger, 1973).

Tannins. Studies of the effects of tea on iron absorption were begun after the observation that absorption from meals with which tea was drunk were lower than expected. Controlled studies revealed the tea does indeed reduce the amount of iron absorbed from bread, a meal of rice and potato and onion soup, or solutions of $FeCl_3$ and $FeSO_4$ (Disler *et al.*, 1975). Table 3–9 shows the mean absorption from the various iron sources when either water or tea was consumed. Drinking tea or other beverages containing tannin may contribute to the development of iron deficiency if the diet consists largely of vegetable foodstuffs. No attempt was made to investigate the mechanism responsible for the interference with absorption, but it was suggested that the tannins were the source. Tannins are known to form colored complexes with ferric iron, and a black discoloration was seen when $FeCl_3$ was added to 200 ml of tea. This complex may interfere with absorption. Preliminary observation by the same authors indicates coffee may have a similar effect.

Table 3-9. Effect of Tea on Iron Absorption from Various Sources[a]

	Percentage absorbed	
Source	Water	Tea
$FeCl_3$	21.7 ± 19.7	6.2 ± 3.9
$FeSO_3$ + ascorbic acid	30.9 ± 19.3	11.2 ± 7.5
	40.9 ± 22.2	13.3 ± 12.0
Iron in bread	10.4 ± 4.4	3.3 ± 3.0
Iron from rice with potato & onion soup + ascorbic acid	10.8 ± 4.1	2.5 ± 1.6
Hemoglobin iron		
Uncooked	14.7 ± 11.4	6.0 ± 4.5
Cooked	13.5 ± 7.0	14.3 ± 9.2

[a] Adapted from Disler *et al.* (1975).

de Alarcon et al. (1979) determined iron absorption in patients with thalassemia major and thalassemia intermedia. As hemoglobin falls throughout the course of the normal transfusion cycle in these patients and the rate of erythopoietin increases, it was felt that absorption of dietary iron might increase. The quality of iron absorbed from the diet by patients with thalassemia major is not usually excessive; however, patients with thalassemia intermedia and other hemolytic anemias may represent a definite risk. Although the number of patients studied was small, inhibition of the absorption of iron by tea was observed in all patients. An iron absorption indicative of iron deficiency was observed both times a patient with thalassemia intermedia was tested (28.2 and 41.1%). Inhibition by tea dropped these values by 55 and 71%.

Phytate. Some researchers have found that the addition of relatively high levels of phytic acid to animal diets had no effect on iron availability, while others found that absorption was substantially reduced (McCance *et al.*, 1943). The absorption of nonheme iron is so sensitive to both the conditions of the small intestine and other dietary components that small differences in experimental design can lead to confusing results. Sharpe *et al.* (1950) found that added sodium phytate reduced iron absorption considerably, while there was no correlation between the phytate content of foods and reduction in iron absorption. There was also no correlation between absorption and the phytate content of soybean seeds (Welch and van Campen, 1975). Further it appeared that immature soybean seeds contained a factor or factors other than phytic acid that depresses iron availability. Liebman and Driskell (1979) fed iron-deficient rats diets in which bread was the source of phytate and constituted 50% of the diet. The remaining foods were varied so that they contained six distinctly different levels of phytate. Total protein in the diet was fairly constant, although the source of the protein varied. The iron content of the liver was similar in all cases, and it would appear that phytate had no effect on iron absorption.

Morris and Ellis (1975, 1976) showed that over 60% of the iron in wheat bran was present as monoferric phytate (MFP). Studies in rats showed that the MFP was bioavailable. Humans and rats have an intestinal phytase which might liberate food iron as the phytate is degraded, although the physiological importance of this enzyme is not known. MFP is soluble at pH 7.0 and above, unlike the phytate complexed with two or more iron atoms. Unleavened bread has a higher phytate content than leavened bread. Phytate is hydrolyzed by moist heat, and this hydrolysis during baking may have a deleterious effect on the total biological availability of the iron in monoferric phytate.

In dogs MFP was about one-half as available as ferrous sulfate at lower doses (equivalent to 1.5 mg) and about one-seventh as available at higher doses (15 mg of iron) when given without food (Lipschitz *et al.*, 1979). When given with food, the MFP iron exchanged completely with the nonheme dietary iron, and absorption was the same as the major pool of inorganic dietary iron from meals of high or low iron availability.

Geophagia. Geophagia has been suggested both as a cause and an effect of iron deficiency anemia. Information on the effect of clay eating on mineral absorption is limited. Oral iron and zinc tolerance tests were performed on 12 patients with iron deficiency anemia who practiced geophagia. Five controls with iron deficiency anemia alone were also studied. In contrast to the patients with iron deficiency anemia only, those who practiced geophagia did not show elevation in serum iron values after a test does. Peroral intestinal biopsies were done in 4 patients presumably all of whom practiced geophagia. The intestinal villi were shortened, blunted, and occasionally fused in addition to cellular infiltration in the lamina propria (Arcasoy *et al.*, 1978).

3.5.4 Fortification Iron

Because of the high prevalence of iron deficiency, fortification of the diet has long been advocated. There are two major problems: what to fortify, and what to fortify with. Table 3–10 lists eight of the most common salts used for fortification. Steinkamp *et al* (1955) studied absorption of

Table 3-10. Relative Availability and Costs of Nutritional Iron Stores[a]

Name	Contained iron (%)	Relative bioavailability (%)	Cost Per Pound of "available" iron ($)
Ferrous sulfate heptahydrate	20.1	100	2.68
Ferrous sulfate dried	32.1	100	1.56
Ferrous fumarate	32.9	102	4.16
Ferric ammonium citrate	17.0	107	12.69
Ferric orthophosphate	28.6	15	21.93
Sodium iron pyrophosphate	14.5	12	58.00
Reduced iron, electrolytic	97.0	50	3.18
Reduced iron, H_2	96.0	32	2.65

[a] Adapted from *Iron Data Sheet*, Mallinckrodt, Inc. St. Louis, Print No. 021-R (1979).

four different iron compounds baked into bread. In spite of the highly variable individual results, it was concluded that there was no difference among the four sources. Several developments raised doubts regarding the interpretation and the results. Therefore, 20 years later more detailed studies were done in which absorption of each alternative salt was compared to the absorption of ferrous sulfate. The least available of the supplements was sodium iron pyrophosphate with a mean absorption ratio relative to ferrous sulfate (ARFS) of 5%. Intermediate was ferric orthophosphate (ARFS = 31%), and still greater availability was noted for reduced iron (ARFS = 95%) (Cook *et al.,* 1973). Using a chick bioassay system, Motzok and co-workers also found biological availability from foods supplemented with ferric orthophasphate and sodium iron phosphate to be low (Motzok *et al.,* 1977). These data are somewhat different from those included in Table 3–10 and illustrate the necessity of testing under the same conditions as actual use. Also, there is a great deal of variability in compounds called *reduced iron* (Waddell, 1974).

Ten commercial variety breads, five enriched with ferrous sulfate, one with reduced iron, one with unspecified iron, and three which were not enriched, were tested recently using rats. Bioavailability compared to absorption of ferrous sulfate added to a low-iron diet varied from 32 to 80%. Interference with absorption was unrelated to phytate or fiber content (Ranhotra *et al.,* 1979).

It also became apparent that the same salt added to different foods would be absorbed differently. Combinations of rice, wheat flour, and milk were fed in six ratios to provide constant nitrogen per day. Iron as ferrous sulfate was added in a constant amount to each of the diets, and absorption ranged from 7.5 to 25.6%. Absorption was higher when the diets contained more rice than when there was a high wheat content (Senchak *et al.,* 1973).

Fe(III)–EDTA has recently been proposed as a salt for iron fortification. Absorption of this salt with six different food vehicles compared with absorption of ferrous sulfate indicates that while absorption from Fe(III)–EDTA remained practically the same, the absorption of ferrous sulfate varied from 2 to 30%. The Fe(III)–EDTA complex is apparently not affected by vegetable foods or milk and because of the small amount needed may be very useful in fortification. About 16% of the iron absorbed from the Fe–EDTA complex is separated during the process of absorption and excreted in the urine. The mechanism of absorption of the EDTA complex is not understood (Martinez-Torres *et al.,* 1979). The EDTA complex has also been added to sugar and is well utilized. Unfortunately, when added to tea, it causes a discoloration (Layrisse *et al.,* 1976a, 1976b).

3.6 Mucosal Factors Affecting Iron Absorption

Thomas *et al.* (1972) reported that treatment of normal rats with phenobarbitone for 5 days increased the intestinal absorption of $^{59}FeSO_4$ compared to the controls (20.4 vs. 9.8%). *In vitro,* phenobarbital treatment also resulted in a significant increase by everted duodenal gut sac but did not affect transport by $^{59}FeSO_4$ by jejunal gut sac. The increased absorption might be due to increased synthesis of the carrier protein. Edwards and Hoke (1975) demonstrated a one-third enhancement of absorption of a 1-μm dose of iron and a 67% increase in the absorption of a 10-μm dose in normal mice. Mice with an X-linked recessive trait for anemia (SLA) showed impaired intestinal iron absorption. The defect is in the transfer of iron from the mucosal cell to the plasma and is associated with accumulation of iron in the duodenal epithelium. These mice fail to respond to the phenobarbitone by increasing absorption of iron, indicating that phenobarbitone acts on the second step of absorption. When these animals were fed an iron- deficient diet, however, the phenobarbitone enhancement of iron absorption took place, indicating that these animals have not lost their ability to alter the amount of iron absorbed from the intestine but rather the ability to regulate iron absorption. There may be two possible explanations: (1) that there is malabsorption in the synthesis of the mucosal transport substance which is structurally and functionally normal and (2) that the regulation of the synthesis of the mucosal transport substance itself is structurally abnormal and less capable of iron binding. Further work with this strain might clarify the regulation of iron absorption.

Rats treated with 2,3,7,8-tetrachlorodibenzo-*p*-dioxin (TCDD), orally or intraperitioneally, absorb significantly more iron *in vivo* than do control rats (Manis and Kim, 1979). The same observations was seen in *in vitro* studies. TCDD decreased calcium transport by gut sacs *in vitro* and does not influence D-galactose or L-protein transport. This would indicate that the compound exerts selective effects on discrete intestinal mechanisms rather than a general metabolic or toxic effect. As suggested by Manis and Kim, further studies are needed to determine if other inducers of mixed-function oxidase systems alter intestinal absorption of minerals and thereby pose a previously unsuspected public health hazard.

Cycloheximide impairs the intestinal transport of iron possibly by interfering with the synthesis of a carrier protein (Yeh and Shels, 1969). Phenobarbital treatment partially reverses the inhibition induced by cycloheximide (4.9 vs. 2.2%).

Since both the plasma level of iron and its transport protein transferrin change in women on oral contraceptive agents, King *et al.* (1978) studied the absorption of stable isotopes of iron, copper, and zinc in young women 19–22 years of age. Absorption was determined by the balance technique with modification since the excreted stable isotopes were activated by neutron activation analysis. Fourteen of the women were on oral contraceptive agents, and 8 were not. All appeared to be in good iron status as determined by hemoglobin, hematocrit, serum iron, and percent saturation values. There was no significant difference between the two groups, and the absorption was 13.5% ± 12.3%. The coefficient of variation for iron absorption was 90%. Norrby *et al.* (1972) also found that there was no change in the absorption of ^{59}Fe before and during treatment of oral contraceptive agents. Their subjects absorbed 50% of theoretical dose.

Alcohol increases absorption of ferric iron in subjects with normal gastric secretion but not in subjects with achlorhydria (Charlton *et al.*, 1964). In rats there is no effect (Tapper *et al.*, 1968).

3.7 Regulation of Iron Absorption

Our present understanding of intestinal iron absorption and its control is still very limited. Older theories are well discussed in some of the reviews mentioned earlier (Forth and Rummel, 1973; Turnbull, 1974). Almost every factor involved in iron absorption has been suggested as the determining factor in regulation at one time or another. One constant observation is that conditions that stimulate erythropoiesis or depletion of iron stores, such as pregnancy, hemorrhage, and iron deficiency, increase iron absorption and that conditions associated with ample stores or decreased erythropoiesis, such as iron overload and large doses of radiation, decrease absorption. Patients with a regenerative anemia, i.e., erythropoiesis and iron stores not affected by anemia or transfusion, absorb significantly more iron when anemic than when the hemoglobin values are restored to near normal by transfusion (Schiffer *et al.*, 1965). Anemic rats also increased absorption when placed in conditions in which the body iron content and the rate of red cell production were unchanged.

Intraluminal factors determine the amount of iron available for absorption. Regulation of absorption by secretions of the gastrointestinal tract were proposed in the 1960's, but experimental data fail to support the concept. Most evidence supports the idea that regulation of absorption is largely by the intestinal epithelial cells. The question remains, however, as to how the cells are made aware of the need for iron. That there was a humoral factor was suggested by a study of the effect of hypoxia on iron absorption in parabiotic rats (Brittin *et al.*, 1968). When one member of the pair was made hypoxic, the partner showed increased absorption of iron. Additional evidence was obtained when iron-replete rats were perfused with serum of iron-deficient animals and absorption increased (MacDermott and Greenberger, 1969). The identity of the humoral factor remains unknown. Serum ferritin, a potential candidate, does not respond quickly enough to changes in dietary iron level but is heat-stable.

There are a number of changes which take place in the mucosa cells during times of need which can be correlated with increased iron absorption. A direct causal relationship has not, however, been established. As mentioned earlier, there are a number of iron-binding proteins present in the mucosa. After administration of a dose of FeCl, the major radioactive components of mucosal cytosol are ferritin, transferrin, and a transferrin-like protein and a nonprotein low-molecular-weight form.

The elution pattern of proteins labeled with ^{59}Fe and the electrophoretic data are different for anemic and iron-overloaded rats (Van Campen, 1974). An iron-binding protein (transferrin-like) having a molecular weight of approximately 80,000 was very apparent in the material obtained by iron-deficient rats but not in the material from overloaded rats. The total protein content of the mucosal cell remains unchanged, while the concentration of iron-binding proteins increases up to 10 times in iron-deficient animals (Forth and Rummel. 1973).

Although the ferritin content of the mucosa increases with iron administration, it is now generally accepted that ferritin accumulates to sequester excess iron present in the cell and does not participate in the regulation of absorption. However, a role for ferritin cannot be ruled out unequivocally. Halliday *et al.* (1978) examined the relationships between serum ferritin and duodenal ferritin in normal subjects and in patients with iron deficiency, secondary iron overload, or idiopathic hemochromatosis (IHC) to investigate the changes of the isoferritins in these various conditions to determine if the isoferritins could play a regulatory role. In conditions of iron overload, duodenal iron was lower at all levels of serum ferritin in comparison with normal and iron-deficient subjects. Purified duodenal ferritin from normal subjects and patients with iron overload conditions showed the same two distinct isoferritins. Two additional isoferritins were detected after oral administration of iron. Control patients and patients with secondary iron overload and the patient with untreated IHC exhibited similar profiles with a predominance of isoferritins of pI 5.25–5.35 and some evidence of a smaller peak at a pI of 5.54–5.56. The pattern was different in patients with iron deficiency who were being treated in that there was a substantial increase in the isoferritins of pI 5.54–5.62, suggesting an increase in more basic isoferritin when inorganic iron is presented to mucosal cells. The authors suggest that the basic isoferritin may play a regulatory role in the control of iron absorption and that more information is needed in this area.

The hypothesis that the complement of iron taken up by the mucosal cells during their formation regulates their subsequent behavior has received considerable support. Conrad *et al.* (1963, 1964) found that radioiron given orally to rats initially resulted in uniformly distributed labeling of the columnar epithelial cells of the villi and that during the next 40 hr the labeled cells moved progressively to the tips of the villi. In contrast, little activity was observed in iron-deficient or iron-loaded animals. The decreased incorporation of radioiron resulted from rapid passage through the cell to the plasma in iron-deficient animals. In animals with iron overload, decreased incorporation was due to lack of uptake. Parenteral iron resulted in the basal cells being labeled first in both iron-overloaded rats and the controls. It was 72 hr before all the cells were labeled. There was very little iron in the mucosal cells of iron-deficient rats (Peters *et al.,* 1971. The authors hypothesized that mucosal cells are imprinted at the time they are formed. This concept is consistent with some data but not adequate to explain other data; for example, acute hemolysis can result in increased absorption, while mucosal cell iron in unchanged (Pirzio-Biroli and Finch, 1960). There is the possibility that it may be the form of the iron rather than the total amount or the internal iron exchange (Worwood *et al.,* 1975) which is important.

Assembling a composite picture of the mechanism and regulation of iron absorption is difficult for several reasons. The mucosal cell has need for iron to function as a metabolically active cell; therefore, iron can enter the cell from the plasma to meet these needs as well as from the lumen. Further studies frequently examine only duodenal cells, while regulatory changes may occur farther down the intestine. Absorption is three-dimensional, not two-dimensional. More data are needed on suckling animals. Finally, conflicting data must be resolved. Figure 3–9 is an attempt at a composite picture of iron absorption which includes some of the normal functions of iron within the cell because all iron must play a role in the overall picture. Mucosal uptake is shown to occur by two pathways, the receptor mechanism being an energy-requiring process. Receptor populations increase more quickly in the distal portion of the gut of iron-deficient animals. No data are available on receptor populations of neonatal animals. Pinocytosis seen in neonatal rats may be a pathological response to exceeding the capacity of the proximal gut, or it may be a capacity seen only in the very immature.

Once within the cell, the iron appears to react with a low-molecular-weight compound such as an amino acid carrier and become part of a

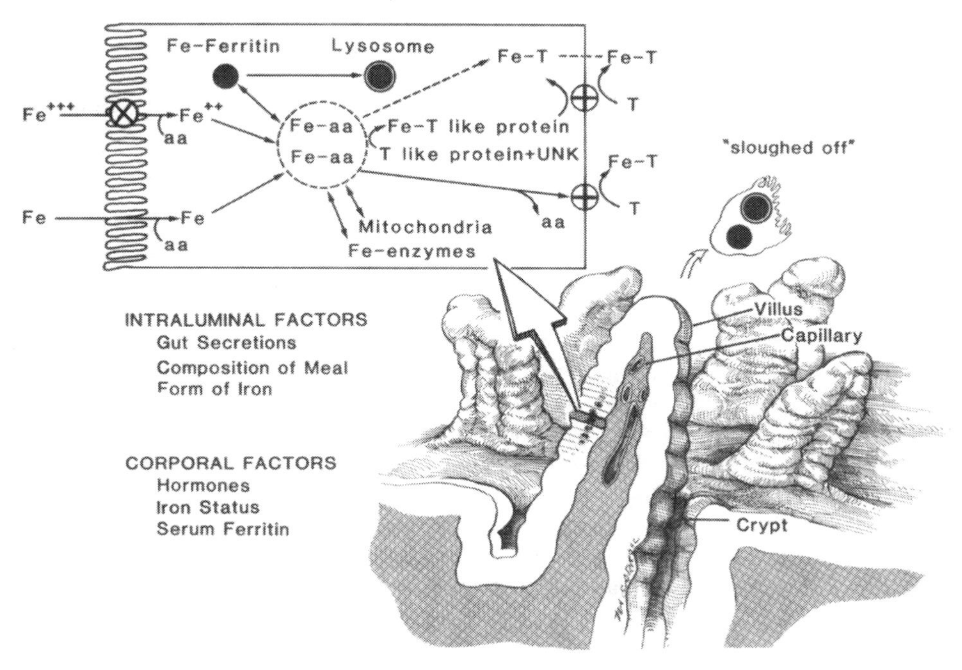

Figure 3-9. Composite of the possible mechanisms and regulation of iron absorption in the intestinal mucosa. aa—Amino acid; T—transferrin.

pool. Much iron remains in the pool only a short period of time before transfer to ferritin. A small amount of iron is quickly attached to a transferrin-like protein for rapid transfer across the serosal surface. An unknown factor in the noncytosol fraction may be needed for the transfer. As the pool is depleted, additional iron is mobilized from the ferritin. Iron not mobilized from ferritin remains in storage until the cell itself is sloughed, making the role of ferritin similar to its role in other cells. Treatment of rats with cycloheximide results, in impaired uptake of iron into the mucosal cell as well as defective transfer (Greenberger and Ruppert, 1966).

Some of the amino-acid-bound iron may participate in the serosal transfer also. Transfer across the serosal surface may or may not require energy. Iron also enters the cell from the serosal side—possibly bound to transferrin, possibly by simple diffusion. Mitochondria accumulate iron for cellular metabolism and do not appear to participate in iron absorption.

Thus, unfortunately, the mechanism remains unclear but complex. It appears that our understanding may not be clarified in the near future since the regulatory mechanism for each type of mucosal uptake may be different, the elusive transferrin-like protein must be further characterized, serosal transfer must be further investigated, etc. On the other hand, studies of the past 10 years in particular have contributed a great deal of information, some of which can be interpreted to indicate that control of regulation may not be a single factor.

3.8 Groups at Risk

From available but inadequate information, over 50 million U.S. citizens would be classified as iron depleted. In developing countries, because iron deficiency is prevalent among both the poor and the rich, it is probably more important, although not quite as detrimental as protein and calorie deficiency. Because anemia does not have overt, specific symptoms, it is not usually recognized by lay persons as a disease. Its symptoms are subtle and chronic. Iron deficiency anemia can impair exercise performance and is potentially an important determinant of productivity. In children, it may impair behavior as well as decrease activity (Pollitt and Leibel, 1976).

In developed countries, groups at risk include pregnant and menstruating women, infants, and children. Very little information is available on the elderly, but there are indications that anemia is also common in this group. In developing countries, the same groups plus adult males would be considered at risk. Approximately 88% of the adult male workers studied recently (Basta *et al.,* 1979) had hookworm infections, and 45% were anemic. Available literature indicates a prevalence approaching 80% in women of the same developing country (Van Veen, 1971). Treatment of the males with elemental iron resulted in improved work performance and morbidity. There was a transient improvement in the placebo group who received income supplementation. Differences in productivity amounted to approximately 20%. Based on stable market conditions and maintaining the same productivity, the benefit of supplementation would be an extra $44 of product for $0.17 of cost for the iron used for supplementation, approximately a 260:1 benefit–cost ratio. This does not include the less tangible benefits on overall health and pride in the ability to provide for one's family. In a developed country where the deficiency is less severe initially, improvement would be more difficult to document.

3.9 Fortification

If one's philosophy is to provide optimal nutrition, then a rational program for food supplementation based on available information of factors which regulate and affect iron absorption must be developed by nutritionists. What to fortify will vary from country to country. Adding minerals to foods is technically more difficult than the addition of vitamins. Flour for bread, hamburger buns, enriched macaroni, and noodles has been enriched in the United States since 1941, yet we continue to see iron deficiency and iron deficiency anemia. There are at least two reasons for this. Some forms of iron were said by the manufacturers to impart a grayish appearance to white foods, so in certain cases salts used for enrichment were those which are less well absorbed. Further, many individuals at high risk do not consume sufficient enriched flour-containing products to meet their needs. Grain products such as crackers, sweet rolls, cakes, and cookies are not always enriched. Sugar has been suggested as a vehicle for fortification in some countries. Unfortunately, there is a color change when fortified sugar is added to tea (Disler *et al.*, 1975). Further, some people object because of the encouragement of poor eating habits. The iron compound used for fortification will have to be compatible with the food, not destroyed during preparation, not deleterious to other foods, and absorbed well by humans. Data on reduced iron, particularly from animal and human assays, do not agree, partly due to the diversity of products all called reduced iron and possibly partly because of different responses between species (Waddell, 1974). Cost is another factor in the decision as to which salt to use (Table 3–10).

The problem of what to fortify with which salt pales beside the problem of how much should be added. The iron must be added at a sufficiently high level to meet the needs of groups at high risk and yet low enough not to be harmful to other population groups. Adult men need only 40% as much iron per 1000 kcal as women. Several alternative supplementation programs (Table 3–11) have been presented by Monsen (1971). The flour enrichment noted in Table 3–11 is higher than the present level of enrichment. It is readily apparent that any fortification program designed to meet the needs of women and children result in excess intake for men. Because of the possible harmful effects of excessive intake of iron, it is imperative that controls be developed, and to develop controls, more data are needed.

There is also the need for better and more complete food tables. With advances in technology and the formulation of new products, we can no longer depend on the amount of iron in a food as grown. We need additional information on the effects of processing both on the total amount and on any changes in absorption characteristics. It may also be that the majority of people can adapt to a wide range of iron intake, while a minority have limited ability to increase their proportionate absorption, to reduce their losses, or both (Elwood, 1966). Before we fortify for optimum nutrition, we need to know more about optimum need; however, in the

Table 3-11. Iron Provided by Various Dietary Supplementation Programs[a]

Sex–Age (years) Group	1965–1966 Survey (% RDA)	Alternative fortification programs: A, 32 mg/lb of white bread, rolls, etc. (%RDA)	B, 20 mg/lb of all baked goods (% RDA)	C, 10 mg/qt of milk (% RDA)	D, 20 mg/lb of all baked goods + 10 mg/qt of milk (% RDA)
Children					
1–2	46[b]	58[b]	59[b]	87	99
Females					
12–14	62	83	85	89	112
18–19	61	81	81	78	98
20–34	63	81	82	74	94
35–54	61	79	78	69[b]	87[b]
Males					
12–14	77	106	105	110	139
18–19	138	189	191	186	239
20–34	179[c]	234[c]	232[c]	212[c]	265[c]
35–54	167	220	221	191	245

[a] Adapted from Monsen (1971).
[b] Lowest in column.
[c] Highest in column.

meantime we need to make judgments on the basis of available information and try to come up with a rational program.

Summary

Lack of iron is by far the most common nutritional deficiency in the United States and presumably the rest of the world. Iron deficiency contributes to impairment of health and substandard performance of millions of persons. To correct this problem, many experts suggest fortification of the food supply. To fortify adequately to meet the needs of the high-risk groups and not to overdose the remainder of the population, more information is needed on absorption of nonheme iron and the factors which affect it. Much progress has been made since the introduction of radioisotopes, but every aspect of the absorption of nonheme iron still needs further investigation.

Techniques have improved greatly, but there is no absolute best for all situations. Whole body counting, while providing excellent data on total absorption cannot be used to study large numbers of subjects and does not detect redistribution of iron within the body. Incorporation of radioiron into red blood cells, the most commonly used technique, allows study of larger numbers of subjects; however, the assumption that there is almost complete incorporation of the iron into red blood cells may not be true for the elderly. All techniques using radioisotopes can be used with infants, children, and pregnant women only very cautiously. Stable isotopes may allow more thorough studies of these groups which are high risk for iron deficiency.

The use of extrinsic tag to study absorption from a whole meal was another landmark which enabled investigators to begin the study of absorption from meals, as commonly eaten. Although there are some problems which must still be overcome, this technique should provide us with much needed information in the near future. At present, studies of absorption of dietary iron have provided sufficient data to suggest that the animal protein and ascorbic acid content of the diet can be used to obtain correction factors when estimating dietary iron intake from food records. More data are needed on the effects of other dietary components.

A survey of studies of iron absorption throughout the life cycle reveals a paucity of information about certain groups and physiological conditions. Absorption is higher during periods of growth, such as pregnancy and early infancy. However, more data are needed on the absorption of iron from human milk as well as on the recurrent suggestion that cow's milk suppresses iron absorption. Preliminary data which suggest that, unlike the small for gestation age infant, the premature infant cannot control iron absorption have ramifications for the development of formulas for these infants. At the other end of the age spectrum, additional studies must be conducted to confirm the finding of adequate absorption but incomplete utilization in the elderly if true additional fortification for these persons would not resolve the basic problem.

Finally, the mechanism and regulation of iron absorption need to be clarified. Much progress has been made in the last 10 years. The concept of the two-pool model for heme and nonheme iron absorption was a big step in unraveling the complex processes of absorption. Studies which combine ultrastructural observations with biochemical data should help resolve some of the ambiguities of present data. Almost every review of the subject ends with the same conclusion: Many problems unfortunately still remain.

References

Abernathy, R. P., Miller, J., Wentworth, J., and Speirs, M., 1965. Metabolic patterns in preadolescent children. XII. Effect of amount and sources of dietary protein on absorption of iron. *J. Nutr.* 85:265–270.

Acheson, L. S., and Schultz, S. G., 1972. Iron influx across the brush border of rabbit duodenum: Effects of anemia and iron loading, *Biochim. Biophys. Acta* 225:479–483.

Amine, E. K., and Hegsted, D. M., 1971. Effect of diet on iron absorption in iron deficiency rats, *J. Nutr.* 101:927–936.

Amine, E. K., and Hegsted, D. M., 1975. Effect of dietary carbohydrates and fats on inorganic iron absorption. *J. Agric. Food Chem.* 23:204–212.

Amine, E. K., Corey, J., Hegsted, D. M., and Hayes, K. C., 1970. Comparative hematology during deficiencies of iron and vitamin A in the rat, *J. Nutr.* 100:1033–1040.

Anderson, T. A., Kim, I., and Fomon, S. J., 1972. Iron status of anemic rats fed iron-fortified cereal–milk diets, *Nutr, Metab.* 14:355–363.

Apte, S. V., and Iyengar, L., 1970. Absorption of dietary iron in pregnancy. *Am. J. Clin. Nutr.* 23:73–77.

Arcasoy, A., Cavdar, A. O., and Babacan, E., 1978. Decreased iron and sinc absorption in Turkish children with iron deficiency and geophagia, *Acta Haematol.* 60:76–84.

Baird, I. M., Podmore, D. A., and Wilson, G. M., 1957. Changes in iron metabolism following gastrectomy and other surgical operations, *Clin. Sci.* 16:463–469.

Baird, I. M., Walters, R. L., and Sutton, D. R., 1974. Absorption of slow-release iron and effects of ascorbic acid in normal subjects and after gastrectomy, *Br. Med. J.* 4:505–508.

Ballas, S. K., 1979. Normal serum iron and elevated total iron-binding capacity in iron deficiency states, *Am. J. Clin. Pathol.* 71:401–403.

Bannerman, R. M., Callender, S. T., and Williams, D. L., 1962. Effect of desferrioxamine and D.T.P.A. in iron overload, *Br. Med. J.* 2:1573–1577.

Bannerman, R. M., Callender, S. T., Hardisty, R. M., and Sephton Smith, R., Iron absorption in thalassaemia, *Br. J. Haematol.* 10:490–495.

Bartels, U., Strandberg Pederson, N., and Jarnum, S. 1978. Iron absorption and serum ferritin in chronic inflammatory bowel disease, *Scand. J. Gastroenterol.* 13:649–656.

Basta, S. S., Soekirman, M. S., Karyadi, D., and Scrimshaw, N. S., 1979. Iron deficiency anemia and the productivity of adult males in Indonesia, *Am. J. Clin. Nutr.* 32:916–925.

Bedard, Y. C., Pinkerton, P. H., and Simon, G. T., 1971. Radioautographic observations in iron absorption by normal mouse duodenum, *Blood* 37:232–245.

Bedard, Y. C., Pinkerton, P. H., and Simon, G. T., 1973. Radioautographic observations on iron absorption by the duodenum of mice with iron overload, iron deficiency and X-linked anemia, *Blood* 42:131–140.

Bedard, Y. C., Pinkerton, P. H., and Simon, G. T., 1976. Uptake of circulating iron by the duodenum of normal mice with altered iron stores including sex-linked anemia. High resolution radioautographic study. *Lab. Invest.* 34:611–615.

Benzie, R. McD., 1963. The influence of age upon the iron content of bone marrow, *Lancet* 1:1074–1075.

Beresford, C. H., Neale, R. J., and Brooks, O. G., 1971. Iron absorption and pyrexia, *Lancet* 1:568–572.

Bezwoda, W., Charlton, R., Bothwell, T., Torrance, J., and Mayet, F., 1978. The importance of gastric hydrochloric acid in the absorption of non-heme food iron. *J. Lab. Clin. Med.* 92:108–116.

Bing, F. C., 1972. Assaying the availability of iron, *J. Am. Diet. Assoc.* 60:114–122.

Bjorn-Rasmussen, E., 1974. Iron absorption from wheat bread: Influence of various amounts of bran, *Nutr. Metab.* 16:101–110.

Bjorn-Rasmussen, E., and Hallberg, L., 1979. Effect of animal proteins on the absorption of food iron in man, *Nutr. Metab.* 23:192–202.

Bjorn-Rasmussen, E., Hallberg, L., and Walker, R. B., 1972. Food iron absorption in man.

I. Isotope exchange between food iron and inorganic iron salts added to food: Studies on maize, wheat and eggs, *Am. J. Clin. Nutr.* 25:317–323.

Bjorn-Rasmussen, E., Hallberg, L., and Walker, R. B., 1973. Food iron absorption in man. II. Isotope exchange of iron between labeled foods and between a food and an iron salt, *Am. J. Clin. Nutr.* 26:1311–1319.

Bjorn-Rasmussen, E., Hallberg, L., Isaksson, B., and Arvidsson, B., 1974. Food iron absorption in man. Applications of the two pool extrinsic tag method to measure heme and non-heme iron absorption from the whole diet, *J. Clin. Invest.* 53:247–255.

Bjorn-Rasmussen, E., Hallberg, L., and Magnussen, B., Svanberg, B., and Arvidsson, B., 1976. Measurement of iron absorption from composite meals, *Am. J. Clin. Nutr.* 29:772–778.

Bothwell, T. H., 1969. Iron overload in the Bantu, in *Iron Metabolism,* F. Gross (ed.), Springer, Berlin, pp. 362–373.

Bothwell, T. H., and Charlton, R. W., 1970. Absorption of iron, *Ann. Rev. Med.* 21:145–156.

Bothwell, T. H., Mallett, B., Oliver, R., and Smith, M. D., 1955. Inability to assess absorption of iron from plasma radioiron curves, *Br. J. Haematol.* 1:352–357.

Bothwell, T. H., Pirzio-Biroli, G., and Finch, C. A., 1958. Iron Absorption I Factors influencing absorption, *J. Lab. Clin. Med.* 51:24–36.

Bowering, J., Masch, G. A., and Lewis, A. R., 1977. Enhancement of iron absorption in iron depleted rats by increasing dietary fat, *J. Nutr.* 107:1687–1693.

Brise, H., and Hallberg, L., 1962. A method for comparative studies of iron absorption in man using two different isotopes, *Acta Med. Scand. Suppl.* 171:7–22.

Britten, G. M., Haley, J., and Brecker, G., 1968. Enhancement of intestinal iron absorption by a humoral effect of hypoxin in parabiotic rats, *Proc Soc. Exp. Biol. Med.* 128:178–184.

Brodan, V., Kuhn, E., Masek, J., Kordar, K., Brodanova, M., and Valek, J., 1968. The influence of concomitant absorption in the digestive system, *Acta Biol. Med. Germ.* 20:597–605.

Brown, E. B., and Justus, B. W., 1958. In vitro absorption of radioiron by everted pouches of rat intestine, *Am. J. Physiol.* 194:319–326.

Brown, E. B., and Rother, M. L., 1963. Studies of the mechanism of iron absorption. II. Influence of iron deficiency and other conditions on uptake by rats, *J. Lab. Clin. Med.* 62:804–816.

Buttner, W., and Muhler, J. C., 1959. Effect of dietary iron on phosphate metabolism, *Proc. Soc. Exp. Biol. Med.* 100:440–442.

Callender, S. T., 1975. Iron deficiency due to malabsorption of food iron, in *Iron Metabolism and Its Disorders,* H. Kief (ed.), Excerpta Medica, Amsterdam, pp. 168–175.

Callender, S. T., and Warner, G. T., 1968. Iron absorption from bread, *Am. J. Clin. Nutr.* 21:1170–1174.

Callender, S. T., Mallett, B. J., and Smith, M. D., 1957. Absorption of hemoglobin iron, *Br. J. Haematol.* 3:186–192.

Callender, W. T., Witts, L. J., Warner, G. T., and Oliver, R., 1966. The use of simple whole body counter for haematological investigations, *Br. J. Haematol.* 12:276–282.

Callender, S. T., Marney, S. R., Jr., and Warner, G. T., 1970. Eggs and iron absorption, *Br. J. Haematol.* 19:657–665.

Carmichael, D., Christopher, J., Hegenauer, J., and Saltman, P., 1975. Effect of milk and casein on the absorption of supplemental iron in the mouse and chick, *Am. J. Clin. Nutr.* 28:487–493.

Celada, A., Rudolf, H., Herreros, V., and Donath, A., 1978. Inorganic iron absorption in subjects with iron deficiency anemia, achylia gatriea and alcoholic cirrhosis using a whole body counter, *Acta Haematol.* 60:182–92.

Chapman, D. G., and Campbell, J. A., 1957. Effect of calcium and phosphorus salts on the utilization of iron anemic rats, Br. J. Nutr. 11:127–133.

Charlton, R. W., Jacobs, P., Seftel, H., and Bothwell, T. H., 1964. Effect of alcohol on iron absorption, *Br. Med. J.* 2:1427–1429.

Charlton, R. W., Derman, D., Skikne, B., Lynch, S. R., Sayers, M. H., Torrance, J. D., and Bothwell, T. H., 1977. Iron stores, serum ferritin and iron absorption, in *Proteins of Iron Metabolism,* E. P. Brown, P. Aisen, J. Fielding, and E. Crichton (eds.), Grune & Stratton, New York, pp. 387–392.

Chirasiri, L., and Izak, G., 1966. The effect of acute hemorrhage and acute hemolysis on the intestinal iron absorption in the rat, *Br. J. Haematol.* 12:611–622.

Conley, M. C., and Hathcock, J. N., 1978. Effects of dietary protein and amino acids on iron utilization by iron-depleted rats, *J. Nutr.* 108:475–480.

Conrad, M. E., 1970. Factors affecting iron absorption, in *Iron Deficiency,* L. Hallberg, G. H. Harwerth, and A. Vannotti (eds.), Academic Press, New York, pp. 87–120.

Conrad, M. E., and Crosby, W. H., 1963. Intestinal mucosal mechanisms controlling iron absorption, *Blood* 22:406–415.

Conrad, M. E., and Schade, S. G., 1968. Ascorbic acid chelates in iron absorption: A role for hydrochloric acid and bile, *Gastroenterology* 55:35–45.

Conrad, M. E., Weintraub, L. R., and Crosby, W. H., 1964. The role of the intestine in iron kinetics, *J. Clin Invest* 43:963–967.

Conrad, M. E., Weintrabu, L. R., Sears, D. A., and Crosby, W. H., 1966. Absorption of hemoglobin iron, *Am. J. Physiol.* 211:1123–1130.

Cook, J. D., 1977. Absorption of food iron, *Fed. Proc Fed Am. Soc. Exp. Biol.* 36:2028–2032.

Cook, J. D., and Lipschitz, D. A., 1977. Clinical Measurements of iron absorption, *Clin. Hematol.* 6:567–581.

Cook, J. D., and Monsen, E. R., 1975. Food iron absorption. I. Use of a semisynthetic diet to study absorption of non-heme iron, *Am. J. Clin. Nutr.* 28:1289–1295.

Cook, J. D., and Monsen, E. R., 1976. Food iron absorption in human subjects. III. Comparison of the effect of animal protein on non-heme iron absorption, *Am. J. Clin. Nutr.* 29:859–867.

Cook, J. D., Brown, G. M., and Valberg, L. S., 1964. The effect of achylia gastrica on iron absorption, *J. Clin. Invest.* 43:1185–1191.

Cook, J. D., Layrisse, M., and Finch, C. A., 1969. The measurement of iron absorption, *Blood* 33:421–429.

Cook, J. D., Palmer, H. E., Pailthorp, K. G., anf Finch, C. A., 1970. The measurement of iron absorption by whole body counting, *Phys. Med. Biol.* 15:467–473.

Cook, J. D., Layrisse, M., Martinez-Torres, C., Walker, R., Monsen, E., and Finch, C. A., 1972. Food iron absorption measured by an extrinsic tag, *J. Clin. Invest.* 51:805–815.

Cook, J. D., Minnich, V., and Moore, C. V., 1973. Absorption of fortification iron in bread, *Am. J. Clin. Nutr.* 26:861–872.

Cook, J. D., Lipschitz, D. A., Miles, L. E. M., and Finch, C. A., 1974. Serum ferritin as a measure of iron stores in normal subjects, *Am. J. Clin. Nutr.* 27:681–687.

Cook, M. B., 1947. Absorption of radioactive iron by children 7–10 years of age, *J. Nutr.* 33:107–119.

Coulson, K. M., Cohen, R. L., Caulson, W. F., and Jelliffe, D. B., 1977. Hematocrit levels in breast-fed American babies, *Clin. Pediatr. Philadelphia* 16:649–651.

Cowan, J. W., Esfahani, M., Salji, J. P., and Azzan, S. A., 1966. Effect of phytate on iron absorption in the rat, *J. Nutr.* 90:423–427.

Crosby, W. H., 1968. Control of iron absorption by intestinal luminal factors, *Am. J. Clin. Nutr.* 21:1189–1193.
Dagg, J. H., Kuhn, I. N., Templeton, F. E., and Finch, C. A., 1967. Gastric absorption of iron, *Gastroenterology* 53:918–922.
Darby, W. J., Hahn, P. F., Kaser, M. M., Steinkamp, R. C., Densen, P. M. and Dallman, P. R., Siimes, M. A. and Stekel, A., 1980. Iron deficiency in infancy and childhood, *Am. J. Clin. Nutr.* 33:86–118.
Dauncey, M. J., Davies, C. G., Shaw, J. C. L., and Urman, J., 1978. The effect of iron supplements and blood transfusion on iron absorption by low birth-weight infants fed pasteurized human breast milk, *Pediatr. Res.* 12:889–904.
Davis, P. S., Luke, C. G., and Deller, P. G., 1966. Reduction of gastric iron binding protein in haematochromatosis. A previously unrecognized metabolic defect, *Lancet* 1:1431–1433.
de Alarcon, P. A., Donovan, M., Forbes, G. B., Landaw, S. A., and Stockman, J. A., 1979. Iron absorption in the thalassemia syndromes and its inhibition by tea, *N. Engl. J. Med.* 300:5–8.
DHEW–USPHS–Health Resources Administration, 1974. *Dietary Intake and Biochemical Findings, Preliminary Findings of First Health and Nutrition Examination Survey. U.S. 1971–72,* National Center for Health Statistics, Rockville, Md.
Disler, P. B., Lynch, S. R., Charlton, R. W., Torrance, J. D., Bothwell, T. H., Walker, P. B., and Mayet, F., 1975a. The effect of tea on iron absorption, *Gut* 16:193–200.
Disler, P. B., Lynch, S. R., Charlton, R. W., Bothwell, T. H., Walker, R. B., and Mayet, F., 1975b. Studies on the fortification of cane sugar with iron and ascorbic acid, *Br. J. Nutr.* 34:141–152.
Dymock, I. W., Godfrey, B. E., and Williams, R., 1971. A comparison of methods for the determination of iron absorption using a whole-body counter, *Phys. Med. Biol.* 16:269–273.
Edwards, J. A., and Hoke, J. E., 1975. Effect of dietary iron manipulation and phenobarbitone treatment on in vivo intestinal absorption of iron in mice with sex-linked anemia, *Am. J. Clin. Nutr.* 28:140–145.
El-Hawary, M. F. S., El-Shobaki, F. A., Kholeif, T., Sakr, R., and El-Bassoussy, M., 1975. The absorption of iron, with or without supplements of single amino acids and of ascorbic acid, in healthy and Fe-deficient children, *Br. J. Nutr.* 33:351–355.
Elwood, P. C., 1966. Utilization of food iron—an epidemiologist's view, *Nutr. Dieta* 8:210–225.
Elwood, P. C., Newton, D., Eakins, J. D., and Brown, D. A., 1968. Absorption of iron from bread, *Am. J. Clin. Nutr.* 21:1162–1169.
Enwonwu, C. O., Monsen, E. R., and Jacobsen, K., 1972. Absorption of iron in protein–calorie deficient rats and immediate effects of re-feeding an adequate protein diet, *Am. J. Dig. Dis.* 17:959–968.
Erlandson, M. E., Walden, B., Stern, G., Hilgartner, M. W., Wehman, J., and Smith, C., 1962. Studies on congenital hemolytic syndromes. IV. Gastrointestinal absorption of iron, *Blood* 19:359–378.
Ezekiel, E., 1967. Intestinal iron absorption by neonates and some factors affecting it, *J. Lab. Clin. Med.* 70:138–149.
Fairbanks, V. T., Fahey, J. L., and Beutler, E., 1971. *Clinical Disorders of Iron Metabolism,* 2nd ed., Grune & Stratton, New York, pp. 65–89.
Finch, C. A., Ragan, H. A., Dyer, I. A., and Cook, J. D., 1978. Body iron loss in animals, *Proc. Soc. Exp. Biol. Med.* 159:335–338.
Forth, W., and Rummel, W., 1973. Iron absorption, *Physiol. Rev.* 53:724–792.
Frannsson, G-B., and Lonnerdal, B, 1980. Iron in human milk, *J. Pediatr.* 96:380–384.

Freiman, H. D., Tauber, S. A., and Tulsky, E. G., 1963. Iron absorption in the healthy aged, *Geriatrics* 18:716–720.

Fritz, J. C., Pla, G. W., Roberts, T., Boehne, J. W., and Hove, E. L., 1970. Biological availability in animals of iron from common dietary sources, *J. Agric. Food Chem.* 18:647–651.

Furugouri, K., 1977. Iron binding substances in the intestinal mucosa of neonatal piglets, *J. Nutr.* 107:487–494.

Gallagher, N. D., Mason, R., and Foley, K. E., 1973. Mechanisms of iron absorption and transport in neonatal rat intestine, *Gastroenterology* 64:438–444.

Garby, L., and Sjolin, S., 1959. Absorption of labeled iron in infants less than three months old, *Acta Paediatr.* 48:*Suppl.* 117:24–28.

Garretson, F. D., and Conrad, M. E., 1967, Starch and iron absorption, *Proc. Soc. Exp. Biol. Med.* 126:304–308.

Gautier du Defaix, H., Puente, R., Vidal, B., Perez, E., and Vidal, H., 1980. Liver storage iron in normal populations of Cuba, *Am. J. Clin. Nutr.* 33:133–136.

Goldberg, A., Lockhead, A. C., and Doggy, J. H., 1963. Histamine fast achlorhydria and iron absorption, *Lancet* 1:848–850.

Goldberg, L., and Smith, J. P., 1960. Vitamin A and E deficiencies in relation to iron overloading in the rat, *J. Pathol. Bacteriol.* 80:173–180.

Gorton, M. K., Hepner, R., and Workman, J. B., 1963. Iron metabolism in premature infants. I. Absorption and utilization of iron as measured by isotope studies, *J. Pediatr.* 63:1063.

Greenberger, N. J., 1973. Effect of antibiotics and other agents on the intestinal transport of iron, *Am. J. Clin. Nutr.* 26:104–112.

Greenberger, N. J., and Ruppert, R. D., 1966. Inhibition of protein synthesis. A mechanism for the production of impaired iron absorption, *Science* 153:315–316.

Greenberger, N. J., Balcerzak, S. P., and Ackerman, G. A., 1969. Iron uptake by isolated intestinal brush borders, changes induced by alteration in iron stores, *J. Lab. Clin. Med.* 73:711–721.

Greenman, J., and Jacobs, A., 1975. The effect of iron stores on iron absorption in the rat: The possible role of circulating ferritin, *Gut* 16:613–616.

Greig, W. A., 1952. The effects of additions of calcium carbonate to the diet of breeding mice, 2 haematology and histopathology, *Br. J. Nutr.* 6:280–294.

Gross, S., 1968. The relationship between milk protein and iron content on hematologic values in infancy, *J. Pediatr.* 73:521–530.

Hahn, P. F., Ross, J. F., Bale, W. F., and Whipple, G. H., 1940. The utilization of iron and the rapidity of hemaglobin formation in anemia due to blood loss, *J. Exp. Med.* 71:731–736.

Hahn, P. F., Carothers, E. L., Darby, W. J., Martin, M., Sheppard, C. W., Cannon, R. O., Beam, A. S., Densen, P. M., Peterson, J. C., and McClellan, G. S., 1951. Iron metabolism in human pregnancy as studied with the radioactive isotope Fe^{59}, *Am. J. Obstet. Gynecol.* 61:477–486.

Hallberg, L., 1974. The pool concept in food iron absorption and some of its implications, *Proc. Nutr. Soc.* 33:285–291.

Hallberg, L., and Bjorn-Rasmussen, E., 1972. Determination of iron absorption from whole diet, a new two-pool model using two radioiron isotopes given as heme and non-heme iron, *Scand. J. Haematol.* 9:193–197.

Hallberg, L., and Solvell, L., 1960. Iron absorption studies, *Acta Med. Scand. Suppl.* 358, 168:1–69.

Hallberg, L., Hallgren, J., Hollender, A., Hogdahl. A. M., and Tibblin, G., 1968. Occurrence of iron deficiency anemia in Sweden, in *Occurrence, Causes and Prevention of Nutritional Anemias, Symp. VI, Swedish Nutrition Foundation,* G. Blix (ed.), Almquist & Wiksell, Uppsala, Sweden, p. 19.

Hallberg, L., Bjorn-Rasmussen, E., Rossander, L., and Suwanik, R., 1979. The measurement of food iron absorption in man. A methodological study on the measurement of dietary non-haem–Fe absorption when the subjects have a free choice of food items, *Br. J. Nutr.* 41:283–289.

Hallberg, L., Bjorn-Rasmussen, E., Howard, L. and Rossander, L., 1979. Dietary heme iron absorption. A discussion of possible mechanisms for the absorption-promoting effect of meat and for the regulation of iron absorption, *Scand. J. Gastroent.* 14:769–779.

Halliday, J. W., and Powell, L. W., 1973. The use of suspensions of isolated rat mucosal cells to study mechanisms of iron absorption, *Clin, Chim Acta* 43:267–276.

Halliday, J. W., Powell, L. W., and Mack, U., 1975. Intestinal iron-binding complexes in iron absorption, in *Proteins of Iron Storage and Transport in Biochemistry and Medicine,* R. R., Crighton (ed.), North-Holland, Amsterdam, pp. 405–410.

Halliday, J. W., Mack, U., and Powell, L. W., 1978. Duodenal ferritin content and structure. Relationship with body iron stores in man, *Arch. Intern. Med.* 138:1109–1113.

Hart, H. V., 1971. Comparison of the availability of iron in white bread, fortified with iron powder, with that of iron naturally present in whole meal bread, *J. Sci. Food Agric.* 22:354–357.

Hazell, T., Ledward, D. A., and Neale, R. J., 1978. Iron availability from meat, *Br. J. Nutr.* 39:631–638.

Hegsted, D. M., Finch, C. A., and Kinney, T. D., 1949. The influence of diet on iron absorption. II. The interrelation of iron and phosphorus, *J. Exp. Med.* 90:147–156.

Heinrich, H. C., 1970. Intestinal iron absorption in man, method of measurements, dose relationships, diagnostic and therapeutic applications, in *Iron Deficiency, Pathogenesis, Clinical Aspects and Therapy,* Hallberg, L., Herwerth, G. H., and Vanotti, A. (eds.), Academic Press, New York, pp. 213–298.

Heinrich, H. C., Bartels, H., Goetze, C., and Schafer, K. H., 1969. Normal range of intestinal iron absorption in newborns and infants, *Klin. Wochenschr.* 47:984–991.

Herndon, J. F., Rice, E. G., Tucker, R. G., Van Loon, E. J., and Greenberger, S., 1958. Iron absorption and metabolism. III. The enhancement of iron absorption in rats by *d*-sorbitol, *J. Nutr.* 64:615–623.

Hoglund, S., and Reizenstein, P., 1969. Studies in iron absorption. V. Effect of gastrointestinal factors on iron absorption, *Blood* 34:496–504.

Howard, J., and Jacobs, A., 1972. Iron transport by rat small intestine in vitro: Effect of body iron status, *Br. J. Haematol.* 23:595–603.

Huebers, H., 1975. Identification of iron binding intermediates in intestinal mucosal tissue of rats during absorption, in *Proteins of Iron Storage and Transport in Biochemistry and Medicine,* R. R. Crighton (ed.), North-Holland, Amsterdam, pp. 381–388.

Huebers, H., Huebers, E., Forth, W., and Rummel, E. W., 1971. Binding of iron to a nonferritin protein in the mucosal cells of normal and iron deficient rats during absorption, *J. Med. Sci.* 10:1141–1144.

Humphreys, J., Walpole, B., and Worwood, M., 1977. Intracellular iron transport in rat intestinal epithelian, biochemical and ultrastructural observations, *Br. J. Haematol.* 36:209–217.

Hussain, R., Walker, R. B., Layrisse, M., Clark, P., and Finch, C. A., 1965. Nutritive value of food iron, *Am. J. Clin. Nutr.* 16:464–471.

Hwang, Y. F., and Brown, E. B., 1965. Effect of desferrioxamine on iron absorption, *Lancet* 1:135–137.

Jacobs, A., 1970. Digestive factors in iron absorption, in *Progress in Gastroenterology, Vol. 2*, G. B. Glass (ed.), Grune & Stratton, New York, pp. 221–235.

Jacobs. A., 1973. The mechanism of iron absorption, *Clin. Haematol.* 2:323–337.

Jacobs, A., 1975. Iron balance and absorption, *Bibl. Nutr. Dieta* 22:61–73.

Jacobs, A. and Greenman, D. A., 1969. Availability of food iron, *Br. Med. J.* 1:673–676.

Jacobs, A. M., and Owen, G. M., 1969. The effect of age on iron absorption, *J. Gerontol.* 24:95–96.

Jacobs, P., Bothwell, T., and Charlton, R. W., 1964. Role of hydrochloric acid in iron absorption, *J. Appl. Physiol.* 19:187–188.

Jansen, C., and Harrill, I., 1977. Intakes and serum levels of protein and iron for 70 elderly women, *Am. J. Clin. Nutr.* 30:1414–1422.

Johnston, F. A., Frenchman, R., and Burroughs, E. D., 1948. The absorption of iron from beef by women, *J. Nutr.* 35:453–465.

Josephs, H. W., 1958. Absorption of iron as a problem in human physiology. A critical review, *Blood* 13:1–54.

Karp, R. J., Haaz, W. S., Starko, K., and Gorman, J. M., 1974. Iron deficiency in families of iron deficient inner-city school children, *Am. J. Dis. Child.* 128:18–20.

Kimber, C. L., Mukherjee, T., and Weller, W. J., 1973. In vitro iron attachment to the intestinal brush border. Effect of iron stores and other environmental factors, *Am. J. Dig. Dis.* 18:781–791.

King, J. C., Reynolds, W. L., and Margen, S., 1978. Absorption of stable isotopes of iron, copper and zinc during oral contraceptive use, *Am. J. Clin. Nutr.* 31:1198–1203.

Klavins, J. V., Kinney, T. D., and Kaufman, N., 1962. The influence of dietary protein on iron absorption, *Br. J. Exp. Pathol.* 43:172–180.

Kuhn, I. N., Layrisse, M., Roche, M., Martinez, C., and Walker, R. B., 1968. Observations on the mechanism of iron absorption, *Am. J. Clin. Nutr.* 21:1184–1188.

Kumpulainen, L. H., and Saukkonen, H. A., 1979. Blood sample $^{59}Fe/^{55}Fe$ activity ratio measurement using a semiconductor detector, *Int. J. Appl. Radiat. Isot.* 30:407–410.

Larsen, L., and Milman, N., 1975. Normal iron absorption determined by means of whole body counting and red cell incorporation of ^{59}Fe, *Acta Med. Scand.* 198:271–274.

Layrisse, M., and Martinez-Torres, C., 1972. Model for measuring dietary absorption of heme iron: Test with a complete meal, *Am. J. Clin. Nutr.* 25:401–411.

Layrisse, M., Martinez-Torres, C., and Roche, M., 1968. Effect of interaction of various foods on iron absorption, *Am. J. Clin. Nutr.* 21:1175–1183.

Layrisse, M., Martinez-Torres, C., Cook, J. D., Walker, R., and Finch, C. A., 1973. Iron fortification of food: Its measurement by extrinsic tag method, *Blood* 41:333–352.

Layrisse, M., Martinez-Torres, C., and Gonzalez, M., 1974. Measurement of the total daily dietary iron absorption by the extrinsic tag model, *Am. J. Clin. Nutr.* 27:152–162.

Layrisse, M., Martinez-Torres, C., Renzy, M., and Leets, I., 1975. Ferritin iron absorption in man, *Blood* 45:689–698.

Layrisse, M., Martinez-Torres, C., Renzi, M., Velez, F., and Gonzalez, M., 1976a. Sugar as a vehicle for iron fortification, *Am. J. Clin. Nutr.* 29:8–18.

Layrisse, M., Martinez-Torres, C., and Renzi, M., 1976b. Sugar as a vehicle for further studies, *Am. J., Clin. Nutr.* 29:274–279.

Levine, P. H., Levine, A. J. and Weintraub, 1972. The role of transferrin in the control of absorption: studies on the cellular level, *J. Lab. Clin. Med.* 80:333–341.

Liebman, M., and Driskell, J., 1979. Dietary phytate and liver iron repletion in iron-depleted rats, *Nutr. Rep. Int.* 19:281–287.

Linder, M. C., and Munro, H. N., 1975. Ferritin and free iron in iron absorption, in *Proteins of Iron Storage and Transport in Biochemistry and Medicine,* R. R., Crighton (ed.), North-Holland, Amsterdam, pp. 395–400.

Linder, M. C., and Munro, H. N., 1977. The mechanism of iron absorption and its regulation, *Fed. Proc. Fed. Am. Soc. Exp. Biol.* 36:2017–2023.

Linder, M. C., Dunn, V., Isaacs, E., Jones, W., Lim, S., Van Volkom, M., and Munro, H. N., 1975. Ferritin and intestinal iron absorption: Pancreatic enzymes and free iron, *Am. J. Physiol.* 228:196–204.

Lipschitz, D. A., Simpson, K. M., Cook, J. D., and Morris, E. R., 1979. Absorption of monoferric phytate by dogs, *J. Nutr.* 109:1154–1160.

Loh, T. T., and Kaldor, I., 1971. Intestinal iron absorption in suckling rats, *Biol. Neonat.* 17:173–186.

Lonnerdal, B., Keen, C. L., Frannsson, G-B., Hambraeus, L., and Hurley, L., 1980. New perspectives on iron supplementation of milk, *J. Pediatr.* 96:242.

Lunn, J. A., Richmond, J., Simpson, J. D., Leask, J. D., and Tothill, P., 1967. Comparison between three radioisotope methods for measuring iron absorption, *Br. Med. J.* 3:331–333.

Lynch, S. R., Becker, D., Seftel, H., Bothwell, T. H., Stevens, K., and Metz, J., 1970. Iron absorption in kwashiorkor, *Am. J. Clin. Nutr.* 23:792–797.

MacDermott, R. P., and Greenberger, N. J., Evidence for a humoral factor influencing iron absorption, *Gastroenterology* 57:117–125.

Mackey, H. M. M., 1928. Anemia in infancy, its prevalency and prevention, *Arch. Dis. Child.* 3:1175–1179.

Magnusson, B. E. O., 1976. Iron absorption after antrectomy with gastroduodenostomy, *Scand. J. Haematol. Suppl.* 26:1–111.

Mahoney, A. W., and Hendricks, D. G., 1976. Effect of dietary iron level on efficiency of converting food iron into hemoglobin in the anemic rat, *Nutr. Metab.* 20:222–227.

Mahoney, A. W., Orden, C. C., Van, C. C., and Hendricks, D. G., 1974. Efficiency of converting food iron into hemoglobin by the anemic rat, *Nutr. Metab.* 17:223–230.

Manis, J., 1971. Intestinal iron-transport defect in the mouse with sex-linked anemia, *Am. J. Physiol.* 220:135–139.

Manis, J., and Kim, G., 1979. Stimulation of iron absorption by polychlorinated aromatic hydrocarbons, *Am. J. Physiol.* 236:E763–768.

Manis, J. G., and Schachter, D., 1962a. Active transport of iron by intestine features of the two-step mechanism, *Am. J. Physiol.* 203:73–80.

Manis, J. G., and Schachter, D., 1962b. Active transport of iron by intestine: Effects of oral iron and pregnancy, *Am. J. Physiol.* 203:81–86.

Martinez-Torres, C., and Layrisse, M., 1971. Iron absorption from veal muscle, *Am. J. Clin. Nutr.* 24:521–540.

Martinez-Torres, C., and Layrisse, M., 1973. Nutritional factors in iron deficiency: Food iron absorption, *Clin. Haemat.* 2:339–352.

Martinez-Torres, C., and Layrisse, M., 1974. Interest for the study of dietary absorption and iron fortification, *World Rev. Nutr. Diet. 19:51–70.*

Martinez-Torres, C., Renzi, M., and Layrisse, M., 1976. Iron absorption by humans from hemosiderin and ferritin, further studies, J. Nutr. 106:128–135.

Martinez-Torres, C., Romano, E. L., Renzi, M., and Layrisse, M., 1979. Fe(III)–EDTA complex as iron fortification. Further studies, *Am. J. Clin. Nutr.* 32:809–816.

Marx, J. J. M., 1979. Normal iron absorption and decreased red cell iron uptake in the aged, *Blood* 53:204–211.

McCance, R. A., Edgecome, C. N., and Widdowson, E. M., 1943. Phytic acid and iron absorption, *Lancet* 2:126–128.

McCance, R. A., and Widdowson, E. M., 1937. Absorption and excretion of iron, *Lancet* 1:680–684.

McMillan, T. J., and Johnston, F. A., 1951. The absorption of iron from spinach by six young women and the effect of beef upon absorption, *J. Nutr.* 44:383–398.

McMillan, J. A., and Oski, F. A., 1978. Reply to letter to the editor, *Pediatrics* 62:441–442.

McMillan, J. A., Landaw, S. A., and Oski, F. A., 1976. Iron sufficiency in breast-fed infants and the availability of iron from human milk, *Pediatrics* 58:686–691.

McMillan, J. A., Oski, F. A., Lourie, G., Tomarelli, R. M., and Landaw, S. A., Iron absorption from human milk, simulated human milk and proprietary formulas, *Pediatrics* 60:896–900.

Mejia, L. A., Hodges, R. E., Arrorace, C., Viteri, F., and Torum, B., 1977. Vitamin A deficiency and anemia in Central American children, *Am. J. Clin. Nutr.* 30:1175–84.

Mejia. L. A., Hodges, R. E., and Rucker, R. B., 1979. Role of vitamin A in the absorption, retention and distribution of iron in the rat, *J. Nutr.* 109:129–137.

Miller, J., and Landes, D. R., 1976a. Modification of iron and copper metabolism by dietary starch and glucose, *Nutr. Repts. Int.* 13:187–191.

Miller, J. and Landes, D. R., Effects of starch, sucrose and glucose on iron absorption by anemic rats. 1976b. *Nutr. Rep. Int.* 14:7–11.

Miller, D. D., and van Campen, D., 1979. A method for the detection and assay of iron stable isotope tracers in blood serum, *Am. J. Clin. Nutr.* 32:2354–2361.

Monsen, E. R., 1971. The need for iron fortification, *J. Nutr. Educ.* 2:152–155.

Monsen, E. R., 1974. Validation of an extrinsic iron label in monitoring absorption of nonheme food iron in normal and iron-deficient rats, *J. Nutr.* 104:1490–1495.

Monsen, E. R., and J. D., Cook, 1976. Food iron absorption in human subjects. IV. The effects of calcium and phosphate salts on the absorption of nonheme iron, *Am. J. Clin. Nutr.* 29:1142–1148.

Monsen, E. R., and Cook, J. D., 1979. Food iron absorption in human subjects. V. Effects of the major dietary constituents of a semisynthetic meal, *Am. J. Clin. Nutr.* 32:804–808.

Monsen. E. R., Hallberg, L., Layrisse, M., Hegsted, D. M., Cook, J. D., Mertz, W., and Finch, C. A., 1978. Estimation of available dietary iron, *Am. J. Clin. Nutr.* 31:134–141.

Moore, C. V., 1968. The absorption of iron from foods, in *Occurrence, Causes* and *Prevention of Nutritional Anemias, Symp. VI, Swedish Nutrition Foundation,* G. Blix (ed.), Almquist & Wiksell, Uppsala, Sweden, pp. 92–103.

Moore, C. V., and Dubach, R., 1951. Observations on the absorption of iron from foods tagged with radioiron, *Trans. Assoc. Am. Physicians* 64:245–256.

Moore, C. V., Arrowsmith, W. R., Welch, J., and Minich, V., 1939. Studies in iron transportation and metabolism. IV. Observations on the absorption of iron from the gastrointestinal tract, *J. Clin. Invest.* 18:553–580.

Moore, C. V., Dubach, V., Minnich, V., and Roberts, H. K., 1944. Absorption of ferrous and ferric radioactive iron by human subjects and by dogs, *J. Clin. Invest.* 23:755–767.

Morgan, E. H., 1974. Transferrin and transferrin iron, in *Iron in Biochemistry and Medicine,* A. Jacobs and M. Worwood, (eds.), Academic Press, New York, pp. 29–37.

Morris, E. R., and Ellis, R., 1975. Isolation of a soluble iron complex from wheat bran and its biological availability to the rat, *Fed. Proc. Fed. Am. Soc. Exp. Biol.* 34:923.

Morris, E. R., and Ellis, R., 1976. Isolation of monoferric phytate from wheat bran and its biological value as an iron source to the rat, *J. Nutr.* 106:753–760.

Morris, E. R., and Greene, F. E., 1972. Utilization of the iron of egg yolk for hemoglobin formation by the growing rat, *J. Nutr.* 102:901–908.

Motzok, I., Davies, M. I., Verma, R. S., and Pennell, M. D., 1977. Biological availability of iron from foods and tonics containing various iron supplements, *Nutr. Rep. Int.* 15:459–467.

Norrby, A., Aybo, G., and Solvell, L., 1972. The influence of combined oral contraceptive on the absorption of iron, *Scand. J. Haematol.* 9:43–47.

Oettinger, L., Mills, W. B., and Hahn, P. F., 1954. Iron absorption in premature and full-term infants, *J. Pediatr.* 45:302–306.

Olszon, E., Isakson, B., Norrby, A., and Solvell, L., 1978. Food iron absorption in iron deficiency, *Am. J. Clin. Nutr.* 31:106–111.

Owen, G. M., Nelson, C. E., and Garry, P. J., 1970. Nutritional status of preschool children: Hemoglobin, Hematocrit and plasma iron values, *J. Pediatr.* 76:761–762.

Palmer, H. E., Cook, J. D., Pailthrop, K. G., and Finch, C. A., 1970. A whole-body counter for precision in vivo measurement of radio-iron, *Phys. Med. Biol.* 15:457–465.

Parmley, R. T., Barton, J. C., Conrad, M. E., and Austin, R. L., 1978. Ultrastructural cytochemistry of iron absorption, *Am. J. Pathol.* 93:707–728.

Pearson, W. N., Reich, M., Frank, H., and Salamat, L., 1967. Effects of dietary iron level on gut iron levels and food absorption in the rat, *J. Nutr.* 92:53–65.

Peters, T., Jr., Apt, L., and Ross, M. F., 1971. Effect of phosphates upon iron absorption studies in normal human subjects and in an experimental model using dialysis, *Gastroenterology* 61:315–322.

Pinkerton, P. H., 1969. Control of iron absorption by the intestinal epithelial cell. Review and hypotherapy, *Ann. Intern. Med.* 70:401–408.

Pirzio-Biroli, G., and Finch, C. A., 1960. Iron absorption. III. Influence of iron stores on iron absorption in normal subjects, *J. Lab. Clin. Med.* 55:216–220.

Pirzio-Biroli, G., Bothwell, T. H., and Finch, C. A., 1958. Iron absorption. II. The absorption of radioiron administered with a standard meal in man, *J. Lab. Clin. Med.* 51:37–48.

Pla, G. W., and Fritz, J. C., 1970. Availability of iron, *J. Assoc. Off. Anal. Chem.* 53:791–800.

Pollack, S. R., Kaufman, M., and Crosby, W. H., 1964. Iron absorption: The effect of an iron deficient diet, *Science* 144:1015–1016.

Pollack, S., Compann, T., and Arcario, A., 1972. A search for a mucosal iron carrier. Identification of mucosal fractions with a rapid turnover of ^{59}Fe, *J. Lab. Clin. Med.* 80:322–332.

Pollitt, E., and Leibel, R. L., 1976. Iron deficiency and behavior, *J. Pediatr.* 88:372–381.

Powell, L., Campbell, C. B., and Wilson, E., 1970. Intestinal mucosal uptake of iron and iron retention in idiopathic hemochromatosis as evidence for a mucosal abnormality, *Gut* 11:727–731.

Price, D. C., Cohn, S. H., Wasserman, L. R., Reizenstein, P. G. and Cronkite, E. P., 1962. The determination of iron absorption and loss by whole body counting, *Blood* 20:517–530.

Ranhotra, G. S., Lee, C., and Gelroth, J. A., 1979. Bioavailability of iron in some commercial variety breads, *Nutr. Rep. Int.* 19:851–857.

Rao, B. S. N., and Prahavathi, T., An in vitro method of predicting the bioavailability of iron from foods, *Am. J. Clin. Nutr.* 31:169–175.

Rao, B. S. N., Prasad, J. S., and Sarthy, C. V., 1977. An animal model to study iron availability from human diets, *Br. J. Nutr.* 37:451–456.

Richmond, V. S., Worwood, M., and Jacob, A., 1972. The iron content of intestinal epithelial cells and its subcellular distribution: Studies on normal, iron overloaded and iron deficient rats, *Br. J. Haematol.* 23:605–614.

Rios, E., Hunter, R. E., Cook, J. D., Smith, N. J., and Finch, C. A. 1975. The absorption of iron as supplements in infant cereal and infant formulas, *Pediatrics* 55:686–693.

Robertson, I. K., and Worwood, M., 1978. Lead and iron absorption from rat small intestine: The effect of dietary iron deficiency, *Br. J. Nutr.* 40:253–260.

Saarinen, W. M., and Siimes, M. A., 1977. Iron absorption from infant milk formula and the optimal level of iron supplementation, *Acta Paediatr. Scand.* 66:719–722.

Saarinen, U. M., and Siimes, M. A., 1979. Iron absorption from breast milk, cow's milk, and iron-supplemented formula: An opportunistic use of changes in total body iron determined by hemoglobin, ferritin and body weight in 132 infants, *Pediatr. Res.* 13:143–147.

Saarinen, U. M., Siimes, M. A., and Dallman, P. R., 1977. Iron absorption in infants: High availability of breast milk iron as indicated by the extrinsic tag method of iron absorption and by the concentration of serum ferritin, *J. Pediatr.* 91:36–39.

Saltman, P., 1965. The role of chelation in iron metabolism, *J. Chem. Educ.* 42:682–687.

Sayer, L., and Finch, C. A., 1953. Determination of iron absorption using two isotopes of iron, *Am. J. Physiol.* 172:372–376.

Sayers, M. H., Lynch, S. R., and Jacobs, P., 1973. The effects of ascorbic acid supplementation on the absorption of iron in maize, wheat and soya, *Br. J. Haematol.* 24:209–218.

Schiffer, L. M., Price, D. C., Cuttner, J., Cohn, S. H., and Cronkite, E. P., 1964. A note concerning the 100 percent value in iron absorption studies by whole body counter, *Blood* 23:757–760.

Schiffer, L. M., Price, D. C., and Cronkite, E. P., 1965. Iron absorption and anemia, *J. Lab. Clin. Med.* 65:316–321.

Schultz, J., and Smith, N. J., 1958a. A quantitative study of the absorption of food iron in infants and children, *Am. J. Dis. Child.* 95:109–119.

Schulz, J., and Smith, N. J., 1958b. Quantitative study of the absorption of iron salts in infants and children, *Am. J. Dis. Child.* 95:120–125.

Scott, D. E., and Pritchard, J. A., 1967. Iron deficiency in healthy young college women, *J. Am. Med. Assoc.* 199:897–900.

Senchak, M. M., Howe, J. M., and Clark, H. E., 1973. Iron absorption by adults fed mixtures of rice, milk and wheat flour, *J. Am. Diet. Assoc.* 62:272–275.

Shahid, M. J., and Haydar, N. A., 1967. Absorption of inorganic iron in thalassemia, *Br. J. Haematol.* 13:713–718.

Sharpe, L. M., Peacock, W. C., Cooke, R., and Harris, R. S., 1950. The effect of phytate and other food factors on iron absorption, *J. Nutr.* 41:433–496.

Sheehan, R. G., and Frenkel, E. P., 1972. The control of iron absorption by the gastrointestinal mucosal cell, *J. Clin. Invest.* 51:224–231.

Siimes, M. A., Vuori, E., and Kuitman, P., 1979. Breast milk iron—a declining concentration during the course of lactation, *Acta Paediatr. Scand.* 68:29–31.

Six, K. M., and Goyer, R. A., 1972. The influence of iron deficiency on tissue content and content and toxicity of ingested lead in the rat, *J. Lab. Clin. Med.* 79:128–136.

Smith, M. D., and Mallet, B., 1957. Iron absorption before and after partial gastrectomy, *Clin. Sci.* 16:23–24.

Steinkamp, R., Dubach, R., and Moore, C. V., 1955. Studies in iron transportation and metabolism, *Arch, Intern. Med.* 95:181–193.

Steinke, F. H., and Hopkins, D. T., 1978. Biological availability to the rat of intrinsic and extrinsic iron with soybean protein isolates, *J. Nutr.* 108:481–489.

Svanberg, B., 1974. Absorption of iron in pregnancy, *Acta Obstet. Gynecol. Scand. Suppl.* 48:1–108.

Tapper, E. J., Bush, S., Ruppert, R. D., and Greenberger, N. J., 1968. Effect of acute and chronic ethanol treatment on the absorption of iron in rats, *Am. J. Med. Sci.* 255:46–52.

Theuer, R. C., Kemmerer, K. S., Martin, W. H., Zoumas, B. L., and Sarett, H. P., 1971. Effect of processing on availability of iron salts in liquid infant formula products. Experimental soy isolate formulas, *Agric. Food Chem.* 19:555–558.

Thomas, F. B., McCullough, F. S., and Greenberger, N. J., 1972. Effects of phenobarbital on the absorption of inorganic and hemoglobin iron in the rat, *Gastroenterology* 62:590–599.

Turnbull, A. L., 1965. The absorption of radioiron given with a standard meal for polya partial gastrectomy, *Clin. Sci.* 28:499–509.

Turnbull, A., 1974. Iron absorption, in *Iron in Biochemistry and Medicine,* A. Jacobs and M. Worwood (eds.), Academic Press, New York, pp. 369–401.

Turnbull, A., Cleton, F., and Finch, C. A., 1962. Iron absorption. IV. The absorption of hemoglobin iron, *J. Clin. Invest.* 41:1897–1907.

Van Campen, D., 1972. Effect of histidine and ascorbic acid on the absorption and retention of ^{59}Fe by iron-depleted rats, *J. Nutr.* 102:165–170.

Van Campen, D., 1974. Regulation of iron absorption, *Fed. Proc. Fed. Am. Soc. Exp. Biol.* 33:100–105.

Van Campen, D., and Gross, C., 1969. Effect of histidine and certain other amino acids on the other absorption of iron-59 by rats, *J. Nutr.* 99:68–74.

Van Veen, M. S., 1971. Some ecological considerations of nutrition problems in Java, *Ecol. Food Nutr.* 1:31–36.

Waddell, J., 1974. The bioavailability of iron sources and their utilization in food enrichment, *Fed. Proc. Fed. Am. Soc. Exp. Biol.* 33:1779–1783.

Warner, G. T., and Oliver, R., 1966. A whole-body counter for clinical measurements utilizing the "shadow-shield" technique, *Phys. Med. Biol.* 2:83–86.

Watanabe, Y., 1974. Iron absorption in low birth weight infants by the whole body counter, *Acta Paediatr. Jpn Overseas Ed.* 16:14–21.

Webb, J., Multani, J. S., Saltman, P., and Gray, H. B., 1973. Spectroscopic and magnetic studies of iron III, *Biochemistry* 12:265–267.

Webling, D., and Holdsworth, E. S., 1966. Bile and the absorption of strontium and iron, *Biochem. J.* 100:661–663.

Welch, R. M., and Van Campen, D. R., 1975. Iron availability to rats from soybeans, *J. Nutr.* 105:253–256.

Wheby, M. S., 1966. Site of iron absorption in man, *Clin. Res.* 14:50.

Wheby, M. S., and Crosby, W. H., 1963. The gastrointestinal tract and iron absorption, *Blood* 22:416–428.

Wheby, M. S., Conrad, M. E., Helberg, S. E., and Crosby, W. H., 1962. The role of bile in control of iron absorption, *Gastroenterology* 42:319–324.

Wheby, M. S., Jones, L. G., and Crosby, W. H., 1964. Studies on iron absorption. Intestinal regulatory mechanism, *J. Clin. Invest.* 43:1433–1442.

Wheby, M. S., Suttle, G. E., and Ford, K. T., 1970. Intestinal absorption of hemoglobin iron, *Gastroenterology* 58:647–654.

Williams, R., Pitcher, C. S., Parsonson, A., and Williams, H. S., 1965. Iron absorption in the relatives of patients with idiopathic hemochromatosis, *Lancet* 1:1243–1246.

Worwood, M., and Jacobs, A., 1971. The subcellular distribution of ^{59}Fe during iron absorption in the rat, *Br. J. Haematol.* 20:587–597.

Worwood, M., and Jacobs, A., 1972. The subcellular distribution of ^{59}Fe in small intestinal mucosa: Studies with normal, iron deficient and iron loaded rats, *Br. J. Haematol.* 20:265–272.

Worwood, M., Jacobs, A., and Cavill, I., 1975. Iron absorption: Regulation by interval iron exchange, in *Protein of Iron Storage and Transport in Biochemistry and Medicine,* R. R. Creighton (ed.), North-Holland, Amsterdam, pp. 401–404.
Yeh, S. D., and Shels, M. E., 1969. Quantitative aspects of cycloheximide inhibition of amino acid incorporation, *Biochem. Pharmacol.* 18:1919–1926.

Chemistry and Metabolism of the Transferrins

4.1 Introduction

The discovery of transferrin-type proteins in egg white and serum proceeded along independent though parallel routes. In their investigation of the composition of egg white, Osborne and Campbell (1900) had occasion to purify a new protein that was very similar to ovalbumin. They wrote that "...since it so closely resembles ovalbumin and is so intimately associated with it, the writer suggests that it be called conalbumin." Conalbumin was purifed by ammonium sulfate precipitation and manipulation of pH. Osborne and Campbell were not aware of the iron-binding properties of conalbumin. Almost half a century later, Schade and Caroline (1944) discovered the antimicrobial properties of raw egg white and showed that such activity could be abolished by iron. They also discovered that the addition of iron to the egg white produced a tan to brownish coloration. Two years later, Alderton *et al.* (1946) showed that the antimicrobial substance of raw egg white was conalbumin. They isolated 95% pure conalbumin by ammonium sulfate precipitation (1.5 *M* ammonium sulfate at pH 3.0), and it accounted for 80% of all conalbumin present in egg white. They estimated that conalbumin accounted for some 10% of all egg white protein.

It is generally accepted that the first definitive work on plasma iron was done by Fontes and Thivolle (1925). They showed that plasma contained small but significant amounts of iron and that this was decreased in experimental anemia. Later, largely through the efforts of Barkan and co-workers, it was shown that iron was nondialyzable and nonultrafiltrable and therefore bound to protein. Barkan also termed such iron, which he called "easily splittable" (leicht abspaltbar), transport iron (Barkan, 1927; Barkan and Schales, 1937). In 1941, Vahlquist reported that electrophoretically iron migrated with albumin (30–50%) and α- and β-globulins (50–70%). None was found associated with γ-globulin. He further discovered that plasma iron remained undialyzable at pH 4.5–10 and that some 50% was lost at pH 3.5–4.0 (Vahlquist, 1941). In 1945, serum iron-binding capacity was determined to be 300 μg of iron per 100 ml. Serum was titrated with iron until it became bipyridyl-positive. That point was taken as that of maximum serum iron-binding capacity (Holmberg and Laurell, 1945). And last but not least, Heilmyer and Plötner in 1937 ascribed a central role to plasma protein-bound iron in the overall iron metabolic system (see Chapter 1).

The iron-binding protein of serum was first purified by Schade and Caroline (1946) from Cohn's fraction IV-3,4 by ammonium sulfate precipitation. It was determined that this preparation inhibited bacterial growth, and the inhibition was abolished by iron. It was also found that some 0.63 μg of iron saturated 1 mg of protein, which indicates that their preparation was about 50% pure. It was calculated that a human being had enough of the iron-binding protein to bind a maximum of 260 μg of iron per 100 ml of plasma.

The first physical data on the plasma (serum) iron-binding protein were published by Oncley *et al.* (1947) and by Cohn (1947) in a review article. The work was done on Cohn's fraction IV (also called β-globulin or β_1-metal-combining globulin). The sedimentation constant was reported to be 5.5 S and the molecular weight 90,000. Cohn believed that this protein transported iron, copper, and zinc. Schade *et al.* (1949), using the same material, reported that the iron-binding protein binds two iron atoms per molecule, and for each iron atom bound, one molecule of sodium bicarbonate was utilized and bound. The maximum absorption of the iron-saturated protein was at 460 nm. The protein was termed siderophilin.

Independently of the American groups who utilized materials obtained through the wartime plasma fractionation projects, the serum iron-binding protein was also isolated in Sweden by Laurell and Ingelman (1947). The Swedish workers used swine plasma as the starting material because of its high TIBC. They utilized both ammonium sulfate and alcohol at low temperatures to precipitate the iron-binding protein "as a beautiful red sediment." Its molecular weight was near 88,000, and it contained 1.4 atoms of iron per molecule of protein. Maximum absorption of the iron-containing protein was 460–470 nm, and in the copper state, the absorption was at 430–440 nm. The term *transferrin* was suggested as an appropriate name for this protein (Holmberg and Laurell, 1947), and this name has been utilized in preference to siderophilin.

The milk iron-binding protein, variously termed lactotransferrin, lactosiderophilin, the red protein, and most recently and perhaps most appropriately lactoferrin, is a relative newcomer to the field of nonheme iron-binding proteins. For years, researchers working on various milk proteins had observed the presence of a red protein in many of their preparations (Groves, 1971), but it was not until the late 1950s and early 1960s that this protein was actually isolated and characterized. Johansson (1958) was one of the first to report on the purification of lactoferrin from human milk. It was a by-product of his α-lactalbumin purification process. He reported that lactoferrin had a salmon-pink color with an absorption maximum at 460–470 nm and that the iron was not easily removable therefrom. In 1960, several reports appeared on lactoferrins from human milk (Johansson, 1960; Montreuil *et al.*, 1960) and bovine milk (Groves, 1960). These initial studies established that the lactoferrins, like transferrins and conalbumin, had a molecular weight of near 90,000 and an absorption maximum of near 465 nm with the absence of a Soret band, the presence of carbohydrate, but, in contrast to transferrin and conalbumin, a loss of iron at pH 2 rather than at pH 5–6. Lactoferrin was reported to bind two iron atoms per molecule of protein, except for Montreuil *et al.* (1960), who reported a binding capacity of six iron atoms per molecule of protein. A question arose as to whether or not human and bovine lactoferrins were related to the respective transferrins. Immunological studies for both human (Montreuil *et al.*, 1960) and bovine (Szuchet-Derechin and Johnson, 1962) lactoferrins has revealed that the two types of proteins are not immunochemically related. It should be remembered, however, that circulating transferrin can and does occur in milks of many mammalian species, and in fact the iron-binding protein of rabbit milk appears to be entirely circulating transferrin. Surprisingly, it accounts for nearly 50% of all rabbit milk proteins (Jordan *et al.*, 1967).

A few words are in order in regard to nomenclature. The egg white, milk, and plasma iron-binding proteins are all termed transferrins or the transferrin class of proteins, even though the term transferrin has been and still is used to designate the iron-binding protein of circulation. Confusion can therefore arise when such a term is used. In this volume, therefore, in this and the following chapters, we shall utilize the term *serotransferrin* to refer to the circulatory iron-binding protein, or alternately, the terms serum or plasma transferrin are used.

4.2 The Levels of Transferrins in Biological Fluids

The criteria one can use to classify a protein as a transferrin are that it is freely soluble in water, that it binds a maximum of two ferric iron atoms per molecule of protein, that it has a molecular weight of near 80,000 and a single polypeptide chain, that the iron-laden form shows a characteristic absorption spectrum in the visible light range with a maximum at about 460 nm, that it has a characteristic EPR spectrum, and in case of serotransferrin, that it can donate its iron to the immature red cells (see Chapter 6). Since the transferrins have probably arisen through a gene duplication process during the course of evolution (Greene and Feeney, 1968), it is possible that a transferrin-like protein with a single iron-binding site and a molecular weight of near 40,000 will be isolated in the future from some primitive animal.

Practically every human biological fluid contains a protein of the transferrin class. The circulation, lymphatic fluid, cerebrospinal fluid, milks of some species, and even some cells contain serotransferrin, whereas lactoferrin is present in all body secretions such as milk, synovial fluid, saliva, and semen and in many cells. Conalbumin is present in the egg white of all birds. It apparently differs from the respective serotransferrin in the nature of the carbohydrate chain only (Williams, 1962). Serotransferrin content in plasma or serum can be approximated from TIBC measurements (see Chapter 1) as follows: Since each serotransferrin molecule can bind a maximum of two gram atoms of iron per mole of protein, the amount of iron bound will be $55.9 \times (2/80{,}000) = 0.0013975$ g of iron per g of protein, or approximately 1.4 μg of iron per mg of protein. If the TIBC of a normal adult male is 350 μg/dl (Table 1-9), then there must be $350/1.4 = 250$ mg of serotransferrin per dl or 2.5 mg/ml. Serotransferrin levels in serum or plasma can also be measured directly by immunochemical means, and they have largely corresponded to the values obtained with TIBC data. The clinical biochemist accepts human serotransferrin levels of 205–374 mg/dl as being "normal." Systemic lupus erythematosus (SLE), acute and chronic inflammation, chronic liver disease, and starvation are accompanied by low serotransferrin levels. Serotransferrin is increased in iron deficiency anemia (Wallach, 1974). Otherwise serotransferrins remain remarkably constant.

Lactoferrin is present in serum in very small amounts, and efforts to correlate its levels with disease processes have failed (Rümke *et al.*, 1971). Its levels are relatively high in various secretions of the mammalian organism, the milk being its major source. In human milk, lactoferrin is a major protein component, accounting for between 0.1 and 0.3 g of a total protein content of 0.5–0.7 g/100 ml (Bezkorovainy, 1977; Nagasawa *et al.*, 1972). The milks of other species contain much less lactoferrin, as, for instance, does cow's milk, whose lactoferrin content is barely one-tenth that of human milk. The various mammalian species thus show varying lactoferrin–serotransferrin ratios in their milks. An extreme situation exists in rabbit milk, which contains large amounts of a protein that differs from serotransferrin by one sialic acid residue only (Baker *et al.*, 1968). It is termed milk transferrin rather than lactoferrin, since lactoferrin by definition differs from serotransferrin both immunologically and with respect to amino acid composition. Rabbit milk contains little if any true lactoferrin. The situation in the rabbit is thus similar to the fowl serotransferrin–conalbumin relationship.

Lactoferrins have been isolated from various milks and their levels therein determined. Unlike serotransferrin, whose levels in blood remain constant unless severe disease is present, milk lactoferrin and serotransferrin levels show a marked variation from one individual to another as well as changes with increasing time of lactation within the same individual. As a rule, milk lactoferrin and serotransferrin levels are highest at the beginning and end of the lactation period. Thus in one sow, the milk serotransferrin level was near 425 μg/ml 1 day following delivery, dropped to about 40 μg/ml between the 5th and 45th day, and then rose to about 60 μg/ml. Lactoferrin values were about 80, 20, and 50 μg/ml, respectively (Masson and Heremans, 1971). The same picture is seen in the human being, where lactoferrin levels in one individual were 5000 μg/ml at the beginning of the lactation period, dropped to about 2000 μg/ml between the 10th and the 160th day, and then rose again to 3000 μg/ml (Figure 4-1) (Lönnerdal *et al.* 1976a). Table 4-1 illustrates lactoferrin content of human milks as a function of time postpartum.

Lactoferrin is present in a variety of biological fluids as well as cells. Of special interest is the presence of lactoferrin in neutrophilic leukocytes (Masson *et al.*, 1969). It has been proposed that lactoferrin, at least in part, plays a role in the development of hypoferremia during infections because of the proliferation of neutrophils and therefore lactoferrin, which can sequester the iron in preference to serotransferrin. Such iron is eventually taken up by the reticuloendothelial cells to be stored there (Malm-

Table 4-1. Lactoferrin Content of Milks of Swedish Mothers as a Function of Time of Lactation

Time of lactation	Lactoferrin content (mg/ml)	Reference
6–10 days	4.5 ± 0.8	Nagasawa *et al.*, 1972
11–60 days	2.1 ± 0.5	Nagasawa *et al.*, 1972
61–197 days	1.6 ± 0.3	Nagasawa *et al.*, 1972
0–0.5 months	3.53 ± 0.54	Hambraeus *et al.*, 1978
0.5–1.5 months	1.94 ± 0.38	Hambraeus *et al.*, 1978
1.5–3.5 months	1.65 ± 0.29	Hambraeus *et al.*, 1978
3.5–6.5 months	1.39 ± 0.26	Hambraeus *et al.*, 1978

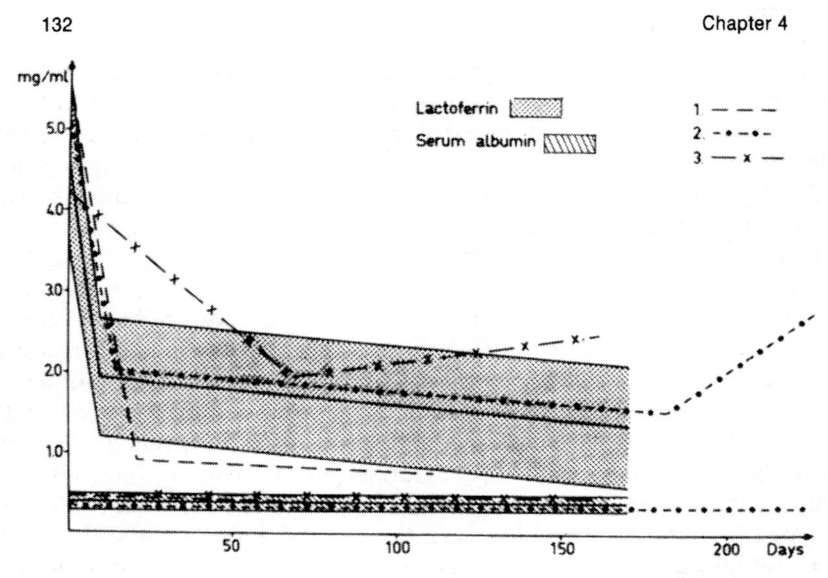

Figure 4-1. Lactoferrin content of milks of three subjects as a function of time following parturition. It is seen that lactoferrin levels are variable, whereas those of serum albumin remain constant at about 0.4 mg/ml. From Lönnerdal, B., Forsum, E., and Hambraeus, L., *Am. J. Clin. Nutr.* 29:1127–1133, 1976, by permission of the American Society for Clinical Nutrition.

quist *et al.*, 1978). Additionally, the lactoferrin content of normal synovial fluid is near 1 μg/ml, whereas in synovial fluid of inflamed joints it is near 26.4 μg/ml. The increase is believed to be due to neutrophil infiltration of the inflamed joint (Bennett *et al.*, 1973). Table 4-2 summarizes serotransferrin and lactoferrin contents of various biological fluids.

Conalbumin is also a major protein component of egg white protein content. Few quantitative data are available on conalbumin content, and some estimates have stated the following: Conalbumin accounts for 10% (Osborne and Campbell, 1900) and 15% (Longsworth *et al.*, 1948) of all hen's egg white protein, and there are 10 g of conalbumin per 1 liter of egg white (Azari and Baugh, 1967). Feeney *et al.* (1960) have analyzed the egg whites of a number of species, and some of their results are given in Table 4-3. Indications are that conalbumin may account for as little as 2% of the dry weight of egg white and for as much as 16%.

4.3 Isolation of the Transferrins

The isolation of serotransferrin does not present any undue difficulty. Most of the older procedures have utilized either Cohn's fraction IV-7

Table 4-2. Mean Serotransferrin and Lactoferrin Values of Various Human and Animal Biological Fluids (μg/ml)

Fluid	Animal	Serotransferrin	Lactoferrin	Reference
Serum	Human	—	<1–3.5	Malmquist, 1972
	Human	—	0.8 ± 40%	Rümke *et al.*, 1971
	Human	2,640	—	Van Eijk and Kroos, 1978
	Human	2,530	—	Weippl *et al.*, 1973
	Human	2,140	—	Hughes, 1972
	Rabbit	2,800 ± 100	—	Van Vugt *et al.*, 1975
	Rabbit	8,400	—	Jordan *et al.*, 1967
	Cattle	3,360	—	Martinsson and Möllerberg, 1973
	Calves	4,500	—	Martinsson and Möllerberg, 1973
	Dog	3,060(1,250–5,450)	—	Altman and Dittmer, 1974
Urine	Human	—	0.3	Haupt and Baudner, 1973
Saliva	Human	—	8	Haupt and Baudner, 1973
Seminal fluid	Human	—	550	Haupt and Baudner, 1973
Bile	Rabbit	16.4 ± 3.0	1600 ± 200	Van Vugt *et al.*, 1975
Amniotic fluid	Human	299(140–445)	—	Larsen *et al.*, 1973
Cerebrospinal fluid	Human	17.2	—	Bleijenberg *et al.*, 1971
Lymph	Dog	2,210(1,440–2,950)	—	Altman and Dittmer, 1974
Tears	Human	100	—	Altman and Dittmer, 1974
	Human	—	1400	Haupt and Baudner, 1973
Colostrum	Human	—	4900 ± 600	Nagasawa *et al.*, 1974
	Human	—	4200 ± 49	Reddy *et al.*, 1977
	Human	—	3300	Hambraeus *et al.*, 1978
	Cattle	409	—	Martinsson and Möllerberg, 1973
Milk	Human	—	1700	Hambraeus, 1977
	Human	25	4000	Masson and Heremans, 1971
	Human	—	1550	Lönnerdal *et al.*, 1976b
	Human	—	3320 ± 72	Evans *et al.*, 1978
	Human	—	2500 ± 65	Reddy *et al.*, 1977
	Rabbit	20,000	Trace	Jordan *et al.*, 1967

(Koechlin, 1952) or fraction IV (Inman *et al.*, 1961) as starting materials. Advantage was also taken of the fact that iron-saturated serotransferrin is about 40 times more soluble in ethanol–water mixtures than is apo-serotransferrin (Inman *et al.*, 1961). With the preparation of the cellulose-based ion exchangers by Sober and Peterson, several procedures for the isolation of serotransferrin using either DEAE– or CM–cellulose were proposed. A two-step procedure was proposed by Bezkorovainy *et al.* (1963) which involved the initial precipitation of a serotransferrin-rich fraction from serum by ammonium sulfate followed by chromatography

Table 4-3. Conalbumin Content of Egg Whites of a Number of Avian Species[a]

Species	Dry weight of egg white (mg/ml)	% Conalbumin
Duck	132	2
Goose	133	4
Chicken	125	12
Golden pheasant	104	13
Lady amherst	108	16
Turkey	124	11
Pigeon	101	9
Rhea	110	3

[a] From Feeney *et al.* (1960).

on DEAE–cellulose. Although serotransferrin preparations obtained by this method appeared homogeneous by the methods of zone electrophoresis and ultracentrifugation, certain impurities amounting to 15% were supposedly found in such preparations by Frenoy *et al.* (1971), who instead proposed a three-step procedure involving the initial precipitation with ammonium sulfate followed by preparative zone electrophoresis and finally chromatography on DEAE–cellulose. Frenoy *et al.* claimed to have obtained preparations that were 90–95% pure.

A two-step procedure for the preparation of rat serotransferrin was proposed by Gordon and Louis (1963), who initially used preparative zone electrophoresis followed by chromatography on DEAE–cellulose. Rat serotransferrin exists in the form of two species, the *fast* (Tf_f) and *slow* (Tf_s). They can be separated electrophoretically. Gordon and Louis isolated both forms and showed that the difference in electrophoretic mobility was not due to sialic acid or amino acid content.

The most widely used methods for the preparation of serotransferrin involve the initial precipitation of the bulk of plasma proteins with Rivanol® (2-ethoxy-6,9-diaminoacridine lactate) followed by at least two additional steps. Some difficulty exists in removing traces of Rivanol® from the final serotransferrin preparations, though potato starch has been found to be quite effective (Sutton and Karp, 1965). The first investigators to use Rivanol® for serotransferrin isolation were Boettcher *et al.* (1958), who made an "...almost pure preparation" of serotransferrin by precipitating the bulk of plasma proteins at pH 8–10 with 3.5 parts of 0.4% Rivanol®, followed by removal of Rivanol® by charcoal, precipitation of γ-globulins at a final ethanol concentration of 25% at pH 6.8 and -6°C, and finally precipitation of the serotransferrin at a final ethanol concentration of 40% at pH 5.8.

Other investigators have combined Rivanol® precipitation of plasma proteins with chromatography on ion exchangers, ammonium sulfate precipitation, and chromatography on various gel filtration media to prepare 90–95% homogeneous serotransferrins from human (Nagler *et al.*, 1962; Roop and Putnam, 1967), swine (Leibman and Aisen, 1967), cattle (Stratil and Spooner, 1971), and sheep (Spooner *et al.*, 1975) sera. A representative scheme used by Roop and Putnam (1967) is shown in Figure 4-2.

Conalbumin, as pointed out, was first isolated at the turn of the century. Since that time, ammonium sulfate precipitation has remained the cornerstone of the various isolation procedures for conalbumin preparation. An exception is the ethanol fractionation procedure devised by Bain and Deutsch (1948), who reported the preparation of "essentially pure" hen's egg conalbumin in the yield of 2 g/liter of egg white.

Ammonium sulfate precipitation followed by alcohol fractionation and finally chromatography on CM–Sephadex was used by Evans and Holbrook (1975). They obtained 4 g of conalbumin from 1 liter of egg white (36 eggs). Duck conalbumin was prepared by a three-step procedure involving initial ammonium sulfate fractionation followed by DEAE– and CM–cellulose chromatography. The yield was about 2.25 g/liter of egg white (Das, 1974). It would appear that the most rapid procedure for the preparation of hen's egg conalbumin was proposed by Azari and Baugh (1967). "Mucin"-free egg white at pH 4.7 was passed over a column of CM–cellulose equilibrated with 0.01 *M* citrate buffer at pH 4.7. Conalbumin was retained and after washing the column, was eluted with 0.3 *M* $(NH_4)_2SO_4$ at pH 8.5. It was precipitated by making the column eluate 65% saturated with ammonium sulfate and adjusting the pH to 6. Conalbumin can be further crystallized in the iron-saturated form from solutions containing $(NH_4)_2SO_4$. The yield was near 7.2 g/liter of egg white at the ammonium sulfate precipitation step and 4 g/liter after the second crystallization.

Bovine milk contains very little lactoferrin, and it appears to be tightly associated with other milk proteins. In human milk, too, lactoferrin prefers to associate with acidic macromolecules, so that such association may be electrostatic in nature (Hekman, 1971), and one may surmise that the same is true for bovine milk lactoferrin. Bovine lactoferrin is usually isolated from crude preparations of other milk proteins. Thus, Groves (1960) isolated lactoferrin from acid casein by first extracting casein with acetate buffer at pH 4.0 and then precipitating the lactoferrin with $(NH_4)_2SO_4$ (lactoferrin precipitated between 40 and 65% saturation with ammonium sulfate at pH 7.6). The final purification step involved chromatography on DEAE–cellulose. Final recovery was 0.74 g from 15 gal of skim milk. Castellino *et al.* (1970) isolated lactoferrin from commercial α-lactalbumin preparations. The procedure was simple: Three grams of

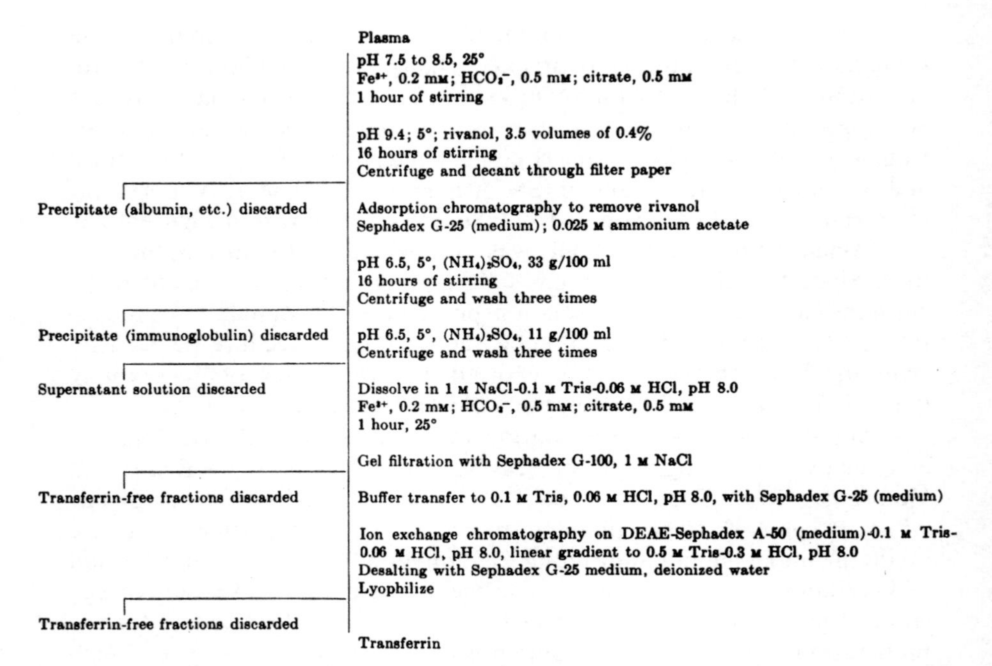

Figure 4-2. Preparation of human serum transferrin from plasma. From Roop and Putnam (1967).

lactalbumin at pH 8.6 were chromatographed on a column of Sephadex G-100, and lactoferrin, amounting to 250 mg, appeared in the void volume.

Human lactoferrin is usually isolated from milk whey. There are several methods available. One of the earlier procedures involves ammonium sulfate fractionation of the whey followed by chromatography on Amberlite XE 64 (Montreuil *et al.*, 1960). A combination of ethanol and ammonium sulfate precipitation steps has also been proposed. It has the advantage of separating all the other milk proteins in addition to lactoferrin (Got, 1965). The most rapid method for the isolation of human lactoferrin is that of Johansson (1969), who first saturated lactoferrin in milk with iron and then added 3 g of dry CM–Sephadex to 1 liter of human milk diluted with two volumes of water. Lactoferrin was bound to the ion exchanger and was then eluted thereform with 2 *M* NaCl at pH 8.0. It was crystallized from a solution of low ionic strength at pH 7.0. A somewhat better yield was obtained by Querinjean *et al.* (1971), who first precipitated the bulk of the milk proteins with 2 *M* ammonium sulfate at pH 7.0. The supernatant was then dialyzed to remove the ammonium sulfate, the lactoferrin present therein was saturated with iron, and CM–Sephadex was added. The lactoferrin was eluted from the ion exchanger with 0.33 *M* NaCl at pH 7.0. The yield was 1.3 g of lactoferrin per liter of milk.

4.4 Physical Properties of the Transferrins

4.4.1 Hydrodynamic Parameters, Molecular Weights, and X-Ray Diffraction

Transferrins are globular proteins that are soluble in aqueous buffers. Serotransferrins and conalbumin are also soluble in water. Molecular weights of the transferrins were originally calculated from hydrodynamic parameters and composition data. Bain and Deutsch (1948) reported a molecular weight for conalbumin of 87,000 on the basis of sedimentation–diffusion measurements. Later, Warner and Weber (1953), on the basis of iron uptake measurements, reported a minimum molecular weight of 38,300 and a true molecular weight of 76,600 for conalbumin. Subsequent physical measurements have confirmed the molecular weight of conalbumin as 77,000.

Early measurements of hydrodynamic properties of serotransferrins had placed the molecular weights close to 90,000 (Oncley *et al.*, 1947; Laurell and Ingelman, 1947; Schutze *et al.*, 1957). From such data, the dimensions of human serotransferrin molecule were calculated to be 37 × 190 Å (Oncley *et al.*, 1947) and those of swine serotransferrin as 35 × 170 Å (Laurell and Ingelman, 1947). As better isolation and analytical procedures became available, it was discovered that the molecular weights of most serotransferrins, determined by physical methods, were around 78,000. The earlier high values were probably due to the presence of serotransferrin aggregates in solutions analyzed. This was particularly reflected by the sedimentation constants. Mann *et al.* (1970) have in fact stated that sedimentation constant values of higher than 5.4 S "can probably be disregarded because of aggregation." The most recent estimate for the molecular weight of the serotransferrin polypeptide chain is near 81,000. This is based on amino acid sequence studies of MacGillivray *et al.* (1977), who have determined that human serotransferrin contains 676 amino acid residues. If the carbohydrate moieties are added to this (two chains with a molecular weight of 2200 each), the molecular weight becomes close to 85,000. X-Ray diffraction studies at 6-Å resolution using human serotransferrin have revealed a molecular weight of 84,884, where orthorhombic crystals with unit cell dimensions of a = 78 Å, b = 94 Å, and c = 112 Å were observed (De Lucas *et al.*, 1978). For swine serotransferrin crystals, X-ray diffraction studies indicated a molecular weight of 81,500 and unit cell dimensions (orthorhombic crystals) of a = 87 Å, b = 88 Å, c = 114 Å (Magdoff-Fairchild and Low, 1970). Perhaps the most extensive investigation on any transferrin by the X-ray diffraction method was done on diferric rabbit serotransferrin at 6-Å resolution (Gorinsky *et al.*, 1979). The molecular dimensions were determined to be 95 × 60 × 50 Å, and a model was constructed to approximate the molecular shape. It had two *lobes*, and both lobes had pronounced clefts (iron-binding sites?) close to the bridge joining them. This model is represented in Figure 4-3.

The molecular dimensions of the transferrins and degrees of their hydration have been investigated by several groups. Bezkorovainy (1966) reported an axial ratio of 1:2 for iron-free and 1:3 for iron-saturated serotransferrin. Effective hydrodynamic volumes (V_e) were 20.3 and 15.4 × 10^4 Å^3, respectively. Rosseneu-Motreff *et al.* (1971) showed that apo-serotransferrin can best be represented by a prolate ellipsoid of revolution with the dimensions 24.6 by 62 Å, an axial ratio of 2.5, and an effective hydrodynamic volume of 15.4 × 10^4 Å^3. Iron-saturated serotransferrin had the dimensions 27.6 by 55.2 × 10^4 Å^3 with an axial ratio of 2.1 and an effective hydrodynamic volume of 16.9 × 10^4 Å^3. More recently, Yeh *et al.* (1979) reported molecular dimensions of 24.6 by 62 Å for conalbumin. They estimated the degree of hydration to be near 0.28 g of water per g of protein. Hydrodynamic parameters and other physical constants of human serotransferrin are given in Table 4-4, and those of lactoferrin, conalbumin, and serotransferrins of other species are given in Table 4-5.

Figure 4-3. Three-dimensional appearance of human serum transferrin showing its bi-lobal structure. From Gorinsky *et al.* (1979).

Table 4-4. Physical Properties of Human Serotransferrin

Reference	$S^{\circ}_{20,w}$ (S unit)	$D_{20,w}$ (cm^2/sec)	[η] (dl/g)	$\bar{V}$ (cc/g)	$E^{1\%}_{280}$	$E^{1\%}_{465}$	Molecular weight	pI
Schultze *et al.*, 1957	6.1	6.2	0.042	0.725	10.9	—	88,000	—
Oncley *et al.*, 1947	5.5	—	—	0.725	—	—	90,000	5.6
Van Eijk *et al.*, 1972	5.4	—	—	—	—	—	83,000	—
Roberts *et al.*, 1966	4.85	5.85	—	—	—	—	73,000	—
Mann *et al.*, 1970	5.25	—	0.040	—	—	—	79,300	—
Leger *et al.*, 1977	5.30	—	0.041	—	—	—	75,000	—
Bezkorovainy *et al.*, 1968	5.35	5.75	0.044	0.723	—	—	82,000	—
Keller and Pennell, 1959	5.00	—	—	—	11.2	0.570	90,000	5.45 (apo) 5.80 (Fe_2)
Bearn and Parker, 1966	5.5	5.4	—	0.725	11.2	0.57	—	5.40 (apo) 5.58 (Fe_2)
Aisen *et al.*, 1967	—	—	—	—	14.1 (Fe_2) 11.4 (apo)	—	81,000	—

Table 4-5. Physical Properties of Transferrins from Various Sources

Protein	$S^{\circ}_{20,w}$ (S unit)	$D^{\circ}_{20,w}$ (cm^2/sec)	[η] (dl/g)	$\bar{V}$ (cc/g)	pI	$E^{1\%}_{280}$	$E^{1\%}_{465}$	Molecular weight	Reference
Human lactoferrin iron saturated	5.02	—	—	0.726	—	14.2	—	82,600	Haupt and Baudner, 1973
	5.25	—	—	—	—	—	0.540	80,000	Johansson, 1969
	4.8	4.6	—	0.735	—	11.7	0.500	89,000–95,000	Montreuil *et al.*, 1960
	5.35	—	0.040	—	—	—	—	77,000	Leger *et al.*, 1977
60% saturated	4.91	5.6	—	—	—	14.6	—	75,100	Querinjean *et al.*, 1971
Iron-free	4.94	—	—	0.716	—	11.2	—	75,100	Querinjean *et al.*, 1971
Bovine lactoferrin, iron saturated	5.55	5.75	—	0.725	7.8	—	0.47	86,100	Groves, 1960
	5.3	—	—	—	—	14.5	—	77,100	Castellino *et al.*, 1970
	—	—	—	0.736	8.0	—	—	93,000	Weiner and Szuchet, 1975
Rabbit serotransferrin, iron saturated	—	—	—	—	—	13.9	0.509	70,000	Baker *et al.*, 1968
	4.6	—	—	—	—	—	—	82,000	Van Eijk *et al.*, 1972
Conalbumin, iron-free	5.4	5.66	—	0.731	6.0	—	—	87,000	Bain and Deutsch, 1948
	5.49	5.72	—	0.721	—	—	—	84,000	Bezkorovainy *et al.*, 1968
Sheep serotransferrin	5.15	—	—	0.734	—	12.5	—	77,500	Guerin *et al.*, 1976
Murine serotransferrin, iron saturated	—	—	—	—	8.7–9.6	—	—	75,000–78,000	Kinkade *et al.*, 1976
Rat serotransferrin, iron saturated	5.1	—	—	0.733	—	—	0.680	67,000	Charlwood, 1963
									Gordon and Louis, 1963
	4.7	—	—	—	—	—	—	83,000	Van Eijk *et al.*, 1972
Cattle serotransferrin	5.18	—	—	—	—	—	—	77,500	Richardson *et al.*, 1973
Hagfish serotransferrin, iron-free	5.80	6.70	—	—	—	10.2	—	75,600	Aisen *et al.*, 1972
Swine serotransferrin, 70% saturated	5.8	5.82	—	0.725	4.4	—	—	88,000	Laurell and Ingelman, 1947
Shark serotransferrin	5.35	—	—	—	9.2	12.0	—	77,500	Got *et al.*, 1967
Tench serotransferrin	4.8	—	—	—	—	—	—	81,000	Van Eijk *et al.*, 1972

4.4.2 Denaturation of the Transferrins

Transferrins, being typical globular proteins, are susceptible to denaturation through a variety of agents. A point of interest is that iron-saturated transferrins are much more stable to denaturation than are the apoproteins. This was first demonstrated by Azari and Feeney (1958, 1961) with respect to thermal and urea denaturation. Thus, iron was not lost from conalbumin even in 8 *M* urea, whereas with apoconalbumin, the iron complex could not be formed in urea concentrations above 2 *M*. The effect of urea and acid on conalbumin was extensively studied by Glazer and McKenzie (1963), who used viscosity, optical rotation, and sedimentation techniques to detect denaturation in conalbumin. It was shown that extensive denaturation of the protein occurred below pH 4.2 and that this was accelerated by urea even at low urea concentrations. In the pH range of 5.9–8.2, apoconalbumin showed no change in urea solutions at 3 *M* concentration. Iron–conalbumin showed no change even at 5 *M* urea concentrations. Denaturation was apparently reversible even if carried out at pH 4.5. Thus, at pH 6.0, native conalbumin had a specific rotation of $-37°$ and a reduced viscosity of 0.040; in 6 *M* urea these figures were $-92°$ and 0.182, respectively, and after diluting the latter to 1.5 *M* urea, the parameters returned to $-35°$ and 0.034, respectively. The *renatured* protein had also recovered 85% of its iron-binding capacity.

Denaturation of conalbumin is apparently accompanied by "swelling" and an increase in random coil content (Yeh *et al.*, 1979). Native conalbumin, assuming a prolate shape, had the molecular dimensions 21 by 68 A, an axial ratio of 0.31, and a solvation value of 0.28 g of water per g of protein. Denaturation in 7.2 *M* urea resulted in a prolate ellipsoid 42 × 84 Å in size with an axial ratio of 0.5 and an apparent solvation value of 4.16 g of water per g of protein. Renatured protein (urea diluted) had molecular dimensions of 22 × 67 Å and a hydration value of 0.36 g of water per g of protein. The α-helix and β-pleated structure contents were 11, 32, and 57% for the native protein. They were 6, 19, and 75% for denatured conalbumin and 12, 29, and 59% for the renatured material. Iron-binding activity lost upon denaturation was completely restored upon diluting the urea.

Similar results were obtained with human serotransferrin (Bezkorovainy and Grohlich, 1967). Iron-free serotransferrin was denatured in 3.8 *M* urea after 1 hr at pH 8.6, whereas iron-saturated serotransferrin was stable in 5.6 *M* urea after 2 days. Native serotransferrin had an intrinsic viscosity of 0.044, whereas in 5.4 *M* guanidine hydrochloride this value was 0.147. Upon denaturation, serotransferrin apparently undergoes an aggregation process (Bron *et al.*, 1968). Lactoferrin was found to be somewhat less stable to denaturation than serotransferrin when in the iron-free state; it was, however, more stable in the iron-saturated state (Teuwissen *et al.*, 1974). The same high degree of stability in iron-saturated lactoferrin was also observed during denaturation by acid (Bezkorovainy G. and Bezkorovainy A., 1978).

The transferrins are readily denatured when modified chemically, especially via carboxyacylation with succinic, maleic, or citraconic anhydrides (Bezkorovainy *et al.*, 1972; Zschocke *et al.*, 1972). Physical parameters and iron-binding properties can be largely restored to those of native proteins by either removing the carboxyacyl groups (in the case of maleyl and citraconyl groups) or by dissolving the modified proteins in concentrated salt solutions, e.g., 2 *N* NaCl.

4.4.3 Quaternary Structure of the Transferrins

Because the transferrins can bind a maximum of two iron atoms for each protein molecule, it was originally suspected that these proteins may consist of two subunits. And indeed Jeppsson (1967) has reported that he could split human serotransferrin into two identical subunits by reduction–alkylation or performic acid oxidation. The molecular weight of the subunits was near 40,000 with an $S^{\circ}_{20,w}$ of 1.24 S and a $D^{\circ}_{20,w}$ of 2.8 × 10^{-7} cm^2/sec. Tryptic digestion followed by two-dimensional chromatography yielded only half the expected peptides (34–38 from a protein with 38 arginine and 68 lysine residues). In the same year, however, Bezkorovainy and Grohlich (1967) reported that the molecular weight of human serotransferrin did not change upon reduction–alkylation in 8 *M* urea and 6 *M* guanidine hydrochloride and in spite of the extensive degree of denaturation remained a single-chain protein. The same results were obtained with sulfitolyzed conalbumin and human serotransferrin, with alkylated conalbumin (Bezkorovainy *et al.*, 1968), and with reduced–alkylated–succinylated serotransferrin and conalbumin (Bezkorovainy *et al.*, 1969).

The single-chain nature of human and rabbit serotransferrin and hen's egg conalbumin was independently determined by Greene and Feeney (1968) using ultracentrifugal techniques on the reduced–alkylated proteins in 8 *M* urea and 6 *M* guanidine hydrochloride. They also suggested that the transferrins might perhaps consist of two identical halves that arose during the course of evolution by a gene-duplicating process. The results of Greene and Feeney and of Bezkorovainy and Grohlich were confirmed in a careful study of Mann *et al.* (1970), who obtained molecular weights of near 78,000 for both the native and reduced–alkylated serotransferrin.

They were also able to recover 55–70 ninhydrin-positive peptides out of a possible 76 following tryptic digestion of serotransferrin. Identical results were obtained with lactoferrin (Querinjean *et al.*, 1971). Thus all transferrins heretofore studied have been proved to consist of a single polypeptide chain.

4.4.4 Secondary Structure of the Transferrins

No extensive X-ray diffraction studies have so far been made on any of the transferrins, so that information on the secondary structure must be gathered from optical rotatory dispersion or circular dichroism studies. Such studies indicated a rather low α-helix content and a substantial β structure in all transferrins studied (Table 4-6). Hen's egg conalbumin had a high random coil content compared to serotransferrin and lactoferrin. The circular dichroic pattern of serotransferrin and lactoferrin is shown in Figure 4-4.

The secondary structure of transferrins does not appear to change much upon the binding of iron (Table 4-6). This is also true for copper

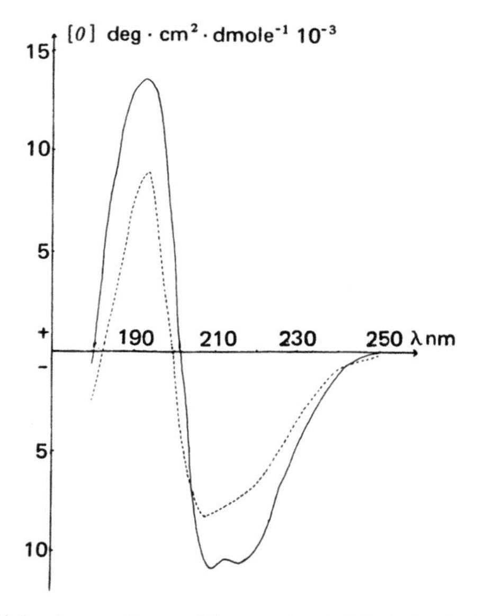

Figure 4-4. Circular dichroism patterns of iron-saturated lactoferrin (solid line) and human serotransferrin (broken line) in the ultraviolet range of the spectrum. From Mazurier *et al.* (1976).

Table 4-6. Secondary Structures of the Transferrins (in % of amino acid residues)

Protein	α helix	β-Pleated sheet	Random coil	Reference
Human serotransferrin				
Iron-free	17	68	15	Mazurier *et al.*, 1976
	17–18	—	—	Nagy and Lehrer, 1972
Iron saturated	23	62	15	Mazurier *et al.*, 1976
Hen's egg conalbumin				
Iron-free	11	32	57	Yeh *et al.*, 1979
	28	32	40	Tan, 1971
Iron saturated	28	32	40	Tan, 1971
Human lactoferrin				
Iron-free	28	64	8	Mazurier *et al.*, 1976
Iron saturated	26	57	17	Mazurier *et al.*, 1976

binding (Tomimatsu and Vickery, 1972). On the other hand, the so-called extrinsic region of the circular dichroic spectrum (i.e., bands due to restricted rotation of certain amino acid side chains such as tyrosyl or tryptophanyl groups) shows considerable alteration following the binding of iron (Figure 4-5) (Nagy and Lehrer, 1972; Mazurier *et al.*, 1976; Tomimatsu and Vickery, 1972; Brown and Parry, 1974). This is to be expected, since various amino acid side chains are either involved in the binding of iron directly or are in some fashion perturbed by such binding. It has been pointed out that the secondary structure of lactoferrin differs considerably from that of serotransferrin, where the α-helix content of lactoferrin is greater than that of serotransferrin or conalbumin and the iron held by lactoferrin, in contrast to that of serotransferrin, is located deeper in the protein's interior (Mazurier *et al.*, 1976).

At this point, then, it is possible to summarize the preceding by stating that the transferrins are single-chain proteins with molecular weights of 77,000–90,000 and containing relatively large amounts of β pleated-sheet structure and relatively little α-helix.

4.5 Metal-Binding Properties of the Transferrins

4.5.1 Metal Ions that Are Bound by the Transferrins

It has been repeatedly stated that the transferrins bind a maximum of two atoms of ferric iron for each transferrin molecule. In doing so, they acquire a salmon-pink color with an absorption maximum at 460–470 nm and the absorbances presented in Tables 4-4 and 4-5. Titration of serotransferrin with iron using the salmon-pink color as an indicator is shown in Figure 4-6. Ferrous iron is not bound to the transferrins, since the color can be discharged by reduction with hydrosulfite, and no color is obtained from ferrous iron and a transferrin unless O_2 is made available (Koechlin, 1952). There is a report asserting that ferrous iron must be bound to the transferrins, albeit very weakly, in order to account for the rapid oxidation of ferrous iron in the presence of transferrin (Bates *et al.*, 1973). However, it has been pointed out that ferrous iron is oxidized to the ferric state much more rapidly in the presence of any chelating agent such as EDTA, nitroloacetate, and citrate and that transferrin does not present any special situation in this regard (Harris and Aisen, 1973). Direct-binding experiments using ultraviolet difference spectroscopy as an assay method have tended to exclude the possibility that the transferrins bind ferrous iron (Gaber and Aisen, 1970).

Transferrins are capable of combining with a number of transition metals other than iron, though ferric iron appears to form the strongest bond with the proteins. The relative stability of the various complexes has been given as $Fe^{3+} > Cr^{3+}$, $Cu^{2+} > Mn^{2+}$ and Co^{2+}, and $Cd^{2+} > Zn^{2+}$

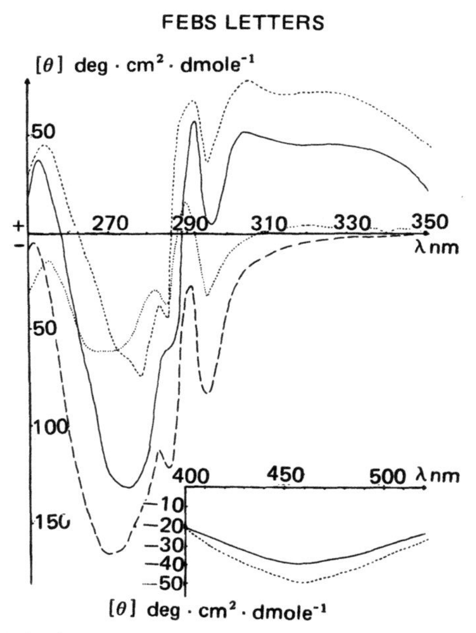

Figure 4-5. Circular dichroism spectra of iron-saturated and iron-free transferrins between 250 nm and 500 nm. Iron-saturated lactoferrin (———), iron-free lactoferrin (— — —), iron-saturated serotransferrin (------), and iron free serotransferrin (⋯⋯). Human source in all cases. From Mazurier *et al.* (1976).

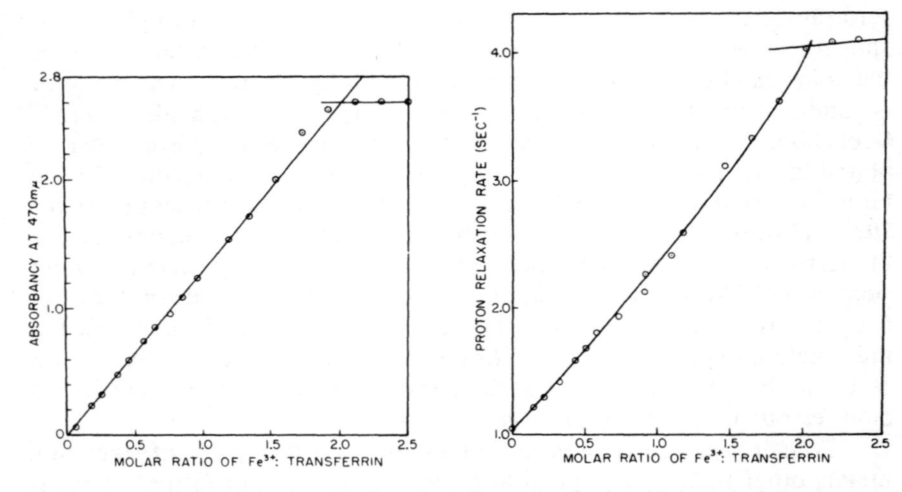

Figure 4-6. Titration of the iron-binding sites of serotransferrin with iron. Left-hand curve uses E_{470} as indicator; right-hand curve uses proton relaxation rate as an indicator of the amount of iron bound. Both methods indicate that serotransferrin binds a maximum of two iron atoms per protein molecule. From Aisen *et al.* (1966).

and Ni^{2+} (Tan and Woodworth, 1969). Other metals that can be bound by the transferrins include scandium (Ford-Hutchinson and Perkins, 1971), vanadium (Cannon and Chasteen, 1975), gallium (Woodworth *et al.*, 1970), platinum (Stjernholm *et al.*, 1978), and the elements of the lanthanide series (Luk, 1971). The various metals give different absorption spectra in the visible portion of the spectrum; e.g., copper–serotransferrin has an absorption maximum at 430 nm, and the zinc complex is colorless. The binding of zinc is of physiological importance, as serotransferrin is involved in zinc absorption and its transport in the portal circulation (Evans, 1976). There are also reports to the effect that serotransferrin is a folate carrier (Markkanen *et al.*, 1972), but this has been disputed (Jacob and Herbert, 1974).

4.5.2 The Binding of "Synergistic" Anions by the Transferrins

In 1949, Schade *et al.* (1949) reported that the typical salmon-pink color of the iron–serotransferrin complex does not develop unless bicarbonate (a synergistic anion) is present in the medium. It is now accepted, however, that carbonate rather than bicarbonate is the true synergistic anion under physiological conditions (Bates and Schlabach, 1973a). In fully saturated transferrins, two molecules of carbonate are bound per molecule of protein, i.e., one per each iron atom. They exchange with the atmospheric CO_2 very slowly (half-life = 20 days), though the process is accelerated in the presence of other anions (Aisen *et al.*, 1973a). Various divalent, trivalent, and tetravalent but not monovalent anions can take the place of carbonate in the iron–transferrin complex. These include EDTA, nitriloacetate, oxalate, and others, though carbonate is by far the most preferred species. In fact carbonate can displace the others from iron-laden serotransferrin. It was at one time believed that the binding of iron by the transferrins could proceed in the absence of the synergistic anions, albeit with an altered resulting spectrum (Aasa *et al.*, 1963; Aisen *et al.*, 1967); however, it has now been well settled that the synergistic anion is an absolute requirement for the binding of iron by the transferrins (Price and Gibson, 1972a; Van Snick *et al.*, 1973) and that in the absence of such anions a *nonspecific* binding of iron by the transferrins (in the form of ferric hydroxide polymers) is the result (Tsang *et al.*, 1975; Bates and Schlabach, 1975). The carbonate apparently participates directly in the binding of iron by the transferrins. Thus, if other synergistic anions replace carbonate in the complex, there is a change in the EPR spectrum, indicating that a change in the iron ligand field had taken place (Pinkowitz and Aisen, 1972). Carbonate is also a ligand in the copper–lactoferrin and copper–serotransferrin complexes (Masson and Heremans, 1968; Zweier and Aisen, 1977) and in the vanadium–serotransferrin complex (Campbell and Chasteen, 1977).

What, then, is the structural relationship among the synergistic anion, iron, and the transferrin molecule? Schlabach and Bates (1975) have addressed this question by testing some 25 synergistic and potentially synergistic anions for their ability to give a colored complex with iron and serotransferrin and therefore to promote the *specific* binding of iron by the protein. They came to the conclusion that a typical synergistic anion must have the general structure given by

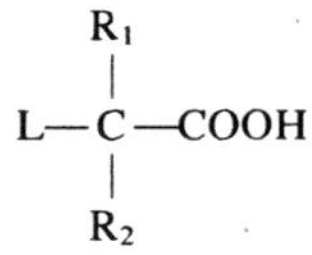

$$\begin{array}{c} R_1 \\ | \\ \mathrm{L{-}C{-}COOH} \\ | \\ R_2 \end{array} \tag{4-1}$$

where L is an electron-withdrawing group. Typically, there are such compounds as

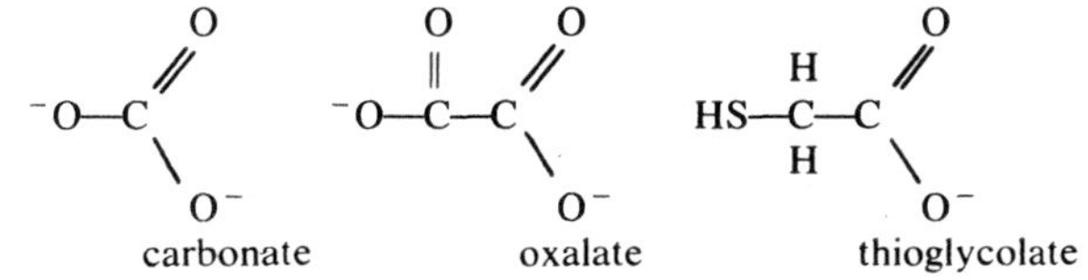

$$\underset{\text{carbonate}}{{}^{-}\mathrm{O{-}C({=}O)O^{-}}} \qquad \underset{\text{oxalate}}{{}^{-}\mathrm{O{-}C({=}O){-}C({=}O)O^{-}}} \qquad \underset{\text{thioglycolate}}{\mathrm{HS{-}CH_2{-}C({=}O)O^{-}}} \tag{4-2}$$

According to Schlabach and Bates, the iron is ligated with the anion through the L grouping and linked to the serotransferrin backbone through electrostatic interactions. It has been suggested that the counterion for the synergistic anion is an arginyl residue, since chemical modification of conalbumin and serotransferrin shows the association of one arginyl group with each iron atom bound (Rogers *et al.*, 1978). The resultant structure, shown in Figure 4-7, has been termed the *interlocking sites model.* In this model, group L must be within 6.3 Å of the iron atom. This proposal received support from a report by Najarian *et al.* (1978), who used EPR spectroscopy with spin-labeled oxalate to show that the distance between the L group and iron was equal to or less than 6 Å. The physiologic significance of the synergistic anion may be its function in the release of iron to immature red cells, to be discussed in Chapter 6.

4.5.3 The Reaction of Iron with the Transferrins

Iron can be bound up with the transferrins and removed therefrom at will under the appropriate conditions without denaturing the protein. It has now been well established that the loading of transferrin with iron must be accomplished either by using ferrous iron (which is immediately oxidized to ferric iron by O_2) or with ferric iron complexed with an appropriate chelating agent, such as citrate, nitriloacetate, etc. If $FeCl_3$ is used, as was the case until the mid-1970s, a considerable proportion of the iron added may be bound to the transferrins in a nonspecific manner, i.e., in the form of ferric hydroxide polymer (Bates and Schlabach, 1973b). This occurs regardless of whether or not carbonate is present in the medium.

Kinetic measurements have indicated that iron from the ferric iron–nitriloacetate complex is incorporated into serotransferrin through a one-step reaction in the absence of carbonate, whereas in the presence of the anion, this occurs as a two-step process (Bates and Wernicke, 1971; Binford and Foster, 1974). In the first case, this may be represented by

$$\begin{gathered}2\ \text{Fe–nitriloacetate} + \text{serotransferrin} \rightarrow \\ (\text{Fe–nitriloacetate})_2\text{–serotransferrin} \\ \Delta H = -10.4\ \text{kcal/mole}\end{gathered} \tag{4-3}$$

In the presence of carbonate, the reaction proceeds as

$$\begin{gathered}2\ \text{Fe–nitriloacetate} + \text{serotransferrin} \rightarrow \\ (\text{Fe–nitriloacetate})_2\text{–serotransferrin} \\ (\text{Fe–nitriloacetate})_2\text{–serotransferrin} + 2\ \text{carbonate} \rightarrow \\ (\text{Fe–carbonate})_2\text{–serotransferrin} + 2\ \text{nitriloacetate}\end{gathered} \tag{4-4}$$

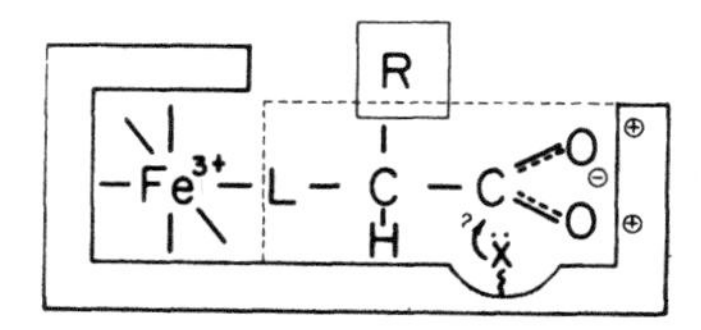

Figure 4-7. Model of the interlocking iron- and synergistic anion-binding sites in the transferrins. Dashed line indicates the confines of the synergistic anion, where L is its "proximal ligand." From Schlabach and Bates (1975).

The overall $\Delta H = -22.50 \pm 0.60$ kcal/mole. The first reaction in equation (4-4) is very rapid and is complete within 0.1 sec, whereas the second reaction is slow, requiring 10 sec for completion. It is clear, however, that enthalpy effects would favor the incorporation of carbonate into the iron–serotransferrin complex over nitriloacetate.

The removal of iron from the iron–transferrin complexes can be accomplished in the presence of citrate by dropping the pH to perhaps 5.0 (Okada *et al.*, 1978). Lactoferrin is an exception, where the pH must be dropped to about 2–3 before iron will come off (Masson and Heremans, 1968).

4.5.4 Are the Two Iron-Binding Sites of the Transferrins Identical?

4.5.4.1 The Equal and Independent Hypothesis

A large volume of information exists on the question as to whether or not the two iron-binding sites of the transferrins are chemically and/or biologically identical. The question of biological (non)identity will be dealt with in Chapter 6, whereas in this section we shall address the chemical properties of the two iron-binding sites. Paradoxically, in spite of all the knowledge gained, the issue of identity or nonidentity is far from settled.

Among the methods that can be utilized to study the properties of iron-binding sites are kinetic and thermodynamic measurements, spectroscopic methods, the determination of various acid–base properties of the transferrins, and chemical modification techniques.

The earliest attempts at looking at this problem were made by Warner and Weber (1953), who used equilibrium dialysis to determine the association constants involved in the iron–conalbumin interaction. They determined that one association constant was much greater than the other and that there was a positive cooperativity between the two iron-binding sites. Aasa *et al.* (1963) and Aisen and Leibman (1968a) found the opposite: There was no difference between the two iron-binding sites of conalbumin and serotransferrin. They suggested that the dialysis times used by Warner and Weber (1953) were too short to establish full equilibrium, which can take as long as 7 days. These workers determined that the pertinent association constants for transferrins was in the vicinity of 10^{23} if iron–citrate was used as the reactant. In the absence of citrate, the figure is near 10^{36}. This is an extremely large association constant, where some 10^4 years would be required to release a single iron atom from serotransferrin if such a release was to be spontaneous (Aisen and Leibman, 1968b). It was furthermore proposed that the two iron-binding sites of transferrins were independent and equivalent.

Subsequent to this, electrophoretic experiments in a Tiselius apparatus showed that at pH 6.7 the transferrins, saturated at various levels with iron, would show three distinct species with mobilities of -1.13×10^{-5}, 1.20×10^{-5}, and -1.50×10^{-5} $cm^2/V/sec$ corresponding to apotransferrins, transferrins with a single iron atom, and the fully iron-saturated transferrins. At the 50% iron saturation level, all three species were present (Aisen *et al.*, 1966). The reason such transferrin species show differences in electrophoretic mobilities is that with the binding of each iron(III) atom three protons are released, three positive charges are added through the entry of each iron atom, and two negative charges are added through the binding of each carbonate anion (Warner and Weber, 1953; Aasa *et al.*, 1963; Aisen *et al.*, 1966). We are thus adding two negative charges for each iron atom bound, and the protein thereby becomes more acidic. Thus, the pIs for iron-free, monoferric, and diferric conalbumin are 6.73, 6.25, and 5.78, respectively (Wenn and Williams, 1968). For serotransferrin the pIs are 5.80 and 5.45 for the iron-free and iron-saturated species, respectively (Keller and Pennell, 1959), or 5.6 and 5.2 for the monoferric and diferric human serotransferrins, respectively (Hovanessian and Awdeh, 1976).

Several successful attempts have been made to prepare the one-iron species of the transferrins by preparative isoelectric focusing technology using transferrin preparations less than 100% saturated with iron (Wenn and Williams, 1968; Hovanessian and Awdeh, 1976; Williams *et al.*, 1970; Van Eijk *et al.*, 1969a). On the basis of such experiments, it was concluded that iron is taken up by the transferrins in two steps:

$$\begin{aligned} \text{apotransferrin} + \text{Fe} &\rightleftharpoons \text{transferrin–Fe} \\ \text{transferrin–Fe} + \text{Fe} &\rightleftharpoons \text{transferrin–Fe}_2 \end{aligned} \tag{4-5}$$

It was also concluded that the coexistence of the monoferric, diferric, and apotransferrin species "...clearly agrees..." with the conclusion of Aasa *et al.* (1963) that "the metal binds randomly to two independent and equivalent sites on the protein" (Wenn and Williams, 1968). However, the same group states that the fact that iron binds randomly to the transferrins "...reflects only the availability of these sites, and we cannot draw conclusions as to the relative values of the dissociation constants" of each site (Williams *et al.*, 1970).

Direct determination of enthalpies of binding of iron to serotransferrin to achieve half saturation from aposerotransferrin and full saturation from half-saturated serotransferrin gave identical values of −10 to −11 kcal/gram atom of iron bound (Binford and Foster, 1974), apparently supporting the equivalent and independent site theory. It should be noted that the binding of iron by lactoferrin is, from a thermodynamic point of view, some 300 times stronger than that of either serotransferrin or conalbumin (Aisen and Leibman, 1972).

The distribution of iron in a transferrin, assuming that it binds randomly to the two iron-binding sites, can be calculated from

$$\begin{aligned}\%\ \text{apotransferrin} &= \frac{(100 - s)^2}{100} \\ \%\ \text{monoferric transferrin} &= 2s\,\frac{100 - s}{100} \\ \%\ \text{diferric transferrin} &= \frac{s^2}{100}\end{aligned} \tag{4-6}$$

(Lane, 1975), where s is percent saturation of transferrin with iron.

4.5.4.2 Evidence for Nonidentity of the Iron-Binding Sites

In contrast to the equivalent and independent site point of view, there exists a body of evidence that tends to indicate that the two iron-binding sites of the transferrins are not equivalent and perhaps not even independent. One of the earlier suggestions in this regard was made by Line *et al.* (1967) on the basis of chemical modification data, where it was found that the tyrosyl residues at the two iron-binding sites were not equally reactive. Further evidence to that effect came from calorimetric measurements, spectroscopic data on transferrins containing iron and other metals, and acid–base properties of these proteins, as described in the following paragraphs.

EPR spectroscopy has been an extremely valuable tool in describing the mode of binding of iron by the transferrins, which includes the definition of the iron-binding ligands as well as a probe into whether or not the two iron-binding sites are equivalent. A paramagnetic metal ion such as Fe^{3+} or Cr^{3+} will give a well-defined angular momentum. When a magnetic field is applied to such a system, the normally degenerate energy levels are lifted so that the new energy level E is given by

$$E = E_0 + m_s g \beta B \tag{4-7}$$

where E_0 is the degenerate energy level, m_s is the spin quantum number, g is a constant for each paramagnetic metal or free radical (g is taken as 2 for Fe^{3+} and Cr^{3+}), β is the Bohr magneton (9.2732×10^{-28}), and B is the applied magnetic field. Now if electromagnetic radiation in the form microwaves, $h\nu$, is applied to a metal or free radical with unpaired electron(s) which is simultaneously being subjected to an external magnetic field B, then whenever $h\nu = g\beta B$, an electron can be promoted from one level to another, resulting in energy absorption. This is termed *resonance*, and it registers as a *signal* in the EPR spectrograph. The position of the signal is characterized by the so-called g' value, defined as

$$g' = \frac{h\nu}{\beta B_{obs}} \tag{4-8}$$

The most typical g' value for the iron–transferrin complex is at approximately 4.3, and it is 1.98 for chromium. The EPR spectra obtained for iron can be fitted to the spin Hamiltonian, which combines the Zeeman energy term ($g\beta BS$) with two distortion factors, as given by

$$H = g\beta BS + D[S_z^2 - \tfrac{1}{3}S(S + 1)] + E(S_x^2 - S_y^2) \tag{4-9}$$

where S is the spin quantum number ($\frac{5}{2}$ for Fe^{3+} and $\frac{3}{2}$ for Cr^{3+}), D is the axially symmetric component, and E "reflects any rhombic departure from axial symmetry." D and E are also termed *zero-field splitting parameters* (Aisen *et al.*, 1969; Pinkowitz and Aisen, 1972). The ratio of E/D, termed λ, determines the ligand field of the paramagnetic atom. For both iron and chromium it is near one-third, implying a rhombic ligand field for both. Typical EPR spectra for three transferrins are shown in Figure 4-8, which indicates some difference between conalbumin and human serotransferrin.

Let us now return to the question of whether or not the two metal-binding sites of the transferrins are identical. The first spectroscopic work showing dissimilarity between the two sites was provided by Aisen *et al.* (1969), who determined a then-novel spectrum of the chromium–serotransferrin complex. They then proceeded to displace the chromium with cobalt and iron over a period of several days. As the lines attributable to chromium disappeared, lines attributable to iron appeared. Intermediate species indicated that the chromium lines were superimposed on those of iron, indicating a nonrandom replacement. Price and Gibson (1972b) later used perchloric acid to perturb iron-binding sites of serotransferrin and conalbumin. It was found that the perturbant did not act uniformly on all the iron present, indicating differences in iron environments.

In a very clever investigation, Aisen *et al.* (1973b) assumed a *priori*

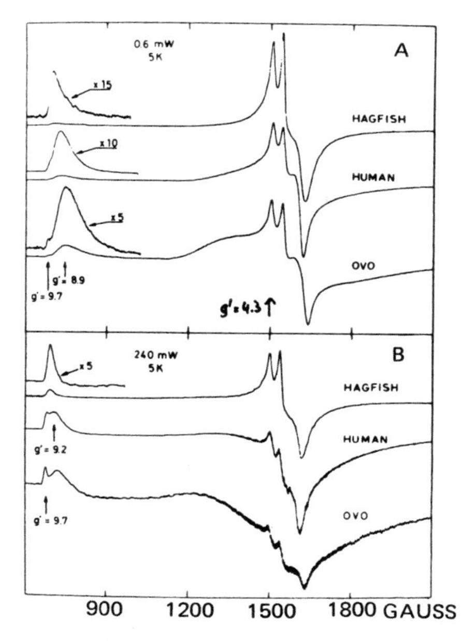

Figure 4-8. Electron paramagnetic resonance spectra of various transferrins saturated with iron. Temperature was 5°K, microwave power was 0.6 mW for group A and 240 mW for group B. From Aasa (1972).

that the two metal-binding sites of conalbumin were unequal and had different metal association constants. Following up on this assumption, gallium (a diamagnetic metal and therefore EPR silent) was first added to conalbumin to saturate one site, say site A, and then the paramagnetic iron was added to saturate the other site, site B. Conversely, iron was added first to saturate site A and then gallium to bind up with site B. EPR spectra of the two types of proteins were dissimilar, but when the spectra were superimposed through a computer simulation program, the result was identical to the spectrum of iron-saturated conalbumin. These relationships are shown in Figure 4-9. The authors concluded that conalbumin "does not bind metal ions in a simple random fashion." Differences in the two binding sites were also shown using vanadyl ions, which have the distinction of giving extremely sharp EPR spectra when combined with the transferrins (Cannon and Chasteen, 1975; Chasteen *et al.*, 1977).

Differences between the two iron-binding sites of the transferrins can be demonstrated via iron binding as a function of pH (Princiotto and Zapolski, 1975; Lestas, 1976). A fully iron-saturated transferrin does not lose its iron in a continuous manner as the pH is dropped. Instead, the

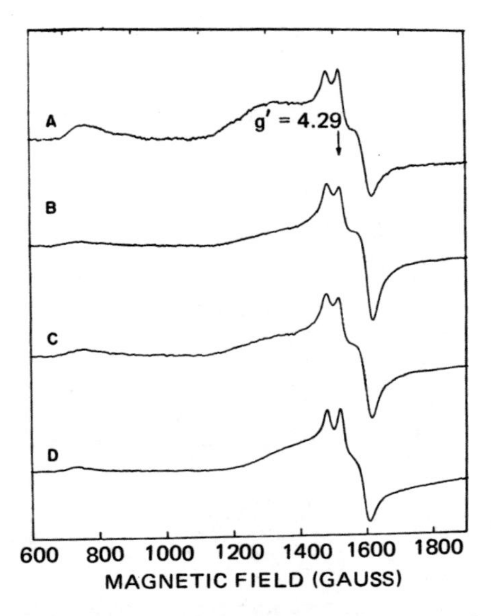

Figure 4-9. Electron paramagnetic spectra indicating nonidentity of the two iron-binding sites of conalbumin. A is conalbumin with one site occupied by iron, B is that with the other site occupied by iron. Note differences in patterns around 1300G ($g' = 4.98$). C is a computer-simulated summation of patterns A and B, and D is iron-saturated conalbumin. From Aisen *et al.* (1973b).

loss is biphasic, where one iron-binding site tends to lose iron between pH 6 and 7, and the other retains iron well below pH 6.0 (Figure 4-10). In addition, if serotransferrin is partially saturated with the iron–nitriloacetate complex, iron is apparently bound preferentially to that site, which retains the iron at the lower pH. The latter site has been designated as site A, whereas the site that loses its iron at pH 6–7 is designated as site B. It is thus possible to label selectively site A with ^{59}Fe if such labeling is done at pH 5–6 (Harris, 1977). Site B can then be labeled with ^{55}Fe or another metal isotope. Moreover, addition of citrate to such selectively labeled transferrin preparations does not lead to a rapid randomization of the label. Selectivity is retained for at least 80 min at 37°. The pH effects are observed with the binding of other metal ions to serotransferrin, as with, e.g., chromium, copper, and vanadium. EPR spectra of selectively labeled serotransferrins utilizing the pH phenomenon in their preparation have been determined, and the two sites have been found to give different EPR spectra (Harris, 1977; Zweier, 1978).

The work of Harris on serotransferrin was extended to conalbumin

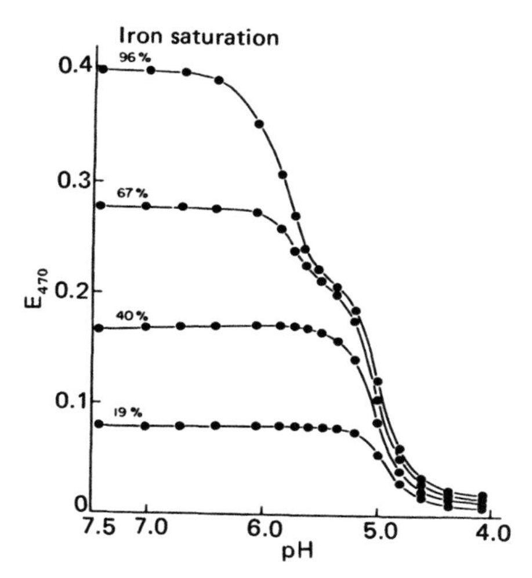

Figure 4-10. Binding of iron by serotransferrin as a function of pH. The amount of iron bound by the protein is assayed by absorption at 470 nm. From Lestas, A. N., *Brit. J. Haematol.* 32:341–350, 1976, by permission of the Blackwell Scientific Publications, Ltd.

by Williams *et al.* (1978), who showed that at alkaline pH values (around pH 8) the iron–nitriloacetate complex in the presence of bicarbonate–carbonate is taken up preferentially by the N-terminal portion of the protein molecule. This was shown at total iron levels of 0.21 to about 1.8 atoms of iron per each conalbumin molecule. When the iron was added at pH 6.0, there was a preferential binding by the C-terminal portion of the protein molecule. These findings were obtained by adding a limited amount of ^{59}Fe–nitriloacetate to conalbumin and then saturating it with ^{55}Fe–nitriloacetate. The protein was then counted to determine the ^{59}Fe–^{55}Fe ratio; then the protein was digested to isolate the C-terminal fragment (see Section 4.9.4 for methodology) and counted. These results were also confirmed by an electrophoresis method developed by Makey and Seal (1976), who discovered that a serotransferrin sample partially saturated with iron (e.g., 60%) can be separated into four bands by electrophoresis in a 6% polyacrylamide gel slab using a Tris–borate–EDTA buffer at pH 8.4 and containing 6 *M* urea. The slowest band corresponded to aposerotransferrin, the fastest to iron-saturated serotransferrin, and the intermediate two bands to serotransferrins saturated at either of the two sites. At a total iron saturation level of 30%, only the slowest three bands were observed. In case of conalbumin, Williams *et al.* (1978) showed that the slowest band was apoconalbumin, the next in mobility was conalbumin with iron occupying only the C-terminal site, the next was conalbumin with iron occupying the N-terminal site, whereas the most rapidly migrating species was the iron-saturated conalbumin.

Milne (1978) developed equations applicable to the binding of iron by the transferrins. If we define p as the probability of iron binding to the N-terminal site, q as that for the C-terminal site ($q = 1 - p$), C as the total amount of iron bound, and x as the N-terminal site fractional saturation, then it is possible to write the following relationship:

$$\frac{q}{p} \ln(1 - x) = \ln(1 + x - C) \qquad (4\text{-}10)$$

If one plots $\ln(1 + x - C)$ against $\ln(1 - x)$ and if there is no interaction between the two sites, a straight line whose slope is q/p should be obtained. In fact, such plots have not been straight lines in situations where the N-terminal site is saturated first (at slightly alkaline pH), indicating an intersite cooperativity. Straight-line plots are obtained if saturation of the transferrin is carried out at pH 6.0 (C-terminal site preferentially) (Williams *et al.*, 1978).

The procedure of Makey and Seal (1976) was also used by Evans and Williams (1978) to study the order of increasing mobilities of the four serotransferrin species. As was the case with conalbumin, in the order of increasing mobilities, the four serotransferrin forms were aposerotransferrin, monoferric serotransferrin with the C-terminal site occupied, monoferric serotransferrin with the N-terminal site occupied, and last the iron-saturated serotransferrin. Likewise, at pH 6 and neutrality, the C-terminal site had a greater avidity for iron than did the N-terminal site. At pH 8.5 the reverse was true. Evans and Williams (1978) identified the C-terminal iron-binding site as site A of Harris (1977), and the N-terminal site was identified as site B. It has been proposed by Frieden and Aisen (1979) that the site A and B designations be abolished and substituted by the C-terminal and N-terminal designations, respectively. Other methods for the separation of transferrins labeled with iron at the N-terminal or the C-terminal site include isoelectric focusing (Van Eijk *et al.*, 1978) and chromatography on DEAE–cellulose (Lane, 1975).

Differences in the two iron-binding sites can also be demonstrated calorimetrically when measuring the enthalpies of denaturation. In conalbumin partially saturated with iron, as the temperature is raised, it is possible to observe four endotherms, presumably corresponding to the denaturation of the iron-free species, each of the two monoferric forms, and the iron-saturated conalbumin. The temperatures at which the four endotherms were observed were 63° for apoconalbumin, 68° and 77° for the two monoferric forms, and 84° for the diferric protein (Donovan and Ross, 1975). Denaturation experiments on separated iron-saturated N-terminal and C-terminal halves of conalbumin (see Section 4.9.4) revealed isotherms at 77° and 86°, respectively. When the two apo fragments were mixed and iron added to saturate half the iron-binding sites present, the denaturation isotherm was located at 77°. This clearly indicates that at pH 7.5 iron is bound preferentially to the site located at the N-terminal region of conalbumin (Evans *et al.*, 1977).

The most convincing evidence for the unequal affinity for iron on part of the N- and C-terminal iron-binding sites of the transferrins would require the determination of their respective iron-binding association constants. This work was executed by Aisen *et al.* (1978), using a combination of equilibrium dialysis methodology and the electrophoretic method of Makey and Seal (1976). Equilibrium dialysis was performed at pH 6.0 and 7.4; then electrophoresis to separate the various serotransferrin forms was performed. By using equations derived for this purpose, association constants with respect to iron were calculated for both sites. If the Fe_a site of Aisen *et al.* (1978) is equated with the C-terminal site and the Fe_b site with the N-terminal site, then at both pH values, the C-terminal site binds iron more avidly than the N-terminal site if iron is offered as the nitriloacetate complex in the presence of carbonate. This was especially evident at pH 6.7, where only the C-terminal site was occupied at all iron concentrations tested. One can represent the pathways of the binding of iron by a transferrin by

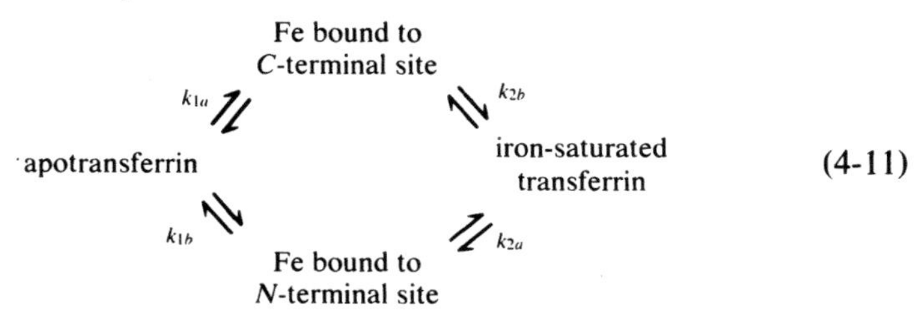

(4-11)

The intrinsic apparent association constants k_{1a}, k_{1b}, etc., are given in Table 4-7. It is seen from this table that not only does the C-terminal iron-binding site have a greater affinity for iron but that there exists a negative cooperativity between the two sites. This becomes clear when one compares k_{1a} with k_{2a} and k_{1b} with k_{2b}: When one site is occupied, the second site becomes occupied by a metal less efficiently than it would have been had the first site remained open. Such a cooperation has also been detected by Williams *et al.* (1978) for conalbumin. Since the two iron-binding sites are only 25 Å apart (Meares and Ledbetter, 1977), such an interaction is not surprising.

Table 4-7. Intrinsic Affinity Constants for the Two Iron-Binding Sites of Human Serotransferrin when Iron Is Offered as the Nitriloacetate Complex in the Presence of Carbonate[a]

pH	k_{1a}	k_{1b}	k_{2a}	k_{2b}
6.7	3×10^{19}	$\leq 1.4 \times 10^{18}$	$\leq 4.8 \times 10^{18}$	$\leq 2.4 \times 10^{17}$
7.4	4×10^{20}	6.8×10^{19}	1.4×10^{20}	2.8×10^{19}

[a] From Aisen *et al.* (1978).

In summary, then, most recent evidence indicates that the two iron-binding sites of the transferrins are neither equivalent nor independent. They can be distinguished by a number of physical and chemical methods, and their intrinsic association constants for iron are significantly different. The most convenient method for their designation appears to be in terms of the C-terminal (greater affinity for iron) and the N-terminal iron-binding sites.

4.6 Distribution of Iron in Human Serum Transferrin *in Vivo*

There are two reports on the physiological distribution of iron in serotransferrin, which is normally one-fourth to one-third saturated with iron. One investigation has shown that some 44% of serotransferrin normally exists in the apo state, that the N-terminal monoferric species accounts for 24% and the C-terminal species for 10%, and that 22% is accounted for by the iron-saturated form (Williams, 1979).

In a more complete report, where the *crossed* electrophoresis method was utilized as an assay procedure to separate the various forms of serotransferrin, the results were somewhat different. They are illustrated in Table 4-8, the main feature of which is that the most abundant serotransferrins were the monoferric species where iron occupied the N-terminal site and aposerotransferrin. These were present in approximately equal amounts when the iron saturation level of serotransferrin was one-fourth to one-third of saturation (Leibman and Aisen, 1979). The distribution of iron in serotransferrin under physiological conditions is thus not random, and the amount of iron present in the C-terminal iron-binding site is not very large in spite of its higher association constant for iron compared to the N-terminal binding site. Clearly, in an *in vivo* situation, there is selective iron uptake by the N-terminal iron-binding site, which in turn may reflect the iron-donating properties of the reticuloendothelial system. As pointed out previously (Chapter 1), the RES is the major source of circulating iron.

4.7 Some Gross Differences between Iron-Free and Iron-Saturated Transferrins

When the apo forms of the transferrins become saturated with iron, their physical and chemical properties undergo certain alterations. Among these are the immunochemical peculiarities of the two proteins. It was shown that antibodies to apoconalbumin will form lighter precipitates with iron-saturated conalbumin than with the homologous protein (Tengerdy *et al.*, 1966) and that in the Ouchterlony system, aposerotransferrin and iron-saturated serotransferrin give lines indicating only partial immunochemical identity (Kourilsky and Burtin, 1968).

In a previous section of this chapter it was pointed out that the binding of each iron atom by a transferrin is accompanied by the release of three protons, but because in addition to each ferric iron atom bound a carbonate ion is also bound, there is an increase in the net negative charge of the protein. The protein thus becomes more acidic and is readily distinguished from its apo counterpart by electrophoresis. When subjected to chromatography on DEAE–cellulose, one would expect the diferric transferrin to be bound more firmly with the ion exchanger. However, the opposite turns out to be true. When partially iron-saturated serotransferrin is subjected to chromatography on DEAE–cellulose, iron-saturated serotransferrin is eluted first, followed by serotransferrin containing a single iron atom per molecule, and aposerotransferrin is eluted last (Lane, 1971). The explanation for this bizarre behavior is that there is a surface charge redistribution as iron is bound to aposerotransferrin, which is apparently also very important in the process of serotransferrin interaction with immature red cells (see Chapter 6).

That the surface of the iron-saturated transferrin is indeed different from that of an apotransferrin has been shown by the methods of hydro-

Table 4-8. Serotransferrin Forms in Normal Human Serum[a]

				% distribution			
Subject	TI (μg/100 ml)	TIBC (μg/100 ml)	% Saturation with iron	Apotransferrin	C-terminal site occupied	N-terminal site occupied	Diferric
1	86	334	26	41	9	41	9
2	60	290	21	34	2	56	10
3	48	310	16	52	2	46	2
4	143	322	44	16	6	47	31
5	111	336	33	45	13	34	8
6	110	322	34	34	13	45	8

[a] From Leibman and Aisen (1979).

gen–tritium exchange and analysis for buried amino acid side chains. Hydrogen is exchanged with the medium more slowly in the iron-saturated serotransferrin and conalbumin than with the apoproteins. Divalent metal ions had a less drastic effect on this property than did the trivalent metals (Ulmer, 1969). Chemical modification procedures have also shown that iron-saturated serotransferrin contains 8 phenolic and 11 amino groups that are unreactive and may therefore be considered as "buried." The corresponding numbers for aposerotransferrin are 1 and 2, respectively (Bezkorovainy and Zschocke, 1974). These data suggest that the iron-saturated serotransferrin is more compact than is aposerotransferrin.

The change in shape accompanying the binding of iron and other metals by the transferrins has also been evaluated by hydrodynamic methods. One of the first studies comparing iron-free and iron-saturated serotransferrins hydrodynamically was done by Bezkorovainy, who found that the sedimentation constants were 5.31 and 5.38 S, respectively, and that the diffusion constants were 5.54×10^{-7} and 5.75×10^{-7} cm^2/sec, respectively, in cacodylate buffer at pH 7.0. It was concluded that the axial ratios were 1:2 and 1:3 for the iron-free and iron-saturated serotransferrins, respectively (Bezkorovainy, 1966). Careful differential sedimentation velocity studies on the two proteins were subsequently carried out by Charlwood (1971), who found sedimentation constants of 5.09 S for aposerotransferrin and 5.16 for iron-saturated serotransferrin in a Tris–HCl buffer at pH 7.2. These findings were confirmed for a number of other transferrins by Jarritt (1976), who also determined differences in Stokes' radii by the gel filtration method. Changes in sedimentation constants and Stokes' radii were similar, though opposite in sign, indicating that these differences were due to conformational changes in protein structure rather than aggregation.

There are a number of other physical–chemical properties of the transferrins that change upon the binding of iron by these proteins. Many have been mentioned in the previous sections of this chapter; others, such as changes in the ultraviolet absorption spectra and the "extrinsic" portion of optical rotatory dispersion and circular dichroism spectra, can be traced directly to the binding of iron by specific amino acid side chains. These effects are described in greater detail in Section 4.8.

4.8 The Iron-Binding Ligands of the Transferrins

Various physical–chemical properties of the transferrins had indicated as long ago as the 1940s that these proteins do not contain heme. Because of the peculiarities of the spectrum of iron-saturated serotransferrin and conalbumin, it was originally proposed that the binding of iron was similar to that found in the hydroxamate-like siderophores (see Chapter 7) (Fiala and Burk, 1949). It was soon discovered, however, that upon the binding of iron by either conalbumin or serotransferrin, three protons per iron atom and two protons per copper atom bound were released into the medium and that these hydrogen ions originated from amino acid side chains with a p*K* value greater than 10. This, of course, pointed to tyrosine as an iron-binding ligand in the transferrins (Warner and Weber, 1953). Wishnia *et al.* (1961) later found that iron-saturated conalbumin had six fewer titrable phenolic groups than did apoconalbumin, thus "...establishing the nature and number of the chelating groups beyond question as three phenolate residues per ferric ion."

Following this classical work, investigations of the nature and number of the iron-binding ligands centered around various spectroscopic techniques, especially the EPR and difference spectroscopy methods, chemical modification procedures, and optical rotation effects due to restricted rotation of the amino acid side chains. It should be remembered that ferric iron has an octahedral ligand field with six ligands in the transferrins and that carbonate occupies one of the ligand positions (see Section 4.5.2). The earliest attempts to define iron-binding ligands of the transferrins using EPR spectroscopy were performed independently by Aasa *et al.* (1963) and Windle *et al.* (1963). Both groups noted that EPR spectra of the transferrin–copper complexes gave hyperfine structures in the $g \perp$ peak which were characteristic of copper–nitrogen model compounds. They postulated that each iron and copper atom is coordinated with two nitrogen residues, since both metals can occupy the same binding site. Aasa *et al.* (1963) pointed out that these nitrogen atoms must belong to imidazoyl residues, because these are the only residues that can exist in the unprotonated state at pH 7.5, the pH of their experiments. Both groups also proposed that each iron atom is also coordinated with three tyrosyl residues. Windle *et al.* (1963) saw no difference in the EPR spectra of conalbumin, serotransferrin, and lactoferrin. By 1963, therefore, on the basis of titration and EPR spectroscopy evidence, it was surmised that each ferric atom in the transferrins was coordinated with three tyrosyl and two histidyl residues and one carbonate residue. This is illustrated in Figure 4-11.

Subsequent spectroscopic studies have attempted to refine the model shown in Figure 4-11 and to provide further evidence for the participation of various ligands in the binding of metals by the transferrins. The participation of the tyrosine residues was confirmed by resonance Raman spectroscopy (Tomimatsu *et al.*, 1976; Carey and Young, 1974), proton magnetic resonance (Woodworth *et al.*, 1970), ultraviolet difference spec-

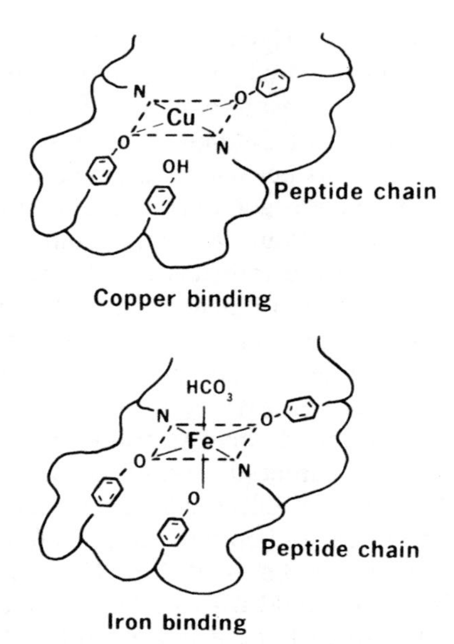

Figure 4-11. Model for the binding of iron and copper by the transferrins, as proposed by Windle and co-workers. The Ns represent histidyl residues. The iron complex is unsymmetric and in a high-spin state. An octahedral model best fits the data available. Copper apparently forms a square planar complex with the transferrins and exists in a symmetrical low-spin state. Reprinted with permission from Windle, J. J., Wiersema, A. K., Clark, J. R., and Feeney, R. E., 1963, Investigation of the iron and copper complexes of avian conalbumins and human transferrins by electron paramagnetic resonance, *Biochemistry* 2:1341–1345. Copyright by the American Chemical Society.

troscopy (Tan and Woodworth, 1969; Tomimatsu and Donovan, 1976; Krysteva *et al.*, 1976), and circular dichroism measurements (Nagy and Lehrer, 1972; Mazurier *et al.*, 1976; Tomimatsu and Vickery, 1972). In addition, such ultraviolet spectroscopy and circular dichroism measurements as well as fluorescence spectroscopy measurements have implicated tryptophan residues in the iron-binding process (Lehrer, 1969). It is unclear whether the tryptophanyl residues are perturbed because of conformational changes in the protein molecule or because they are directly involved as metal-binding ligands. The latter possibility is not necessarily favored by most workers in the field (Tomimatsu and Donovan, 1976; Lehrer, 1969; Brown and Parry, 1974). Figure 4-5 shows the CD spectra in the 250–350-nm region of serotransferrin and lactoferrin and their iron-saturated counterparts. As can be seen, considerable differences exist between the apoproteins and the iron-saturated species. The effects around 250 and 320 nm are attributable to some disulfide groupings and the fact that their dihedral angle may change upon the binding of iron by the transferrins. The bands in the 290–300-nm region are attributable to tryptophan perturbations and those in the 275–290-nm region to tyrosyl groups. These spectra do not tell, however, which of the perturbed side chains are directly involved in metal coordination. Recent EPR spectroscopy experiments have rather conclusively demonstrated the involvement of histidyl residues in the binding of copper by serotransferrin (Zweier *et al.*, 1979).

There is one report that claims to have resolved the EPR signals of the two metal-binding sites of serotransferrin and conalbumin (Mazurier *et al.*, 1977). The two sites appear to be quite different, as can be seen from Table 4-9, and, moreover, it was claimed that sites A of serotransferrin and lactoferrin were very similar chemically, whereas site B differed in regard to their response to chemical modification as well as in regard to their EPR signals. This work is clearly only the forerunner of what it is essential to accomplish: the complete elucidation of the chemical structure of the iron-binding sites of serotransferrin, conalbumin, and lactoferrin.

Although the various forms of spectroscopic methodology have dominated the thinking and approaches toward the elucidation of metal-binding ligands in the transferrins, it should be recognized that such methodology, in spite of its many advantages from the protein chemist's point of view, has often failed to provide us with definite answers in regard to the identity and quantity of the iron-binding ligands in transferrins. It is therefore of value to take a look at another approach, the concept of chemical modification of specific amino acid residues, which strives to chemically alter specific amino acid side chains and correlate such modification with a loss (or lack) of biological activity such as binding of metal ions. The spectroscopists have severely criticized such an approach, allegedly because a loss of biological activity following chemical modification of certain amino acid side chains may have nothing to do with the function of the amino acid side chain in question but may, instead, reflect a subtle change in the conformation of the protein (e.g., Woodworth *et al.*, 1970; Aisen and Brown, 1975). This criticism is of course quite valid if the chemical modification studies were not carefully monitored for changes in the secondary and tertiary structures of the proteins. Such monitoring may be performed by the measurement of, for instance, the CD spectra of native and modified proteins. When such precautions are taken, interpretation of chemical modification data on transferrins and other biologically active proteins becomes less ambiguous, and valuable information could be gathered therefrom that could then complement spectroscopic data.

Table 4-9. Metal-Binding Ligands of Various Transferrins as Revealed by Spectroscopic and Chemical Modification Methodologies (It is understood that one molecule of carbonate is bound for each metal atom.) (in ligands per metal atom)

Protein	Metal bound	Methodology	No. of histidyl groups	No. of tyrosyl groups	Reference
Serotransferrin	Cu	EPR, titration	2	2	Aasa *et al.*, 1963
	Fe	EPR, titration	2	3	Aasa *et al.*, 1963
Serotransferrin, lactoferrin, conalbumin	Fe	EPR, titration	2	3	Windle *et al.*, 1963
	Cu	EPR, titration	2	2	Windle *et al.*, 1963
Serotransferrin	Cu	EPR	—	2	Zweier and Aisen, 1977[a]
Conalbumin	Divalent	Ultraviolet difference spectroscopy	—	1	Tan and Woodworth, 1969
	Trivalent	Ultraviolet difference spectroscopy	—	2	Tan and Woodworth, 1969
Serotransferrin					
Site A	Cu	EPR	1	—	Mazurier *et al.*, 1977
Site B	Cu	EPR	3–4	—	Mazurier *et al.*, 1977
Serotransferrin, lactoferrin	Fe	Spectrophotometric titration	—	3	Teuwissen *et al.*, 1972
Serotransferrin	Fe	Tetranitromethane, bromoacetate	2	3	Line *et al.*, 1967
Serotransferrin, conalbumin	Fe	*N*-acetylimidazole	—	3	Komatsu and Feeney, 1967
Conalbumin	Fe	Iodination	—	2	Phillips and Azari, 1972
Serotransferrin, conalbumin	Fe	Ethoxyformylation, photooxidation	2	—	Rogers *et al.*, 1977
Serotransferrin	Fe	Diethylpyrocarbonate	1.5	—	Krysteva *et al.*, 1975
Lactoferrin	Fe	Diethylpyrocarbonate	0.5	—	Krysteva *et al.*, 1975

[a] One water molecule was an additional ligand.

The advantages of the chemical modification approach include such capability as focusing upon specific amino acid side chains. EPR methodology, for instance, has been unable until recently (Zweier *et al.*, 1979) to be specific as to which particular side chains provided the nitrogen atoms associated with metals. Another advantage of the chemical modification technique is that it can provide quantitative information on how many amino acid side chains are involved in the binding of metals. Such information is more often than not unavailable from spectroscopic data. Ideally, when chemical modification of a transferrin is performed, one would monitor, in addition to any changes in the secondary and tertiary structures of protein, either the amount or iron remaining bound or the amount of iron that the modified protein is capable of binding, depending on whether iron-saturated or iron-free transferrin is used. In any case, one should also monitor the specificity of metal binding through visible-range absorption spectroscopy (maximum at 460–470) and/or EPR spectroscopy. Very often the chemically modified transferrin may continue to bind the appropriate amount of iron, but its spectrum changes from the normal 460 nm to perhaps 430 nm. This clearly indicates that the active site of the transferrin has been somehow damaged, and interpretation of results then becomes quite difficult.

The involvement of tyrosine residues in the binding of iron by the transferrins was investigated with tetranitromethane, which nitrates the phenolic ring structure (Line *et al.*, 1967; Tsao *et al.*, 1974c; Teuwissen *et al.*, 1973); by iodination (Phillips and Azari, 1972); and by acetylation of the hydroxyl group with *N*-acetylimidazole (Komatsu and Feeney, 1967). In all cases, modification of the tyrosyl residues abolished the capacity of the proteins to bind iron, though modification of the iron-saturated protein did not result in a loss of iron. Iron therefore served to protect certain tyrosyl residues against modification, and this amounted to some two to three tyrosyl groups per iron atom bound (see Table 4-9). Iodination of serotransferrin also suggested the involvement of tryptophanyl groups in the binding of iron (Phillips and Azari, 1972), though this could not be confirmed through modification with the very specific reagents 2-hydroxy-5-nitrobenzyl bromide, 2-nitrophenyl sulphenyl chloride, and dimethyl-(2-hydroxy-5-nitrobenzyl)-sulfonium bromide (Ford-Hutchinson and Perkins, 1972).

Amino groups of the transferrins were modified by a variety of reagents including succinic anhydride, trinitrobenzene sulfonic acid, and reductive methylation (Zschocke *et al.*, 1972) and with acetic anhydride and cyanate (Buttkus *et al.*, 1965), but no involvement of these groups in the binding of iron could be shown. The same was true for carboxyl groups, which were modified with water-soluble carbodiimides (Bezkorovainy and Grohlich, 1970). Histidyl residues of serotransferrin were first modified by Line *et al.* (1967) using bromoacetic acid at near neutrality. It was found that aposerotransferrin lost its ability to combine with iron as a result of this, and in a later communication it was reported that iron-saturated serotransferrin lost its iron as a result of the reaction (Bezkorovainy and Grohlich, 1971) (Figure 4-12). It was concluded that histidyl residues were essential for the binding of iron by serotransferrin. Rogers *et al.* (1977) came to the same conclusion by using ethoxyformic anhydride and by photooxidizing the histidyl residues in both conalbumin and serotransferrin. Krysteva *et al.* (1975) found somewhat puzzling results when diethylpyrocarbonate was used as the histidyl residue modifying reagent. Each iron atom in serotransferrin was associated with 1.5 histidyl residues, whereas in lactoferrin this figure was only 0.5. The involvement of histidyl residues in the iron-binding phenomenon of lactoferrin clearly requires further work.

Some isolated attempts have been made to identify the iron-binding site of transferrins using chemical modification procedures. When iron-saturated serotransferrin is succinylated, certain sections of the protein, presumably at or near the iron-binding sites, are protected against succinylation. When iron is removed from such modified protein and antibodies to it made in rabbits, the antiserum contains antibodies to succinyl groups (which can be removed by any fully succinylated protein) and antibodies specific for the active sites (Zschocke and Bezkorovainy,

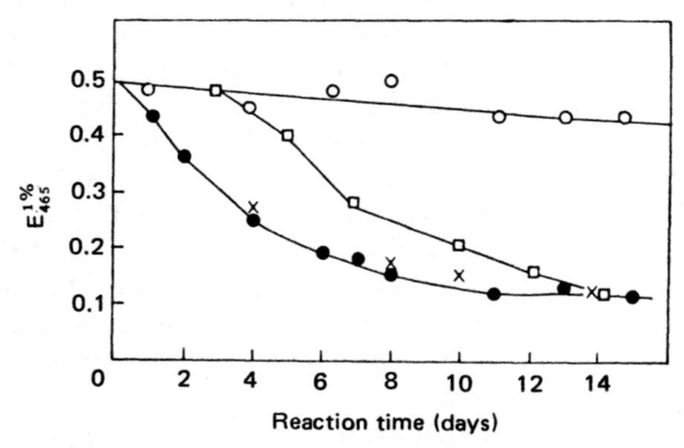

Figure 4-12. The effect of bromoacetic acid on the iron-binding properties of human serotransferrin. Open circles, the iron-saturated protein without the reagent; filled circles, iron-saturated protein in the presence of the reagent; crosses, the modified protein with residual iron removed, and then reconstituted with the protein; squares, iron-free serotransferrin to which iron was added following the reaction with the reagent. From Bezkorovainy and Grohlich (1971).

1970). Such an antiserum clearly has the potential for identifying the iron-binding portions of serotransferrin when and if such can be prepared by, e.g., proteolytic digestion.

We may conclude this section by stating that the exact nature and number of iron-binding ligands at the two iron-binding sites of the various transferrins are still unclear. Not only is it possible that such ligands are different for the two sites of the same protein molecule, but we may also have qualitative and quantitative differences among the metal-binding sites of serotransferrin, conalbumin, and lactoferrin. Unambiguous answers to these questions can only be provided through the determination of the complete primary structure of the transferrins and the identification of the iron-binding segments therein.

4.9 Primary Structure of the Transferrins

The composition of the various transferrins has been studied by numerous authors, and the data are summarized in Tables 4-10 to 4-13. It is seen that all the transferrins contain a full complement of amino acids as well as carbohydrate. They are therefore glycoproteins. It is now agreed that valine, alanine, and glycine are the N terminals of human serotransferrin, hen's egg conalbumin, and human lactoferrin, respectively, and that proline is the C-terminal group of human serotransferrin.

4.9.1 Structure and Significance of the Carbohydrate Moiety of the Transferrins

The physiological significance of the carbohydrate components of the transferrins is at present unclear. Carbohydrate is apparently not required for either the binding of iron or the delivery of iron to immature red cells by serotransferrin (Kornfeld, 1968). The primary purpose of the carbohydrate, like that of other glycoproteins, may be in cell recognition processes and/or in the clearance mechanism. Nevertheless, carbohydrate content provides us with the means of recognizing and distinguishing different types of transferrins, such as hen's egg conalbumin and chicken serotransferrin. Table 4-14 summarizes some of the more careful studies on carbohydrate composition of various transferrins.

Jamieson (1965) was the first investigator to report information on the status of the carbohydrate moiety of human serotransferrin. He digested the protein with pronase and isolated the carbohydrate-containing fragments. Their molecular weights ranged from 3100 to 3800, with car-

Table 4-10. Amino Acid and Carbohydrate Compositions of Human Serotransferrin Preparations (in residues per molecule of protein)

	Reference[a]						
Component	1	2	3	4	5	6	7
Lys	72	68	60	51	49	52	66
Arg	27	30	27	24	23	23	29
His	23	22	20	19	17	17	18
Asx	72	91	79	81	71	75	84
Thr	32	34	28	30	25	28	30
Ser	42	43	39	40	35	37	40
Glx	66	66	63	58	53	54	63
Pro	28	39	31	30	36	37	32
Gly	53	58	48	54	46	46	52
Ala	60	66	56	59	51	56	59
Val	48	49	43	43	40	39	39
Met	9	9	8	4	7	8	9
Ile	15	17	15	14	14	13	15
Leu	65	70	57	57	52	55	64
Tyr	24	29	22	24	24	27	26
Phe	32	33	28	28	27	30	30
½-Cys	43	40	39	35	30	35	36
Trp	11	8	—	9	—	—	—
GlcNAc	—	8	8	—	—	8	—
Hexose	10	12	12	—	—	10	—
Sialic acid	4	4	4	—	—	4	—
Fucose	—	—	6	—	—	—	—
N-terminus	—	—	—	—	Val	Val	Val
C-terminus	—	—	—	—	—	Cys	Pro

[a] References and notes:

1. Frenoy *et al.* (1971), molecular weight 90,000.
2. Roop and Putnam (1967), molecular weight 90,000.
3. Van Eijk *et al.* (1972), molecular weight 80,000.
4. Mann *et al.* (1970), molecular weight 76,600.
5. Sutton and Brew (1974b), molecular weight 76,600.
6. Montreuil and Spik (1975), molecular weight 75,200; galactose-4 residues, mannose-6 residues.
7. Bezkorovainy *et al.* (1968) and Sly and Bezkorovainy (1974), molecular weight 82,000.

bohydrate accounting for two-thirds of that. Asparagine was determined to be the amino acid with which the carbohydrate was linked. Jamieson concluded that serotransferrin contained two oligosaccharide chains, each with a molecular weight of 2200–2400 daltons.

In a subsequent publication Jamieson *et al.* (1971) reported the complete structure of the carbohydrate chains, which they determined to be identical. The purification process that they used was able to remove all amino acids but asparagine from the oligosaccharide fragment preparations. The structure, shown in equation (4-12), provides for 4 moles of sialic acid, 4 moles of galactose, 8 moles of *N*-acetylglucosamine, and 8

Table 4-11. Composition of Various Conalbumin Preparations Isolated from Hen's Egg White (in residues per molecule of protein)

Component	Reference[a]				
	1	2	3	4	5
Lys	63	56	49	63	63
His	11	12	14	13	11
Arg	29	28	34	33	32
Asx	68	72	68	75	76
Thr	31	33	28	35	36
Ser	39	40	30	43	44
Glx	61	67	65	71	66
Pro	22	26	30	29	29
Gly	43	52	49	59	51
Ala	52	55	55	53	53
Val	43	52	52	45	44
Met	9	8	10	12	10
Ile	19	24	30	25	23
Leu	39	49	50	49	48
Tyr	17	20	21	21	20
Phe	21	26	29	26	26
½-Cys	19	31	41	21	28
Trp	15	12	—	18	—
Hexosamine	—	6	—	5	—
Hexose	—	4	—	4	—
Sialic acid	—	—	—	0	—
N-terminus	—	—	Ala	Ala	—

[a] References and notes:
1. Azari and Baugh (1967), molecular weight 76,600.
2. Tsao *et al.* (1974c), molecular weight 76,600.
3. Williams (1974), molecular weight 77,000.
4. Williams (1962), molecular weight 77,000.
5. Bezkorovainy *et al.* (1968), molecular weight 76,600.

Table 4-12. Compositions of Serotransferrins from Various Sources (in residues per molecule)

	Species and reference[a]																
	Rabbit				Bovine		Sheep		Rat		Tench,	Shark,	Frog,	Turtle,	Baboon,	Horse,	Swine,
Component	1	2	3	4	5	6	7	8	9	4	4	12	10	10	11	13	14
Lys	46	54	52	61	56	47	62	50	50	57	61	60	72	60	53	44	41
His	19	18	16	20	17	12	17	19	15	22	18	18	16	16	19	14	12
Arg	25	26	24	29	23	15	22	23	23	33	29	26	20	28	25	23	23
Asx	66	70	70	85	81	74	92	81	65	94	85	70	68	81	78	73	78
Thr	33	22	20	32	32	26	38	35	34	44	34	33	37	54	29	32	25
Ser	40	39	34	59	38	33	43	41	39	57	55	40	39	67	42	39	35
Glx	60	61	61	74	59	50	65	62	54	51	79	60	65	82	61	60	59
Pro	36	35	33	36	28	21	29	29	33	39	34	30	37	34	31	42	36
Gly	46	46	45	56	44	41	57	52	53	71	62	40	49	67	45	42	35
Ala	48	50	52	62	49	45	61	57	51	65	58	54	55	61	59	50	48
Val	47	37	42	50	35	30	43	31	36	63	44	33	37	57	38	44	41
Met	5	5	6	7	10	7	7	6	4	5	10	6	8	8	9	4	5
Ile	17	14	15	9	21	14	19	16	19	26	32	34	23	31	12	16	17
Leu	50	55	58	70	47	41	54	47	55	70	49	60	50	62	60	54	53
Tyr	22	21	23	22	20	15	24	23	19	22	36	25	21	24	20	21	16

Phe	25	25	25	32	27	22	29	29	31	39	29	30	27	31	27	20	23
½-Cys	34	34	32	33	34	40	36	31	15	40	32	8	25	18	38	33	33
Trp	9	9	9	—	—	8	8	—	8	—	—	—	5	9	—	9	9
GlcNAc	5–6	8	—	6	—	4	—	—	3	5	5	8	—	—	9	9	14
Hexose	9	9	—	6	—	7	—	—	—	7	10	18	—	—	13	10	17
Sialic acid	4	4	—	4	—	4	—	—	—	1	2	0	2	4	3	4	7–8
Fucose	0	—	—	1	—	—	—	—	—	1	1	4	—	—	—	0	3–4
N-terminus	—	—	—	—	—	Asp	Ser	—	—	—	—	Ala, Val	—	—	Val	—	—
C-terminus	—	—	—	—	—	—	—	—	—	—	—	Val	—	—	—	—	—

[a] References and notes:

1. Hudson *et al.* (1973), molecular weight 77,000; galactose-4, mannose-5.
2. Strickland and Hudson (1978), molecular weight 77,000; galactose-4, mannose-5.
3. Baker *et al.* (1968), molecular weight 70,000.
4. Van Eijk *et al.* (1972); molecular weights assumed were 80,000 for rat, rabbit, and tench (fish) serotransferrins.
5. Richardson *et al.* (1973), molecular weight 77,500.
6. Brock *et al.* (1978), molecular weight 74,000.
7. Guerin *et al.* (1976), molecular weight assumed 77,500.
8. Spooner *et al.* (1975), molecular weight 77,000.
9. Gordon and Louis (1963), molecular weight 67,000.
10. Palmour and Sutton (1971); molecular weights used were 92,000 for turtle and 76,000 for frog serotransferrins.
11. Bezkorovainy and Grohlich (1974), molecular weight 77,000.
12. Got *et al.* (1967), molecular weight 77,500.
13. Hudson *et al.* (1973), molecular weight 77,000; galactose-4, mannose-6.
14. Hudson *et al.* (1973), molecular weight 77,000; galactose-8, mannose-9.

Table 4-13. Composition of Various Lactoferrins (residues per molecule)

Component	Species and reference[a]								
	Human, 1	Human, 2	Human, 3	Human, 4	Bovine, 5	Bovine, 6	Bovine, 7	Bovine, 8	Murine, 9
Lys	39	56	41	41	49	42	60	47	55
His	10	12	10	10	10	10	11	9	10
Arg	38	53	41	40	36	32	39	32	39
Asx	65	81	64	71	64	71	68	62	61
Thr	33	35	28	30	34	39	35	35	40
Ser	46	52	42	54	40	45	44	45	50
Glx	69	88	64	62	66	73	74	65	71
Pro	33	42	31	26	32	31	35	27	33
Gly	49	66	49	53	48	43	50	45	51
Ala	56	72	61	59	64	59	66	62	60
Val	43	50	37	37	44	43	41	40	45
Met	5	6	4	6	5	4	3	4	5
Ile	16	20	14	14	16	17	14	15	17

Leu	54	66	54	54	62	61	64	62	57
Tyr	20	28	19	21	20	19	28	18	18
Phe	28	37	28	32	26	25	30	24	24
½-Cys	26	32	25	26	36	28	34	28	28
Trp	11	1	—	13	15	9	15	14	10
Hexosamine	8	—	—	8	12	10–11	—	9.5	5
Hexose	10	—	—	10	24	20–22	—	26	—
Sialic acid	3	—	—	3	1	1	—	9.5	—
Fucose	2	—	—	2	—	—	—	0.7	—
N-terminus	Gly	—	Gly	None	—	Ala	—	Ala	Lys
C-terminus	Arg	—	—	HSer	—	—	—	Thr	—

[a] References and notes:

1. Querinjean *et al*. (1971) and Bluard-Deconinck *et al*. (1978), molecular weight 76,400; mannose-6, galactose-4.
2. Blanc *et al*. (1963), molecular weight 88,000.
3. Bezkorovainy and Grohlich (1974), molecular weight 77,000.
4. Montreuil and Spik (1975), molecular weight 77,000; mannose-6, galactose-4.
5. Groves (1960) and Gordon *et al*. (1963), molecular weight 86,100.
6. Castellino *et al*. (1970), molecular weight 77,100; mannose-15 to -16, galactose-5 to -6.
7. Blanc *et al*. (1963), molecular weight 80,000.
8. Cheron *et al*. (1977), molecular weight 77,100; mannose-24, galactose-2, glucosamine-8.4, galactosamine-1.2.
9. Kinkade *et al*. (1976), molecular weight 78,000.

Table 4-14. Carbohydrate Content of Various Transferrins (moles of sugar per mole of protein)

Protein	Component							Reference
	Gal	Man	GlcNAc	GalNAc	Fucose	Sialic acid	Chains	
Human serotransferrin	4	8	8	0	—	4	2	Jamieson *et al.*, 1971
	4	6	8	—	—	4	2	Spik *et al.*, 1975
	6.0	6.2	8.1	—	—	3–4	2	Graham and Williams, 1975
Pig serotransferrin	4	6	8	—	1	2	1.23	Graham and Williams, 1975
Rabbit serotransferrin	4	6	8	—	—	4	2	Strickland and Hudson, 1978
	2	3	4	—	—	1–2	1	Leger *et al.*, 1978
Hen's egg conalbumin	—	4	8	—	—	—	1	Williams, 1968
Human lactoferrin	4	4	8	—	2	4	2	Spik and Mazurier, 1977
Chicken serotransferrin	2	2	3	—	—	1–2	1	Williams, 1968
Cattle serotransferrin	3.4–3.6	4.2–4.5	4.3–4.4	—	—	Present	2	Graham and Williams, 1975
Bovine lactoferrin	24	2	8.4	1.2	0.7	9.5	—	Cheron *et al.*, 1977

moles of mannose for each mole of serotransferrin, with a galactose to mannose ratio of 1 : 2.

$$
\begin{array}{l}
\text{sialic acid} \xrightarrow{2-6} \text{Gal} \xrightarrow{1-3(4)} \text{GlcNAc} \xrightarrow{1-3} \text{Man} \\
\qquad\qquad\qquad\qquad\qquad\qquad\qquad \downarrow 1-2(4) \\
\qquad\qquad\qquad\qquad\qquad\qquad\qquad \text{Man} \\
\qquad\qquad\qquad\qquad\qquad\qquad\qquad \downarrow 1-2(4) \\
\qquad\qquad\qquad\qquad\qquad\qquad\qquad \text{Man} \\
\qquad\qquad\qquad\qquad\qquad\qquad\qquad \downarrow 1-2(4,\ 6) \\
\qquad\qquad\qquad\qquad\qquad\qquad 1-3\ \ \text{Man} \xrightarrow[1-3(4)]{} \text{GlcNAc} \rightarrow \text{Asn} \\
\qquad\qquad\qquad\qquad\qquad\qquad\qquad \uparrow \\
\text{sialic acid} \xrightarrow[2-6]{} \text{Gal} \xrightarrow[1-3(4)]{} \text{GlcNAc} \xrightarrow[1-3(4)]{} \text{GlcNAc}
\end{array}
\tag{4-12}
$$

The oligosaccharide structure of serotransferrin was also investigated by Spik *et al.* (1975), who confirmed the previous finding that two carbohydrate chains were present therein. It was also determined that one of the chains was located in the so-called N-terminal domain of the proteins, whereas the other was in the C-terminal domain (but see the results of MacGillivray *et al.*, 1977, in Section 4.9.3). Both chains were linked to asparagine through a 4-*N*-(acetamido-2-deoxy-β-D-glucopyranosyl)-L-asparagine bond. The structure given by Spik *et al.* (1975) accounted for a molecular weight of 2202 daltons, and its structure is

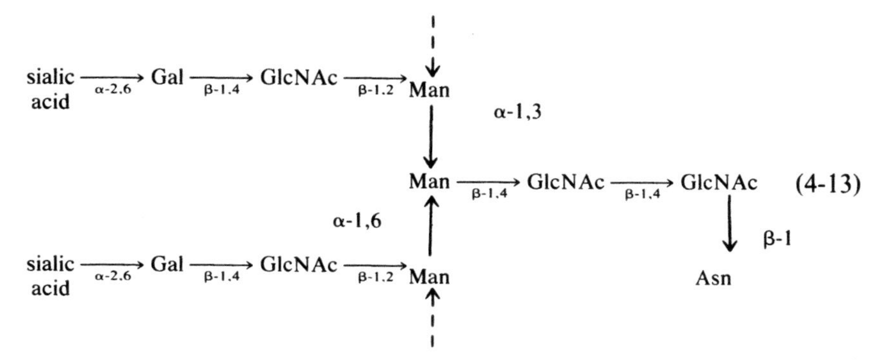

The structure presented in equation (4-13) was subsequently confirmed using proton magnetic resonance spectroscopy (Dorland *et al.*, 1977) and mass spectroscopy (Karlsson *et al.*, 1978) techniques. On the basis of electrophoretic studies, Wong *et al.* (1978) have proposed that there may be an additional sialic acid—Gal—GlcNAc unit attached to mannose residues [marked by a dashed arrow in equation (4-13)]. Thus a transferrin molecule may contain as many as five sialic acid residues.

Lactoferrin apparently has the same oligosaccharide chain structure with the exception that there is a fucose residue attached to the glucosamine residue in position 3 with an α-1,6 or an α-1,3 linkage (Spik and Mazurier, 1977).

Rabbit serotransferrin has also been the subject of a number of investigations. As early as in 1968, it was shown that serum and milk transferrins differed from each other by one sialic acid residue, the milk transferrin containing less sialic acid (Baker *et al.*, 1968). Strickland and Hudson (1978) determined that again there are two identical oligosaccharide chains for each rabbit serotransferrin molecule, and glycopeptides they isolated following tryptic digestion of the protein had molecular weights of near 3000. The structure of the carbohydrate moiety was identical to that of human serotransferrin as determined by Spik *et al.* (1975) (Strickland *et al.*, 1979).

On the other hand, Leger *et al.* (1978), by using proton magnetic resonance spectroscopy, came to the conclusion that rabbit serotransferrin has but one oligosaccharide chain per molecule and that its structure was similar to that of human serotransferrin. Three types of oligosaccharide chains were recognized, differing only in position and content of sialic acid: an oligosaccharide unit containing two terminal sialic acid molecules, a unit containing a sialic acid residue on one branch of the *biantennary structure*, and a unit containing a sialic acid residue on the other branch.

Hen's egg conalbumin differs from hen's egg serotransferrin in the structure of their carbohydrate units only. Both contain a single oligosaccharide chain, which in conalbumin consists only of mannose and *N*-acetylglucosamine. In serotransferrin, there is also galactose and sialic acid (Table 4-14). Both conalbumin and serotransferrin of the hen share the same messenger RNA in the chicken oviduct system (Lee *et al.*, 1978). Preliminary results indicate that the oligosaccharide structure of hen's egg conalbumin may be one or both of the following (Spik *et al.*, 1979):

$$
\begin{array}{llllll}
 & \text{(GlcNAc)} & & & & \\
 & \downarrow{\scriptstyle \beta\text{-}1,4} & & & & \\
\text{GlcNAc} \xrightarrow{\beta\text{-}1,2} & \text{Man} & & & & \\
 & \searrow{\scriptstyle \alpha\text{-}1,3} & & & & \\
 & \text{GlcNAc} \xrightarrow{\beta\text{-}1,4} & \text{Man} \xrightarrow{\beta\text{-}1,4} & \text{GlcNAc} \xrightarrow{\beta\text{-}1,4} & \text{GlcNAc} \xrightarrow{\beta\text{-}1} & \text{Asn} \\
 & \nearrow{\scriptstyle \alpha\text{-}1,6} & & & & \\
\text{GlcNAc} \xrightarrow{\beta\text{-}1,2} & \text{Man} & & & & \\
 & \uparrow{\scriptstyle \beta\text{-}1,4} & & & & \\
 & \text{(GlcNAc)} & & & &
\end{array}
\qquad (4\text{-}14)
$$

The GlcNAc in parentheses indicates the possible position(s) of the additional GlcNAc residues.

4.9.2 Cyanogen Bromide Fragmentation of the Transferrins

Following the establishment that the transferrins consist of but a single polypeptide chain, a proposal was put forth suggesting that these proteins may be composed of two identical or nearly identical halves, which may have arisen through a gene duplication process (Greene and Feeney, 1968). This hypothesis was tested by several investigators by subjecting the transferrins to fragmentation with cyanogen bromide (CNBr), which splits those peptide bonds, where methionine contributes carboxyl residues. Since human serotransferrin contains 7–9 methionine residues (Table 4-10), one would expect to get 8–10 different cyanogen bromide peptides if the serotransferrin structure were not a duplicate. If it were, half the expected number of different cyanogen bromide fragments would be obtained.

In 1971 there was a report describing the isolation of only three CNBr fragments from conalbumin (eight methionine residues), and, moreover, each fragment was isolated in twice the molar quantity with respect to the starting material. The molecular weights of the fragments were 21,000, 9400, and 7000, accounting for 74,800 daltons (Phillips and Azari, 1971). It was concluded that conalbumin consisted of two identical halves. Fol-

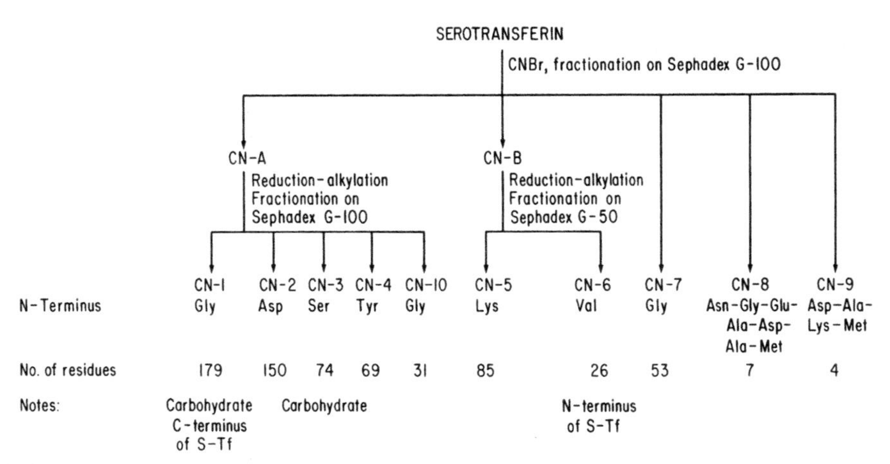

Figure 4-13. The cleavage of human serotransferrin into fragments by cyanogen bromide. Derived largely from work described by MacGillivray *et al.* (1977), Sutton and Brew (1974a), and Sutton *et al.* (1975).

lowing this, human serotransferrin was subjected to CNBr cleavage coupled with reduction–alkylation of the disulfide bonds, and it was shown that at least six different fragments could be obtained thereby. The molecular weights were 26,000, 16,000, 15,000, 9000, 6000, and 4000 (N-terminus). These added up to 76,000 daltons. It was concluded that human serotransferrin cannot consist of a duplicate structure, though limited regions of identical amino acid sequences could not be excluded (Bezkorovainy and Grohlich, 1973). Later it was shown that at least eight fragments could be obtained from conalbumin by CNBr fragmentation (Bezkorovainy and Grohlich, 1974), and, finally, the results of the paper reporting duplicate structure in conalbumin were withdrawn (Tsao *et al.*, 1974a). Lactoferrin also yielded seven unique CNBr peptides as expected from its six methionyl residues. Their molecular weights ranged from 1468 to 25,000, and they accounted for 74,969 daltons (Mazurier *et al.*, 1974). It thus appears that the evidence is overwhelmingly against a simple duplicate structure in the transferrin group of proteins.

4.9.3 Amino Acid Sequences of Cyanogen Bromide Fragments of Human Serotransferrin

Cyanogen bromide fragmentation has proved to be the method of choice for the primary structure determination of human serotransferrin. Sutton and Brew (1974a) and Sutton *et al.* (1975) reported on the preparation of nine CNBr fragments in a quantity sufficient for sequence studies using essentially the scheme in Figure 4-13. The N-terminal fragment (CN-6) was sequenced (Sutton and Brew, 1974b), and its amino acid sequence, as well as that obtained by others, is shown in Table 4-15. Most of the other CNBr fragments of human serotransferrin, or portions thereof, have also been sequenced (Sutton *et al.*, 1975; Jollès *et al.*, 1974; MacGillivray *et al.*, 1977), and the CNBr peptides have been satisfactorily assembled:

$$\begin{array}{l}(\mathrm{NH_2{-}CH\text{-}6})\text{–}(\mathrm{CN\text{-}5})\text{–}(\mathrm{CN\text{-}2})\text{–}(\mathrm{CN\text{-}7})\text{–}(\mathrm{CN\text{-}9})\text{–}(\mathrm{CN\text{-}4})\text{–}\\ \mathrm{Val}\\ \qquad(\mathrm{CN\text{-}8})\text{–}(\mathrm{CN\text{-}3})\text{–}(\mathrm{CN\text{-}10})\text{–}(\mathrm{CN\text{-}1{-}COOH})\\ \qquad\qquad\qquad\qquad\qquad\qquad \mathrm{Pro}\end{array} \tag{4-15}$$

All transferrins are homologous proteins; i.e., they probably arose from the same ancestral gene. The high degree of homology in the N-terminal regions of these proteins can be observed from Table 4-15. Other areas of homology between serotransferrin and lactoferrin have also been observed (Metz-Boutigue *et al.*, 1978). Especially interesting are peptides

Table 4-15. N-Terminal Amino Acid Sequences of Human Serotransferrin and Lactoferrin and Hen's Egg Conalbumin

Protein[a]	Reference	Sequence
		4 — 10 — 15
Tf	Sutton and Brew, 1974b	NH_2—Val-Pro-Asp-Lys——Thr-Val——Arg-Trp——Ala-Val-Ser-Glu-His-Glu-
	Brune *et al.*, 1978	NH_2—Val-Pro-Asp-Lys——Thr-Val——Arg-Trp-Cys-Ala-Val-Ser-Glu-His-Glu-
	Jollès *et al.*, 1976	NH_2—Val-Pro-Asp-Lys——Thr-Val——Arg-Trp-Cys-Ala-Val-Ser-Glu-His-Glu-
Lf	Jollès *et al.*, 1976	NH_2—Gly-Arg-Arg-Arg-Arg-Ser-Val——Gln-Trp-Cys-Ala-Val-Ser-Gln-Pro-Glu-
C	Jollès *et al.*, 1976	NH_2—Ala-Pro-Pro-Lys——Ser-Val-Ile-Arg-Trp-Cys-Thr-Ile-Ser-Ser-Pro-Glu-
		18 — 22 — 26
Tf		Ala-Thr-Cys-Lys-Ser-Glu-Cys-Phe-Arg-Asp-
		Ala-Thr-Lys-Cys-Glu-Ser——Phe-Arg-Asp-
		Ala-Thr-Lys-Cys-Gln-Ser——Phe-Arg-Asp-
Lf		Ala-Thr-Lys-Cys-Phe-Gln——Trp-Gln——
C		Gln-Lys-Lys-Cys-Asn-Asn——Leu-Arg-Asp-

[a] Abbreviations: Tf, serotransferrin; Lf, lactoferrin; C, conalbumin.

containing carbohydrate and those containing cystine. The latter are of course largely conserved during the course of evolution. Table 4-16 shows the amino acid sequences of some glycopeptides isolated by Graham and Williams (1975) and Kingston and Williams (1975). Homologies have also been noted in regard to cystine-containing peptides of human serotransferrin and lactoferrin (Bluard-Deconinck *et al.*, 1974). A number of cystine-containing peptides have also been isolated from conalbumin, though their homologies with respect to peptides isolated from human serotransferrin have not been reported (Elleman and Williams, 1970).

The most interesting aspect of amino acid sequence study in the transferrins is the demonstration of some internal homology in these proteins. When the amino acid sequences of residues 1–339 are compared to those of residues 340–676, a number of regions with identical amino acid sequences may be identified (Figure 4-14) (MacGillivray *et al.*, 1977; MacGillivray and Brew, 1975). Such internal homology has most likely originated from gene duplication and gene fusion processes during the course of evolution. It will be remembered that a proposal had been made by Greene and Feeney (1968) to the effect that the transferrins are products of a gene duplication process and should consist of two identical halves. This has not proved to be entirely correct, since following such gene duplication and fusion events, numerous point mutations have also taken place. Lactoferrin also contains regions of internal homology (Metz-Boutigue *et al.*, 1978). It is interesting to note, however, that the two carbohydrate chains of serotransferrin are not situated in the N-terminal and the C-terminal domains, as might be expected from a gene duplication process and as determined by Spik *et al.* (1975). The MacGillivray *et al.* (1977) model locates the carbohydrate chains in the C-terminal region attached to asparagine residues 415 and 608.

4.9.4 Fragments of the Transferrins Containing a Single Iron-Binding Site

In 1974, Williams described the cleavage of 30% iron-saturated conalbumin into a number of small peptides and an iron-containing fraction with a molecular weight of near 35,000 by trypsin or chymotrypsin digestion (Williams, 1974). The iron-containing fraction was separated by isoelectric focusing into two components, isolelectric points 5.9 and 6.3. In all other respects the two components were identical: Both had a single iron-binding site, an identical molecular weight of near 35,000, identical amino acid composition, and a single-chain structure; both were apparently derived from the N-terminal section of conalbumin; and both con-

Table 4-16. Amino Acid Sequences of Glycopeptides Isolated from Some Transferrins[a]

Protein[b]	Sequence
HTf	CHO -Cys-Gly-Leu-Val-Pro-Val-Leu-Ala-Glu-Asn-Tyr-Asn-Lys-Ser-
BTf	CHO -Arg-Val-Tyr-Asn-Ser-Ser-
STf	CHO -Asp-Asp-Ser-Ser-Arg-Lys-Asn-Arg-Ser-Leu-Thr-Val-Gly-Glu-
C	CHO -Ala-Gly-Trp-Val-Ile-Pro-Met-Gly-Leu-Ile-His-Asn-Arg-Thr-Gly-Thr-Cys-Asn-Phe-

[a] Compiled from Graham and Williams (1975) and Kingston and Williams (1975).
[b] Abbreviations: HTf, human serotransferrin; BTf, bovine serotransferrin; STf, swine serotransferrin; C, conalbumin-hen serotransferrin; CHO, carbohydrate.

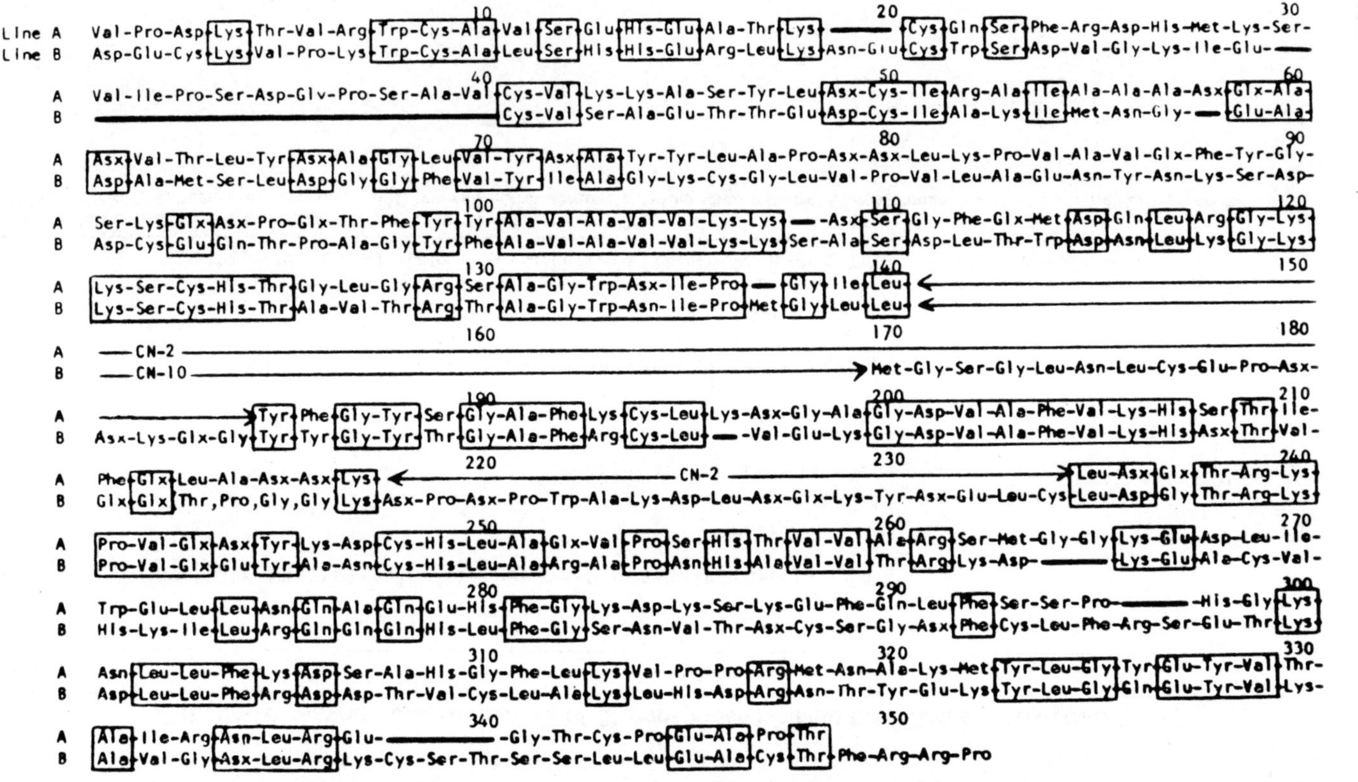

Figure 4-14. Partial amino acid sequence of human serotransferrin, showing internal homologies (boxed residues). A is the N-terminal domain of the protein (residues 1–339), B is the C-terminal domain of the protein (residues 340–676). N-terminus is valine, C-terminus is proline. Reprinted by permission from MacGillivray, R. T. A., Mendez, E., and Brew, K., 1977, Structure and evolution of human serum transferrin, *in Proteins of Iron Metabolism*, E. B. Brown, P. Aisen, J. Fielding, and R. R. Crichton (eds.), Grune and Stratton, New York, pp. 133–141. Copyright by Grune and Stratton, Publ.

tained no carbohydrate. The C-terminal fragment could be prepared from conalbumin by either a digestion of the iron-saturated protein with subtilisin or by digestion of the 30% iron-saturated protein with trypsin. However, contrary to the method for N-terminal fragment preparation, the conalbumin was treated with iron at pH 5, where only the C-terminal site becomes occupied by iron. Both the N- and C-terminal fragments could donate iron to rabbit reticulocytes. It is noteworthy that fragments isolated by Williams possessed a single polypeptide chain and possessed only internal disulfide linkages. This indicates that conalbumin consists of two independent domains each containing an iron-binding site, a structure very similar to that of serotransferrin as shown in Figure 4-3. An iron-containing fragment was also prepared by Tsao *et al.* (1974b) from conalbumin using CNBr as the fragmenting agent. Its molecular weight was 36,000, and it apparently consisted of at least two polypeptide chains. Williams speculated that this fragment was representative of the C-terminal portion of the protein (Williams, 1975).

Both the N- and C-terminal fragments capable of binding iron were prepared from bovine serotransferrin fully saturated with iron using trypsin digestion (Brock *et al.*, 1976, 1978). Amino acid composition was nearly identical in both fragments, though carbohydrate was localized in the C-terminal domain. Although whole bovine serotransferrin donated iron to rabbit reticulocytes, the monoferric fragments apparently failed to do so. Amino acid analyses and peptide maps for the two fragements were quite different.

Human serotransferrin has also been separated into the C- and N-terminal fragments, each carrying one iron-binding site (Evans and Williams, 1978). The former was prepared from 30% iron-saturated serotransferrin, where iron was added in the form of nitriloacetate at either pH 5.5 or 8.5, and then the unoccupied N-terminal portion was digested away. The N-terminal fragment was prepared by adding iron to 30% saturated protein at pH 8.5 in the form of ferric citrate. The C-terminal fragment, which had a molecular weight of near 43,000 and carried carbohydrate, carried the iron-binding site designated as site A by Harris (1977). The N-terminal fragment had a molecular weight of 36,000 and contained carbohydrate and thus was consistent with the MacGillivray *et al.* (1977) model. The two fragments had similar amino acid compositions; however, their peptide maps were quite different.

Human lactoferrin was first used to prepare an iron-carrying fragment by Line *et al.* (1976). Fully iron-saturated lactoferrin was digested with pepsin at pH 3, yielding a homogeneous iron-containing fragment with a molecular weight of 33,000–39,500. It carried carbohydrate, and its N-terminal residue was alanine. Bluard-Deconinck *et al.* (1978) later showed that this fragment represented the C-terminus of lactoferrin. They were also able to isolate both the N-terminal and C-terminal lactoferrin fragments from 30% iron-saturated lactoferrin by trypsin digestion and to show that, contrary to the situation with human serotransferrin, the carbohydrate was equally distributed between the two domains of the protein molecule. The data on iron-containing fragments obtained from the various transferrins are summarized in Table 4-17.

The fact that single-chain iron-containing fragments can be produced from all transferrins indicates that these proteins consist of semi-independent structural entities, which have been termed the N- and C-terminal domains. This is to be expected if one accepts the hypothesis that the transferrins arose through a gene duplication process in the course of evolution. It is also clear from these studies that the degree to which such domains become denatured under a given set of conditions may not be and probably is not the same. For instance, the fact that iron-saturated lactoferrin can yield a C-terminal fragment through extensive digestion with pepsin at pH 3 indicates that its N-terminal domain may be much more susceptible to acid denaturation than is its C-terminal domain (Bezkorovainy G. and Bezkorovainy A., 1978). Moreover, such results reiterate the findings made in the late 1950s to the effect that iron-containing transferrins are much less susceptible to denaturation and digestion by proteolytic enzymes than are the iron-free species. The availability of transferrin fragments carrying but a single iron-binding site makes it possible to investigate the physical–chemical properties of each site without the interfering effects of the other. The interpretation of the results, be they of spectroscopic or chemical modification nature, now becomes much less ambiguous than was previously possible. Moreover, the possible biological differences between the two iron-binding sites of serotransferrin (see Chapter 6) may at last be satisfactorily proved to exist or not to exist by the use of fragments carrying a single iron-binding center. Although the results in this regard with bovine serotransferrin fragments have been disappointing, the situation may prove to be different with the fragments of other species.

4.10 Microheterogeneity of the Transferrins

Multiplicity of components observed with homogeneous preparations of the transferrins in electrophoretic, isoelectric focusing, or various chromatography systems may be due to genetically controlled amino acid substitutions, slight differences in carbohydrate content, or as yet unexplained reasons. Among the last group of transferrins are the well-characterized rat serotransferrin and cattle transferrins. Rat serotrans-

Table 4-17. Preparation of Iron-Binding Fragments of the Transferrins

Protein	Conditions used	Molecular weight of fragment	N-terminus	Notes	Reference
Conalbumin	30% saturated with iron; digested with trypsin or chymotrypsin	35,000	Ala	N-Terminal portion of conalbumin; tryptic and chymotryptic fragments differ only in pIs: 6.3 and 5.9.	Williams, 1974
	Fe saturated; subtilisin digestion	31,000–36,000	—	C-Terminal portion of conalbumin; contains carbohydrate; two components; pIs 5.45 and 5.61.	Williams, 1975
	30% saturated with iron; prepared at pH 5, then digested with trypsin	35,500–44,500	—	C-Terminal portion of conalbumin; contains carbohydrate; pI 5.45.	Williams, 1975
Bovine serotransferrin	Fe saturated; digested with trypsin	38,500	—	C-Terminal fragment; contains carbohydrate.	Brock *et al.*, 1976, 1978
		32,000	Asp	N-Terminal fragment; contains no carbohydrate.	Brock *et al.*, 1976, 1978
Conalbumin	CNBr	36,000	—	Two polypeptide chains.	Tsao *et al.*, 1974b
Human serotransferrin	30% saturated with iron–nitriloacetate, pH 5.5 or pH 8.5	43,000	—	C-Terminal fragment; contains carbohydrate and acid-stable Fe-binding site.	Evans and Williams, 1978
	30% saturated with iron–citrate at pH 8.5	36,000	Val	N-Terminal fragment; contains no carbohydrate.	Evans and Williams, 1978
Human lactoferrin	Fe saturated; digested with pepsin at pH 3.0	33,000–39,500	Ala	Contains carbohydrate.	Line *et al.*, 1976
	Fe saturated; digested with pepsin at pH 3.0	40,000	Ala	C-Terminal fragment; contains carbohydrate; arg is C-terminal amino acid.	Bluard-Deconinck *et al.*, 1978
	30% saturated with iron; digested with trypsin	40,000	—	C-Terminal fragment; contains carbohydrate; arg is C-terminl amino acid.	Bluard-Deconinck *et al.*, 1978
		40,000	Gly	N-Terminal fragment; contains carbohydrate.	Bluard-Deconinck *et al.*, 1978

ferrin migrates in the form of two components in gel electrophoresis. These have been separated by isoelectric focusing and are termed Tf_s (slow) and Tf_f (fast). Their isoelectric points are 5.9 and 5.7, respectively (Okada *et al.*, 1979). There is a controversy as to whether or not the two isotransferrins are biologically identical (see Chapter 6).

Bovine serotransferrins exist in the form of several genetic variants, the most common of which are TfA, TfD_1, TfD_2, and TfE. All give at least five bands in starch gels (Stratil and Spooner, 1971). Several individual bands have been isolated from homozygous phenotypes and have been shown to differ in regard to their sialic acid contents. If sialic acid is removed from the five-band phenotypes, two bands remain in each of the genetic variants (Stratil and Spooner, 1971; Hatton *et al.*, 1977). It is therefore clear that sialic acid contributes in part to the microheterogeneity of homozygous bovine serotransferrins, though another factor is also operative. The two asialo bands were examined in regard to molecular weights, amino acid content, and peptide maps and were found to be identical in every regard tested (Richardson *et al.*, 1973). It thus appears that the basis for the microheterogeneity of rat serotransferrin and for the different bands observed in homozygous genetic variants of bovine serotransferrins remains to be discovered.

There are several reports dealing with transferrin heterogeneity based on differences in carbohydrate content. The case with bovine homozygous genetic variants has already been mentioned. Human serotransferrin of type C and fully saturated with iron can be separated into three components by DEAE–cellulose chromatography (Wong *et al.*, 1978; Regoeczi *et al.*, 1977). Such separation is based on sialic acid content. The fractions isolated (termed TfC-4-6, TfC-5-6, and TfC-6-6) had 0.99, 1.33, and 1.63% sialic acid, respectively, and had sialic acid to galactose ratios of 0.82, 1.02, and 0.97, respectively. These accounted for 6, 62, and 32% of the serotransferrin examined. These data suggest that in terms of the structure worked out by Spik *et al.* (1975) (Section 4.9.1), TfC-4-6 contains three sialic acid residues, TfC-5-6 contains four residues, and TfC-6-6 may represent a serotransferrin with a fifth *N*-acetylneuraminyl-α-2,6-*N*-acetyllactosamine unit attached to a mannose residue. It was found that TfC-4-6, TfC-5-6, and TfC-6-6 are not the products of the separation process but are instead produced by the biosynthetic mechanism in the liver.

Hen's egg conalbumin has been separated into the major and minor components by isoelectric focusing. The major component had a pI of 5.78, whereas the minor component had a pI of 5.62 in the iron-saturated state. Isoelectric focusing of whole conalbumin partially saturated with iron revealed the presence of six components: three corresponding to the diferric, monoferric, and apoconalbumin (pIs of 5.78, 6.25, and 6.73, respectively) of the major component and the corresponding bands of the minor component (pIs of 5.62, 6.05, and 6.50, respectively). It was determined that both components of conalbumin had identical iron-binding constants, amino acid composition, and peptide maps (Williams and Wenn, 1970).

The two components of hen's egg white conalbumin apparently differ in the structure of their carbohydrate moieties (Iwase and Hotta, 1977). The two known forms of this protein as well as a third component (fractions II, III, and IV, present in a ratio of 17 : 56 : 17) were separated on columns of concanavalin A–Sepharose, which have affinities for terminal hexoses. All three differed in carbohydrate content but not amino acid content. The classical minor component (fraction II) had 1.6% mannose, 0.3% galactose, and 2.5% glucosamine, and the structure of its oligosaccharide chain was deduced to be largely Asn–$(GlcNAc)_3$–$(Man)_4$–$(GlcNAC)_2$. The carbohydrate content of fraction III, the classical major component, was 2.3% mannose, 0.1% galactose, and 4.4% glucosamine, with an oligosaccharide structure largely Asn–$(GlcNAc)_2$–$(Man)_3$–$(GlcNAc)_3$. Fraction IV, which also had the electrophoretic mobility of the classical major component, had 3.5% mannose, 0.6% galactose, and 3.5% glucosamine. Its oligosaccharide chains were apparently a mixture of at least the preceding two types of structures. It should be noted that both of the structures proposed by Iwase and Hotta (1977) could fit the structures deduced by Spik *et al.* (1979) (Section 4.9.1).

Human serotransferrins as well as serotransferrins of other species can exist in the form of genetically controlled polymorphs. In human beings, at least 22 different alleles have been identified (Kühnl and Spielmann, 1978), and the variant transferrins are "inherited as simple Mendelian codominant traits" (Giblett, 1969). There are thus 253 phenotype possibilities. Variant transferrins differ from each other by single amino acid substitutions and not by molecular weights as is the case with haptoglobins.

The most abundant allele in all populations studied is the TfC gene, and the homozygous phenotype TfCC occurs in between 90 and 100% of any given population. Other transferrin variants have been classified on the basis of whether they migrate electrophoretically faster or slower than TfCC: the B types are faster migrating, and the D types are the slower ones. They have been given the subscripts 1, 2, 3, etc., but as intermediately migrating forms are discovered, this nomenclature becomes inadequate, and it has been suggested that classification be made on the basis of locality, e.g., $TfBC_{Atlanti}$ (Giblett, 1969). Some human serotransferrin variant electropherograms are reproduced diagrammatically in Figure 4-15.

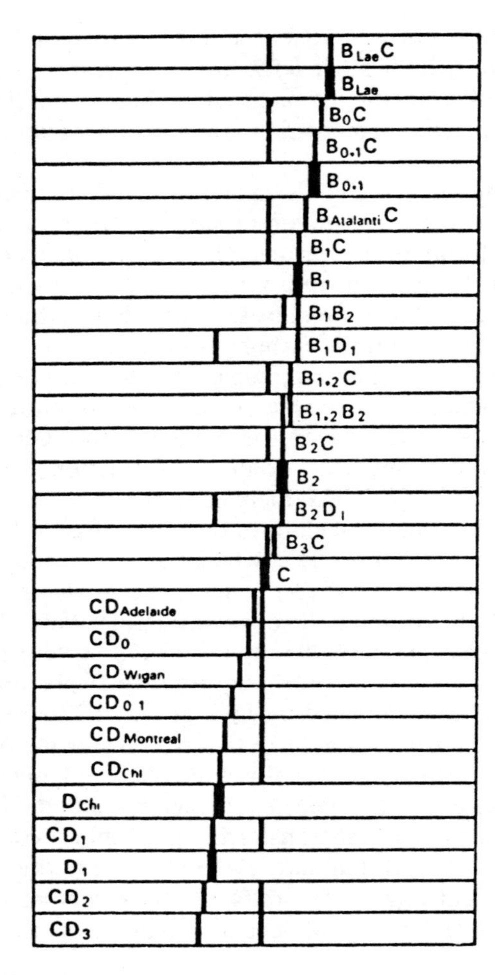

Figure 4-15. Diagrammatic plot of electrophoretic mobilities of the various human serotransferrin variants. From Giblett, E. R., 1969, *Genetic Markers in Human Blood*, F. A. Davis, Philadelphia, pp. 139–159, by permission of the Blackwell Scientific Publications, Ltd.

Gene frequencies of the serotransferrin variants are given in Table 4-18. Further, of 2221 Caucasian blood donors in the eastern United States, 2 had the B_1 gene (TfB_1), 15 had the B_2 gene (TfB_2), and 10 had the D_1 gene (TfD_1). This translates into an apparent TfCC phenotype frequency of 0.9878 (Roop *et al.*, 1968). In populations other than Caucasian, the TfB and TfD gene frequencies may be greater: The D variants occur largely in the Negroes (4.5% as per Roop *et al.*, 1968), Australian aborigines, and Chinese (especially D_{Chi}), whereas the B variants occur mostly in Caucasians and American Indians.

There is a possibility that the TfC gene may exist in the form of two variants, TfC_1 and TfC_2, so that three phenotypes are possible: TfC_1C_1, TfC_1C_2, and TfC_2C_2 (Kühnl and Spielmann, 1978). These were detected by the method of isoelectric focusing. Gene frequencies for the German population studied ($n = 942$) was given at 0.8195 for TfC_1 and at 0.172 for TfC_2. In regard to gene frequencies, this is similar to "a new serum protein polymorphism" ascribed to transferrin, where gene frequencies were reported as 0.81 and 0.19 ($n = 132$, Danish population) (Thymann, 1978). This polymorphism was also detected by isoelectric focusing. It is not known whether the polymorphisms discovered by isoelectric focusing are due to amino acid sequence differences or differences in carbohydrate content described earlier in this section.

Of the known amino acid substitutions in the genetic variants of human serotransferrin, we have an aspartic acid residue in TfC substituted by a glycine residue in the D_1 variant (Wang and Sutton, 1965), and a histidyl residue of TfC is substituted by an arginyl residue in the D_{Chi} variant (Wang *et al.*, 1967). It has been proposed that these substitutions are located in the CN-7 fragment (residues 264–318) of the serotransferrin molecule (Sutton *et al.*, 1975).

Genetic polymorphism of the serotransferrins has been described for numerous mammalian and nonmammalian species; however, since, similarly to the human variants, these do not exhibit any differences in terms of their iron-binding or biological properties, discussion of their pecul-

Table 4-18. Serotransferrin Phenotypes and Gene Frequencies in Various Caucasian Populations[a]

		Phenotypes[b]			Gene frequencies[c]		
Population	Sample	*CC*	*BC*	*DC*	TfC	Tf*B*	Tf*D*
Rome	319	0.9875	0.0031	0.0094	0.99368	0.00156	0.00473
Greece	2050	0.9956	0.0015	0.0029	0.99780	0.00075	0.00145
Sweden	2395	0.9896	0.0092	0.0012	0.99479	0.00462	0.00060
England	139	0.9860	0.0150	0.0020	0.99297	0.00755	0.00101
Finland	3893	0.9553	0.0089	0.0124	0.97739	0.00450	0.00063
Iceland	402	1.0000	—	—	1.00000	—	—
Canada	425	0.9880	0.0120	—	0.99398	0.00602	—
United States	471	0.9870	0.0130	—	0.99348	0.00652	—
Poland	252	0.9683	0.0317	—	0.98402	0.01592	—

[a] From Serafini and Serra (1968).
[b] *BC* and *DC* include all *B* and *D* phenotypes (heterozygous).
[c] Tf*B* and Tf*D* include all *B* and *D* alleles.

iarities will be deferred to a future date. No genetic polymorphisms have been detected in hen's egg conalbumin or the lactoferrins.

4.11 Metabolism of the Transferrins

Many tissues of the human organism are capable of biosynthesizing serotransferrin; however, because of its size and the rate of the biosynthectic process, the liver is by far the most important source of this protein. The contribution of each tissue to serotransferrin biosynthesis is summarized in Table 4-19 (Morgan, 1969). Conalbumin is biosynthesized in the oviduct magnun, which can be cultured, and conalbumin biosynthesis thus studied *in vitro* (Palmiter, 1972). The messenger RNA codings for conalbumin and chicken serotransferrin are identical, and they have a molecular weight of close to 1×10^6 (3200 nucleotides). Both the estrogens and progesterone can increase dramatically the biosynthesis of conalbumin, while only estrogens have a modest effect on serotransferrin biosynthesis in the liver. The two genes, though apparently identical in structure, seem to be under separate control mechanisms (Lee *et al.*, 1978). Conalbumin and chicken serotransferrin are biosynthesized in the form of preconalbumin and preserotransferrin; that is immediately following translation, they carry at their N-terminal segment a 19-amino-acid peptide, the structure of which is (Thibodeau *et al.*, 1978)

*f*Met–Lys–Leu–Ile–Leu–Cys–Thr–Val–Leu–Ser– (4-16)
Leu–Gly–Ile–Ala–Ala–Val–Cys–Phe–Ala–

This peptide is apparently lost as the protein is secreted into the extracellular fluid. Whether this situation also occurs in mammalian systems is still uncertain.

Table 4-19. Biosynthesis of Serotransferrin by Various Tissues in the Rat[a]

Tissue	^{14}C concentration in microsome serotransferrin 20 min after injection of ^{14}C-leucine into rats (counts/min/organ)	^{14}C concentration of tissue slice serotransferrin 2 hr following incubation with ^{14}C-leucine (counts/min/g)
Liver	4600 ± 1400	1920 ± 260
Spleen	139 ± 25	2130 ± 440
Kidney	90 ± 11	61 ± 20
Intestine	67 ± 7.6	100 ± 22
Lung	69 ± 11	—
Bone marrow	45 ± 5.1	2200 ± 740

[a] From Morgan (1969).

Serotransferrin turnover has been studied in a number of animals. The disappearance of ^{131}I- or ^{125}I-labeled serotransferrin from the bloodstream carries a biphasic character (Figure 4-16), the rapid phase representing an equilibration of the injected material with extravascular serotransferrin and the slower phase representing serotransferrin degradation. In the human being, in one study the latter was 7.58 days, and the former was 16–75 hr (eight subjects). The extravascular serotransferrin represented some 60% of the total extracellular pool (17 g). Serotransferrin degraded per day averaged 23.75 mg/kg of body weight, or 1.557 g/day in an average adult male (Katz, 1961). Others have estimated the half-life of human serotransferrin to be 7–9 days (Morton *et al.*, 1976). In the rabbit, the half-life of serotransferrin has been reported to be 4.2 days (Shepp *et al.*, 1973) and 10.5 days (Van Eijk *et al.*, 1969b). In the rat, absolute serotransferrin turnover was estimated at 35 mg/day with a total serotransferrin pool of about 115 mg (Morgan, 1966).

The liver appears to be a major site for serotransferrin catabolism (Morgan, 1966). The removal of sialic acid from serotransferrin increases its rate of catabolism but not nearly as much as for other glycoproteins such as ceruloplasmin or orosomucoid. In certain human subjects, the half-life of native serotransferrin was 8 days, but that of asialoserotransferrin was 6.4 days. In rabbits the two half-lives differed by only 15% (Regoeczi *et al.*, 1974). Glycoproteins are cleared more rapidly after losing sialic acid because a carbohydrate-specific lectin exists on the surface of the liver cells which recognizes the penultimate sugar in the carbohydrate moiety and thereby promotes the endocytosis of the modified protein. Asialoserotransferrin apparently combines with the lectin as efficiently as any other asialoglycoprotein, but its endocytosis proceeds less rapidly (Regoeczi *et al.*, 1978). The reason for this is believed to be the fact that the hepatocyte lectin recognizes the sequence fucosyl-α-1,3-*N*-acetylglucosaminyl-X. Serotransferrin does not possess such a sequence, whereas human lactoferrin does (see Section 4.9.1), and contrary to the situation with serotransferrin, asialolactoferrin is rapidly cleared from the bloodstream by the liver (Prieels *et al.*, 1978).

The control of transferrin biosynthesis is apparently mediated through total iron body stores and may be linked to the ferritin biosynthetic mechanism (see Chapter 5). The starting point for such observations is that in iron deficiency, in the absence of protein metabolism disorders, transferrin biosynthesis is markedly increased, whereas that of ferritin is decreased. If iron-deficient livers are perfused, there is an increased output of serotransferrin with a concomitant increase in TIBC. However, increasing serum iron concentrations will immediately stop the increased output of transferrin *in vivo*, though transfusing the iron-deficient animal

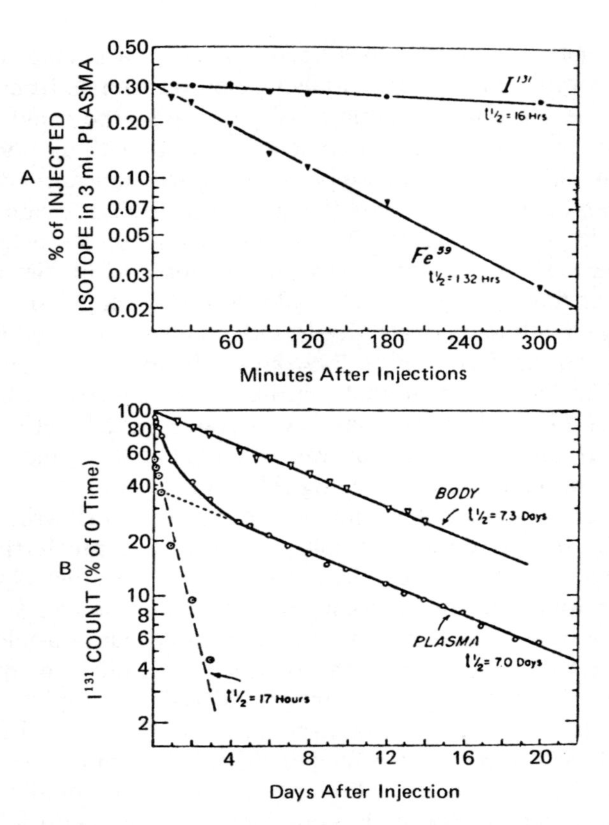

Figure 4-16. Clearance of labeled serotransferrin from human circulation. Graph A represents sampling taken over a 5-hr period (rapid phase of the clearance curve), and B represents the slower phase of the clearance curve. From Katz (1961).

with red cells (without raising the plasma iron levels) will not accomplish this (Morton and Tavill, 1977). These data were used to argue against an earlier proposal of Morgan (1969) that the relative hypoxia associated with anemia may be the factor responsible for the increased output of serotransferrin in iron deficiency anemia. Cycloheximide stops the increased output of serotransferrin by the liver in iron deficiency anemia; hence serotransferrin output is due to a *de novo* biosynthesis of transferrin. It has been proposed that either a "depletion of intrahepatic ferritin may be a major stimulus to transferrin synthesis" or, alternately, a nonferritin liver iron pool of an as yet unidentified nature may be involved (Morton and Tavill, 1978).

Summary

Some transferrins were known to exist since the turn of this century but were not purified until the late 1940s or early 1950s. Transferrins are present in many biological fluids, the major sources being the bloodstream (serotransferrin), milk (lactoferrin), and hen's egg white (conalbumin). The isolation of each class of transferrins can be accomplished by classical protein isolation procedures.

The transferrins are typically globular proteins, are single chain in nature, and have molecular weights of close to 80,000. Their physical properties such as the axial ratios, CD, EPR, and ultraviolet and visible range spectra change upon the binding of iron, and the binding of iron protects the proteins against denaturation and proteolytic digestion. The transferrins bind a maximum of two iron atoms per molecule of protein and in addition one molecule of carbonate, which is called the synergistic anion, per each iron atom. The two iron-binding sites are not chemically equal; i.e., they have different association constants with respect to iron. Additionally, one site is less prone to losing iron at slightly acidic pH values than is the other. The C-terminal iron-binding site is the more avid iron binder, though in the circulation the N-terminal monoferric protein predominates. The amino acid side chains involved in the binding of iron are the tyrosyl and histidyl residues.

The primary structures of transferrins have been investigated by preliminary fragmentation with cyanogen bromide followed by the isolation of the various fragments produced. Serotransferrin apparently contains 676 amino acid residues, and there appears to be a considerable degree of internal homology present, consistent with the concept that the transferrins arose as a result of a gene duplication process in the course of evolution. The two carbohydrate chains of serotransferrin molecule are biantennary in nature and are attached to the C-terminal domain of the molecule. All transferrins can be cleaved under the appropriate conditions into half molecules containing but a single iron-binding site. The transferrins are therefore believed to exist in the form of two semi-independent (bilobal) N-terminal and C-terminal domains.

The transferrins often exhibit various degrees of microheterogeneity, which may be based on differences in carbohydrate content, point mutations, or other as yet poorly understood reasons.

Human serotransferrin is biosynthesized largely in the liver and is apparently degraded there as well. Its biosynthetic rate is controlled by and is inversely proportional to the iron levels in the liver. Conalbumin biosynthesis takes place in the oviduct, and the gene coding for its structure is identical to that coding for the structure of chicken serotransferrin. Gene expression is, however, under separate control mechanisms.

References

Aasa, R., 1972. Re-interpretation of the electron paramagnetic resonance spectra of transferrins, *Biochem. Biophys. Res. Commun.* 49:806–812.

Aasa, R., Malmstrom, B. G., Saltman, P., and Vanngard, T., 1963. The specific binding of iron(III) and copper(II) to transferrin and conalbumin, *Biochim. Biophys. Acta* 75:203–222.

Aisen, P., and Brown, E. B., 1975. Structure and function of transferrin, *Prog. Hematol.* 9:25–56.

Aisen, P., and Leibman, A., 1968a. The stability constants of the Fe^{3+} conalbumin complexes, *Biochem. Biophys. Res. Commun.* 30:407–413.

Aisen, P., and Leibman, A., 1968b. Citrate-mediated exchange of Fe^{3+} among transferrin molecules, *Biochem. Biophys. Res. Commun.* 32:220–226.

Aisen, P., and Leibman, A., 1972. Lactoferrin and transferrin: A comparative study, *Biochim. Biophys. Acta* 257:314–323.

Aisen, P., Leibman, A., and Reich, H. A., 1966. Studies on the binding of iron by transferrin and conalbumin, *J. Biol. Chem.* 241:1666–1671.

Aisen, P., Aasa, R., Malmstrom, B. G., and Vanngard, T., 1967. Bicarbonate and the binding of iron to transferrin, *J. Biol. Chem.* 242:2484–2490.

Aisen, P., Aasa, R., and Redfield, A. G., 1969. The chromium, manganese, and cobalt complexes of transferrin, *J. Biol. Chem.* 244:4628–4633.

Aisen, P., Leibman, A., and Sia, C.-L., 1972. Molecular weight and subunit structure of hagfish transferrin, *Biochemistry* 11:3461–3464.

Aisen, P., Leibman, A., Pinkowitz, R. A., and Pollack, S., 1973a. Exchangeability of bicarbonate specifically bound to transferrin, *Biochemistry* 12:3679–3684.

Aisen, P., Lang, G., and Woodworth, R. C., 1973b. Spectroscopic evidence for a difference between the iron-binding sites of conalbumin, *J. Biol. Chem.* 248:649–653.

Aisen, P., Leibman, A., and Zweier, J., 1978. Stoichiometric and site characteristics of the binding of iron to human transferrin, *J. Biol. Chem.* 253:1930–1937.

Alderton, G., Ward, W. H., and Fevold, H. L., 1946. Identification of the bacteria-inhibiting iron-binding protein of egg white as conalbumin, *Arch. Biochem. Biophys.* 11:9–13.

Altman, P. L., and Dittmer, D. S., 1974. *Biology Data Book,* Vol. III, 2nd ed., FASEB, Bethesda, Md.

Azari, P., and Baugh, R. F., 1967. A simple and rapid procedure for preparation of large quantities of pure ovotransferrin, *Arch. Biochem. Biophys.* 118:138–144.

Azari, P. R., and Feeney, R. E., 1958. Resistance of metal complexes of conalbumin and transferrin to proteolysis and to thermal denaturation, *J. Biol. Chem.* 232:293–299.

Azari, P. R., and Feeney, R. E., 1961. The resistance of conalbumin and its iron complex to physical and chemical treatments, *Arch. Biochem. Biophys.* 92:44–52.

Bain, J. A., and Deutsch, H. F., 1948. Separation and characterization of conalbumin, *J. Biol. Chem.* 172:547–555.

Baker, E., Shaw, D. C., and Morgan, E. H., 1968. Isolation and characterization of rabbit serum and milk transferrins. Evidence for difference in sialic acid content only, *Biochemistry* 7:1371–1378.

Barkan, G., 1927. Eisenstudien. 3. Mitteilung. Die Verteilung des leicht abspaltbaren Eisens zwischen Blutkörperchen und Plasma und sein Verhalten unter experimentellen Bedingungen, *Hoppe-Seyler's Z. Physiol. Chem.* 171:194–221.

Barkan, G., and Schales, O., 1937. Chemischer Aufbau und physiologische Bedeutung des "leicht abspaltbaren" Bluteisens, *Hoppe-Seyler's Z. Physiol. Chem.* 248:96–116.

Bates, G. W., and Schlabach, M. R., 1973a. A study of the anion binding site of transferrin, *FEBS Lett.* 33:289–292.

Bates, G. W., and Schlabach, M. R., 1973b. The reaction of ferric salts with transferrin, *J. Biol. Chem.* 248:3228–3232.

Bates, G. W., and Schlabach, M. R., 1975. The non-specific binding of Fe^{3+} to transferrin in the absence of synergistic anions, *J. Biol. Chem.* 250:2177–2181.

Bates, G. W., and Wernicke, J., 1971. The kinetics and mechanism of iron(III) exchange between chelates and transferrin. IV. The reaction of transferrin with iron(III) nitriloacetate, *J. Biol. Chem.* 246:3679–3685.

Bates, G. W., Workman, E. F., and Schlabach, M. R., 1973. Does transferrin exhibit ferroxidase activity?, *Biochem. Biophys. Res. Commun.* 50:84–90.

Bearn, A. G., and Parker, W. C., 1966. Transferrin, *in Glycoproteins,* A. Gottschalk (ed.), Elsevier, Amsterdam, p. 415.

Bennett, R. M., Eddie-Quartey, A. C., and Holt, P. J. L., 1973. Lactoferrin—an iron binding protein in synovial fluid, *Arthritis Rheum.* 16:186–190.

Bezkorovainy, A., 1966. Comparative study of metal-free, iron-saturated and sialic acid-free transferrins, *Biochim. Biophys. Acta* 127:535–537.

Bezkorovainy, A., 1977. Human milk and colostrum proteins, *J. Dairy Sci.* 60:1023–1037.

Bezkorovainy, A., and Grohlich, G., 1967. The behavior of native and reduced–alkylated human transferrin in urea and guanidine–HCl solutions, *Biochim. Biophys. Acta* 147:497–510.

Bezkorovainy, A., and Grohlich, D., 1970. Modification of carboxyl groups of transferrin, *Biochim. Biophys. Acta* 214:37–43.

Bezkorovainy, A., and Grohlich, D., 1971. Imidazole groups of iron-saturated transferrin, *Biochem. J.* 123:125–126.

Bezkorovainy, A., and Grohlich, D., 1973. Cyanogen bromide fragments of human serum transferrin, *Biochim. Biophys. Acta* 310:365–375.

Bezkorovainy, A., and Grohlich, D., 1974. Comparative study of several proteins of the transferrin class, *Comp. Biochem. Physiol. B* 47:787–797.

Bezkorovainy, A., and Zschocke, R. H., 1974. Structure and function of transferrins. I. Physical, chemical, and iron-binding properties, *Arzneim. Forsch.* 24:476–485.

Bezkorovainy, A., Rafelson, M. E., and Likhite, V., 1963. Isolation and partial characterization of transferrin from normal human plasma, *Arch. Biochem. Biophys.* 103:371–378.

Bezkorovainy, A., Gerbeck, C. M., and Grohlich, D., 1968. Some physical–chemical properties of reduced–alkylated and sulfitolysed human serum transferrin and hen's egg conalbumin, *Biochem. J.* 110:765–770.

Bezkorovainy, A., Zschocke, R. H., and Grohlich, D., 1969. Some physical–chemical properties of succinylated transferrin, conalbumin, and orosomucoid, *Biochim. Biophys. Acta* 181:295–304.

Bezkorovainy, A., Grohlich, D., and Babler, B. J., 1972. Reversible acylation of human serum transferrin, *Physiol. Chem. Phys.* 4:535–542.

Bezkorovainy, G., and Bezkorovainy, A., 1978. Denaturation of human lactoferrin in acidic solution, *Trans. Ill. State Acad. Sci.* 71:326–330.

Binford, J. S., and Foster, J. C., 1974. Calorimetry of the transfer of Fe(III) from nitriloacetate to apotransferrin in the presence and in the absence of bicarbonate, *J. Biol. Chem.* 249:407–412.

Blanc, B., Bujard, E., and Mauron, J., 1963. The amino acid composition of human and bovine lactotransferrins, *Experientia* 19:299–303.

Bleijenberg, B. G., Van Eijk, H. G., and Leijnse, B., 1971. The determination of non-heme iron and transferrin in cerebrospinal fluid, *Clin. Chim. Acta* 31:277–281.

Bluard-Deconinck, J.-M., Masson, P. L., Osinski, P. A., and Heremans, J. H., 1974. Amino acid sequence of cysteic acid peptides of lactoferrin and demonstration of similarities between lactoferrin and transferrin, *Biochim. Biophys. Acta* 365:311–317.

Bluard-Deconinck, J.-M., Williams, J., Evans, R. W., Van Snick, J., Osinski, P. A., and Masson, P. L. 1978. Iron-binding fragments from N-terminal and C-terminal regions of human lactoferrin, *Biochem. J.* 171:321–327.

Boettcher, E. W., Kistler, P., and Nitschmann, Hs., 1958. Method of isolating the β_1-metal-combining globulin from human blood plasma, *Nature (London)* 181:490–491.

Brock, J. H., Arzabe, F., Lampreave, F., and Pineiro, A., 1976. The effect of trypsin on bovine transferrin and lactoferrin, *Biochim. Biophys. Acta* 446:214–225.

Brock, J. H., Arzabe, F. R., Richardson, N. E., and Deverson, E. V., 1978. Characterization of monoferric fragments obtained by tryptic cleavage of bovine transferrin, *Biochem. J.* 171:73–78.

Bron, C., Blanc, B., and Isliker, H., 1968. Etude electrophoretique de la denaturation de transferrine humaine par l'uree, *Biochim. Biophys. Acta* 154:61–69.

Brown, E. M., and Parry, R. M., 1974. A spectroscopic study of bovine lactoferrin, *Biochemistry* 13:4560–4565.

Brune, J. E. L., Martin, S. R., Boyd, B. S., Palmour, R. M., and Sutton, H. E., 1978. Human transferrin: The sequence and cystine bridges of the N-terminal region, *Tex. Rep. Biol. Med.* 36:47–61.

Buttkus, H., Clark, J. R., and Feeney, R. E., 1965. Chemical modification of amino groups of transferrins: Ovotransferrin, human serum transferrin, and human lactoferrin, *Biochemistry* 4:998–1005.

Campbell, R. F., and Chasteen, N. D., 1977. An anion binding study of vanadyl(IV) suman serotransferrin. Evidence for direct linkage to the metal, *J. Biol. Chem.* 252:5996–6001.

Cannon, J. C., and Chasteen, N. D., 1975. Non-equivalence of the metal binding sites in vanadyl-labeled human serum transferrin, *Biochemistry* 14:4573–4577.

Carey, P. R., and Young, N. M., 1974. The resonance Raman spectrum of the metalloprotein ovotransferrin, *Can. J. Biochem.* 52:273–280.

Castellino, F. J., Fish, W. W., and Mann, K. G., 1970. Structural studies on bovine lactoferrin, *J. Biol. Chem.* 245:4269–4275.

Charlwood, P. A., 1963. Ultracentrifugal characteristics of human, monkey, and rat transferrins, *Biochem. J.* 88:394–398.

Charlwood, P. A., 1971. Differential sedimentation–velocity and gel-filtration measurements on human apotransferrin and iron–transferrin, *Biochem. J.* 125:1019–1026.

Chasteen, N. D., White, L. K., and Campbell, R. F., 1977. Metal site conformational states of vanadyl(IV) human serotransferrin complexes, *Biochemistry* 16:363–368.

Cheron, A., Mazurier, J., and Fournet, B., 1977. Fractionnement chromatographique et etudes sur la microheterogeneite de la lactotransferrine de vache preparee par un procede original, *C. R. Acad. Sci. Ser. D* 28:585–588.

Cohn, E. J., 1947. Chemical, physiological, and immunological properties and clinical uses of blood derivatives, *Experientia* 3:125–136.

Das, B. R., 1974. Immunochemical studies on duck (*Anas Platyrhynchos*) ovotransferrin—

isolation, isoelectric fractionation, and antigen–antibody reaction, *Proc. Soc. Exp. Biol. Med.* 146:795–802.

De Lucas, L. J., Suddath, F. L., Gams, R. A., and Bugg, C. E., 1978. Preliminary X-ray study of crystals of human transferrin, *J. Mol. Biol.* 123:285–286.

Donovan, J. W., and Ross, K. D., 1975. Iron binding to conalbumin. Calorimetric evidence for two distinct species with one bound iron atom, *J. Biol. Chem.* 250:6026–6031.

Dorland, L., Haverkamp, J., Schut, B. L., Vliegenthart, J., Spik, G., Strecker, G., Fournet B., and Montreuil, J., 1977. The structure of the asialo-carbohydrate units of human serotransferrin as proven by 360 MHz proton magnetic resonance spectroscopy, *FEBS Lett.* 77:15–20.

Elleman, T. C., and Williams, J., 1970. The amino acid sequence of cysteic acid-containing peptides from performic acid-oxidized ovotransferrin, *Biochem. J.* 116:515–535.

Evans, G. W., 1976. Transferrin function in zinc absorption and transport, *Proc. Soc. Exp. Biol. Med.* 151:775–778.

Evans, R. W., and Holbrook, J. J., 1975. Differences in the protein fluorescence of the two iron(III)-binding sites of ovotransferrin, *Biochem. J.* 145:201–207.

Evans, R. W., and Williams, J., 1978. Studies of the binding of different iron donors to human serum transferrin and isolation of iron-binding fragments from the N- and C-terminal regions of the protein, *Biochem. J.*173:543–552.

Evans, R. W., Donovan, J. W., and Williams, J., 1977. Calorimetric studies on the binding of iron and aluminum to the amino and carboxyl-terminal fragments of hen ovotransferrin, *FEBS Lett.* 83:19–22.

Evans, T. J., Ryley, H. C., Neale, L. M., Dodge, J. A., and Lewarne, V. M., 1978. Effect of storage and heat on antimicrobial proteins in human milk, *Arch. Dis. Child.* 53:239–241.

Feeney, R. E., Anderson, J. S., Azari, P. R., Bennett, N., and Rhodes, M. B., 1960. The comparative biochemistry of avian egg white proteins. *J. Biol. Chem.* 235:2307–2311.

Fiala, S., and Burk, D., 1949. On the mode of iron binding by siderophilin, conalbumin, hydroxylamine, aspergillic acid, and other hydroxamic acids, *Arch. Biochem. Biophys.* 20:172–175.

Fontes, G., and Thivolle, L., 1925. Sur la teneur du serum en fer non hemoglobinique et sur sa diminution au cours de l'anemie experimenteale, *C. R. Seances Soc. Biol. Paris* 93:687–689.

Ford-Hutchinson, A. W., and Perkins, D. J., 1971. The binding of scandium ions to transferrin *in vivo* and *in vitro, Eur. J. Biochem.* 21:55–59.

Ford-Hutchinson, A. W., and Perkins, D. J., 1972. Chemical modifications of the tryptophanyl groups of transferrin, *Eur. J. Biochem.* 25:415–419.

Frenoy, N., Goussault, Y., and Bourrillon, R., 1971. Preparation de la transferrine du serum humain, *Clin. Chim. Acta* 32:243–249.

Frieden, E., and Aisen, P., 1980. Forms of iron transferrin, *Trends in Biochemical Sciences (TIBS)* 5:xl.

Gaber, B. P., and Aisen, P., 1970. Is divalent iron bound to transferrin? *Biochim. Biophys. Acta* 221:228–233.

Giblett, E. R., 1969. *Genetic Markers in Human Blood,* F. A. Davis Company, Philadelphia, pp. 135–159.

Glazer, A. N., and McKenzie, H. A., 1963. The denaturation of proteins. IV. Conalbumin and iron III–conalbumin in urea solution, *Biochim. Biophys. Acta* 71:109–123.

Gordon, A. H., and Louis, L. N., 1963. Preparation and properties of rat transferrin, *Biochem. J.* 88:409–414.

Gordon, W. G., Groves, M. L., and Basch, J. J., 1963. Bovine milk "red protein": Amino acid composition and comparison with blood transferrin, *Biochemistry* 2:817–820.

Gorinsky, B., Horsbaugh, C., Lindley, P. F., Moss, D. S., Parkar, M., and Watson, J. L., 1979. Evidence for the bilobal nature of differic rabbit plasma transferrin, *Nature (London)* 281:157–158.

Got, R., 1965. Fractionnement des proteines du lactoserum humain, *Clin. Chim. Acta* 11:432–441.

Got, R., Font, J., and Goussault, Y., 1967. Etude sur une transferrine de selacien, le grande roussette (*Scyllium Stellare*), *Comp. Biochem. Physiol.* 23:317–327.

Graham, I., and Williams, J., 1975. A comparison of glycopeptides from the transferrins of several species, *Biochem. J.* 145:263–279.

Greene, F. C., and Feeney, R. E., 1968. Physical evidence for transferrins as single polypeptide chains, *Biochemistry* 7:1366–1371.

Groves, M. L., 1960. The isolation of a red protein from milk, *J. Am. Chem. Soc.* 82:3345–3350.

Groves, M. L., 1971. Minor milk proteins and enzymes, *in Milk Proteins*, Vol. III, H. A. McKenzie (ed.), Academic Press, New York, p. 368.

Guerin, G., Vreeman, H. J., and Nguyen, T. C., 1976. Preparation et caracterisation physicochimique partielle de la transferrine serique ovine, *Eur. J. Biochem.* 67:433–445.

Hambraeus, L., 1977. Proprietary milk versus human breast milk in infant feeding, *Pediatr. Clin. North Am.* 24:17–36.

Hambraeus, L., Lönnerdal, B., Forsum, E., and Gebre-Medkin, M., 1978. Nitrogen and protein components of human milk, *Acta Paediatr. Scand.* 67:561–565.

Harris, D. C., 1977. Different metal-binding properties of the two sites of human transferrin, *Biochemistry* 16:560–564.

Harris, D. C., and Aisen, P., 1973. Facilitation of Fe(II) autooxidation by Fe(III) complexing agents, *Biochim. Biophys. Acta* 329:156–158.

Hatton, M. W. C., Regoeczi, E., Wong, K.-L., and Kraay, G. J., 1977. Bovine serum transferrin phenotypes AA, D_1D_1, D_2D_2, EE: Their carbohydrate composition and electrophoretic multiplicity, *Biochem. Genet.* 15:621–640.

Haupt, H., and Baudner, S., 1973. Isolierung und Kristallisation von Lactoferrin aus Human-Kolostrum, *Hoppe-Seyler's Z. Physiol. Chem.* 354:944–948.

Hekman, A., 1971. Association of lactoferrin with other proteins, as demonstrated by changes in electrophoretic mobility, *Biochim. Biophys. Acta* 251:380–387.

Holmberg, C. G., and Laurell, C.-B., 1945. Studies on the capacity of serum to bind iron. A contribution to our knowledge of the regulation mechanism of serum iron, *Acta Physiol. Scand.* 10:307–319.

Holmberg, C. G., and Laurell, C.-B., 1947. Investigation of serum copper. I. Nature of serum copper and its relation to the iron-binding protein of human serum, *Acta Chem. Scand.* 1:944–950.

Hovanessian, A. G., and Awdeh, Z. L., 1976. Gel isoelectric focusing of humanserum transferrin, *Eur. J. Biochem.* 68:333–338.

Hudson, B. G., Ohno, M., Brockway, W. J., and Castellino, F. J., 1973. Chemical and physical properties of serum transferrin from several species, *Biochemistry* 12:1047–1053.

Hughes, N. R., 1972. Serum transferrin and ceruloplasmin concentrations in patients with carcinoma, melanoma, sarcoma, and cancers of hematopoietic tissues, *Aust. J. Exp. Biol. Med. Sci.* 50: 97–107.

Inman, J. K., Coryell, F. C., McCall, K. B., Sgouris, J. T., and Anderson, H. D., 1961. A large-scale method for the purification of human transferrin, *Vox Sang.* 6:34–52.

Iwase, H., and Hotta, K., 1977. Ovotransferrin subfractionation dependent upon carbohydrate chain differences, *J. Biol. Chem.* 252:5437–5443.

Jacob, E., and Herbert, V., 1974. Evidence against transferrin as a binder of either vitamin B_{12} or folic acid, *Blood* 43:767–768.

Jamieson, G. A., 1965. Studies on glycoproteins. II. Isolation of the carbohydrate chains of human transferrin, *J. Biol. Chem.* 240:2914–2920.

Jamieson, G. A., Jett, M., and De Bernardo, S. L., 1971. The carbohydrate sequence of the glycopeptide chains of human transferrin, *J. Biol. Chem.* 246:3686–3693.

Jarritt, P. H., 1976. Effect of iron on sedimentation–velocity and gel filtration behavior or transferrins from several vertebrates, *Biochim. Biophys. Acta* 453:332–343.

Jeppsson, J. O., 1967. Subunits of human transferrin, *Acta Chem. Scand.* 21:1686–1694.

Johansson, B. G., 1958. Chromatographic separation of lactalbumin from human milk whey on calcium phosphate columns, *Nature (London)* 181:996–997.

Johansson, B. G., 1960. Isolation of an iron-containing red protein from human milk, *Acta Chem. Scand.* 14:510–512.

Johansson, B. G., 1969. Isolation of crystalline lactoferrin from human milk, *Acta Chem. Scand.* 23:683–684.

Jollès, J., Charet, P., Jollès, P., and Montreuil, J., 1974. Sequence studies concerning human serum transferrin: The primary structure of two cyanogen bromide fragments, *FEBS Lett.* 46:276–280.

Jollès, J., Mazurier, J., Boutigue, M.-H., Spik, G., Montreuil, J., and Jollès, P., 1976. The N-terminal sequence of human lactotransferrins: Its close homology with the amino-terminal region of other transferrins, *FEBS Lett.* 69:27–31.

Jordan, S. M., Kaldor, I., and Morgan, E. H., 1967. Milk and serum iron and iron-binding capacity in the rabbit, *Nature (London)* 215:76–77.

Karlsson, K.-A., Pascher, I., Samuelsson, B. E., Finne, J., Krusius, T., and Rauvala, H., 1978. Mass spectrometric sequence study of the oligosaccharide of human transferrin, *FEBS Lett.* 94:413–417.

Katz, J. H., 1961. Iron and protein kinetic studies by means of doubly labeled human crystalline transferrin, *J. Clin. Invest.* 40:2143–2152.

Keller, W., and Pennell, R. B., 1959. Sterilization of plasma components by heat. 1. β_1-metal combining protein, *J. Lab Clin. Med.* 53:638–645.

Kingston, I. B., and Williams, J., 1975. The amino acid sequence of a carbohydrate-containing fragment of hen ovotransferrin, *Biochem. J.* 147:463–472.

Kinkade, J. M., Kendall-Miller III, W. W., and Segars, F. M., 1976. Isolation and characterization of murine lactoferrin, *Biochim. Biophys. Acta* 446:407–418.

Koechlin, B. A., 1952. Preparation and properties of serum and plasma proteins. XXVIII. The β_1-metal-combining protein of human plasma, *J. Am. Chem. Soc.* 74:2649–2653.

Komatsu, S. K., and Feeney, R. E., 1967. Role of tyrosyl groups in metal binding properties of transferrins, *Biochemistry* 6:1136–1141.

Kornfeld, 1968. The effects of structural modification on the biologic activity of human transferrin, *Biochemistry* 7:945–954.

Kourilsky, F. M., and Burtin, P., 1968. Immunochemical difference between iron-saturated and unsaturated human transferrin, *Nature (London)* 218:375–377.

Krysteva, M. A., Mazurier, J., Spik, G., and Montreuil, J., 1975. Comparative study on histidine modification by diethylpyrocarbonate in human serotransferin and lactotransferrin, *FEBS Lett.* 56:337–340.

Krysteva, M. A., Mazurier, J., and Spik, G., 1976. Ultraviolet difference spectral studies of human serotransferrin and lactotransferrin, *Biochim. Biophys. Acta* 453:484–493.

Kühnl, P., and Spielmann, W., 1978. Transferrin: Evidence for two common subtypes of TfC allele, *Hum Genet.* 43:91–95.

Lane, R. S., 1971. DEAE–cellulose chromatography of human transferrin. The effect of increasing iron saturation and copper(II) binding, *Biochim. Biophys. Acta* 243:193–202.

Lane, R. S., 1975. Differences between human Fe_I–transferrin molecules, *Br. J. Haematol.* 29:511–520.

Larsen, B., Snyder, I. S., and Galash, R. P., 1973. Transferrin concentration in human amniotic fluid, *Am. J. Obstet. Gynecol.* 117:952–954.

Laurell, C.-B., and Ingelman, B., 1947. The iron-binding protein of swine serum, *Acta Chem. Scand.* 1:770–776.

Lee, D. C., McKnight, G. S., and Palmiter, R. D., 1978. The action of estrogen and progesterone on the expression of the transferrin gene. *J. Biol. Chem.* 253:3494–3503.

Leger, D., Verbert, A., Loucheux, M.-H., and Spik, G., 1977. Etude de la masse moleculaire de la lactotransferrine et de la serotransferrine humaine, *Ann. Biol. Anim. Biochim. Biophys.* 17:737–747.

Leger, D., Tordera, V., Spik, G., Dorland, L., Haverkamp, J., and Vliegenthart, J. F. G., 1978. Structure determination of the single glycan of rabbit serotransferrin by methylation analysis and 360 MHz ^1H NMR spectroscopy, *FEBS Lett.* 93:255–260.

Lehrer, S. S., 1969. Fluorescence and absorption studies of the binding of copper and iron to transferrin, *J. Biol. Chem.* 244:3613–3617.

Leibman, A. J., and Aisen, P., 1967. Preparation of single crystals of transferrin, *Arch. Biochem. Biophys.* 121:717–719.

Leibman, A., and Aisen, P., 1979. Distribution of iron between the binding sites of transferrin in serum: Methods and results in normal human subjects, *Blood* 53:1058–1065.

Lestas, A. N., 1976. The effect of pH upon human transferrin: Selective labelling of the two iron-binding sites, *Br. J. Haematol.* 32:341–350.

Line, W. F., Grohlich, D., and Bezkorovainy, A., 1967. The effect of chemical modification on the iron-binding properties of human transferrin, *Biochemistry* 6:3393–3402.

Line, W. F., Sly, D. A., and Bezkorovainy, A., 1976. Limited cleavage of human lactoferrin with pepsin, *Int. J. Biochem.* 7:203–208.

Longsworth, L. G., Cannan, R. K., and McInnes, D. A., 1948. Quoted by Bain, J. A., and Deutsch, H. F., 1948. Separation and characterization of conalbumin, *J. Am. Chem. Soc.* 172:547–555.

Lönnerdal, B., Forsum, E., and Hambraeus, 1976a. A longitudinal study of the protein, nitrogen, and lactose contents of human milk from Swedish well-nourished mothers, *Am. J. Clin. Nutr.* 29:1127–1133.

Lönnerdal, B., Forsum, E., and Hambraeus, L., 1976b. Protein content of human milk. I. A transversal study of Swedish normal material, *Nutr. Rep. Int.* 13:125–134.

Luk, C. K., 1971. Study of the nature of the metal-binding sites and estimate of the distance between the metal-binding sites in transferrin using trivalent lanthanide ions as fluorescent probes, *Biochemistry* 10:2838–2843.

MacGillivray, R. T. A., and Brew, K., 1975. Transferrin: Internal homology in the amino acid sequence, *Science* 190:1306–1307.

MacGillivray, R. T. A., Mendez, E., and Brew, K., 1977. Structure and evolution of serum transferrin, *in Proteins of Iron Metabolism,* E. B. Brown, P. Aisen, J. Fielding, and R. R. Crichton (eds.), Grune & Stratton, New York, pp. 133–141.

Magdoff-Fairchild, B., and Low, B. W., 1970. Preliminary X-ray crystallographic study of porcine transferrin, *Arch. Biochem. Biophys.* 138:703–705.

Makey, D. G., and Seal, U. S., 1976. The detection of four molecular forms of human transferrin during the iron binding process, *Biochim. Biophys. Acta* 453:250–256.

Malmquist, J., 1972. Serum lactoferrin in leukemia and polycythemia vera, *Scand. J. Haematol.* 9:305–310.

Malmquist, J., Hansen, N. E., and Karle, H., 1978. Lactoferrin in haematology, *Scand. J. Haematol.* 21:5–8.

Mann, K. G., Fish, W. W., Cox, A. C., and Tanford, C., 1970. Single-chain nature of human serum transferrin, *Biochemistry* 9:1348–1354.

Markkanen, T., Virtanen, S., Himanen, P., and Pajula, R.-L., 1972. Transferrin, the third carrier protein of folic acid activity in human serum, *Acta Haematol.* 48:213–217.

Martinsson, K., and Möllerberg, L., 1973. On the transferrin concentration in blood serum of growing calves and in bovine colostrum, *Zbl. Vet. Med.* 20:277–284.

Masson, P. L., and Heremans, J., 1968. Metal-combining properties of human lactoferrin (red milk protein). I. The involvement of bicarbonate in the reaction, *Eur. J. Biochem.* 6:579–584.

Masson, P. L., and Heremans, J. F., 1971. Lactoferrin in milk from different species, *Comp. Biochem. Physiol. B* 39:119–129.

Masson, P. L., Heremans, J. F., and Schonne, E., 1969. Lactoferrin, an iron-binding protein in neutrophilic leukocytes, *J. Exp. Med.* 130:643–658.

Mazurier, J., Spik, G., and Montreuil, J., 1974. Isolation and characterization of the cyanogen bromide fragments from human lactotransferrin, *FEBS Lett.* 48:262–265.

Mazurier, J., Aubert, J.-P., Loucheux-Lefevre, M.-H., and Spik, G., 1976. Comparative circular dichroism studies of iron-free and iron-saturated forms of human serotransferrin and lactotransferrin, *FEBS Lett.* 66:238–242.

Mazurier, J., Lhoste, J.-M., Spik, G., and Montreuil, J., 1977. The two metal-binding sites of human serotransferrin and lactotransferrin, *FEBS Lett.* 81:371–375.

Meares, C. F., and Ledbetter, J. E., 1977. Energy transfer between terbium and iron bound to transferrin: Reinvestigation of the distance between metal binding sites, *Biochemistry* 16:5178–5180.

Metz-Boutigue, M.-H., Jollès, J., Mazurier, J., Spik, G., Montreuil, J., and Jollès, P., 1978. Structural studies concerning human lactotransferrin: Its relatedness with human serum transferrin and evidence for internal homology, *Biochimie* 60:557–561.

Milne, J. G. C., 1978. Derivation of equations relating the fractional saturation of N-terminal and C-terminal sites to the number of bound iron atoms per molecule of ovotransferrin, *Biochem. J.* 173:541–542.

Montreuil, J., and Spik, G., 1975. Comparative studies of carbohydrate and protein moieties of human serotransferrin and lactotransferrin, *in Proteins of Iron Storage and Transport in Biochemistry and Medicine,* R. R. Crichton, (ed.), North-Holland, Amsterdam, pp. 27–38.

Montreuil, J., Tonnelat, J., and Mullet, S., 1960. Preparation et propertes de la lactosiderophiline (lactotransferrine) du lait de femme, *Biochim. Biophys. Acta* 45:413–421.

Morgan, E. H., 1966. Transferrin and albumin distribution and turnover in the rat, *Am. J. Physiol.* 211:1486–1494.

Morgan, E. H., 1969. Factors affecting the synthesis of transferrin by rat tissue slices, *J. Biol. Chem.* 244:4193–4199.

Morton, A. G., and Tavill, A. S., 1977. The role of iron in the regulation of hepatic transferrin synthesis, *Br. J. Haematol.* 36:383–394.

Morton, A. G., and Tavill, A. S., 1978. The control of hepatic iron uptake: Correlation with transferrin synthesis, *Br. J. Haematol.* 39:497–507.

Morton, A., Hamilton, S. M., Ramsden, D. B., and Tavill, A. S., 1976. Studies on regulatory factors in transferrin metabolism in man and the experimental rat, *in Plasma Protein Turnover,* R. Bianchi, G. Mariani, and A. S. McFarlane (eds.), University Park Press, Baltimore, pp. 165–175.

Nagasawa, T., Kiyosawa, I., and Kuwahara, K., 1972. Amounts of lactoferrin in human colostrum and milk, *J. Dairy Sci.* 55:1651–1659.

Nagasawa, T., Kiyosawa, I., and Takase, M., 1974. Lactoferrin and serum albumin of human casein in colostrum and milk, *J. Dairy Sci.* 57:1159–1163.

Nagler, A. L., Kochwa, S., and Wasserman, L. R., 1962. Improved isolation of purified siderophilin from individual sera, *Proc. Soc. Exp. Biol. Med.* 111:746–749.

Nagy, B., and Lehrer, S. S., 1972. Circular dichroism of iron, copper, and zinc complexes of transferrin, *Arch. Biochem. Biophys.* 148:27–36.

Najarian, R. C., Harris, D. C., and Aisen, P., 1978. Oxalate and spin-labeled oxalate as probes of the anion binding site of human transferrin, *J. Biol. Chem.* 253:38–42.

Okada, S., Rossman, M. D., and Brown, E. B., 1978. The effect of acid pH and citrate on the release and exchange of iron on rat transferrin, *Biochim. Biophys. Acta* 543:72–81.

Okada, S., Jarvis, B., and Brown, E. B., 1979. *In vivo* evidence for the functional heterogeneity of transferrin-bound iron. V. Isotransferrins: An explanation of the Fletcher–Huehns phenomenon in the rat, *J. Lab. Clin. Med.* 93:189–198.

Oncley, J. L., Scatchard, G., and Brown, A., 1947. Physical–chemical characteristics of certain of the proteins of normal human plasma, *J. Phys. Chem.* 51:184–198.

Osborne, T. B., and Campbell, G. F., 1900. The protein constituents of egg white, *J. Am. Chem. Soc.* 22:422–450.

Palmiter, R. D., 1972. Regulation of protein synthesis in chick oviduct, *J. Biol. Chem.* 247:6450–6461.

Palmour, R. M., and Sutton, H. E., 1971. Vertebrate transferrins. Molecular weights, chemical compositions and iron-binding studies, *Biochemistry* 10:4026–4032.

Phillips, J. L., and Azari, P., 1971. On the structure of ovotransferrin. Isolation and characterization of the cyanogen bromide fragments and evidence for a duplicate structure, *Biochemistry* 10:1160–1165.

Phillips, J. L., and Azari, P. 1972. Iodination of ovotransferrin and its iron complex. Extent of involvement of tyrosine phenolic groups in the iron binding, *Arch. Biochem. Biophys.* 151:445–452.

Pinkowitz, R. A., and Aisen, P., 1972. Zero-field splittings of iron complexes of transferrins, *J. Biol. Chem.* 247:7830–7834.

Price, E. M., and Gibson, J. F., 1972a. A re-interpretation of bicarbonate-free ferric transferrin E.P.R. spectra, *Biochem. Biophys. Res. Commun.* 46:646–651.

Price, E. M., and Gibson, J. F., 1972b. Electron paramagnetic resonance evidence for a distinction between the two iron-binding sites in transferrin and in conalbumin, *J. Biol. Chem.* 247:8031–8035.

Prieels, J.-P., Pizzo, S. V., Glasgow, L. R., Paulson, J. C., and Hill, R., 1978. Hepatic receptor that specifically binds oligosaccharides containing fucosyl $\alpha 1 \rightarrow 3$ *N*-acetylglucosamine linkages, *Proc. Natl. Acad. Sci. U.S.A.* 75:2215–2219.

Princiotto, J. V., and Zapolski, E. J., 1975. Difference between the two iron-binding sites of transferrin, *Nature (London)* 255:87–88.

Querinjean, P., Masson, P. L., and Heremans, J. F., 1971. Molecular weight, single-chain structure, and amino acid composition of human lactoferrin, *Eur. J. Biochem.* 20:420–425.

Reddy, V., Bhaskaran, C., Raghuramulu, N., and Jagedeesan, V., 1977. Antimicrobial factors in human milk, *Acta Paediatr. Scand.* 66:229–232.

Regoeczi E., Hatton, M. W. C., and Wong, K.-L., 1974. Studies on the metabolism of asialotransferrins: Potentiation of the catabolism of human asialotransferrin in the rabbit, *Can. J. Biochem.* 52:155–161.

Regoeczi, E., Wong, K.-L., Ali, M., and Hatton, M. W. C., 1977. The molecular components of human transferrin type C, *Int. J. Pept. Protein Res.* 10:17–26.

Regoeczi, E., Taylor, P., Hatton, M. W. C., Wong, K.-L., and Koj, A., 1978. Distinction between binding and endocytosis of human asialotransferrin by the rat liver, *Biochem. J.* 174:171–178.

Richardson, N. E., Buttress, N., Feinstein, A., Stratil, A., and Spooner, R. L., 1973. Structural studies on individual components of bovine transferrin, *Biochem. J.* 135:87–92.

Roberts, R. C., Makey, D. G., and Seal, U. S., 1966. Human transferrin. Molecular weight and sedimentation properties, *J. Biol. Chem.* 241:4907–4913.

Rogers, T. B., Gold, R. A., and Feeney, R. E., 1977. Ethoxyformylation and photooxidation of histidines in transferrins, *Biochemistry* 16:2299–2305.

Rogers, T. B., Børresen, T., and Feeney, R. E., 1978. Chemical modification of the arginines in transferrins, *Biochemistry* 17:1105–1109.

Roop, W. E., and Putnam, F. W., 1967. Purification and properties of human transferrin *C* and a slow moving genetic variant, *J. Biol. Chem.* 242:2507–2513.

Roop, W. E., Roop, B. L., and Putnam, F. W., 1968. Transferrin variants among blood donors, *Vox Sang.* 14:255–257.

Rosseneu-Motreff, M. Y., Soetewey, F., Lamote, R., and Peeters, H., 1971. Size and shape determination of apotransferrin and transferrin monomers, *Biopolymers* 10:1039–1048.

Rümke, P., Visser, D., Kwa, H. G., and Hart, A. A. M., 1971. Radioimmunoassay of lactoferrin in blood plasma of breast cancer patients, lactating and normal women; prevention of false high levels caused by leakage from neutrophile leucocytes *in vitro*, *Folia Med. Neerl.* 14:156–168.

Schade, A. L., and Caroline, L., 1944. Raw hen egg white and the role of iron in growth inhibition of *Shigella dysenteriae*, *Staphylococcus aureus*, *Escherichia coli*, and *Saccharomyces cerevisiae*, *Science* 100:14–15.

Schade, A. L., and Caroline, L., 1946. An iron-binding component in human blood plasma, *Science* 104:340–341.

Schade, A. L., Reinhart, R. W., and Levy, H., 1949. Carbon dioxide and oxygen complex formation with iron and siderophilin, the iron-binding component of human plasma, *Arch. Biochem. Biophys.* 20:170–172.

Schlabach, M. R., and Bates, G. W., 1975. The synergistic binding of anions and Fe^{3+} by transferrin, *J. Biol. Chem.* 250:2182–2188.

Schultze, H. E., Heide, K., and Müller, H., 1957. *Über Transferrin/Siderophilin*, *Behringwerkemitt*, No. 32, pp. 25–48.

Serafini, N. A., and Serra, A., 1968. Plasma transferrin phenotype and gene frequencies in the population of Rome, *Humangenetik.* 6:142–147.

Shepp, M., Yamada, H., Berenfeld, M., and Gabuzda, T. G., 1973. ^{125}I–transferrin turnover in rabbits with haemolytic anaemia, *Br. J. Haematol.* 24:261–266.

Sly, D. A., and Bezkorovainy, A., 1974. Carboxy-terminal residue of human serum transferrin, *Physiol. Chem. Phys.* 6:171–177.

Spik, G., and Mazurier, J., 1977. Comparative structural and conformational studies of polypeptide chain, carbohydrate moiety and binding sites of human serotransferrin and lactotransferrin, *in Proteins of Iron Metabolism*, E. B. Brown, P. Aisen, J. Fielding and R. R. Crichton (eds.), Grune & Stratton, New York, pp. 143–151.

Spik, G., Bayard, B., Fournet, B., Strecker, G., Bouquelet, S., and Montreuil, J., 1975. Studies on glycoconjugates. LXIV. Complete structure of two carbohydrate units of human serotransferrin, *FEBS Lett.* 50:296–299.

Spik, G., Fournet, B., and Monteuil, J., 1979. Etude de la structure du glycanne de l'ovotransferrine de poule, *C. R. Acad. Sci. Ser. D.* 288:967–970.

Spooner, R. L., Oliver, R. A., Richardson, N., Buttress, N., Feinstein, A., Maddy, A. H., and Stratil, A., 1975. Isolation and partial characterization of sheep transferrin, *Comp. Biochem. Physiol. B* 52:515–522.

Stjernholm, R., Warner, F. W., Robinson, J. W., Ezekiel, E., and Katayama, N., 1978. Binding of platinum to human transferrin, *Bioinorg. Chem.* 9:277–280.

Stratil, A., and Spooner, R. L., 1971. Isolation and properties of individual components of cattle transferrin: The role of sialic acid, *Biochem. Genet.* 5:347–365.

Strickland, D. K., and Hudson, B. G., 1978. Structureal studies on rabbit transferrin: Isolation and characterization of the glycopeptides, *Biochemistry* 17:3411–3418.

Strickland, D. K., Hamilton, J. W., and Hudson, B. G., 1979. Structure of the tryptic glycopeptide isolated from rabbit transferrin, *Biochemistry* 18:2549–2554.

Sutton, H. E., and Karp, G. W., Jr., 1965. Adsorption of rivanol by potato starch in the isolation of transferrins, *Biochim. Biophys. Acta* 107:153–154.

Sutton, M. R., and Brew, K., 1974a. Purification and characterization of the seven cyanogen bromide fragments of human serum transferrin, *Biochem. J.* 139:163–168.

Sutton, M. R., and Brew, K., 1974b. The sequence of residues 1–26 of human serum transferrin, *FEBS Lett.* 40:146–152.

Sutton, M. R., MacGillivray, R. T. A., and Brew, K., 1975. The amino acid sequences of three cystine-free cyanogen bromide fragments of human serum transferrin, *Eur. J. Biochem.* 51:43–48.

Szuchet-Derechin, S., and Johnson, P., 1962. Red proteins from bovine milk, *Nature (London)* 194:473–474.

Tan, A. T., 1971. Circular dichroism properties of conalbumin and its iron and copper complexes, *Can. J. Biochem.* 49:1071–1075.

Tan, A. T., and Woodworth, R. C., 1969. Ultraviolet difference spectral studies of conalbumin complexes with transition metal ions, *Biochemistry* 8:3711–3716.

Tengerdy, C., Azari, P., and Tengerdy, R. P., 1966. Immunochemical reactions of conalbumin and its metal complexes, *Nature (London)* 211:203–204.

Teuwissen, B., Masson, P. L., Osinski, P., and Heremans, J. F., 1972. Metal-combining properties of human lactoferrin. The possible involvement of tyrosyl residues in the binding sites. Spectrophotometric titration, *Eur. J. Biochem.* 31:239–245.

Teuwissen, B., Masson, P. L., Osinski, P., and Heremans, J. F., 1973. Metal-combining properties of human lactoferrin. The effect of nitration of lactoferrin with tetranitromethane, *Eur. J. Biochem.* 35:366–371.

Teuwissen, B., Schanck, K., Masson, P. L., Osinski, P. A., and Heremans, J. F., 1974. The denaturation of lactoferrin and transferrin by urea, *Eur. J. Biochem.* 42:411–417.

Thibodeau, S. N., Lee, D. C., and Palmiter, R. D., 1978. Identical precursors for serum transferrin and egg white conalbumin, *J. Biol. Chem.* 253:3771–3774.

Thymann, M., 1978. Identification of a new serum protein polymorphism as transferrin, *Humangenet.* 43:225–229.

Tomimatsu, Y., and Donovan, J. W., 1976. Spectroscopic evidence for perturbation of tryptophan in Al(III) and Ga(III) binding to ovotransferrin and human serum transferrin, *FEBS Lett.* 71:299–302.

Tomimatsu, Y., and Vickery, L. E., 1972. Circular dichroism studies of human serum transferrin and chicken ovotransferrin and their copper complexes, *Biochim. Biohphys. Acta* 285:72–83.

Tomimatsu, Y., Kint, S., and Scherer, J. R., 1976. Resonance Raman spectra of iron(III)–, copper(II)–, cobalt(III)–, and manganese(III)–transferrins and of bis(2.4,6-trichlorophenolato)diimidazole copper(II) monohydrate, a possible model for copper(II) binding to transferrins, *Biochemistry* 15:4918–4924.

Tsang, C. P., Boyle, A. J. F., and E. H. Morgan, 1975. Mossbauer spectra of bicarbonate-free ferric transferrin complex, *Biochim. Biophys. Acta* 386:32–40.

Tsao, D., Azari, P., and Phillips, J. L., 1974a, On the structure of ovotransferrin. I. Isolation and characterization of cyanogen bromide fragments. Reevaluation of the primary structure, *Biochemistry* 13:397–403.

Tsao, D., Morris, D. H., Azari, P., Tengerdy,R. P., and Phillips, J., 1974b. On the structure of ovotransferrin. II. Isolation and characterization of a specific iron-binding fragment after cyanogen bromide cleavage, *Biochemistry* 13:403–407.

Tsao, D., Azari, P., and Phillips, J. L., 1974c. On the structure of ovotransferrin. III. Nitration of iron-ovotransferrin and distribution of tyrosines involved in iron-binding activity, *Biochemistry* 13:408–413.

Ulmer, D. D., 1969. Effect of metal binding on the hydrogen–tritium exchange of conalbumin, *Biochim. Biophys. Acta* 181:305–310.

Vahlquist, B. C. S., 1941. Das Serumeisen. Eine pädiatrisch-klinische und experimentelle Studie, *Acta Paediatr. Suppl.* 5) 28.

Van Eijk, H. G., and Kroos, M. J., 1978. The iron status in healthy individuals aged from 18–25 years, *Folia Haematol.* (Leipzig) 105:93–95.

Van Eijk, H. G., Vermaat, R. J., and Leijnse, B., 1969a. The isoelectric fractionation and properties of rabbit transferrin, *FEBS Lett.* 3:193–194.

Van Eijk, H. G., Penders, T. J., and Leijnse, B., 1969b. Fractionation and properties of rabbit transferrin, *Arch. Int. Pharmacodyn. Ther.* 178:481–482.

Van Eijk, H. G., van Dijk, J. P., van Noort, W. L., Leijnse, B., and Monfoort, C. H., 1972. Isolation and analysis of transferrins from different species, *Scand. J. Haematol.* 9:267–270.

Van Eijk, H. G., van Noort, W. L., Kroos, M. J., and van der Heul, C., 1978. Analysis of the iron-binding sites of transferrin by isoelectric focusing, *J. Clin. Chem. Clin. Biochem.* 16:557–560.

Van Snick, J. L., Masson, P. L., and Heremans, J. F., 1973. The involvement of bicarbonate in the binding of iron by transferrin, *Biochim. Biophys. Acta* 322:231–233.

Van Vugt, H., van Gool, J., Ladiges, N. C. J. J., and Boers, W., 1975. Lactoferrin in rabbit bile: Its relation to iron metabolism, *Q. J. Exp. Physiol.* 60:79–88.

Wallach, J., 1974. *Interpretation of Diagnostic Tests,* 2nd ed., Little, Brown, Boston, pp. 14, 15, 61.

Wang, A.-C., and Sutton, H. E., 1965. Human transferrins *C* and *D*: Chemical difference in a peptide, *Science* 149:435–437.

Wang, A.-C., Sutton, H. E., and Howard, P. N., 1967. Human transferrins *C* and D_{Chi}: An amino acid difference, *Biochem. Genet.* 1:55–59.

Warner, R. C., and Weber, I., 1953. The metal-combining properties of conalbumin. *J. Am. Chem. Soc.* 75:5094–5101.

Weiner, R. E., and Szuchet, S., 1975. The molecular weight of bovine lactoferrin, *Biochim. Biophys. Acta* 393:143–147.

Weippl, G. Pantlitschko, M., and Priebe, H., 1973. Normalwerte und Verteilung von Transferrin beim Erwachsenen, *Blut* 17:376–383.

Wenn, R. V., and Williams, J., 1968. The isoelectric fractionation of hen's egg ovotransferrin, *Biochem. J.* 108:69–74.

Williams, J., 1962. A comparison of conalbumin and transferrin in the domestic fowl, *Biochem. J.* 83:355–364.

Williams, J., 1968. A comparison of glycopeptides from the ovotransferrin and serum transferrin of the hen, *Biochem. J.* 108:57–67.

Williams, J., 1974. The formation of iron-binding fragments of hen ovotransferrin by limited proteolysis, *Biochem. J.* 141:745–752.

Williams, J., 1975. Iron-binding fragments from the carboxyl-terminal region of ovotransferrin, *Biochem. J.* 149:237–244.

Williams, J., 1979. "The distribution of iron in human transferrin under physiological conditions," Abstract C, Fourth International Conference on Proteins of Iron Metabolism, Davos, Switzerland, April 17–21, 1979.

Williams, J., and Wenn, R. V., 1970. Radioactive peptide "maps" of the major and minor components of hen's egg ovotransferrin, *Biochem. J.* 116:533–535.

Williams, J., Phelps, C. F., and Lowe, J. M., 1970. Ovotransferrin with one iron atom, *Nature (London)* 226:858–859.

Williams, J., Evans, R. W., and Moreton, K., 1978. The iron-binding properties of hen ovotransferrin, *Biochem. J.* 173:535–542.

Windle, J. J., Wiersema, A. K., Clark, J. R., and Feeney, R. E., 1963. Investigation of the iron and copper complexes of avian conalbumins and human transferrins by electron paramagnetic resonance, *Biochemistry* 2:1341–1345.

Wishnia, A., Weber, I., and Warner, R. C., 1961. The hydrogen ion equilibria of conalbumin, *J. Am. Chem. Soc.* 83:2071–2080.

Wong, K.-L., Debaune, M. T., Hatton, M. W. C., and Regoeczi, E., 1978. Human transferrin, asialotransferrin, and the intermediate forms, *Int. J. Protein Res.* 12:27–37.

Woodworth, R. C., Morallee, K. G., and Williams, R. J. P., 1970. Perturbations of the proton magnetic resonance spectra of conalbumin and siderophilin as a result of binding Ga^{3+} or Fe^{3+}, *Biochemistry* 9:839–842.

Yeh, Y., Iwai, S., and Feeney, R., 1979. Conformations of denatured and renatured ovotransferrin, *Biochemistry* 18:882–889.

Zschocke, R. H., and Bezkorovainy, A., 1970. Some immunochemical properties of succinylated human transferrin and related proteins, *Biochim. Biophys. Acta* 200:241–246.

Zschocke, R. H., Chiao, M. T., and Bezkorovainy, A., 1972. The function of amino groups in the binding of iron by transferrin, *Eur. J. Biochem.* 27:147–152.

Zweier, J. L., 1978. An electron paramagnetic resonance study of single site copper complexes of transferrin, *J. Biol. Chem.* 253:7616–7621.

Zweier, J. L., and Aisen, P., 1977. Studies of transferrin with the use of Cu^{2+} as an electron paramagnetic resonance spectroscopic probe, *J. Biol. Chem.* 252:6090–6096.

Zweier, J., Aisen, P., Peisach, J., and Mims, W. B., 1979. Pulsed electron paramagnetic resonance studies of the copper complexes of transferrin, *J. Biol. Chem.* 254:3512–3515.

Chemistry and Biology of Iron Storage

5

5.1 Distribution of Ferritin and Hemosiderin

It is said that nearly 20% of all iron in the human organism is associated with the protein ferritin and hemosiderin, which represent the storage forms of iron. Ferritin, but not hemosiderin, is water soluble. For this reason, ferritin has been rather thoroughly characterized, whereas little information is as yet available on the properties of hemosiderin. The latter can be visualized both via the light and electron microscopes and appears in the form of granular structures. Such granules react with Perls' solution [$K_4Fe(CN)_6$] to give the Prussian blue color. Ferritin also reacts with Perls' solution; however, under the light microscope, ferritin appears as a diffuse blue coloration. Ferritin particles can be discerned via the electron microscope. Ferritin is present in the cytosol and the lysosomes and to a much lesser extent in other subcellular structures. All available evidence indicates that insofar as the liver is concerned, hemosiderin is normally located almost exclusively in the Kupffer cells, whereas ferritin is present largely in the parenchymal cells. Both hemosiderin and ferritin are also present in other tissues of the mammalian organism (Arora *et al.*, 1970), and ferritin is found in both vertebrate and invertebrate animals as well as in some plants and even in microorganisms (Richter, 1978).

Hemosiderin is best observed in the iron-overloaded organism, though it occurs in normal cells as well. It has been postulated that hemosiderin is nothing more than a degraded or denatured and polymerized form of ferritin; hence its iron to protein ratio is greater than that of ferritin. The atomic structures of the iron oxide cores of ferritin and hemosiderin cannot be distinguished from each other by the methods of X-ray diffraction, Mössbauer spectroscopy, and electron microscopy (Fischbach *et al.*, 1971). A less popular view holds that hemosiderin is nothing more than iron-laden mitochondria (McKay and Fineberg, 1964). The formation of hemosiderin has been followed histologically in animals that were in the process of being loaded with iron. One first observes the formation of aggregated water-soluble ferritin crystals that appear to be granular in nature and have the ability to give the Prussian blue color. This is follwed by the appearance of water-insoluble amorphous Prussian-blue-staining granules that are typically hemosiderin in nature. Hemosiderin granules first appear in the parenchymal cells of the iron-loaded rabbit when the ferritin content is 20–30 times normal (150–200 mg of iron per liver, normal weight of rabbit liver being 100 g). Iron loading with iron–dextran, ferrous sulfate, or whole blood transfusions leads to iron deposition and the appearance of hemosiderin primarily in the parenchymal cells, whereas loading with iron oxide results in iron accumulation in the Kupffer cells. This perhaps reflects the phagocytic origin of the Kupffer cell iron (Sturgeon and Shoden, 1969). These relationships are illustrated in Table 5–1.

Hemosiderin was probably observed by numerous histologists of the nineteenth century; however, it was E. Neumann who in 1888 first distinguished the granular Prussian blue positive pigment from others and coined the term *hemosiderin*. The supposition was that hemosiderin originated from hemoglobin and thus contained heme. Ferritin, on the other

Table 5-1. Results of Loading Rabbits with Iron[a]

	Iron injected (mg)						
	0	50	100	200	400	800	1600
	Iron oxide						
Total liver iron (mg)	7	38	49	94	213	495	1055
Water-soluble iron (mg)[b]	6	20	28	65	152	354	534
Water-insoluble iron (mg)[b]	1	18	21	28	61	141	521
Parenchymal cell granules	0	0	0	0	±	+ +	+ + +
Kupffer cell granules	0	+	+	+	+ +	+ +	+ + +
	Iron dextran						
Total liver iron (mg)	7	30	54	98	210	408	1048
Water-soluble iron (mg)[b]	6	28	51	88	191	290	494
Water-insoluble iron (mg)[b]	1	2	3	10	19	118	554
Parenchymal cell granules	0	0	0	±	+	+ + +	+ + + +
Kupffer cell granules	0	0	0	±	±	±	±

[a] From Sturgeon and Shoden (1969).
[b] Water-soluble iron = ferritin; water-insoluble iron = hemosiderin.

hand, was apparently first purified in 1894 by Otto Schmiedeberg, a pharmacologist first at Dorpat and then at Strasbourg. He gave it the name *ferratin*, but his preparations contained only 6% iron. The protein was crystallized in 1937 by V. Laufberger, who found its iron content to be near 20%. Laufberger coined the presently used term *ferritin*. Much of the earlier physical–chemical characterization of ferritin that followed was done by S. Granick and P. F. Hahn (Granick, 1946a, 1951). Unfortunately, little if any information is available on the physical–chemical nature of hemosiderin, as its purification is yet to be accomplished if one accepts the view that hemosiderin is a derivative of ferritin. This chapter is therefore devoted almost exclusively to the chemistry and biology of ferritins, which have been the subject of extensive investigations in the past three decades. A number of review articles of recent vintage have summarized our knowledge on this interesting protein (Crichton, 1973a, 1973b, 1973c; Harrison, 1977; Hershko, 1977; Jacobs and Worwood, 1975a; Munro and Linder, 1978; Richter, 1978).

5.2 Isolation of Ferritin

Isolation procedures for ferritin have not changed substantially since its crystallization in 1937 and subsequent modifications of the isolation procedures by Granick in the 1940s. Horse spleen has been the favorite source of ferritin because of the relative abundance of ferritin in this organ and the ease with which it crystallizes.

One of the simpler procedures, described by Pape *et al.* (1968a), involves the homogenization of the spleen, heating the homogenate to 80°C, and then filtering while hot. The filtrate is adjusted to pH 4.6, the precipitate removed by centrifugation, and ferritin crystallized from the supernatant by the addition of cadmium sulfate to a final concentration of 5%. A somewhat more complicated version of the Laufberger–Granick procedure was proposed by Drysdale and Munro (1965), who followed up the pH 4.6 treatment step with a precipitation of ferritin by the addition of ammonium sulfate to the supernatant to 35–50% of saturation. The precipitated ferritin was further purified by chromatography on ion exchangers and/or gel filtration media. Others have omitted the acid precipitation step following the heating of the homogenate (Harris, 1978; Hauser, 1969): Following heating of the spleen homogenate, the precipitated proteins are removed by centrifugation or filtration, and 300 g of $(NH_4)_2SO_4$ per liter of supernatant is added to the supernatant. The precipitated ferritin-rich fraction is further purified by ultracentrifugation or ion-exchange or gel filtration chromatography. Similar procedures were used to isolate porcine (May and Fish, 1977), rabbit (Van Kreel *et al.*,

1972), and human (Powell *et al.*, 1975a) ferritins. A variation of the classical isolation procedure is shown in diagram form in Figure 5–1. If the isolation procedure involves crytallization with 5% $CdSO_4$, the final product may contain as much as 2% Cd. Removal of the Cd may be affected by dialyzing against sodium bisulfite solutions without affecting the iron content (Hegenauer *et al.*, 1979).

Ferritin isolated by the preceding procedures has an average iron content of 20%, which comes to an average of about 1600 atoms of Fe per ferritin molecule (assuming a molecular weight of 450,000 for ferritin). In fact, ferritin preparations contain a spectrum of molecules, differing from each other on the basis of iron content. This may be shown by ultracentrifugation as in Figure 5–2. Hence, to obtain a homogeneous ferritin preparation, it is sometimes desirable to remove the iron and prepare the iron-free protein termed *apoferritin*. This is usually accomplished by reducing the ferric iron of ferritin to the ferrous state followed by dialysis. The most commonly used reducing agents are sodium dithionite or thioglycollic acid solutions containing the ferrous iron chelator αα′-bipyridyl.

Heterogeneity of raw ferritin preparations is also evident upon polyacrylamide or other molecular sieve-type gel electrophoresis (Figure 5–3). This has been shown to be due to the presence of dimers, trimers, and other oligomers of ferritin. It is seen from Figure 5–3 that human ferritin is especially prone to forming oligomers and that the plant ferritin seems to be entirely monomeric. Dimeric ferritin has been isolated and its properties determined (Niitsu and Listowsky, 1973b). Dimers were found to be quite stable and linked through disulfide bonds.

Most ferritin isolation procedures have involved an initial heating step, and there has been concern expressed as to whether or not ferritins prepared by such methods are biologically identical to the nonheated species. Bertrand and Harris (1979) isolated horse spleen ferritin by methods both involving and not involving heat treatment and found that iron was mobilized by reducing agents equally well from both types of ferritins. The conclusion was that the heating step does not damage the ferritin molecule.

5.3 Physical–Chemical Properties of the Ferritins

5.3.1 Gross Structure of Ferritin

Ferritin has been described as a relatively spherical and compact molecule with an outside diameter of 124–130 Å. The interior of the

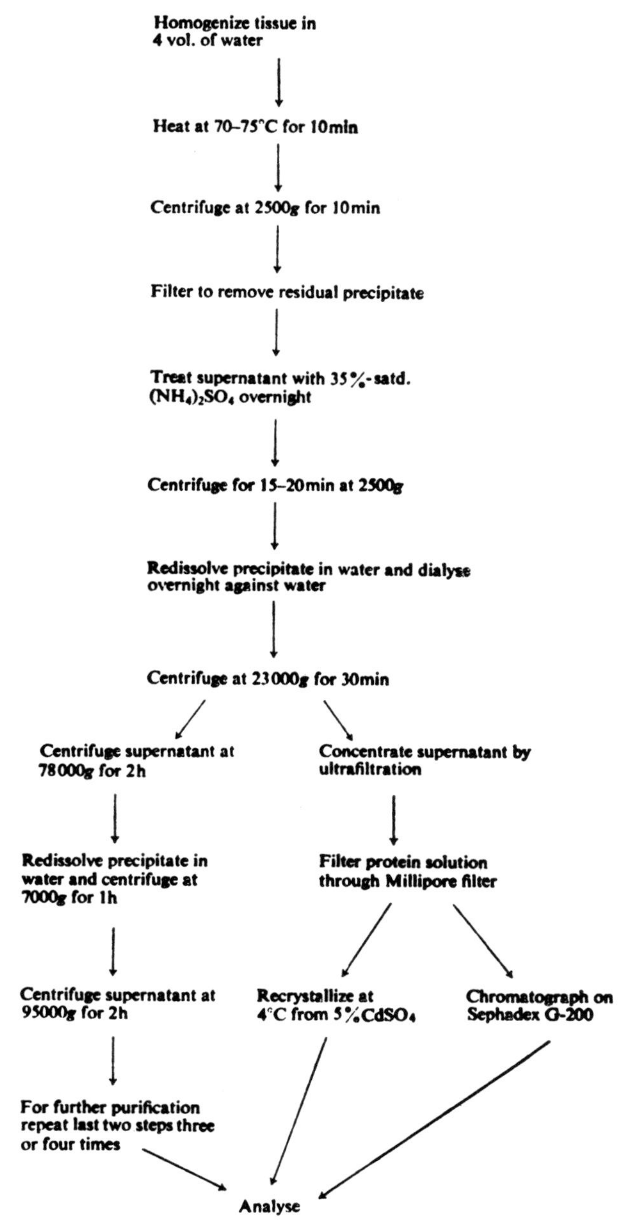

Figure 5-1. A procedure for the isolation of horse spleen ferritin, From Crichton *et al.* (1973a).

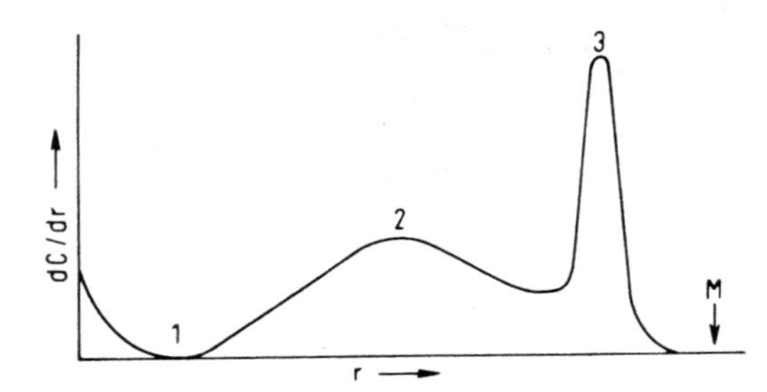

Figure 5-2. Ultracentrifugal pattern of native horse spleen ferritin. M represents the meniscus, 3 is apoferritin, 2 is ferritin of intermediate density, and 1 is ferritin with nearly the maximum amount of iron. From Crichton (1973c).

apoferritin molecule is hollow, the inside diameter of the sphere being 70–80 Å. The shell is about 25 Å thick. Six channels, each 9–12 Å in diameter and situated symmetrically throughout the apoferritin molecule, connect the hollow interior with the outside environment. The channels provide a pathway through which iron and other small molecules such as the bioflavinoids can enter and exit from the apoferritin molecule (Hoare *et al.*, 1975; Harrison, 1977).

The hollow interior of the apoferritin molecule may be filled by the so-called *iron core* or *iron micelle*, its diameter being some 74 Å. There are apparently no specific iron-binding ligands similar to those of transferrin in the ferritin molecule. Instead, iron in ferritin is present in the form of a crystalline aggregate with the composition formula of $(FeOOH)_8(FeOOPO_3H_2)$. Iron core thus contains ferric hydroxide and

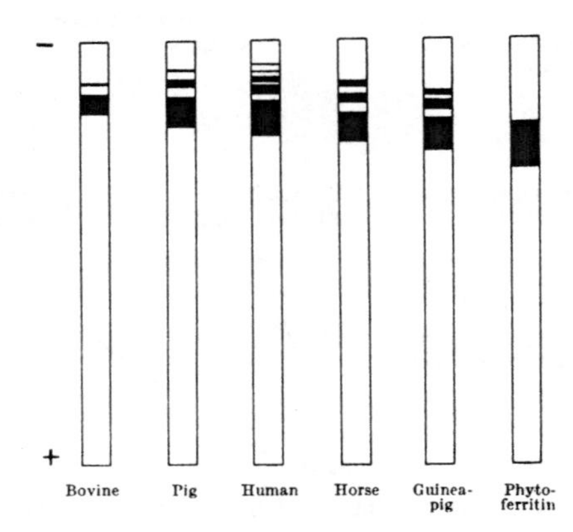

Figure 5-3. Polyacrylamide gel electrophoresis of native ferritins from various sources. From Zamiri and Mason (1968).

ferric phosphate, and iron accounts for about 57% of the iron core weight. X-Ray diffraction patterns for intact ferritin are identical to those of the "naked" iron cores prepared by treating iron-laden ferritin with 1 *N* NaOH. This presumably indicates that the iron core exists as a semi-independent entity in association with the apoferritin molecule. Nevertheless, careful electron microscopic observations have revealed some association between the iron core and its apoferritin envelope (Massover, 1978).

Treatment of ferritin with NaOH also removes some 80% of the phosphate present therein without of course affecting the X-ray diffraction pattern. The conclusion was therefore made that ". . . the phosphate is not an essential structural component of the core, nor does it seem to contribute significantly to the diffraction pattern" (Crichton, 1973c). In reconstitution experiments involving iron, apoferritin, and phosphate, it was found, however, that the reconstituted material could be made to resemble the native ferritin only if phosphate was added last to the preformed iron–apoferritin complex. It was concluded that the phosphate was normally located on the surface of the iron core and probably did not contribute to the integral structure of the iron core crystal (Treffry and Harrison, 1978). A schematic drawing of the iron-laden ferritin molecule is represented in Figure 5–4.

X-Ray diffraction studies of ferritins have revealed patterns very unlike those of any known inorganic iron oxides or hydroxides. The unit cell apparently had a hexagonal symmetry with cell dimensions of a = 2.94 and c = 9.4 Å in one study (Harrison *et al.*, 1967) and a = 11.79 and c = 9.90 Å in another (Crichton, 1973b). The latter study concluded that there were 32 FeOOH molecules per unit cell and that the iron core of fully iron-laden ferritin could accommodate 140 unit cells for a total of about 4500 iron atoms per ferritin molecule. The volume of the iron core would thus be 170,000 $Å^3$.

5.3.2 Molecular Weight of Ferritin and Apoferritin

Ferritin isolated from any source will consist of a spectrum of molecules differing from each other in the amount of iron that they hold. Consequently, a ferritin preparation will show a spectrum of molecular weights, as indicated by the ultracentrifugal pattern in Figure 5–2. The reference point is of course the molecular weight of apoferritin. For the horse spleen material this has been determined to be near 465,000 as early as in 1944 by A. Rothen, while other investigators have established figures somewhat larger (e.g., 480,000 for horse spleen apoferritin and 503,700

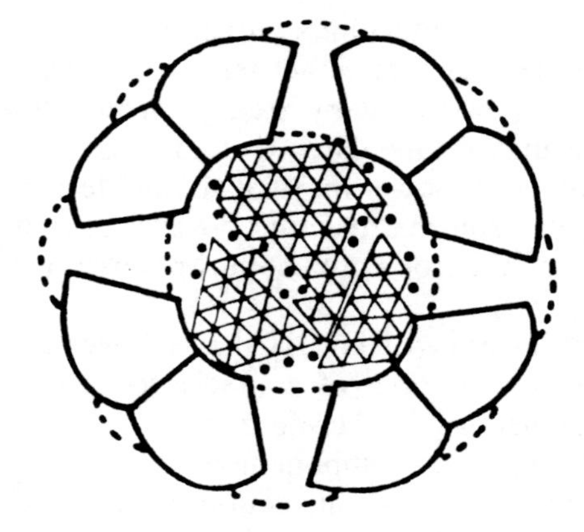

Figure 5-4. A cross section of the ferritin molecule showing subunits, channels leading to the interior of the molecule, the iron crystalloid core (triangular lattices), and phosphate (black dots). From Treffry and Harrison (1978).

for the porcine spleen material) or smaller (e.g., 430,000 for horse spleen apoferritin). Table 5–2 summarizes some molecular-weight data for a number of apoferritin preparations. It is generally agreed that the molecular weight of horse spleen apoferritin should be taken as 450,000.

Normally, horse spleen ferritin contains an average of up to 20% iron when isolated by one of the classical procedures. Such preparations may, of course, be fractionated to yield ferritins of lower or higher iron content. Thus, Niitsu and Listowsky (1973a) subjected horse spleen ferritin and rat liver ferritin to density gradient centrifugation in sucrose and $CsCl_2$ gradients and determined iron distribution therein. The major components, from a density point of view, were apoferritin (10–15%), ferritin containing 1800–2100 iron atoms/molecule in the case of horse spleen and 2000–2400 atoms/molecule in the case of rat liver ferritins (over 50%), and a very heavy fraction containing 3500 and more iron atoms/molecule. The rest (30–35%) were ferritins with an iron content of between 50 and 1800 iron atoms/ferritin molecule. This is illustrated in Figure 5–5.

Ferritins from other tissues or other animals may contain more or less iron. For instance, human serum ferritin contained an average of 5.1% iron, whereas the liver ferritins from the same population contained 16% iron (Zuyderhoudt *et al.*, 1978a). The maximum amount of iron that can be accommodated by an apoferritin molecule amounts to some 4500 atoms, which translates into 442,000 daltons for the iron core, assuming that iron constitutes 57% of the iron core by weight. If it is assumed that apoferritin has a molecular weight of 450,000, then iron-laden ferritin can have a molecular weight of 892,000.

A convenient technique of relating molecular weights of the ferritins to their iron content is by looking at their iron–nitrogen ratios (Fe–N) (Hauser, 1969). The nitrogen content of apoferritin is 16.3%. The normal horse spleen ferritin preparation with 20% iron has an Fe–N = 1.80, whereas a fully iron-laden molecule would have an Fe–N = 3.4. We can then write an expression relating Fe–N to molecular weight as follows:

$$\text{molecular weight} = 450{,}000 + 130{,}000\ (\text{Fe–N})$$

5.3.3 Subunits of the Ferritins

X-Ray and electron microscope studies of the late 1950s and early 1960s suggested that the ferritins might consist of a number of subunits. On the basis of X-ray diffraction data on apoferritin, P. Harrison suggested that apoferritin might consist of 20 identical subunits (Harrison, 1963; Hofmann and Harrison, 1963). Degradation of ferritin into subunits was achieved by first denaturing the protein and then treating it with sodium dodecylsulfate. Later, dissociation was also established with 67% acetic acid (Harrison and Gregory, 1968), and the apoferritin molecule could be reassembled upon the removal of the dissociating agent. The molecular weights of the subunits were estimated to be 23,000, and they were deemed to be identical.

Table 5-2. Molecular Weights and Other Parameters of Apoferritins from Various Sources

Material	Mol. weight	Sedimentation constant (S units)	Reference
Porcine spleen	503,000[a]	17.5	May and Fish, 1977
Horse spleen	480,000[b]	—	Harrison, 1963
	430,000–470,000[c]	17.3	Richter and Walker, 1967
	443,000[a]	—	Crichton, 1972
	465,000[d]	—	Crichton, 1973b
	440,000–465,000[a]	—	Bjork and Fish, 1971
	443,000[e]	—	Vulimiri *et al.*, 1975
	444,000[f]	—	Crichton and Bryce, 1970
	443,000[a]	17.12	Crichton *et al.*, 1973a
Rat liver	493,000[e]	—	Vulimiri *et al.*, 1975
Rat heart (slow)	531,000[e]	—	Vulimiri *et al.*, 1975
Rat heart (fast)	626,000[e]	—	Vulimiri *et al.*, 1975

[a] Sedimentation equilibrium.
[b] X-Ray diffraction.
[c] Light scattering.
[d] Sedimentation–diffusion.
[e] Polyacrylamide gel electrophoresis.
[f] Chemical data.

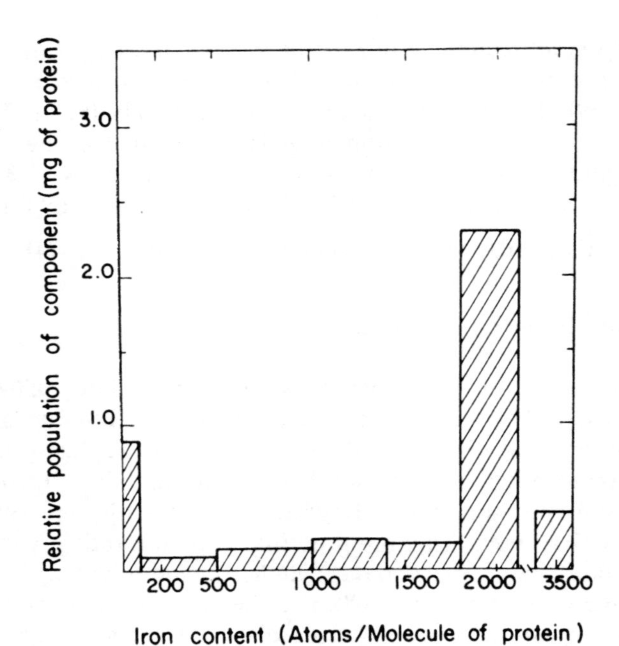

Figure 5-5. Distribution of iron in horse spleen ferritin. It will be noted that the most abundant species contains around 1800–2100 iron atoms per molecule. From Niitsu and Listowsky (1973a).

With the advent of more sophisticated technology, a number of investigators, most notably R. R. Crichton and his colleagues, examined the quaternary structure of ferritin by the method of polyacrylamide gel electrophoresis as well as by the more classical techniques of ultracentrifugation and gel filtration using guanidine–HCl as the dissociating agents. The conclusion reached by Crichton's group (Crichton and Bryce, 1970; Bryce and Crichton, 1971) as well as by others (Bjork and Fish, 1971) was that ferritin consisted of identical subunits with molecular weights of 18,500 each and that each ferritin molecule consisted of 24 such subunits. This model of ferritin is illustrated in Figure 5–6. It may be mentioned at this point that, as far as is known, all ferritins have the same basic subunit structure, though they may differ in regard to amino acid content and immunochemical properties. It should perhaps be even more surprising that ferritins of different animal species are related immunochemically, which may be illustrated by the partial identity patterns in the Ouchterlony system using horse spleen and human liver ferritins (Fine and Harris, 1963).

The mode of interaction among the various subunits to give an extremely stable complex such as ferritin was investigated by the methods of chemical modification and a careful study of conditions required for ferritin dissociation (Crichton, 1973c). It was pointed out that dissociation of ferritin into subunits by urea is pH dependent: At pH 7.0, even 8 *M* urea was not effective, whereas at acidic pH values, dissociation took place. Moreover, dissociation could be accomplished by acid alone only if pH was dropped to about 1.5. Dissociation with guanidine–hydrochloride was also pH dependent: At pH 7, guanidine–hydrochloride failed to dissociate apoferritin, whereas at pH 4.5, dissociation was complete (Listowsky *et al.*, 1972). Further, chemical modification of lysine residues in apoferritin showed that 5 of the 9 groups per subunit (m.w. 18,500) were unavailable for maleylation, and 11 carboxyl groups of 22 were unavailable for reaction with glycineamide in the presence of a water-soluble carbodiimide (Wetz and Crichton, 1976). The conclusion can be reached that electrostatic bonds are extremely important in maintaining the integrity of the ferritin molecule.

In addition to the hydrophilic residues, the tyrosyl groups of ferritin are also involved in its subunit interactions. Thus, whereas in native apoferritin only one tyrosyl residue is available for nitration, upon denaturation all five tyrosyl groups can be modified. Munro and Linder (1978) have summarized the function of the various amino acid side chains in ferritin subunit interactions by Figure 5–7.

Figure 5-6. Assembly of subunits in the complete ferritin molecule. From Crichton (1973c).

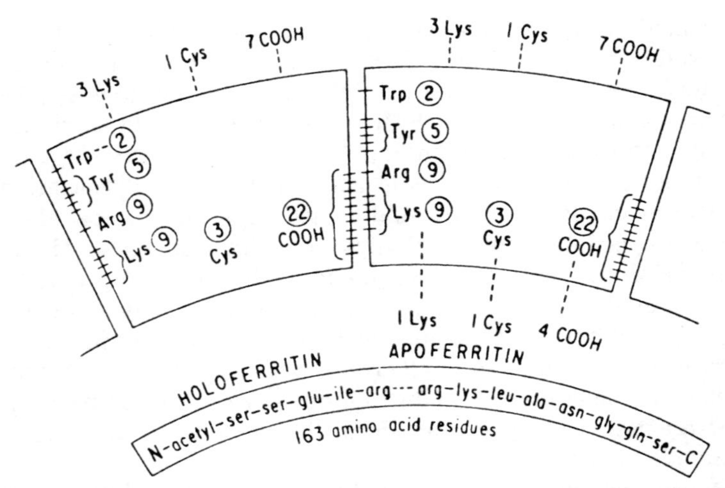

Figure 5-7. Subunit interactions in the ferritin molecule. Note the amino acid residues located at subunit interfaces, where the carboxyl residues of one subunit form salt linkages with the arginyl and lysyl residues of the adjacent subunit. Both the iron-containing (holoferritin) and iron-free (apoferritin) ferritins contain the same amount of residues exposed to the outside environment (3 lys, 1 cys, 7 carboxyl), whereas the apoferritin also has free amino acid residues exposed to the interior of the molecule. These residues are apparently associated with the iron "core" in holoferritin. Each ferritin subunit is shown as having 163 amino acid residues. From Munro and Linder (1978).

Although the existence of the 18,000–19,000-dalton subunits in ferritin is now generally accepted, there are persistent reports that indicate that ferritin may also contain subunits of smaller molecular weight. Smith-Johanssen and Drysdale (1969) were first to report the presence of 12,000-dalton fragments following sodium dodecyl sulfate treatment of apoferritin. Subsequently, Niitsu *et al.* (1973) demonstrated that sodium dodecyl sulfate–polyacrylamide gel electropherograms of human, rat, and rabbit liver ferritins showed four types of subunits: m.w. 19,000 (labeled IV) (this was the major component of all ferritins), m.w. 15,000 (III) (this was absent from human liver ferritin), m.w. 10,000–11,000 (II), and m.w. 7000–8000 (I). It was also shown that the smaller subunits could be produced from the 19,000-dalton species by repeated electrophoresis and recovery. It was concluded that native apoferritin may consist of more than 24 subunits. This has been confirmed by other laboratories (see Munro and Linder, 1978). Current thinking is that the smaller subunits demonstrated by gel electrophoresis in several ferritin preparations arise through a very specific proteolysis of the basic 18,000–19,000-dalton subunit, though the exact physiological significance of such proteolysis is as yet unclear (Collet-Cassart and Crichton, 1975).

A further development in regard to the subunit structures of ferritins is the rather long-standing observation that ferritins from different tissues of the same animal may behave differently upon gel electrophoresis and isoelectric focusing. It is said that Richter (1965) was the first to observe the phenomenon of isoferritins when he analyzed ferritins from human liver and HeLa cells by gel electrophoresis (Drysdale *et al.*, 1977). To qualify for the status of "isoferritin," the difference in electrophoretic or isoelectric focusing mobilities must not be due to aggregation or oligomer formation from the basic molecule, nor can it be due to differences in iron content.

Although certain differences may be observed among tissue isoferritins by the method of polyacrylamide gel electrophoresis (e.g., Figure 5–8), the most revealing technique in this regard is isoelectric focusing. By using this method, it is possible to see dramatic differences among human liver, spleen, heart, and kidney ferritins (Figure 5–9). Similar differences may be observed among the various tissue ferritins of horses, rats, rabbits, and other animals (Drysdale, 1977; Drysdale *et al.*, 1977; Van Kreel *et al.*, 1972). The isoelectric points of human ferritins range from 4.7 to 5.8, those of rat ferritins are from 5.1 to 5.9, and for horse ferritins, the range is 4.1–5.1. The isoferritins may also differ in regard to their immunochemical properties, as can be seen from the precipitin curves in Figure 5–10.

The molecular basis for the existence of tissue isoferritins appears

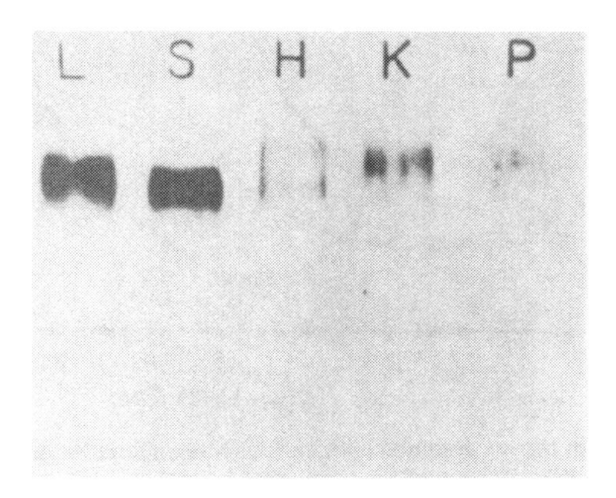

Figure 5-8. Polyacrylamide gel electrophoresis of various human tissue ferritins. Dimers and oligomers are absent, having been removed by gel chromatography. Note especially the two bands in heart ferritin (H). Other ferritins are from the liver (L), spleen (S), kidney (K), and pancreas (P). From Powell, L. W., Alpert, E., Isselbacher, K. J., and Drysdale, J. W., *Brit. J. Haematol.* 30:47–55, 1975, by permission of the Blackwell Scientific Publications Ltd.

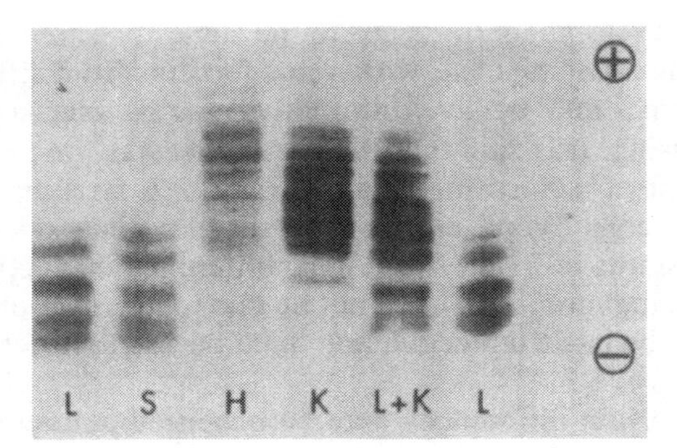

Figure 5-9. Isoelectric focusing of human ferritins from various tissues: liver (L), spleen (S), heart (H), and kidney (K). L + K is a mixture of the liver and kidney ferritins. From Powell, L. W., Alpert, E., Isselbacher, K. J., and Drysdale, J. W., *Brit. J. Haematol.* 30:47–55, 1975, by permission of the Blackwell Scientific Publications Ltd.

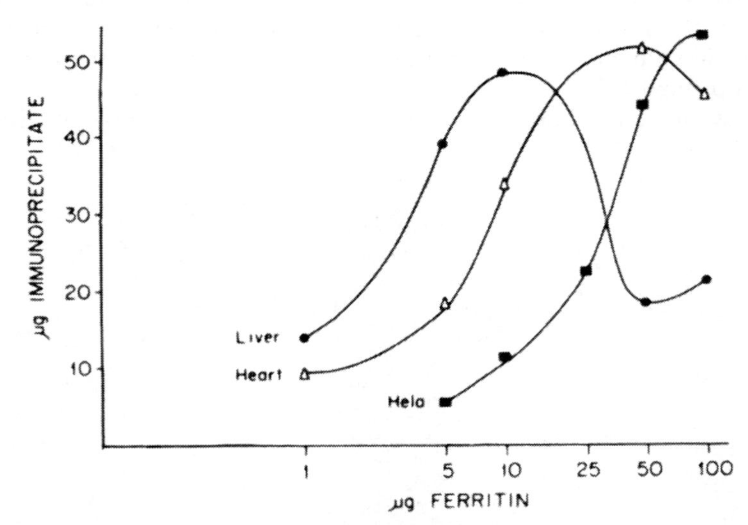

Figure 5-10. Precipitin curves obtained with various human ferritins and human liver ferritin antiserum raised in rabbits. Reprinted by permission from Hazard, J. T., Yokota, M., Arosio, P., and Drysdale, J. W., 1977, Immunologic differences in human isoferritins: implications for immunologic quantitation of serum ferritin, *Blood* 49:139–146. Copyright by Grune and Stratton, Publ.

to rest on the fact that ferritins contain not one but two basic subunits of very similar molecular weights. Ferritins of all species examined could be separated into the two types of subunits by sodium dodecyl sulfate gradient–pore gel electrophoresis. The two types of subunits have molecular weights of 21,000, designated the H subunit, and of 19,000, designated the L subunit (Arosio *et al.*, 1978). This may be illustrated by figure 5–11, where it is seen that substantial differences exist between the isoelectric focusing behavior of the various isoferritins in a number of animals. However, all isoferritins shown are separable into just two subunits with very close electrophoretic mobilities in the sodium dodecyl sulfate gels. Drysdale has then proposed that the various tissue isoferritins have different ratios of the H and L subunits. Because the subunits have differing molecular weights, ferritins constructed from them should also have different molecular weights. Thus, it is possible to have an apoferritin molecule constructed of the H subunits only, which would have a molecular weight near 500,000. On the other hand, a homopolymer of the L subunits would have a molecular weight of near 460,000. Besides the two homopolymers, it is possible to have 23 heteropolymers. In case of the human material, the H homopolymer has a pI of 4.6 and is therefore relatively acidic, whereas the L homopolymer is relatively basic with a pI of 5.7. The heteropolymers will have pIs intermediate to the two extremes. Generally, isoferritins containing a preponderance of the H subunits are found in malignant tumors and in heart, whereas isoferritins with a preponderance of the L subunits are found in the spleen, liver, and iron overload cases. Tissue isoferritin patterns observed are then believed to be due to the presence of mixtures of ferritin molecules each containing a peculiar H–L-subunit ratio, and by the virtue of differences in their pI values, they proceed to separate in isoelectric focusing systems. The subunit distribution in various ferritins is summarized in Figure 5–12.

One report has hinted that the subunit picture in ferritin may be even more complicated than heretofore suspected (Lavoie *et al.*, 1978). Ferritin from human liver was dissociated into subunits in 9 *M* urea at pH 2.5 and then reduced and alkylated and subjected to a two-dimensional procedure involving isoelectric focusing and electrophoresis. The H subunit was separable into four components, termed H_1, H_2, H_3, and H_4, whereas the L subunit was separable into five components, termed L_1, L_2, L_3, L_4, and L_5. In terms of acidity, the following order of decreasing acidity was observed: H_4 (most acidic), H_3, H_2, L_5, H_1, L_4, L_3, L_2, L_1. Relative proportions of the subunits from human liver ferritin were L_3, 0.37; L_4, 0.24; H_2, 0.13; H_1, 0.10; L_5, 0.05; L_2, 0.04; H_3, 0.03; and $H_4 = L_1$, 0.02. Whether or not this newly discovered heterogeneity of ferritins is artifactual remains to be determined.

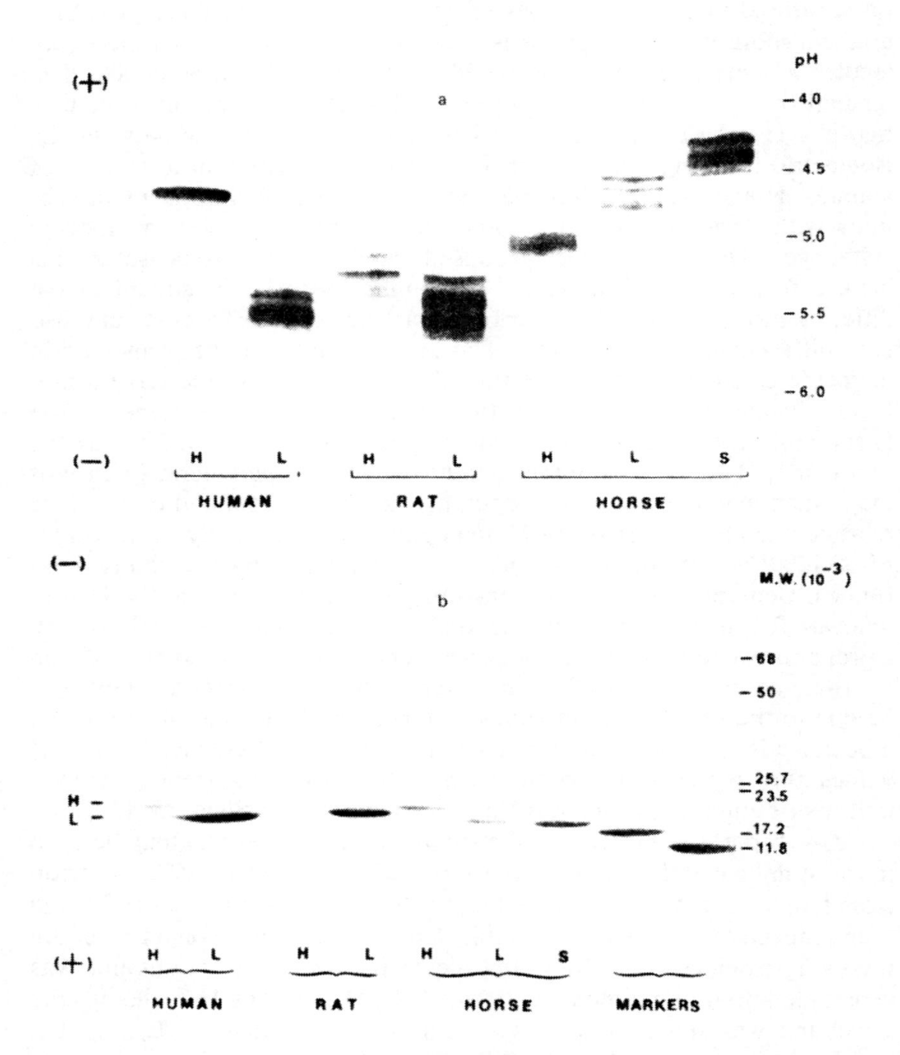

Figure 5-11. (a) Isoferritins from liver (L), heart (H), and spleen (S) of various animals separated by isoelectric focusing. (b) Same materials as in (a) above dissociated into subunits. Note that though a complex multiband system is present in the native proteins, all dissociate into only two types of subunits. From Drysdale (1977).

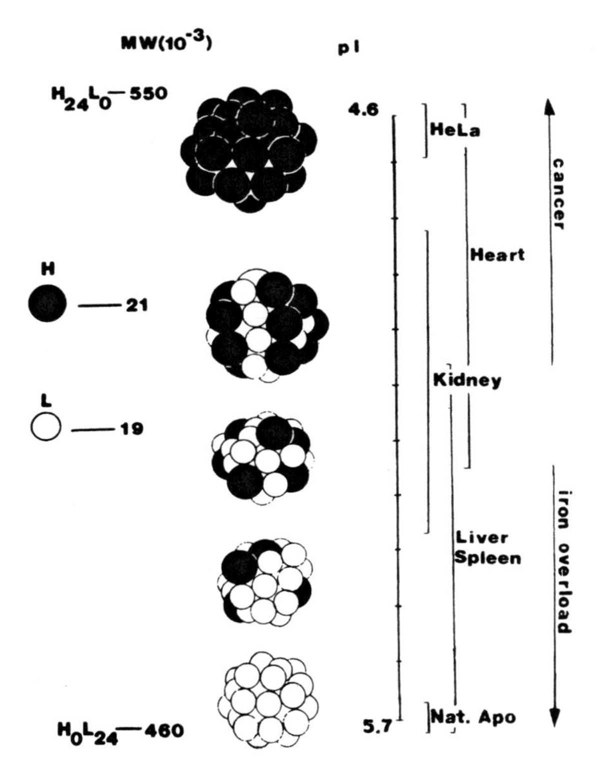

Figure 5-12. Schematic representation of the molecular basis for the existence of tissue isoferritins as proposed by J. W. Drysdale. Salient features of this model are the existence of two types of subunits, H (mol. wt. 21,000) and L (mol. wt. 19,000), and their association in different proportions in the various human tissues. From Drysdale (1977).

There is at least one group that does not subscribe to the multiple-subunit view of ferritin structure (Bryce *et al.*, 1978). It is asserted that there is but a single 18,500-dalton subunit in the ferritin molecule but that this can be cleaved by specific proteases into the 11,000- and the 7000–8000-dalton subunits (subunits II and I, respectively, of Niitsu *et al.*, 1973). A dimerization of subunit II would produce a 22,000-dalton subunit, the subunit of Arosio *et al.* (1978). A dimerization of the 7000–8000-dalton subunit would produce a 15,000-dalton particle (subunit III of Niitsu *et al.*, 1973). Other investigators have also expressed skepticism in regard to some aspects of the isoferritin structure (Russell and Harrison, 1978).

5.3.4 Distribution of Iron among Ferritin Oligomers and Isoferritins

Iron is not distributed evenly in ferritin monomers and oligomers, nor is it distributed evenly among the various tissue isoferritins. Niitsu and Listowsky (1973b) subjected ferritin fractions isolated by sucrose density centrifugation on the basis of iron content to polyacrylamide gel electrophoresis, which can separate ferritin preparations into monomers, dimers, and trimers. The fraction containing largely apoferritin consisted entirely of monomer species, whereas the heavy fraction containing 2200–3000 iron atoms/molecule consisted largely of dimers. Table 5–3 shows the monomer and dimer contents of ferritins with intermediate iron content. The high-iron ferritins thus tend to form dimers and oligomers. It has been suggested that dimer and oligomer formation is the first step in the conversion of ferritin into hemosiderin. An interesting clinical confirmation for the iron content–oligomer relationship was provided by Samarel and Bern (1978), who procured a number of human spleens from which ferritin with differing amounts of iron was isolated. The iron–protein ratio varied from 0.137 down to 0.045. Patients with high iron contents in their ferritins showed the presence of ferritin dimers, trimers, and oligomers. Those with low iron–protein ratios showed only trace amounts of dimers and no trimers or oligomers.

When the various isoferritins from human livers and spleens were examined for iron, it was found that those isoferritins with more basic isoelectric points bound the least amount of iron and consisted largely of the L subunit, whereas the more acidic isoferritins bound more iron and consisted largely of the H subunits. These relationships are illustrated in Table 5–4. The reason for the unequal distribution of iron among the various isoferritins may be that the more acidic subunits somehow facilitate iron deposition to a greater extent than do the more basic L subunits (Bomford *et al.*, 1978). When apoferritins are prepared from the various

Table 5-3. Iron–Protein Relationships in Sucrose Gradient Centrifugation Fractions of Horse Spleen Ferritin[a]

Iron content (atoms of Fe/molecule)	Monomer (mg of protein)	Dimer (mg of protein)
Less than 100	8.5	0
100–500	2.7	1.3
1000–1500	2.8	2.4
1500–2200	17.7	19.8
2200–3000	1.3	9.7

[a] After Niitsu and Listowsky (1973b).

Table 5-4. Iron Content of Isoferritins of Human Liver[a]

pI of isoferritin	Iron to protein ratio	Iron uptake, Half of max. achieved (min)	Subunits
5.34	0.24	5	H
5.43	0.22	—	H ⟩⟩ L
5.57	0.18	—	H ⟩ L
5.67	0.14	7	L ⟩ H
5.74	0.10	10	L ⟩⟩ H
5.81	0.07	—	L

[a] Adapted from Wagstaff *et al.* (1978).

isoferritins and incubated with ferrous iron, it is again the more acidic apoferritins with a preponderance of the H subunits which take up iron fastest (Wagstaff *et al.*, 1978). It is interesting to note that human heart apoferritin, which consists of very acidic isoferritin species, achieves half of its maximum iron content in about 2 min in iron uptake experiments. Yet when heart ferritin from a thalassemic patient with iron overload is examined, the iron uptake time is increased to 4 min. It has been found that in such iron overload ferritins, there is a greater preponderance of the more basic isoferritins than in normal ferritins (Wagstaff *et al.*, 1978). The lower iron content of iron overload ferritins was also noted as early as 1970 by Linder-Horowitz *et al.* (1970) in regard to rat heart and kidney but not liver ferritin. It would thus appear that iron overload induces the greater degree of L-subunit biosynthesis and that the resulting isoferritins, at least in the heart and kidney, are less efficient iron binders than are normal ferritins.

The findings described in this section in regard to human spleen, liver, and heart ferritins are not universal, as opposite results have been obtained with horse tissue ferritins (Russell and Harrison, 1978). The most acidic ferritins were found in the spleen, followed by those of liver, kidney, and finally heart (most basic). Among the horse liver and spleen isoferritins, the more basic species had a greater iron content than did the more acidic species. Apoferritins prepared from the more basic ferritins were able to bind iron at a faster rate than did apoferritins prepared from the acidic ferritins. No correlations of iron content or iron-binding rates were made with the relative content of the H or the L subunits. It is thus clear that in view of the differences in the chemical properties of horse and human isoferritins, it would be impossible to predict the relative iron-binding affinities of tissue ferritins from other animals. Such information must be gathered experimentally on a case-by-case basis.

5.3.5 Composition and Primary Structure of the Ferritins

Amino acid compositions of various ferritins are given in Table 5–5. It will be seen that ferritins of different tissues from the same animal have differing amino acid contents, and this is especially evident in the case of horse heart ferritin on the one hand and the spleen and liver ferritins on the other. This undoubtedly reflects differences in isoferritin composition. Amino acid contents of the H and L subunits are given in Table 5–6.

In addition to amino acids, ferritin contains carbohydrate. There is evidence from at least two laboratories that the subunits of ferritin are not equal with respect to carbohydrate composition, whereas a third laboratory has not detected such differences. Carbohydrate composition of horse spleen ferritin is shown in Table 5–7. It is surprising that mannose is found in ferritin, a tissue protein. It is generally accepted that only plasma proteins contain mannose. It is perhaps even more surprising that glucose is found in ferritin, since few glycoproteins contain this sugar. It has been speculated that there may be at least two types of oligosaccharide chains in ferritin: one containing fucose, mannose, galactose, and hexosamine and the other containing only glucose. It has also been speculated that much of the microheterogeneity of ferritin observed by isoelectric focusing may be due to differences in carbohydrate content of ferritin molecules (Cynkin and Knowlton, 1977).

The amino acid sequences of ferritin subunits are still unknown. Undoubtedly, one reason for this is the uncertainty as to whether or not the subunits of ferritin are identical, and if they indeed are not, then exactly how many different types of subunits are in existence. An N-terminal residue of horse spleen ferritin was found to be *N*-acetylserine, and the N-terminal pentapeptide was found to be *N*-acetylseryl-seryl-glutaminyl-isoleucylarginine (Suran, 1966). A C-terminal peptide was shown to have the structure arginyl-lysyl-leucyl-alanyl-asparaginyl-glutamylserine (C-terminus) (Mainwaring and Hofmann, 1968). Rat liver ferritin apparently has the same tripeptide at its N-terminus as does the horse spleen protein (Huberman and Barahona, 1978).

At the time of the writing of this chapter, "of the 165 amino acids in the (ferritin) sequence, some 85% have been determined" (Crichton *et al.*, 1979). The strategy is to cleave the protein with CNBr, which yields 3–4 peptides, or to cleave the maleylated protein with trypsin, which yields 11–12 peptides. The former procedure depends on the presence of methionyl groups and the latter on arginyl groups in ferritin (Crichton *et al.*, 1977). The individual peptides are then sequenced by the Edman degradation procedure. It is said that horse spleen and liver ferritins, though

Table 5-5. Amino Acid Compositions of Apoferritins of Various Sources (in moles/18,500 g, unless otherwise indicated)

Amino Acid	Source[a]													
	1	2	3	4	5	6	7	8	9	10	11	12	13	14
Lysine	10.8	8.7	8.8	13.5	10.4	10.4	8.7	10	10	10	10	13	8.9	13.3
Arginine	12.0	9.5	9.0	6.7	9.3	8.4	7.6	10	10	8	9	7	6.7	8.9
Histidine	8.7	5.8	6.5	8.4	6.9	5.4	5.0	7	6	6	7	6	5.0	9.7
Aspartate	22.0	17.3	17.9	18.4	19.3	19.2	19.0	20	20	21	19	21	30.3	20.9
Threonine	6.2	5.5	5.6	7.7	6.1	6.2	7.6	7	7	8	6	8	4.0	5.0
Serine	9.8	9.0	8.9	11.2	7.7	9.3	7.8	9	9	12	8	15	14.7	9.7
Glutamate	26.3	23.9	25.3	19.9	22.3	23.9	24.5	26	25	24	24	24	28.9	27.6
Proline	2.6	2.8	3.1	6.4	3.8	2.9	5.7	4	5	6	4	4	3.1	6.0
Glycine	11.6	9.9	10.2	10.2	10.8	10.1	11.0	12	11	12	10	16	11.0	9.6
Alanine	14.2	14.0	13.4	12.7	13.8	13.7	15.3	14	14	14	13	15	14.7	15.4
Valine	8.5	6.9	6.9	7.4	6.0	6.3	7.2	7	7	8	7	10	12.8	17.4
Methionine	3.6	2.8	2.5	4.8	2.7	2.9	1.9	2	2	2	3	4	3.0	3.9
Isoleucine	4.8	3.5	3.6	4.9	3.8	2.5	4.7	3	3	3	4	6	7.4	6.9
Leucine	21.8	25.0	24.2	20.3	23.3	23.4	24.0	23	24	19	23	17	15.8	17.4
Tryptophan	2.1	2.1	2.1	—	2.2	2.2	—	—	—	—	2	4	1.6	1.7
Tyrosine	7.1	5.0	4.1	4.8	4.4	6.0	5.2	3	4	5	6	5	6.0	6.5
Pnenylalanine	9.3	7.3	7.7	7.3	7.0	6.9	6.0	7	7	6	7	7	8.4	9.5
½-Cystine	2.0	2.9	2.6	—	1.7	1.5	—	—	—	—	1	0	—	—

[a] Notes and references:

1. Porcine spleen, m.w. 20,000; May and Fish (1977).
2. Horse spleen; Crichton *et al.* (1973b).
3. Horse liver; Crichton *et al.* (1973b).
4. Horse heart; Crichton *et al.* (1977).
5. Human spleen; Crichton *et al.* (1973b).
6. Human liver; Crichton *et al.* (1973b).
7. Human heart; Crichton *et al.* (1977).
8. Rat spleen; Crichton *et al.* (1975).
9. Rat liver; Crichton *et al.* (1975).
10. Rat intestinal mucosa; Crichton *et al.* (1975).
11. Guinea pig spleen; Crichton *et al.* (1975).
12. Tadpole red blood cells; Crichton *et al.* (1975).
13. Pea; Crichton *et al.* (1977, 1978).
14. Lentil; Crichton *et al.* (1977, 1978).

Table 5-6. Amino Acid Composition of the H and L Subunits from Various Sources (in moles/mole of subunit)[a]

	Horse spleen		Human liver	
Amino acid[b]	H	L	H	L
Lys	11	9	11	10
Arg	9	9	9	9
His	6	4	7	5
Asp	19	16	22	16
Thr	7	5	7	7
Ser	9	8	12	9
Glu	23	22	24	21
Gly	12	9	11	15
Ala	15	14	13	13
Val	8	7	7	7
Met	3	3	3	2
Ile	6	3	6	5
Leu	20	25	20	19
Tyr	6	6	8	5
Phe	7	7	7	6

[a] From Arosio *et al.* (1978). Molecular weights taken as 21,000 and 19,000 for the H and L subunits, respectively.
[b] Proline, $\frac{1}{2}$-cystine, and tryptophan were not determined.

showing extensive homologies, do not have the same amino acid sequence, and the amino acid sequence of the plant ferritins differs from that of mammalian ferritins to such an extent that only a small degree of homology can be detected.

It is noteworthy that the primary structure determination of ferritins is being carried out primarily with the assumption that the subunits in a

Table 5-7. Monosaccharide Content of Horse Tissue Ferritins (in moles/19,000 g)

			Cynkin and Knowlton, 1977		
Monosaccharide	Shinjyo *et al.*, 1975, spleen	Lavoie *et al.*, 1977, spleen	Spleen	Liver	Heart
Glucosamine	0.125	2.3	—	—	—
Galactosamine	0.125	1.1	—	—	—
Glucose	None	1.6	1.6	1.6	3.0
Galactose	0.3	1.3	0.42	0.30	0.58
Mannose	0.7	0.9	0.25	0.38	0.88
Fucose	0.42	None	0.30	0.38	0.79
Sialic acid	Very low	None	Very low	—	—

given tissue ferritin are all identical. As was seen in Section 5.3.3, this assumption may be invalid. Although it is quite likely that the H- and L-type ferritin subunits share extensively in amino acid sequences, it is nevertheless necessary to study the amino acid sequences of the H and L subunits individually.

5.3.6 Secondary and Tertiary Structure of Ferritin Subunits

Ferritins and apoferritins have been examined by both the X-ray diffraction and optical rotation procedures, and the results have yielded information on the physical properties of the subunits as opposed to the entire molecule as a whole. The subunits are approximately cylindrical in shape, some 27 × 54 Å in size, and each subunit contains four parallel rods, 34–42 Å long, and a shorter rod nearly perpendicular to the longer ones (Banyard *et al.*, 1978). All five rods are believed to represent regions of α helix. Some of these regions are only 7 Å apart at the nearest point, which means that they must contain much glycine (Hoare *et al.*, 1975). Alpha helix accounts for some 60% of the 165 amino acid residues present in a ferritin subunit (Harrison *et al.*, 1979).

Circular dichroism (CD) studies on ferritin and apoferritin have shown that the uptake of iron by the latter does not change its CD spectrum. The secondary structures of ferritin and apoferritin are thus identical (Listowsky *et al.*, 1972). A complex series of bands is observed in the 260–300-nm range, reflecting the involvement of tyrosyl, tryptophanyl, and possibly other side chains in the secondary and tertiary structures of ferritin (Listowsky *et al.*, 1972; Wood and Crichton, 1971). Interestingly, ferritin is much more easily susceptible to denaturation with guanidinium chloride than is apoferritin, showing extensive degree of unfolding in 7 *M* guanidinium hydrochloride at pH 7. Apoferritin showed only a partial degree of denaturation under these conditions. Apparently, any situation that dissociates ferritin or apoferritin into subunits will also bring about denaturation of the protein. The CD spectra of ferritin, apoferritin, and denatured apoferritin are shown in Figure 5–13.

Secondary structure parameters calculated from CD data indicate that native proteins contain some 50% α helix, 37% pleated sheet, and only 11% disordered structure. Partial denaturation of apoferritin (e.g., in 7 *M* guanidinium chloride at pH 7.5) affects largely the pleated-sheet structure and not the α helix: Under these conditions, apoferritin contains 45% α helix, 7% pleated sheet, and 48% random coil (Listowsky *et al.*, 1972).

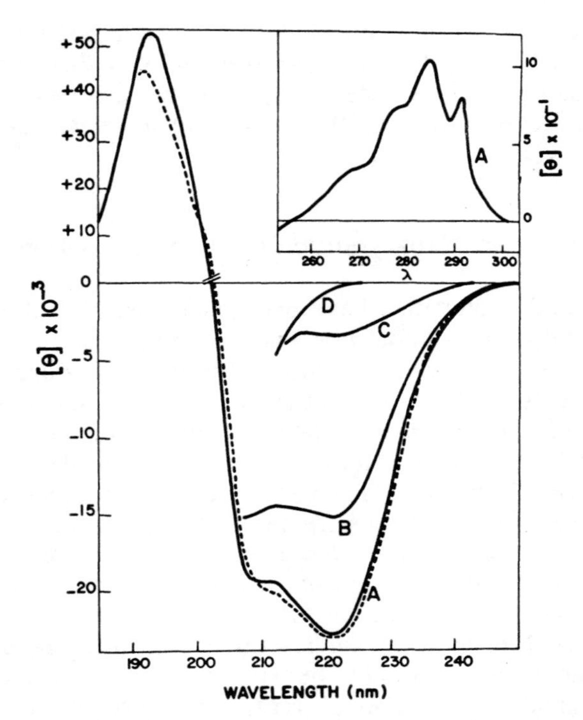

Figure 5-13. Circular dichroism spectra of horse spleen ferritin and apoferritin in 0.1 *M* NaCl at pH 7 (curve A); curve B is apoferritin in 7 *M* guanidinium hydrochloride at pH 7.5; curve C is a high-density ferritin fraction in 7 *M* guanidinium hydrochloride at pH 7.5; curve D is apoferritin in 7 *M* guanidinium hydrochloride at pH 4.5. Reprinted with permission from Listowsky, I., Blauer, G., Englard, S., and Betheil, J. J., 1972, Denaturation of horse spleen ferritin in aqueous guanidinium chloride solutions, *Biochemistry* 11:2176–2182. Copyright by the American Chemical Society.

5.4 Metabolism of Ferritin

The most remarkable phenomenon in the process of ferritin metabolism is that the net content of tissue ferritins can be increased by increasing body iron levels. Traditionally, this has been investigated by injecting iron into rats or other small animals and then evaluating either the incorporations of ^{14}C–amino acids into liver ferritin or the net increase in liver ferritin content. A net increase in liver or any other tissue ferritin can be accomplished by either increasing its rate of biosynthesis, its rate of degradation, or both.

In 1949, Granick postulated that the net increase in tissue ferritin levels in animals treated with iron was primarily due to a retardation of ferritin catabolism. This view was arrived at on the basis of *in vitro* experiments on the susceptibility of iron-laden ferritin to digestion by proteolytic enzymes: Presumably, because of increased iron levels, more apoferritin is converted into ferritin, and the latter, being less susceptible to proteolysis than apoferritin, is retained in the tissues for a longer period of time. This question was further investigated by Fineberg and Greenberg (1955a), who used labeled amino acids to study ferritin biosynthesis in guinea pigs. They not only found increased ferritin levels in iron-treated animals, but they also discovered that there was a four- to fivefold increase in ^{14}C–amino acid incorporation into liver ferritin. The concluded that the increase in tissue ferritin in iron-treated guinea pigs was due to an increase in the *de novo* biosynthesis of ferritin.

These basic studies were extended by Drysdale and Munro (1966) as well as by other investigators. It was confirmed that the iron-loaded animal incorporated ^{14}C–amino acids into liver ferritin much faster than did the normal animal (Figure 5–14); however, the apparent half-life of ferritin was lowered: Normally ferritin has a half-life of about 72 hr as measured by a single-pulse ^{14}C–leucine incorporation, whereas in animals injected with a single dose of iron, as well as animals injected repeatedly with iron, the half-life was near 12 hr (Figure 5–15). It should be noted, however, that in the iron-loaded animals, ^{14}C–ferritin stabilized at a rather high level, which was especially evident in the case of animals treated

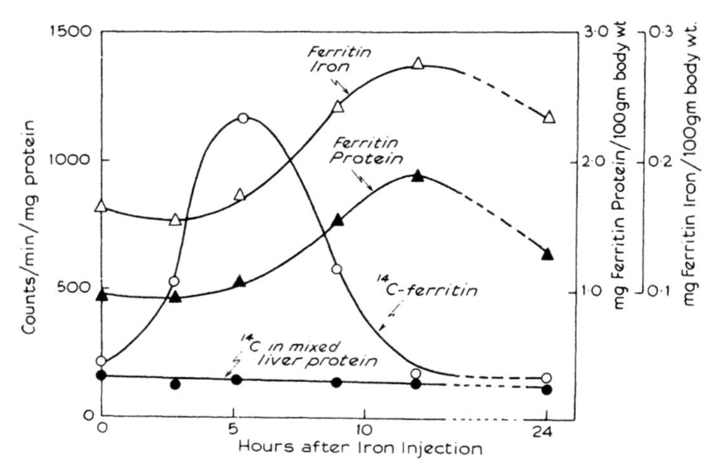

Figure 5-14. Uptake of ^{14}C into rat liver ferritin following a single dose of ^{14}C–leucine injection into rats. Animals were injected with 400 μg iron per 100 g of body weight, then the amino acid was given at 1, 3, 7, 10, and 22 hours following the iron. Rats were sacrificed 2 hr after the amino acid injection. From Drysdale and Munro (1966).

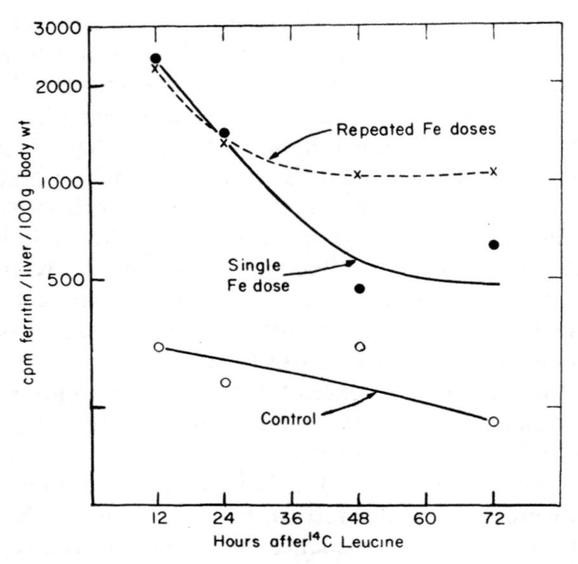

Figure 5-15. Turnover of rat liver ferritin following ^{14}C–leucine injection as a function of the mode of iron administration. From Drysdale and Munro (1966).

repeatedly with iron. This and other evidence led Drysdale and Munro (1966) to propose that the administration of iron to animals results in both the increase of *de novo* ferritin biosynthesis and an inhibition of its breakdown.

There has been a long-standing controversy as to how the "compleat" ferritin is biosynthesized: Is apoferritin made first, followed by the accumulation of iron within its cavity, or is the iron oxide micelle formed first, followed by the assembly of ferritin subunits around the micelle? The latter view was proposed by Pape *et al.*, (1968a); however, previous and subsequent experiments have tended to support the view that apoferritin is assembled first, followed by the incorporation of iron. Earlier evidence for this mechanism was provided by Fineberg and Greenberg (1955b) and Drysdale and Munro (1966), among others. These workers subjected ferritins isolated from animals treated with iron and "pulsed" with a ^{14}C–amino acid to ultracentrifugation and determined that ^{14}C was associated first with the "lighter" apoferritin. Animals sacrificed later showed ^{14}C in the "heavier" ferritin fractions. More recently, this was confirmed by Lee and Richter (1977b). These investigators were able to develop radioimmunoassay methods to distinguish ferritin subunits from the intact protein. When ^{59}Fe–^{14}C ratios were measured in ferritin and ferritin subunits of liver cells that were fed ^{59}Fe and ^{14}C–leucine in tissue culture, the completed ferritin showed a much greater ratio than did the unassembled subunits, contrary to what would be predicted from the theory of Pape *et al.* (1968a). The small amount of iron associated with the unassembled subunits was termed *nonspecific*.

What, then, is the mechanism of control of ferritin biosynthesis by iron on a molecular level? A basic discovery toward the elucidation of this novel system was provided by Drysdale and Munro (1966), who found that actinomycin does not abolish the effect of iron on ferritin biosynthesis. This indicates that iron acts on a posttranslational event in the mechanism of ferritin biosynthesis. This also tends to indicate that a stable ferritin subunit messenger RNA exists in the cytoplasm of the responsive cell and that unless large amounts of iron are available, this messenger RNA is not available for ferritin biosynthesis. Munro's laboratory has indeed discovered that rat liver cells contained a large quantity of *free* (nonribosome) ferritin messenger RNA. In iron-treated animal liver cells, the free messenger RNA was sharply lowered with a concomitant increase in ribosome-bound ferritin messenger RNA. Quantitatively, normal liver cells contain about 50% of the ferritin messenger RNA bound up with ribosomes, whereas the rest is inactive or free RNA. In iron-treated animals, the ribosome-bound ferritin messenger RNA accounts for almost 100% of all cellular ferritin messenger RNA (Munro and Linder, 1978). It has been proposed that ferritin messenger RNA that is normally present in the free or inactive form is "repressed" by ferritin subunits attached to the 5′ end of the messenger RNA. When large amounts of iron are made available to the cell, the existing apoferritin molecules are immediatly filled with iron, thus forcing the unassembled ferritin subunits, including those attached to the free messenger RNA, to aggregate and form apoferritin (this assumes that the unassembled ferritin subunits are in an equilibrium with apoferritin). The result is that the messenger RNA becomes derepressed and available for combination with ribosomes and ferritin biosynthesis (Zähringer *et al.*, 1976); see Figure 5–16. In support of this hypothesis, it has been found that normal rat liver cells and hepatoma cells contain a pool of cytosol ferritin subunits which are available for binding with the messenger RNA (Lee *et al.*, 1975). Moreover, it was found that the administered iron is bound up with ferritin only and not with messenger RNA or ribosomes (Fagard *et al.*, 1978).

It is well known that the cytoplasmic protein synthesizing machinery can be divided into two parts: the cytosol free ribosome system and the endoplasmic reticulum-bound system (microsomes). The former is supposed to represent the machinery for making protein for the cell's internal use, whereas the latter makes protein for export. Ferritin is synthesized by both systems, though the bulk, over 80%, is made by free ribosomes

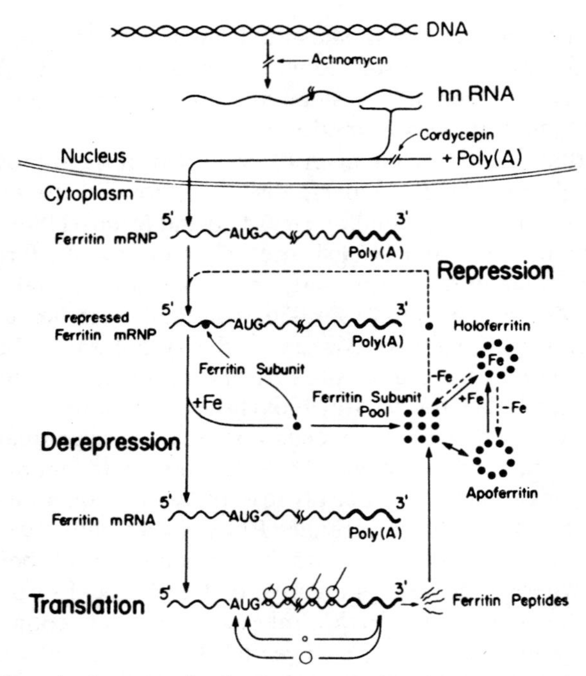

Figure 5-16. A hypothesis accounting for the control of ferritin biosynthesis by tissue iron levels. From Zähringer *et al.* (1976).

and the rest by microsomes (Lee and Richter, 1977a). An attractive possibility is that the microsomes are primarily responsible for making serum ferritin (see Section 5.6.4) and/or ferritin-containing carbohydrate. It has also been proposed that the two ferritin biosynthetic pathways may each lead to a different ferritin subunit. These and other proposals on the origin of isoferritins, the carbohydrate moiety of ferritin, and serum as opposed to tissue ferritin must await experimental substantiation. The involvement of the different subcellular particles in ferritin biosynthesis is illustrated in Figure 5–17.

Not much information is available on the biologic degradation of ferritin or what controls it. One can mention that injected ferritin is very rapidly cleared from the bloodstream by the liver, the half-lives ranging from about 8 min for kidney ferritin to about 16 min for serum ferritin in rats. It has been proposed that the carbohydrate content of ferritin determines, in part, its half-life: The more carbohydrate a ferritin contains, the longer its half-life after parenteral administration (Halliday *et al.*, 1979).

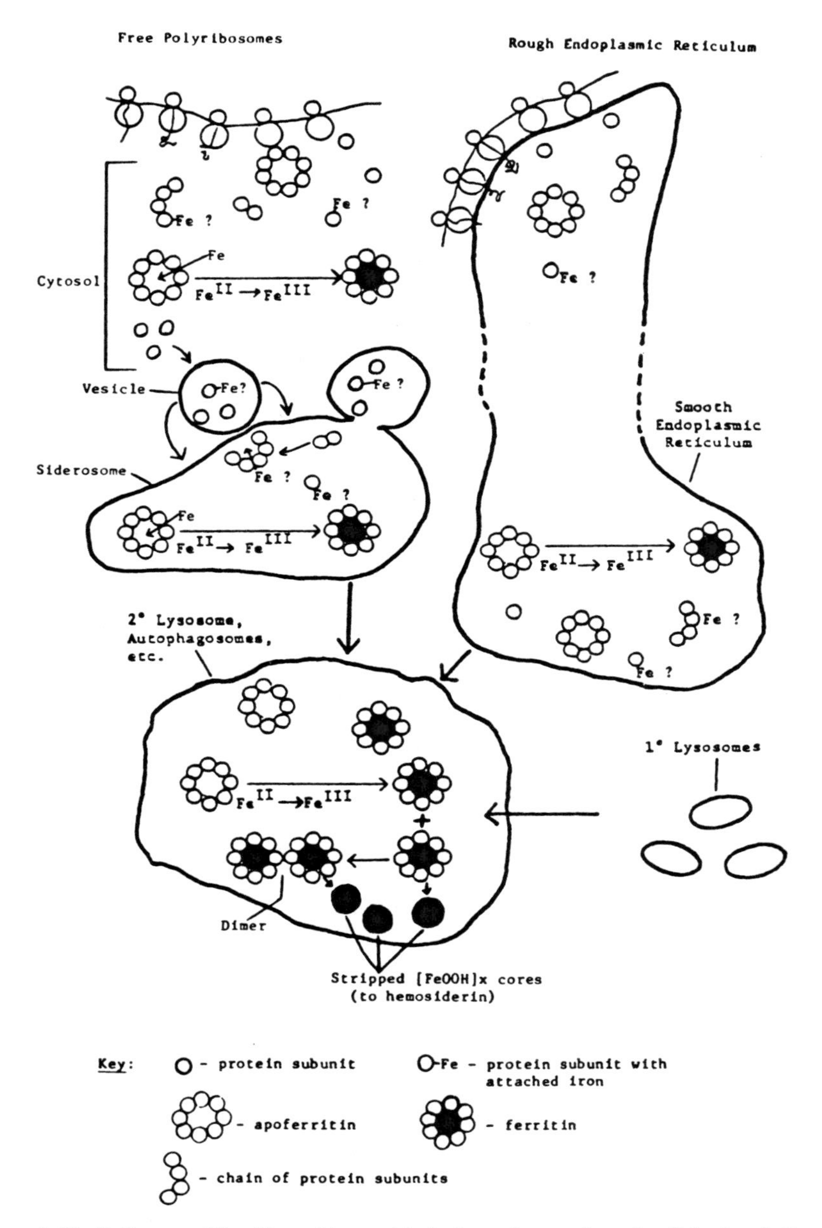

Figure 5-17. Pathways of ferritin and hemosiderin formation on the subcellular level. From Richter (1978).

5.5 Iron Uptake and Release by Ferritin and Apoferritin

It is generally agreed that *in vivo* the formation of apoferritin precedes any specific association of ferritin subunits with iron. The exact mechanism of iron uptake by apoferritin is difficult to study in the intact animal or cell, and investigations have relied on *in vitro* experiments using purified apoferritin, ferritin, and other compounds in an attempt to simulate *in vivo* conditions.

It was proposed as early as in 1955 by H. J. Bielig and E. Bayer that iron uptake and release by ferritin involves a reduction–oxidation cycle: Apoferritin or partially filled ferritin takes up ferrous iron, which is oxidized in the ferritin interior. To be removed, iron must once again be reduced to the ferrous state. These proposals have been confirmed by other workers in subsequent years, though variations on this theme have been observed in several instances. The most widely accepted theory of how iron micelles are formed in the ferritin cavity was formulated by Pauline Harrison and her colleagues (Macara *et al.*, 1972; Harrison *et al.*, 1975; Harrison, 1977): Iron in the ferrous state enters the apoferritin cavity through one of the six channels connecting the cavity with the outside. Ferric iron may also be incorporated into ferritin, albeit at a much slower rate than ferrous iron (Hoy and Harrison, 1976). In the ferritin cavity, the ferrous iron combines with the appropriate amino acid side chains, which tends to catalyze its oxidation to the ferric state in the presence of an oxidizing agent, including O_2 (Treffry *et al.*, 1979). Normally, in *in vitro* experiments, KIO_3–$Na_2S_2O_3$ is utilized as the oxidizing system. As iron acquires the ferric oxidation state, it forms the typical iron oxide crystalloid. The iron oxide micelle may then act as a crystal growth nucleus without any further involvement of the amino acid side chains.

Mobilization of iron from ferritin may be effected by ferrous iron chelating agents such as 1,10-phenanthroline (Hoy *et al.*, 1974a), though this is not the most efficient and certainly not a physiological way to accomplish this. However, in such model systems, the chelating agent will remove iron from the surface of the ferric oxide micelle on a last in–first out basis (Hoy *et al.*, 1974b). An explanation of why 1,10-phenanthroline or other ferrous iron chelators may work in abstracting iron from ferritin was provided by Mazur *et al.* (1955), who have detected small but significant amounts of ferrous iron in ferritin, most likely associated with free sulfhydryl groups present therein. They postulated that such ferrous iron is in equilibrium with the bulk of ferritin's iron, which, of course, is present in the ferric state. Iron may leave ferritin by becoming reduced to the ferrous state, providing the reducing agent can negotiate the channels leading to the interior of the ferritin molecule. This is represented diagrammatically in Figure 5–18.

The hypothesis of Harrison and colleagues predicts that iron uptake rates by ferritin should depend on the availability of iron-binding sites and subsequently on the surface area of the iron oxide micelle. As a corollary, iron uptake and release should not be constant but should depend on the iron content of the existing ferritin molecule. These predictions have been verified experimentally: When iron is added to apoferritin in proportions of 200 to 1000 iron atoms per one protein molecule, a sigmoidal iron uptake curve is obtained, as in Figure 5–19. The initial slow portion (zero order with respect to iron) apparently represents the combination of iron with the amino acid side chains in the ferritin interior (nucleation stage), whereas the rapid phase (growth stage) represents crystal growth on the surface of the iron oxide micelle. When ferritin fractions with varying iron content were incubated with iron or an iron chelator, the iron uptake or release was greatest with ferritin containing some 1200 iron atoms/molecule (Harrison *et al.*, 1974) (Figure 5–20). Computer simulation of the system has shown that the iron oxide micelle has the highest "free" surface area available for crystal growth when close to 1200 iron atoms are present.

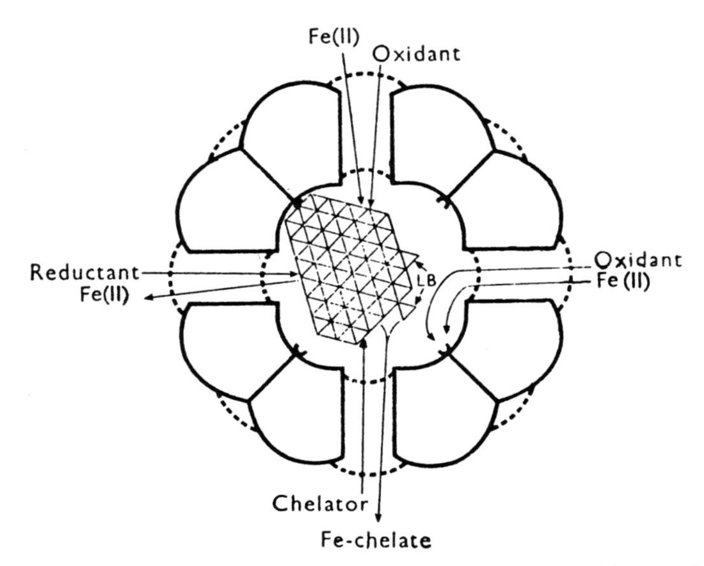

Figure 5-18. Commonly accepted model for the uptake and release of iron by ferritin. U indicates iron oxide initiation sites associated with amino acid side chains (cf. Figure 5-6). LB indicates "loosely bound iron". From Harrison *et al.* (1974).

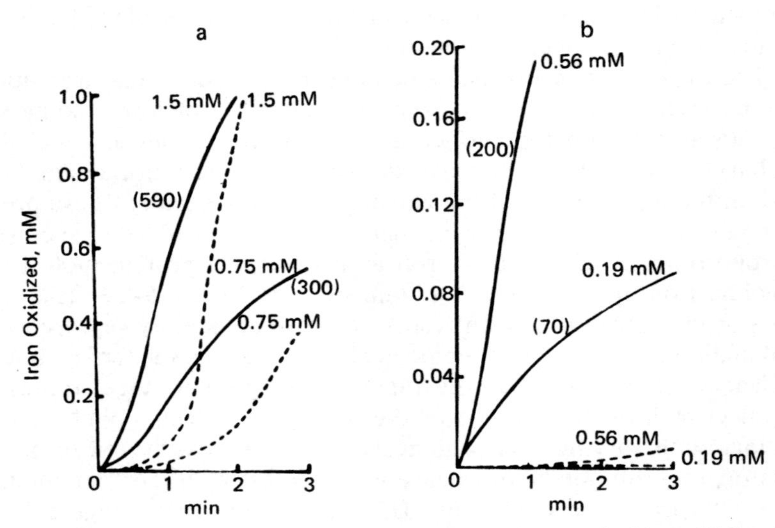

Figure 5-19. Oxidation of iron in the presence of horse spleen apoferritin (solid lines) and serum albumin (dashed lines) at two different ferrous iron concentrations. From Harrison *et al.* (1975).

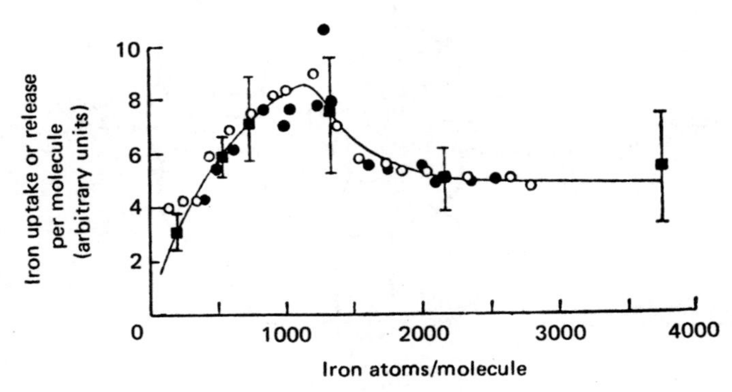

Figure 5-20. Degree of iron uptake and loss by ferritin with varying iron concentrations. Filled circles, using KIO_3–$Na_2S_2O_3$; open circles, O_2; filled squares, loss of iron to 1,10-phenanthroline. From Harrison *et al.* (1974).

In an attempt to test the computer results *in vivo*, iron was injected into rats in small doses (insufficient to stimulate increased ferritin biosynthesis). Ferritins were then isolated from livers at various times after the iron administration and separated into populations according to their total iron content. As in the *in vitro* experiments, iron was incorporated maximally into ferritin that had about one-third of its maximum iron content. Indications are, therefore, that the hypothesis of Harrison and coworkers is operative *in vivo* (Hoy and Harrison, 1976).

The nature of the amino acid side chains responsible for the initiation of iron deposition in apoferritin was investigated by chemical modification and by the inhibitory action of zinc. Modification of the carboxyl but not of the cysteinyl residues abolished the oxidase activity of apoferritin (Wetz and Crichton, 1976). Moreover, zinc was able to likewise abolish the oxidase activity of apoferritin (Macara *et al.*, 1973). The inhibition was of the competitive type. Since zinc is known to be sequestered by imidazole residues, it was concluded that the histidyl groups of apoferritin are involved in the process of iron uptake.

We have briefly touched upon the question of iron release from ferritin in conjunction with the Harrison hypothesis. In their studies, Harrison and collaborators used ferrous iron chelators to remove iron from ferritin, and others have used ferric iron chelators as well, which do not require a preceding reduction step. Apparently the most effective chelating agent was nitriloacetic acid followed by ethylenediaminetetraacetic acid. Citrate was rather ineffective (Pape *et al.*, 1968b). Although such chelators are quite useful in model studies, they do not represent the physiologically occurring mechanism of iron mobilization from ferritin. It is believed that *in vivo* iron is mobilized from ferritin by reduction, and the most effective reducing agents appear to be the flavinoids, especially reduced riboflavin and $FMNH_2$ (Sirivech *et al.*, 1974). It was found that at 200 μM concentration, $FMNH_2$ could release 100% of the iron in an 0.2% ferritin solution in about 12 min, whereas dithionite at 400 μM concentration required 4 hr to accomplish this. Ascorbate removed only 18% of the iron in 4 hr, and glutathione and cysteine were even less effective.

Dognin and Crichton (1975) investigated whether or not iron mobilization by reducing agents depends on the iron content of ferritin and found that at constant ferritin (protein) concentration, the most rapid iron release using 5 mM cysteine occurred when 800–1400 iron atoms were present in each ferritin molecule. Jones *et al.* (1978) subsequently found that the flavinoids mobilized iron from ferritin most rapidly when the latter contained 1200 iron atoms/ferritin molecule. These results are thus consistent with those obtained by Harrison's group, which studied iron release by 1,10-phenanthroline and iron uptake by ferritin from ferrous iron salts.

It should be noted that, size-wise, reduced flavinoids can without much difficulty penetrate the channels leading into the ferritin interior, as they must in order to effect iron reduction. However, there seems to exist a hindrance to the passage of anionic substances through such channels, including $FMNH_2$ and $FADH_2$. Thus, whereas in regard to reducing synthetic iron cores, reduced riboflavin, $FADH_2$, and $FMNH_2$ are identical and the rate-limiting factor is only the redox potential in question, with respect to reducing iron micelles in ferritin, reduced riboflavin is much more effective than are $FADH_2$ and $FMNH_2$. In case of reduced riboflavin, the rate-limiting factor in regard to its ability to reduce ferritin iron is its redox potential (−208 mV), whereas for $FMNH_2$ and $FADH_2$, whose redox potentials are −216 and 219 mV, respectively, the rate-limiting factor is their ability to negotiate the channels into the ferritin interior (Jones *et al.*, 1978).

An *in vivo* system for the mobilization of iron from ferritin has been proposed by Sirivech *et al.* (1974), which includes an enzyme that regenerates reduced flavinoids from NADH. The entire system is termed ferrireductase, and it has been isolated from a number of vertebrate livers. Oxygen is apparently not required. The ferrireductase system is

$$\begin{array}{ccccc} \mathrm{NADH} + \mathrm{H}^{+} & & \mathrm{FMN} & & 2\,\mathrm{Fe}^{2+} + \mathrm{ferritin}\,(\mathrm{Fe}^{3+})_{n-2} \\ \downarrow & \rightleftarrows & \uparrow\downarrow & \rightleftarrows & \uparrow \\ \mathrm{NAD} & & \mathrm{FMNH}_2 & & \mathrm{ferritin}\,(\mathrm{Fe}^{3+})_{n} \end{array} \qquad (5\text{–}1)$$

Crichton (1973d) has summarized a number of findings in regard to ferritin function and has concluded that in every tissue the control of iron metabolism rests upon the relative rates of iron influx into and the iron efflux from the tissue's ferritin. In other words, the rate of oxidation of ferrous iron by ferritin itself during iron uptake and the level of activity of ferrireductases will to a large extent control iron fluxes in the organism. Further work on the uptake of iron by ferritin and the ferrireductases, including *in vivo* studies, is necessary before this model of iron metabolism can be substantiated.

5.6 Ferritin and Tissue Iron Metabolism

In the previous sections of this chapter we have examined some physical–chemical properties of ferritins as well as their metabolism. In this section, certain mammalian tissues and organs are examined with respect to iron metabolism, and the role of ferritin therin is explored. Among other things, in this section we shall deal with the mode of iron transport into or out of cells, the biology of iron overload, and the physiological significance of serum ferritin.

Several attempts have been made to integrate our knowledge in regard to iron metabolism in the "typical cell," most notably by Munro and Linder (1978) and Trump and Berezesky (1977). The latter model is reproduced in Figure 5–21. It is generally agreed that iron enters the cell while bound to serotransferrin by the process of pinocytosis. The iron is abstracted from the serotransferrin, probably by the action of a ferrireductase in conjunction with a reducing agent. The serotransferrin is returned to the circulation via the original pinosome. A similar procedure in hepatocytes presumably internalizes iron associated with the hemoglobin–haptoglobin complex, ferritin, hemopexin, and iron–dextran complexes. The iron normally abstracted from serotransferrin is now in the ferrous state and joins the cytoplasmic "chelatable Fe^{2+} pool," which may include iron bound to small-molecular-weight protein carriers, amino acids, and peptides. The exact nature of this pool has not yet been determined. The ferrous iron may then combine with apoferritin, which is also present in the cytoplasm, or it may be utilized in the biosynthesis of iron-containing enzymes and proteins such as the cytochromes and myoglobin. The ferritin formed may become sequestered into an autophagosome, which, by acquiring degradative enzymes from primary lysosomes, may become a secondary lysosome with the result that the protein becomes digested and hemosiderin emerges.

The ferritin may also lose its iron by reduction as described in the preceding section, thus releasing ferrous iron into the chelatable Fe^{2+} pool, and such iron may eventually find its way into the circulation. Presumably, iron transport into the vascular space takes place by simple diffusion, the driving force being the chemical potential difference. This can be demonstrated in model systems, where ferritin carrying one-third to one-half of its maximum iron-binding capacity will donate its iron to serotransferrin carrying physiological amounts of iron if the two are separated by a semipermeable membrane and if a reducing agent (e.g., ascorbate) and an iron chelator (e.g., citrate or ATP) are also present (Harris, 1978). Another reason chemical potential favors the movement of ferrous iron into the circulation is because the circulatory ferrous iron pool is nonexistent: As soon as ferrous iron enters the bloodstream, it becomes oxidized to the ferric state by O_2 in the presence of the powerful ferroxidase I, also called ceruloplasmin. The ferric iron is then immediately sequestered by serotransferrin. Figure 5–22 illustrates the operation of the ferroxidase system.

There is ample amount of evidence, mainly from Frieden's labora-

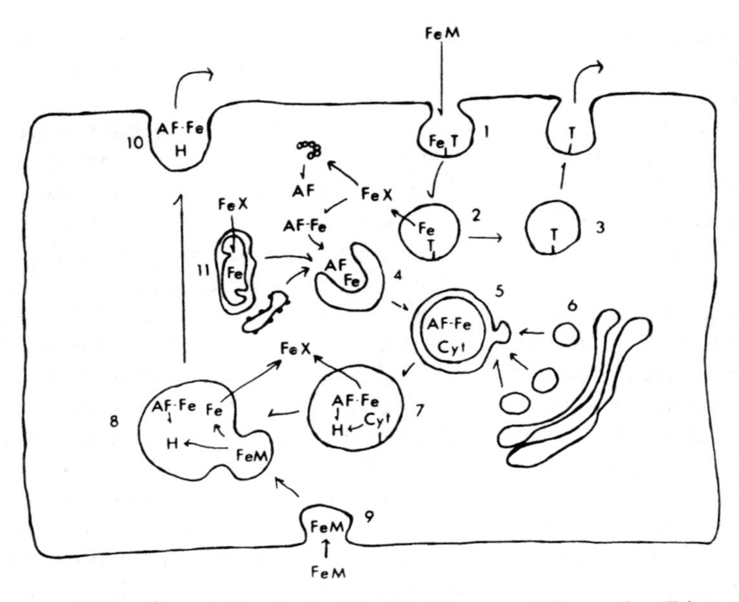

Figure 5-21. Cell iron metabolism is visualized by Trump and Berezesky. T is serotransferrin, AF is apoferritin, X is an iron binder of an unknown nature. 1 indicates iron uptake by pinocytosis, 2 is the internalized vacuole carrying iron-laden transferrin, and 3 is the vacuole returning serotransferrin to the extracellular space following the loss of iron. The iron thus released from serotransferrin combines with apoferritin forming ferritin. Ferritin molecules fuse with cisternae of the endoplasmic reticulum (step 4), which results in the formation of bodies (5) containing ferritin and portions of the cytoplasm. These fuse with primary lysosomes (6) to form secondary lysosomes (7). Ferritin is digested therein to form hemosiderin H (8). Macromolecular iron (FeM) enters the secondary lysosomes directly via pinocytosis (step 9). Reprinted by permission from Trump, B. F., and Berezesky, I. K., 1977, A general model of intracellular iron metabolism, *in Proteins of Iron Metabolism*, E. B. Brown, P. Aisen, J. Fielding, and R. R. Crichton (eds.), Grune & Stratton, New York, pp. 359–364. Copyright by Grune and Stratton, Publ.

tory, that on the molecular level the main function of ceruloplasmin, the copper-containing serum protein, is to oxidize ferrous iron (see reviews by Frieden, 1973, and Frieden and Hsieh 1976). It has been known since the 1920s that copper-deficient animals become anemic and that their serum iron levels become severely depressed even though their total body iron stores may be normal. Such animals are also ceruloplasmin deficient. However, both the red blood cell picture and ceruloplasmin levels can be restored by the administration of copper (Marston and Allen, 1967; Planas and Frieden, 1973). The most dramatic effects are obtained, however, when serum iron levels are measured following ceruloplasmin in-

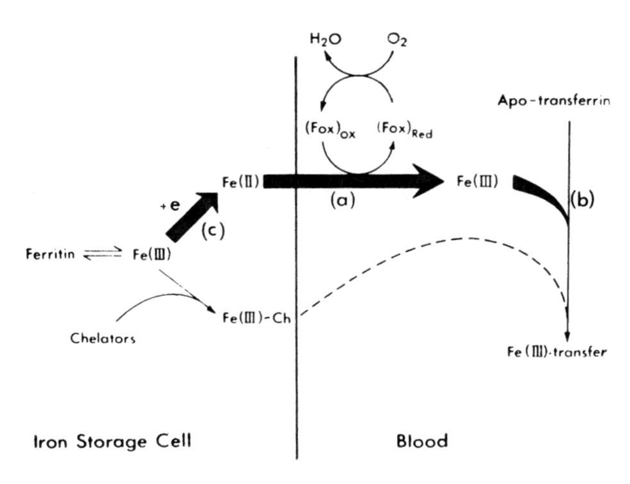

Figure 5-22. The role of ceruloplasmin (ferroxidase I) in iron metabolism. F_{ox} stands for ferroxidase I. From Osaki *et al.* (1971).

fusion in copper-deficient animals. Such effects are presented in Table 5–8.

As useful as the preceding general models may be, they must not be indiscriminately applied to every tissue, as each tissue, organ, or cell line has its own peculiarities with respect to iron metabolism. It is therefore necessary to examine each tissue separately before any specific conclusions about its iron metabolism can be made. The tissue studied most thoroughly has been the liver. This occurred because of its pivotal role in the primary and secondary iron overload disorders. Iron biochemistry in the liver is dealt with in the next section.

Table 5-8. Increase in Serum Iron Levels in Copper-Deficient Pigs Following Infusion of Ceruloplasmin and Copper Sulfate (in μg/100 ml)[a]

Time after injection (min)	Ceruloplasmin injected	Copper sulfate injected
5	18	3
15	26	2
30	84	3
60	136	14
240	225	15

[a] Adapted from Ragan *et al.* (1969).

5.6.1 Iron Metabolism in the Liver

American males carry an average of 600–800 mg of *storage* iron in their livers, with a "normal" maximum of 1500 mg. This translates to an average of 0.1% of the liver's dry weight being accounted for by iron (Barry, 1974). In the human being, of about 35 mg of iron turned over per day, some 4–5 mg enter and leave the liver (0.09 mg of iron per 100 ml of whole blood per day). In patients with iron deficiency anemia, liver iron uptake may amount to 0.01 mg/100 ml/day, and in iron overload states, this figure may be as high as 0.45 mg/100 ml/day (Hershko, 1977). In the rat, on the other hand, total liver *depot* iron is 153 ± 7 μg/g of liver, with ferritin being 499 ± 32 μg/g of liver. About 20% of the iron turned over by the rat per day (171 μg/kg/day) involves the liver, of which 50% involves ferritin (Mazur *et al.*, 1960; Zuyderhoudt *et al.*, 1978b).

The mechanism of iron uptake by liver cells has been studied by a variety of methods. Such uptake experiments involve only the parenchymal cells, as the Kupffer cells do not take up serotransferrin-bound iron. The first experiments attempted utilized liver slices (Saltman *et al.*, 1956a, 1956b). It was found that in such systems liver cells accumulated iron against iron concentration gradients when ferric ammonium citrate was used. This process was not an active one, and iron accumulation occurred because of its sequestration inside the cell, perhaps by ferritin. Moreover, it appeared that the intactness of the cell membrane was not a prerequisite for such iron uptake. Although the work of Saltman and colleagues was criticized because they did not use physiological iron donors, more recent results using isolated liver hepatocytes have corroborated these early findings: Iron presented in the form of ferric and ferrous citrate, as well as that bound to serotransferrin, was taken up equally well by cells with *intact* membranes (as revealed by the trypan blue die exclusion method) as by cells whose membranes were damaged. The use of isolated rat hepatocytes has also shown that the most effective iron uptake by the liver occurs when the iron is presented in the ferrous form. This is followed by ferric citrate, and the uptake of iron from serotransferrin is the slowest of the three (Grohlich *et al.*, 1977; Verhoef *et al.*, 1978; Grohlich *et al.*, 1979). The uptake of iron was dependent on iron concentration in the medium, and the same results were obtained with perfused livers (Zimelman *et al.*, 1977). A large portion of the iron taken up into cytosol from ferrous and ferric citrate, as well as from serotransferrin, is found in ferritin. This was ascertained using isolated rat hepatocytes in suspension (Grohlich *et al.*, 1979), Chang cells in tissue culture (White and Jacobs, 1978), and rat liver cells in tissue culture (Beamish *et al.*, 1975).

Iron taken up from ferric citrate and nitriloacetate by the Chang cells was found largely in the membrane fraction (White and Jacobs, 1978).

The Fletcher–Huehns hypothesis (see Chapter 6) was tested in regard to liver cells both *in vivo* and *in vitro* with conflicting results. It is agreed, however, that diferric serotransferrin is a better iron donor with respect to liver cells than is monoferric serotransferrin, a situation identical to that of the immature red cells and other tissues. *In vivo* experiments in rats showed that monoferric serotransferrin prepared by isoelectric focusing was a poorer iron donor to the liver than was diferric transferrin, the difference being some 2.5-fold in normal animals (Christensen *et al.*, 1978). Identical results were obtained with cells in tissue culture (Beamish *et al.*, 1975). The affinity of liver cells for either site A or site B of the serotransferrin molecule was tested *in vivo* by injecting into rats serotransferrin selectively labeled in the two sites by ^{59}Fe and ^{55}Fe. Should the Fletcher–Huehns hypothesis be operative in regard to liver cells, the ^{59}Fe–^{55}Fe ratio in the liver should deviate from 1. This was not found to be the case, the ratio actually being close to 1 (Pootrakul *et al.*, 1977). Rat cells in tissue culture could not distinguish between the two sites either (Beamish *et al.*, 1975). On the other hand, in hepatocytes in suspension, ^{59}Fe–serotransferrin preincubated with reticulocytes was a better iron donor than was transferrin labeled randomly to the same extent with ^{59}Fe (Verhoef *et al.*, 1978).

There has been much effort expended to determine whether or not hepatocyte membranes contain specific serotransferrin receptors akin to those found in erythroid tissues. In 1971, Fletcher proposed that because liver cells interacted preferentially with diferric serotransferrin as opposed to its monoferric counterpart, liver cell membranes should carry specific serotransferrin receptors (Fletcher, 1971). Later, Van Bockxmeer *et al.* (1975) reported on the extraction of macromolecular material containing serotransferrin from the membranes of various tissues, including liver, and concluded that hepatocytes probably carried transferrin receptors on their plasma membranes. Subsequent work on serotransferrin-bound iron by isolated hepatocytes gave complex results: At low iron saturation levels (10–30%), saturation kinetics in the Michaelis–Menten sense were observed, as might be expected if a receptor were present on the hepatocyte surface. However, with serotransferrin iron saturation levels of over 30%, no such behavior was noted (Figure 5–23). One is then left with the conclusion that the hepatocyte membrane may contain a receptor specific for perhaps monoferric but not diferric serotransferrin, the latter then being handled by the process of simple diffusion into the cell (Grohlich *et al.*, 1979).

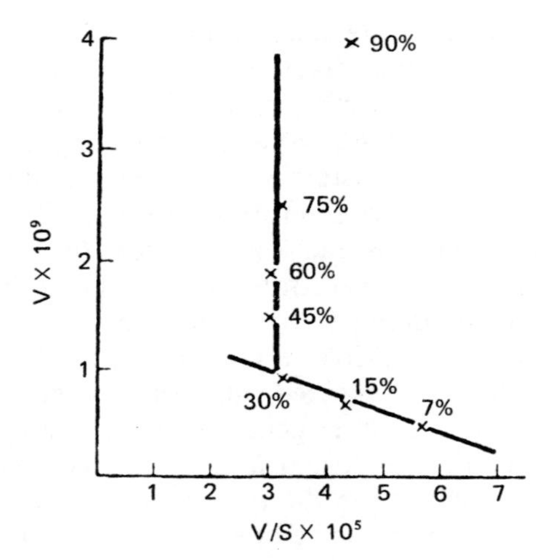

Figure 5-23. Eadie–Hofstee plot of iron uptake by isolated rat hepatocytes from serotransferrin saturated with iron at various levels. *V* is defined as mol per l of iron incorporated into 10^7 cells per min. *S* is defined as iron concentration in the medium in mol per l. Note that whereas iron concentrations were varied (*S*), per transferrin concentrations remained constant. Reprinted with permission from Grohlich, D., Morley, C. G. D, and Bezkorovainy, A., 1979, Some aspects of iron uptake by rat hepatocytes in suspension, *Int. J. Biochem.* 10:797–802, Copyright 1979, Pergamon Press, Ltd.

We can now address the question of whether or not serotransferrin is internalized by the hepatocyte during the process of iron uptake, as has been suggested by the models of Munro and Linder (1978) and Trump and Berezesky (1977). Precious little evidence in either direction is available. In a report by Milsom and Batey (1979), however, it is argued that no pinocytosis of native iron-laden serotransferrin takes place when iron is taken up by the rat hepatocyte. Serotransferrin labeled with both ^{125}I and ^{59}Fe was injected into rats, and their livers were then separated into subcellular particles and counted. The lysosome fraction contained little label, contrary to what would be expected if serotransferrin were internalized by the cell by the process of pinocytosis. In addition, whereas liver ^{125}I reached its maximum some 5 min after the injection, ^{59}Fe reached its maximum only after 16 hr and then remained constant for 14 days. Most of the ^{59}Fe was found in the cytoplasm associated with ferritin. Denatured iron-laden serotransferrin was, however, taken up into the cell interior by the process of pinocytosis.

The release of iron by the liver cells in suspension may be affected by serotransferrin and various small-molecular-weight iron chelators. Hypoxia increased iron release by liver cells, whereas hyperoxia decreased it (Baker *et al.*, 1977). In perfused dog liver, iron was not appreciably mobilized when the organ was perfused with a solution of aposerotransferrin or citrate. However, large amounts of iron left the liver when ceruloplasmin but not copper was added to the perfusion medium (Osaki and Johnson, 1969; Osaki *et al.*, 1971). This aspect of liver metabolism is therefore consistent with Frieden's hypothesis on the function of ceruloplasmin as a ferroxidase. It should be mentioned that the origin of mobilized iron in liver perfusates may be both the parenchymal and Kupffer cells.

5.6.2 Iron Metabolism in the Reticuloendothelial System

The phagocytic reticuloendothelial system (RES) cells are seen largely in the spleen, bone marrow, and the liver in the form of the Kupffer cells. It acquires iron from damaged and presumably normal senescent red cells, converts the heme moiety to bilirubin, and eventually returns the iron to circulation in the form of serotransferrin-bound ferric iron. RES cells will not remove iron from serotransferrin.

Iron metabolism in the RES is apparently ascorbate dependent. In scorbutic guinea pigs, spleen storage iron is greatly increased and consists largely of hemosiderin, whereas in the liver, storage iron is in fact decreased, most of such decrease being due to lowered ferritin levels (Lipschitz *et al.*, 1971). In scorbutic human beings suffering from iron overload, plasma iron levels are low, contrary to their vitamin-C-replete cohorts. Upon the administration of ascorbic acid, plasma iron levels in these patients return to the high levels normally seen in iron overload cases. It has been proposed that ascorbic acid is required for the return of iron into the plasma and that ascorbic acid replaces the ferrireductase system found in other tissues. It is interesting to note that today the most severe cases of scurvy are found in iron overload cases, so that it may be argued that iron overload per se may create ascorbic acid deficiency (Hershko, 1977). It would seem that a form of a vicious circle may be set in motion by iron overload, especially in malnourished individuals: Increased tissue iron levels destabilize ascorbic acid, which in turn prevents the return of iron back into the circulation, thus making iron unavailable to the hemopoietic system. It will be remembered that the RES is quantitatively the most important source of plasma iron.

The cells of the RES are apparently supplied with serotransferrin receptors on their plasma membranes. This conclusion was reached from studies on alveolar macrophages of rabbits, which can serve as a convenient model system for the RES. ^{125}I–serotransferrin was bound to the macrophages and was eluted from them by unlabeled serotransferrin. Such binding of serotransferrin was temperature and energy dependent (Wyllie, 1977).

Iron acquired by the RES apparently enters a *labile* iron pool, part of which is immediately channeled into the circulation (half-life in dogs is 30 min) and a part of which goes into storage, presumably in combination with ferritin (half-life is 7 days). In dogs, the *early* and *late* systems of iron return are approximately equal in volume, whereas in man and rat, some 70% of an administered iron dose is accounted for by the early system and 30% by the late system. These ratios can change dramatically: In the case of hemorrhage or other conditions bringing about an accelerated degree of erythropoiesis, practically all the iron may be handled via the early phase and thus immediately returned to the circulation. Increased body iron stores, on the other hand, shift iron return to the late phase. In infections, inflammations, and neoplasm conditions as well as following endotoxin injections, plasma iron levels drop, which is due to both the movement of iron into the parenchymal cells as well as an increased degree of iron retention by the RES. These conditions in fact stimulate iron incorporation into ferritin of the RES, thereby shifting iron return to the late phase (Torrance *et al.*, 1978). See Chapter 7 for further details. Normally, the availability of iron-binding sites on the circulating serotransferrin molecules and the size of blood-*exchangeable* iron pool will determine the amount of iron the RES will release into the circulation.

A proposal has been made to the effect that idiopathic hemochromatosis (see Section 5.6.3) may be due to an inability by the RES to store iron; i.e., there is a lesion in the operation of the late phase of iron return to plasma. The result is an excessive release of iron into the circulation, which in turn results in an increased iron uptake rate by the liver cells (Fillet *et al.*, 1975).

5.6.3 Iron Overload

The classical iron overload syndrome was first described by Troisier in Paris in 1871 and was termed *bronze diabetes* by him, because the patient had died of diabetes mellitus but had bronze coloration in all his tissues. Today we recognize that iron overload patients may develop other complications as well, including congestive heart failure, hepatic cirrhosis and fibrosis, osteoporosis, and various endocrine gland abnormalities. In fact, hemochromatosis has been implicated as an "easily missed cause" in many cases of cirrhosis, diabetes, and heart failure (Giorgio, 1973). The bronze skin and tissue coloration observed is due to melanin accumulation (Cumming, 1978).

It is possible to classify iron overload syndromes on the basis of their etiology into primary and secondary disorders. The latter may be acquired from dietary sources or blood transfusions. The organs most often accumulating iron are the liver and the RES, and different varieties of iron overload will deposit iron primarily in either the liver, the RES, or both. Iron overload involving largely the RES is considered to be benign, whereas that involving the parenchymal system (liver, pancreas, and other organs) is deemed to be dangerous and likely to elicit potentially fatal complications.

The primary iron overload syndrome called hemochromatosis involves largely the liver, heart, and various endocrine organs. It does not involve the RES. Although it is inherited, the mode of such inheritance is not definitely known. It is suspected that an autosomal dominant mechanism may be involved (Barry, 1974). The primary lesion in the inherited hemochromatosis is also unknown, and for that reason it is often referred to as *idiopathic* hemochromatosis. We do know that iron absorption from dietary sources in hemochromatotic patients is increased, amounting to some 3 mg/day, whereas in the normal individual it is 1.5 mg/day. It has been proposed that idiopathic hemochromatosis is due to a breakdown of the mucosal block system, which normally prevents the absorption of unneeded iron by the organism (Crosby *et al.*, 1963). Patients with idiopathic hemochromatosis, usually men 50–70 years of age, have a total body iron count of 15–50 g, the average being about 20 g. Iron may account for 2–4% of the dry weight of the liver in such cases. The disease begins to develop from the time growth stops, so that assuming an accumulation of 1.5 mg/day (3 mg absorbed minus 1.5 mg, which is the amount excreted daily), the patient accumulates about 0.55 g/year or about 16 g of iron in a span of 30 years when the disease becomes symptomatic. Women are protected against the disease because of their monthly menses.

Another symptom associated with the increased iron absorption by the intestinal mucosa of the hemochromatotic patient is the increased iron uptake by the liver. When hemochromatotic patients whose iron status was reduced to normal by therapy (normally phlebotomy) were given trace amounts of ^{52}Fe, they took up $6.9 \pm 1.2\%$ of the administered dose into the liver. The figure for normal adult men was $1.42 \pm 0.59\%$ (Batey *et al.*, 1978). Iron uptake by the liver in both normal subjects and hemochromatotic patients is dependent on serum iron levels and the degree of serotransferrin saturation with iron. A third phenomenon characteristic of the idiopathic hemochromatosis condition is the failure of the RES to retain iron and the inordinately high rate of iron release into circulation (Charlton *et al.*, 1973; Hershko, 1977). As mentioned in the previous section, the early phase of iron return by the RES completely overshadows the late phase in idiopathic hemochromatosis. Plasma iron levels in hemochromatosis are extremely high and often approach TIBC levels. This is of course due to increased iron absorption and iron output by the RES. Treatment generally involves phlebotomy, whereby 200–250 mg of iron may be removed with each 500 ml of blood. The development of the disease over the life cycle of a typical patient has been graphed by Powell *et al.* (1978) and is shown in Figure 5–24. For further information on idiopathic hemochromatosis, the reader may consult reviews by Hershko (1977), MacDonald (1969), and Pollycove (1978).

An iron overload syndrome can also be brought about by repeated blood transfusions. In one instance, the aplastic anemia, the overload is due entirely to the destruction of foreign red cells, and iron overload involves the RES only. In such a case, the overload may be considered to be "benign." In thalassemia syndromes, excess iron is due to the hemolysis of endogenous red cells as well as the destruction of foreign red cells acquired through transfusion therapy. The large amount of iron entering the circulation apparently overwhelms the capability of serotransferrin to bind iron, and some 10–20% of plasma iron in the thalassemic serum is bound to serotransferrin nonspecifically and/or to serum albumin. This iron is cleared very rapidly by the liver; hence iron overload in thalassemia involves both the RES and the parenchymal cells of the liver (Hershko and Rachmilewitz, 1975). A useful model system for such nonserotransferrin plasma iron has been worked out by White and Jacobs (1978) and involves the Chang cells in tissue culture and iron–nitriloacetate as the iron donor.

The dietary iron overload syndrome has been observed mainly in certain areas of the world where environmental conditions have favored the development of this disease. In most instances, dietary iron overload is associated with high alcohol intake, as alcohol can almost double the rate of iron absorption in humans. Iron levels in many wines is frequently high, and populations consuming such wines have a high rate of dietary iron overload. Likewise, South African natives who brew their beer in cast iron vats have a high incidence of iron overload. It has also been claimed that in certain Western nations that fortify their bread with iron the incidence of iron overload has been on the increase.

In nutritional iron overload as well as in idiopathic hemochromatosis, the examination of tissue ferritins by isoelectric focusing reveals that isoferritins of all tissues are identical; i.e., they are of the liver–spleen type. There is thus a loss of the faster, more acidic isoferritins. In treated idiopathic hemochromatosis, where iron stores are at a normal level albeit temporarily, the isoferritin patterns return to normal (Powell *et al.*, 1975b)). No change in tissue isoferritin patterns is observed in transfusional iron overload (Jacobs and Worwood, 1975b).

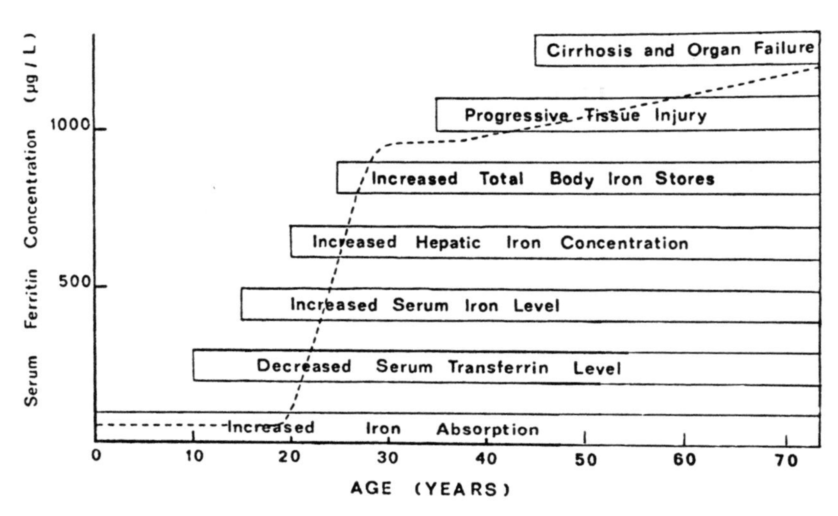

Figure 5-24. Chronology of the development of primary hemochromatosis in man. From Powell *et al.* (1978).

One may now inquire as to the distribution of iron in the tissues of the iron-overloaded human being or experimental animal. There is a tremendous age and case-by-case variation; however, in summary it may be stated that hemosiderin makes up a large proportion of storage iron in the hepatocytes of such patients. Hemosiderin deposits in the spleen are also somewhat increased in idiopathic hemochromatosis. Ferritin levels are, of course, very high (Pollycove, 1978). In animals with artificially induced short-term iron overload using iron–dextran and ferric citrate, most of the iron is found in the cell cytoplasm, i.e., ferritin. Total liver iron in such rats was 7.61 µg/kg of dry weight, whereas controls had only 0.4 µg/mg. Iron was increased in both the parenchymal and Kupffer cells (Bankowsky *et al.*, 1979). In an earlier but more inclusive study, iron was administered to rats in various doses, and animals were sacrificed at various times following iron loading. Nonheme iron was examined in both the parenchymal and Kupffer cells, and the results are shown in Table 5–9. The most important conclusion that can be derived from this study is that the parenchymal cells store iron largely in the form of ferritin but can also store iron in other nonheme (hemosiderin?) forms if the dose administered is very large. On the other hand, the Kupffer cells make little if any ferritin, and the main iron storage form therein is granular and probably hemosiderin in nature (van Wyk *et al.*, 1971). A similar result is also shown in Table 5–1. Attractive as these studies are in terms of providing an understanding of human iron metabolism, it must still be

Table 5-9. Distribution of Iron in Iron-Loaded Rats[a]

Treatment of rats (via injection)	Total iron given (mg)	Total nonheme iron in liver		Whole liver ferritin iron ($\mu g/10^7$ nuclei)	Heme iron ($\mu g/10^7$ nuclei)	Kupffer cell iron ($\mu g/10^7$ nuclei)	
		mg	$\mu g/10^7$ nuclei			Ferritin iron	Total nonheme iron
Control	0	1.0	6.4 ± 1.3	5.3 ± 1.2	1.5	0.15 ± 0.09	0.23 ± 0.33
Hemoglobin	16	3.0	17.3 ± 2.6	12.9 ± 2.5	1.5	1.1 ± 0.6	7.1 ± 4.4
Iron–dextran, one injection							
After 1 week	25	7.8	37.9 ± 7.2	25.3 ± 3.7	1.4	1.6 ± 0.8	27.8 ± 13.9
After 18–20 weeks	25	8.3	49.5 ± 4.9	34.9 ± 2.2	1.6	1.7 ± 0.8	34.8 ± 13.9
Iron–dextran, multiple injections; over 16-week period	140	58.6	290 ± 33	73 ± 5	—	Scant	Very heavy

[a] Adapted from van Wyk *et al.* (1971).

remembered that human beings develop the primary iron overload disease slowly, over a period of 30 years or longer, with a total average excess iron accumulation of 20 g. If it is assumed that an "average" patient weighs 70 kg, this then represents 0.29 g of iron per kg of body weight. In the work of van Wyk *et al.* (1971), the group of rats (average weight, 182.5 g) receiving a total of 140 mg of iron retained some 42% of the iron administered over a 16-week period. This represents 0.16 g/kg of body weight and is very roughly equivalent to the human situation.

5.6.4 Serum Ferritin

The presence of ferritin in serum was discovered only recently following the development of the radioimmunoassay techniques. The interest in serum ferritin is of course based on its potential use as a diagnostic tool, especially in estimating total iron stores in human beings, thus making it possible to detect iron deficiency or overload syndromes.

A graph relating serum ferritin levels with total body storage iron is shown in Figure 5–25. It is seen from this that adult females and males have about 35 and 90 μg of ferritin per liter of serum (Saarinen and Siimes, 1979), whereas newborns and infants show much higher values. Other

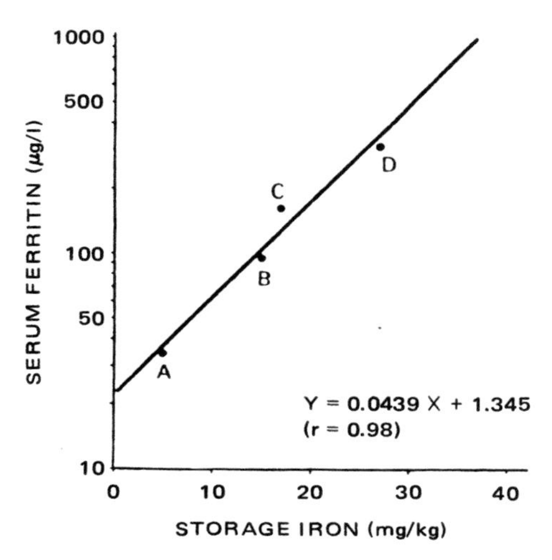

Figure 5-25. Relationship between total body iron stores and serum ferritin levels. A, adult females; B, adult males; C, newborn infants; D, infants 2 weeks old. From Saarinen and Siimes (1979).

authors have given values of 56 and 123 μg/liter for adult females and males, respectively (Jacobs and Worwood, 1975b), and other widely divergent data exist (Engel and Pribor, 1978). A rough estimate of 1 μg of ferritin per liter of serum for every 8 mg of storage iron has also been used.

The source of human serum ferritin is apparently the spleen; at least the isoferritin pattern of serum ferritin is similar to that of spleen and liver ferritin. It is very low in iron content and may be considered to represent spleen apoferritin (Jacobs and Worwood, 1975b). Ferritin administered intravenously into human beings is cleared very rapidly by the liver parenchymal cells (see Section 5.4). Thus, even though it has been the custom to think of serum ferritin as being representative of the total body iron stores, it may be more precise to think of it as representing the reticuloendothelial system iron stores.

It appears that the determination of serum ferritin is most useful in detecting iron deficiency states, since lowered serum ferritin levels become manifest before lowered hemoglobin levels are detected. It has been estimated that iron deficiency is present in 50% of U.S. pregnant women, 30% of preschool children, 12% of adult women, and 3% of adult males (Engel and Pribor, 1978). Thus ferritin determination has been advocated most vigorously in regard to prenatal care, where serum ferritin has been found to be more sensitive to iron deficiency states than serum iron or hemoglobin determinations (Ances *et al.*, 1979). In regard to children, ferritin determinations have also proved quite useful in assessing total body iron stores. Complications arise, however, in cases of infection or neoplastic disease, which tend to give high serum ferritin values (Siimes *et al.*, 1974). One can perhaps conclude that the utility of serum ferritin measurements in detecting iron deficiency states resides in combining these with measurements of total hemoglobin and of serotransferrin iron saturation levels. Yet, invariably, ferritin determinations result in the identification of a larger number of iron deficiency states than do either hemoglobin or serotransferrin iron saturation studies. This phenomenon is illustrated for a population of Canadian Indians in Table 5–10 (Valberg *et al.*, 1979).

There has been a controversy in regard to the utility of serum ferritin assays as diagnostic tools in detecting idiopathic hemochromatosis and, more importantly, in detecting precirrhotic iron overload. The data of one group of researchers have indicated that ferritin assays are a safe and reliable way of detecting preclinical hemochromatosis, where false diagnoses accounted for only 4.8% of the cases, whereas false positives and negatives were obtained in 34% of the cases when serum iron concentrations were used and in 33% of the cases when serotransferrin sat-

Table 5-10. Iron Status in Canadian Indians[a]

Subject group	n	Serum ferritin (ng/ml)		Transferrin saturation with iron (%)		Hemoglobin (g/100 ml)	
		Median	% Abnormal	Median	% Abnormal	Median	% Abnormal
Children, 1–4 yr	85	12	40	19	29	12.6	5
Children, 5–9 yr	200	17	22	20	25	13.0	3
Adolescents, 10–19 yr							
Male	167	17	37	21	21	14.0	10
Female	197	13	47	20	26	13.4	9
Adults, 20–54 yr							
Male	230	54	11	25	7	15.6	6
Female	314	24	34	20	23	13.8	9
Adults, 54 yr							
Male	100	68[b]	3	27	8	15.2	17
Female	94	59	11	22	16	14.2	7
Pregnant females	34	14	56	20	26	12.6	9

[a] Adapted from Valberg *et al.* (1979).
[b] High serum ferritin values in an elderly population was also observed by Loria *et al.* (1979).

uration levels were used (Powell *et al.*, 1978; Powell and Halliday, 1978). Others have claimed that false diagnosis is much more frequent when using serum ferritin assays and have recommended that serum ferritin results not be relied upon for the diagnosis of preclinical hemochromatosis (Crosby, 1977). There is of course no controversy in that in full-blown symptomatic hemochromatosis and secondary iron overload serum ferritin levels are extremely high. Serum ferritin isoferritin patterns in idiopathic hemochromatosis follow the pattern of tissue isoferritins: There is a loss of the acidic isoferritins (pI 5.02–5.06) and their replacement by the more basic ones (pI 5.54–5.62). When the disease is treated (albeit symptomatically), the serum isoferritin pattern returns to normal (i.e., there is a reappearance of the acidic isoferritins).

It has been pointed out that a high serum ferritin does not necessarily indicate a high body iron level, as ferritin can apparently leak out of damaged liver and other tissue cells in the absence of iron overload. In addition, serum ferritin levels have been found to be high in various types of leukemias, Hodgkin's disease, and other neoplasms (Jacobs and Worwood, 1975b). In the various types of cancers given in Table 5–11 (data collected by Hazard and Drysdale, 1977), there is apparently a characteristic distribution of the liver-type (basic, pI ca. 5.6) and the HeLa- or heart-type ferritins (acidic, pI ca. 5.1) in serum. In cases where the serum is suspected of containing considerable amounts of the HeLa-type ferritin,

Table 5-11. Serum Ferritin Levels in Various Types of Cancer and Other Disorders (in μg/liter)[a]

Disease	Liver type	HeLa type[b]	HeLa–liver
Normal	77 ± 48	—	—
Hemochromatosis	5,000	190	0.04
	950	—	—
Teratoblastoma	1,550	1000	0.6
Glioma	130	—	—
Ovarian carcinoma	62	430	7.1
Breast cancer	205	500	2.4
	330	800	2.4
	190	350	1.8
Pancreatic carcinoma	320	600	1.9
	600	1900	3.1
	460	500	1.1
Hepatoma	560	850	1.5
Gall bladder carcinoma	550	2700	5.0
Colon carcinoma	220	1500	6.7
Stomach carcinoma	750	1750	2.4
	118	235	2.0
	550	1050	1.9
Lung carcinoma	630	1750	2.8
Multiple myeloma	2,630	1350	0.5
Acute lymphoblastic leukemia	3,800	165	0.04
	34,670	130	0.004
Acute myeloblastic leukemia	1,675	—	—
	1,866	135	0.07
Pancreatitis	1,170	740	0.6

[a] Adapted from Hazard and Drysdale (1977).
[b] HeLa cells are a neoplastic line of cells originally derived from a tumor of the uterus.

it is necessary to use radioimmunoassay techniques involving antisera to both the liver- and HeLa-type ferritins. Normally, serum ferritin is assayed using antisera to liver–spleen-type ferritin. The immunologic differences among tissue ferritins are considerable and have been discussed thoroughly by Hazard *et al.* (1977).

Ferritin levels in serum have also been correlated with the rates of iron absorption by the gastrointestinal tract. Of course a direct relationship exists between serum ferritin levels and the rate of iron absorption as well as between iron absorption rates and bone marrow nonheme iron stores. An inverse relationship exists between serum ferritin and bone marrow iron stores (Bezwoda *et al.*, 1979). These relationships are illustrated in Figure 5–26. It would be attractive to think of serum ferritin (which represents the level of reticuloendothelial system iron stores) as being the actual messenger controlling the rate of either iron uptake by

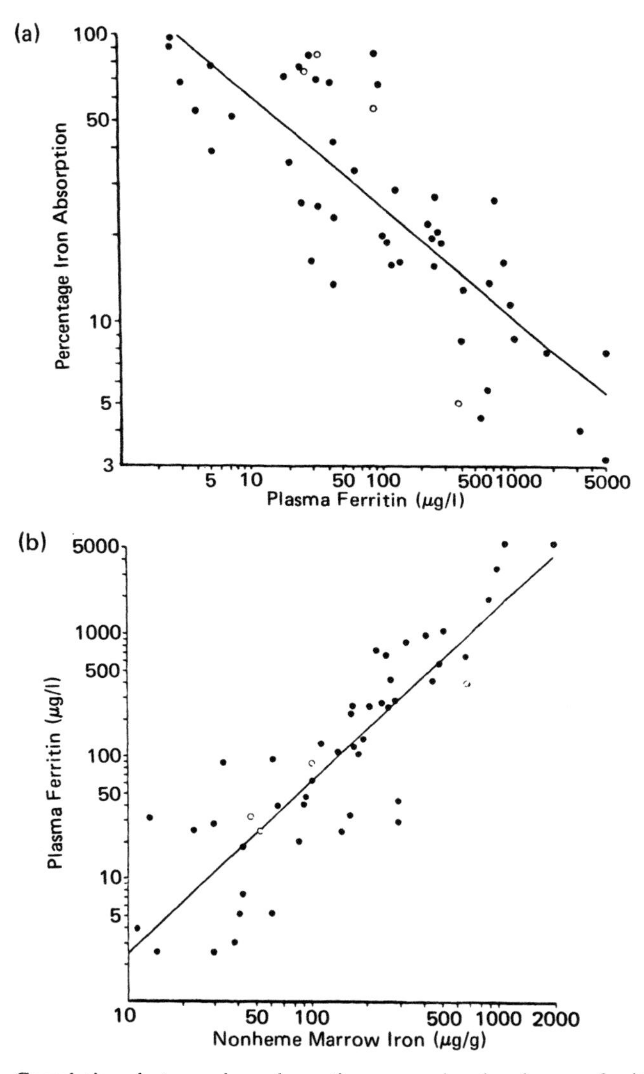

Figure 5-26. Correlations between iron absorption on one hand and serum ferritin and nonheme iron bone marrow stores on the other. (a) Correlation with serum ferritin. (b) Correlation with iron stores. Open circles, patients with primary hemochromatosis. Reprinted with permission from Bezwoda, W. R., Bothwell, T. H., Torrance, J. D., MacPhail, A. P., Charlton, R. W., Kay, G., and Levin, J., 1979, The relationship between marrow iron stores, plasma ferritin concentrations, and iron absorption, *Scand. J. Haematol.* 22:113–120, Copyright 1979, Munksgaard International Publishers, Ltd., Copenhagen, Denmark.

the intestinal messenger controlling the rate of either iron uptake by the intestinal mucosal cells or the release of iron by these cells into the circulation. No direct evidence for such serum ferritin function is yet available, and in fact it has been found that the infusion of ferritin into the circulation of the rat did not result in any changes in iron absorption (Munro and Linder, 1978).

In closing the discussion of serum ferritin, it would be of interest to present some clinical biochemistry data on serum ferritin and other pertinent parameters found in a varity of hematologic disorders. Such a correlation appeared in a review chapter by Pollycove (1978) and is given in Figure 5–27.

5.6.5 Intestinal Mucosal Cells and Iron Metabolism

Duodenal intestinal mucosal cells are the primary locus for iron absorption in the mammalian organism, and the brush borders of these cells are responsible for more than half of such iron uptake. The gross aspects of iron absorption by the organism were dealt with in Chapter 3, and this section will highlight the function of ferritin and other iron-binding proteins in the iron absorption process.

An average adult male ingests some 16 mg of iron per day but absorbs into the circulation only 0.9–1.5 mg. Much of the daily iron intake does find its way into the mucosal cell, but the bulk of it remains there and is eventually lost through the normal sloughing off process of the intestinal villi. This retention of iron by the mucosal cells, that is, the withholding of iron from the circulation, or, more precisely, the control of iron release into the circulation, has often been referred to as the mucosal block. The primary component of the mucosal block mechanism is believed to be ferritin, whose function in the control of iron absorption was postulated more than 30 years ago (e.g., Granick, 1946b). This model states that relatively large amounts of iron become absorbed daily by the duodenal mucosal cells but that the bulk of it becomes sequestered by the ferritin present therein. Presumably the ferritin is in an equilibrium with a *chelatable* ferrous iron pool of the mucosal cell cytosol and may release its iron to this pool and thence into the circulation should the organism demand it. This concept was later revised (Crosby, 1963), and it was proposed that uptake of iron by the mucosal cell ferritin was an irreversible event and that, in effect, ferritin-bound iron of the intestinal mucosa was in a holding pattern waiting to be excreted. The operation of the mucosal block was effectively demonstrated by Charlton *et al.* (1965): Two hours following a dose of oral radioiron to experimental animals, 35% of the

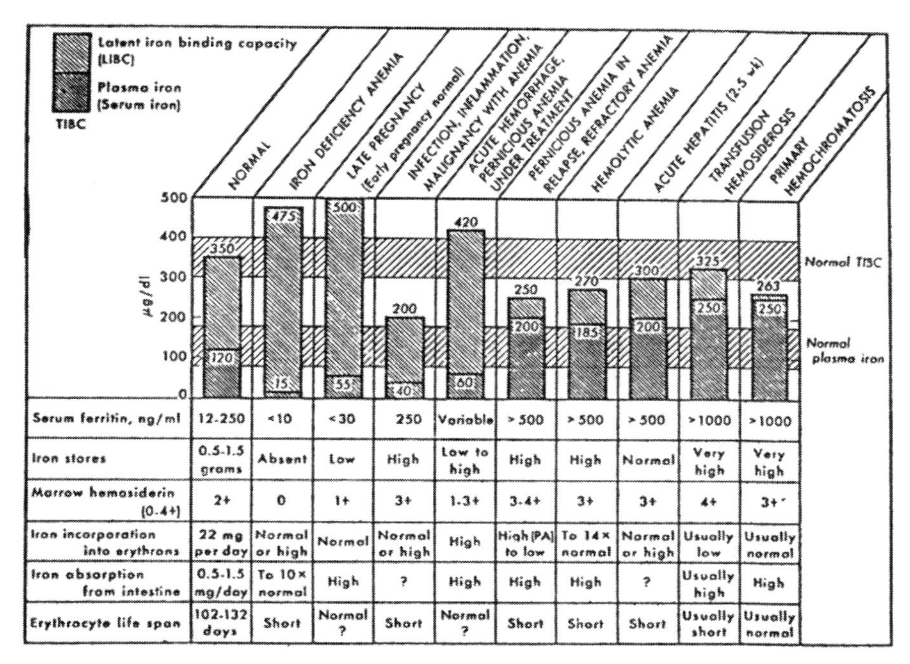

	NORMAL	IRON DEFICIENCY ANEMIA	LATE PREGNANCY (Early pregnancy normal)	INFECTION, INFLAMMATION, MALIGNANCY WITH ANEMIA	ACUTE HEMORRHAGE, PERNICIOUS ANEMIA UNDER TREATMENT	PERNICIOUS ANEMIA IN RELAPSE, REFRACTORY ANEMIA	HEMOLYTIC ANEMIA	ACUTE HEPATITIS (2-5 wk)	TRANSFUSION HEMOSIDEROSIS	PRIMARY HEMOCHROMATOSIS
Serum ferritin, ng/ml	12-250	<10	<30	250	Variable	>500	>500	>500	>1000	>1000
Iron stores	0.5-1.5 grams	Absent	Low	High	Low to high	High	High	Normal	Very high	Very high
Marrow hemosiderin (0-4+)	2+	0	1+	3+	1-3+	3-4+	3+	3+	4+	3+
Iron incorporation into erythrons	22 mg per day	Normal or high	Normal	Normal or high	High	High (PA) to low	To 14× normal	Normal or high	Usually low	Usually normal
Iron absorption from intestine	0.5-1.5 mg/day	To 10× normal	High	?	High	High	High	?	Usually high	High
Erythrocyte life span	102-132 days	Short	Normal ?	Short	Normal ?	Short	Short	Short	Usually short	Usually normal

Figure 5-27. Some biochemical and hematologic parameters in a variety of blood and other disorders. From *Metabolic Basis of Inherited Disease* by M. Pollycove. Copyright 1978 by McGraw-Hill Book Company. Used with the permission of McGraw-Hill Book Company.

dose was in the intestinal mucosa and 3.3% in the remainder of the organism. In 24 hr, the latter had risen to 4.7%, whereas the mucosal radioiron was now negligible, apparently lost through the process of cell exfoliation.

It is well settled that increases in dietary iron will induce the biosynthesis of more mucosal cell ferritin. Conversely, in iron deficiency states, practically no ferritin is present in these cells, indicating the virtual absence of the mucosal block mechanism.

At this point it may be asked what exactly constitutes the total iron pool of the intestinal mucosal cell. Rat mucosa contains about 100 μg or iron per g of tissue, some 25 μg of which is heme iron and 5 μg ferritin iron (Munro and Linder, 1978). About 50% of all mucosal iron is present in *particulate* form (Yoshino *et al.*, 1977). The particulate fraction, in turn, consists of ferritin and an unidentified iron-containing *peak one*, which has a molecular weight of near 10^6 and a density of 1.15–1.16 g/cc. Another nonferritin iron-carrying compound isolated from guinea pig intestinal mucosa is the mucosal transferrin or the so-called gut iron-binding protein (GIBP) (Pollack and Lasky, 1977). This protein is similar to serotransferrin in many respects; however, it consists of two subunits and also differs from serotransferrin in regard to amino acid composition and peptide maps. Apparently GIBP is present in other tissues as well. It has been reported that iron is bound to a non-ferritin-soluble component (GIBP?) before becoming incorporated into ferritin following an oral dose of radioiron (Charlton *et al.*, 1965).

What, then, controls the rate of iron absorption in mammals? It is well established that absorption is increased in iron deficiency states and is lowered to almost nothing in cases of iron overload (with the exception of idiopathic hemochromatosis). Mucosal ferritin levels follow this pattern as well: Normally, rat mucosa contains 5 μg of ferritin iron per g of tissue, the iron-overloaded rat may have some 40 μg of ferritin iron per g, and iron-deficient animals have less than 1 μg of ferritin iron per g (Munro and Linder, 1978). The molecular basis for the control of iron absorption has been under investigation for many years, yet this point is today as unclear as ever. It was proposed as far back as the 1940s that the level of serotransferrin saturation with iron may be the signal for changes in iron absorption. *In vivo* studies by Taylor and Gatenby (1966) in human beings have correlated iron absorption with serotransferrin saturation levels, which varied between 45% in a normal student down to 4% in a patient with bleeding piles. Correlations of iron absorption with such hematologic parameters as hemoglobin levels, total serum iron, and hematocrit were not as good. On the other hand, Levine *et al.* (1972) found no correlation between iron absorption and the level of serotransferrin saturation with iron in rats.

More recent experiments have demonstrated that mucosal cell response to altered dietary iron is rather immediate—the lag period is less than 24 hr in mice. Thus, the mucosal cells respond by adjusting iron absorption before the total body iron stores have an opportunity to readjust to the dietary iron change. This indicates that perhaps the control mechanism is local in nature, as had been suggested earlier by Charlton *et al.* (1965). It has been proposed that the channeling of iron into the biosynthesis of the heme-containing cytochrome P-450 represents the actual control mechanism: The biosynthetic rate of cytochrome P-450 was parallel to iron absorption changes and bore a direct relationship to dietary iron levels (Hegenauer *et al.*, 1977). This proposal needs further amplification.

It is clear that information on iron absorption that can be obtained from whole animals is limited in scope and that experimentation must eventually be expanded to the cellular and even molecular levels. This would involve the isolation of free intestinal mucosal cells and an investigation of the mode of iron assimilation and control mechanisms in the isolated cells either in tissue culture or suspension. Attempts to characterize iron metabolism in such isolated cells have been made and involve either the uptake or release of iron. Levine *et al.* (1972) reported that isolated rat mucosal epithelial cells labeled with ^{59}Fe *in vivo* were able to lose iron much faster in the presence of serum and especially in an aposerotransferrin-enriched serum than in plain incubation medium. Moreover, there was binding of aposerotransferrin and a weaker binding of iron-saturated serotransferrin to such cells, suggesting the presence of serotransferrin receptors on the cell surfaces. The presence of an intracellular ferroxidase promoting the uptake of mucosal cell iron by circulating serotransferrin has been reported by at least two authors (Manis, 1973; Topham, 1978). The ferroxidase in question, in contrast to ceruloplasmin, apparently facilitates the intracellular oxidation of iron, which then finds its way into the extracellular space containing serotransferrin (there is no evidence that serotransferrin penetrates the mucosal cell during iron absorption, though this is possible).

Iron uptake by isolated mucosal cells was studied by at least two groups with conflicting results. In one case (Savin and Cook, 1978) it was found that iron uptake (from $FeCl_3$–ascorbate) took place by simple diffusion, that it was proportional to the amount of iron in the medium, and that in time-course studies it showed a biphasic character. Its uptake was not inhibited by any metabolic inhibitors. Radioiron was released from such cells much more readily into serum containing aposerotransferrin than into medium containing albumin, thus confirming the data of Levine *et al.* (1972). Such iron uptake by the mucosal cells is reminiscent of the uptake of ferrous iron observed in rat hepatocytes by Grohlich *et al.* (1979). In another study, however, iron (in the form of ferric nitriloacetate) was found to be taken up by the human mucosal cells via an active process. The uptake reached a maximum in 5 min, and saturation kinetics was a prominent feature of the process. It was implied that an iron receptor exists on the brush borders of the mucosal cells (Cox and Peters, 1979). A lactoferrin receptor was also identified in the human brush borders (Cox *et al.*, 1979), which has important implications in infant nutrition (see Chapter 4).

Summary

The principal forms of iron storage in the mammalian organism are ferritin and hemosiderin. Hemosiderin is a partially degraded ferritin. Ferritin is a nearly spherical glycoprotein consisting of 24 subunits. Most authorities accept the notion that there are two types of ferritin subunits, the H and the L with molecular weights of 21,000 and 18,500, respectively. Different tissues have ferritins with various combinations of these subunits, and these are termed isoferritins. They are best separated by isoelectric focusing. Each tissue has a characteristic isoferritin pattern, the extremes being the spleen type, consisting largely of the L-type subunits with a relatively basic character, and the heart type, consisting largely of the H-type subunits with a relatively acidic character.

The ferritin molecule, if stripped of iron, is called apoferritin. Apoferritin can bind iron in the form of $(FeOOH)_8(FeO \cdot OPO_3H_2)$ micelles, the maximum being about 4500 iron atoms/ferritin molecule. Normally, there is a spectrum of ferritin molecules containing varying amounts of iron, the average being about 1800–2400, so that iron accounts for about 20% of the weight of the ferritin molecule. Iron is incorporated into ferritin in the ferrous state and is oxidized inside the ferritin cavity to the ferric state. To leave ferritin, iron must be reduced back to the ferrous state by bioflavinoids.

Biosynthesis of ferritin occurs in all tissues and is influenced in a positive fashion by iron. Iron exerts its effects on the posttranslational level by derepressing a stable ferritin messenger RNA. It is an important mediator of iron metabolism in a variety of tissues, including liver, the reticuloendothelial system, the blood, and the small intestine. It is generally agreed that iron may enter the cell in the ferrous state by some method of pinocytosis to join the so-called *chelatable iron pool*. The iron may enter ferritin, participate in the biosynthesis of a variety of heme proteins, or be delivered to the circulation. In the last case, it is oxidized to the ferric state by ceruloplasmin and is immediately bound by serotransferrin. Liver ferritin is greatly increased in iron overload cases, and much of it is converted to hemosiderin. Iron overload involving the liver, such as occurs in idiopathic hemochromatosis, is dangerous because it elicits cirrhosis and fibrosis. On the other hand, iron overload in the reticuloendothelial system, mostly in the form of hemosiderin, is relatively benign. Ferritin is also found in the serum, and its levels therein reflect total body iron stores, especially those of the reticuloendothelial system. Ferritin has been implicated as a component of the so-called mucosal block mechanism operative in the intestinal mucosal cells. It acts to sequester iron entering the mucosal cell but not required by the organism. Ferritin of all tissues accounts for some 20% of all body iron in the human being.

References

Ances, I. G., Granados, J., and Baltazar, M., 1979. Serum ferritin as an early determinant of decreased iron stores in pregnant women, *South. Med. J.* 72:591–592.

Arora, R. S., Lynch, E. C., Whitley, C. E., and Alfrey, C. P., 1970. The ubiquity and significance of human ferritin, *Tex. Rep. Biol. Med.* 28:189–196.

Arosio, P., Adelman, T. G., and Drysdale, J. W., 1978. On ferritin heterogeneity. Further evidence for heteropolymers, *J. Biol. Chem.* 253:4451–4458.

Baker, E., Vicary, F. R., and Huehns, E. R., 1977. Iron mobilisation from isolated rat hepatocytes, *in Proteins of Iron Metabolism*, E. B. Brown, P. Aisen, J. Fielding, and R. R. Crichton (eds.), Grune & Stratton, New York, pp. 327–334.

Bankowsky, H. L., Carpenter, S. J., and Healey, J. F., 1979. Iron and the liver, *Arch. Pathol. Lab. Med.* 103:21–29.

Banyard, S. H., Stammers, D. K., and Harrison, P. M., 1978. Electron density map of apoferritin at 2.8 Å resolution, *Nature (London)* 271:282–284.

Barry, M., 1974. Progress report: Iron and the liver, *Gut* 15:324–334.

Batey, R. G., Pettit, J. E., Nicholas, A. W., Sherlock, S., and Hoffbrand, A. V., 1978. Hepatic iron clearance from serum in treated hemochromatosis, *Gastroenterology* 75:856–859.

Beamish, M. R., Keay, L., Okigaki, T., and Brown, E. B., 1975. Uptake of transferrin-bound iron by rat cells in tissue culture, *Br. J. Haematol.* 31:479–491.

Bertrand, M. L., and Harris, D. C., 1979. Insensitivity of the ferritin iron core to heat treatment, *Experientia* 35:300.

Bezwoda, W. R., Bothwell, T. H., Torrance, J. D., MacPhail, A. P., Charlton, R. W., Kay, G., and Levin, J., 1979. The relationship between marrow iron stores, plasma ferritin concentrations, and iron absorption, *Scand. J. Haematol.* 22:113–120.

Bjork, I., and Fish, W. W., 1971. Native and subunit molecular weights of apoferritin, *Biochemistry* 10:2844–2848.

Bomford, A., Berger, M., Lis, Y., and Williams, R., 1978. The iron content of human liver and spleen isoferritins correlates with their isoelectric point and subunit composition, *Biochem. Biophys. Res. Commun.* 83:334–341.

Bryce, C. F. A., and Crichton, R. R., 1971. The subunit structure of horse spleen apoferritin. I. The molecular weight of the subunit, *J. Biol. Chem.* 246:4198–4205.

Bryce, C. F. A., Magnusson, C. G. M., and Crichton, R. R., 1978. A reappraisal of the electrophoretic patterns obtained from ferritin and apoferritin in the presence of denaturants, *FEBS Lett.* 96:257–262.

Charlton, R. W., Jacobs, P., Torrance, J. D., and Bothwell, T. H., 1965. The role of the intestinal mucosa in iron absorption, *J. Clin. Invest.* 44:543–555.

Charlton, R. W., Bothwell, T. H., and Seftel, H., 1973. Dietary iron overload, *Clin. Haematol.* 2:383–403.

Christensen, A. C., Huebers, H., and Finch, C. A., 1978. Effect of transferrin saturation on iron delivery in rats, *Am. J. Physiol.* 235:R18–R22.

Collet-Cassart, D., and Crichton, R. R., 1975. Structural studies on horse spleen apoferritin, *in Proteins of Iron Storage and Transport in Biochemistry and Medicine*, R. R. Crichton (ed.), North-Holland, Amsterdam, pp. 185–192.

Cox, T. M., and Peters, T. J., 1979. The kinetics of iron uptake *in vitro* by human duodenal mucosa: Studies in normal subjects, *J. Physiol.* 289:469–478.

Cox, T. M., Mazurier, J., Spik, G., Montreuil, J., and Peters, T. J., 1979. Iron binding proteins and influx of iron across the duodenal brush border: Evidence for specific lactotransferrin receptors in the human intestine, *Biochim. Biophys. Acta* 588:120–128.

Crichton, R. R., 1972. The subunit structure of apoferritin and other eicosamers, *Biochem. J.* 126:761–764.

Crichton, R. R., 1973a. The biochemistry of ferritin, *Br. J. Haematol.* 24:677–680.

Crichton, R. R., 1973b. Ferritin, *Struct. Bonding (Berlin)* 17:67–134.

Crichton, R. R., 1973c. Structure and function of ferritin, *Angew. Chem. Int. Ed. Engl.* 12:57–65.

Crichton, R. R., 1973d. A role for ferritin in the regulation of iron metabolism, *FEBS Lett.* 34:125–128.

Crichton, R. R., and Bryce, C. F. A., 1970. Molecular weight estimation of apoferritin subunits, *FEBS Lett.* 6:121–124.

Crichton, R. R., Eason, R., Barclay, A., and Bryce, C. F. A., 1973a. The subunit structure of horse spleen apoferritin: The molecular weight of the oligomer and its stability to dissociation by dilution, *Biochem. J.* 131:855–857.

Crichton, R. R., Millar, J. A., Cumming, R. L. C., and Bryce, C. F. A., 1973b. The organ specificity of ferritin in human and horse liver and spleen, *Biochem. J.* 131:51–59.

Crichton, R. R., Huebers, H., Huebers, E., Collet-Cassart, D., and Ponce, Y., 1975. Comparative studies on ferritin, *in Proteins of Iron Storage and Transport in Biochemistry and Medicine*, R. R. Crichton (ed.), North-Holland, Amsterdam, pp. 193–200.

Crichton, R. R., Collet-Cassart, D., Ponce-Ortiz, Y., Wauters, M., Roman, F., and Paques, E., 1977. Ferritin: Comparative structural studies, iron deposition, and mobilization, *in Proteins of Iron Metabolism*, E. B. Brown, P. Aisen, J. Fielding, and R. R. Crichton (eds.), Grune & Stratton, New York, pp. 13–22.

Crichton, R. R., Ponce-Ortiz, Y., Koch, M. H. J., Parfait, R., and Stuhrmann, H. B., 1978. Isolation and characterization of phytoferritin from pea (*Pisum sativum*) and lentil (*Lens esculenta*), *Biochem. J.* 171:349–356.

Crichton, R. R., Heusterspreute, M., Collet-Cassart, D., Wustefeld, C., Ponce-Ortiz, Y., Magnusson, C. G., and Schank, K., 1979. "The primary structure of apoferritins from plant and animal origins," Abstract E, Fourth International Conference on Proteins of Iron Metabolism, Davos, Switzerland, April 17–21, 1979.

Crosby, W. H., 1963. The control of iron balance by the intestinal mucosa, *Blood* 22:441–449.

Crosby, W. H., 1977. Normal serum ferritin in precirrhotic hemochromatosis, *N. Engl. J. Med.* 296:1116.

Crosby, W. H., Conrad, M. E., and Wheby, M. S., 1963. The rate of iron accumulation in iron storage disease, *Blood* 22:429–440.

Cumming, R. L. C., 1978. Disorders of iron metabolism, *Practitioner* 221:184–192.

Cynkin, M. A., and Knowlton, M., 1977. Studies on the carbohydrate components of ferritin, *in Proteins of Iron Metabolism*, E. B. Brown, P. Aisen, J. Fielding, and R. R. Crichton (eds.), Grune & Stratton, New York, pp. 115–120.

Dognin, J., and Crichton, R. R., 1975. Mobilisation of iron from ferritin fractions of defined iron content by biological reductants, *FEBS Lett.* 54:234–236.

Drysdale, J. W., 1977. Ferritin phenotypes: Structure and metabolism, *in Ciba Foundation Symposium 51, Iron Metabolism*, Elsevier–North-Holland, Amsterdam, pp. 41–67.

Drysdale, J. W., and Munro, H. N., 1965. Small scale isolation of ferritin for the assay of the incorporation of ^{14}C-labelled amino acids, *Biochem. J.* 95:851–858.

Drysdale, J. W., and Munro, H. N., 1966. Regulation of synthesis and turnover of ferritin in rat liver, *J. Biol. Chem.* 241:3630–3637.

Drysdale, J. W., Adelman, T. G., Arosio, P., Casareale, D., Fitzpatrick, P., Hazard, J. T., and Yokota, M., 1977. Human isoferritins in normal and disease states, *Semin. Hematol.* 14:71–88.

Engel, R. H., and Pribor, H. C., 1978. Serum ferritin: A convenient measure of body iron stores, *Lab. Manage.* Oct.: 31–34.

Fagard, R., Tan, H. V., and Saddi, R., 1978. Control of ferritin biosynthesis: Are ribosomes the target of iron? *Biochimie* 60:517–520.

Fillet, G., Marsaglia, G., and Finch, C. A., 1975. Idiopathic hemochromatosis. Abnormality in RBC transport of iron by the reticuloendothelial system (RES), *Blood* 46:1007.

Fine, J. M., and Harris, G., 1963. Electrophoretic and immunological studies of horse and human ferritin, *Clin. Chim. Acta* 8:794–798.

Fineberg, R. A., and Greenberg, D. M., 1955a. Ferritin biosynthesis. II. Acceleration of synthesis by the administration of iron. *J. Biol. Chem.* 214:97–106.

Fineberg, R. A., and Greenberg, D. M., 1955b. Ferritin biosynthesis. Apoferritin, the initial product, *J. Biol. Chem.* 214:107–113.

Fischbach, F. A., Gregory, D. W., Harrison, P. M., Hoy, T. G., and Williams, J. M., 1971. On the structure of hemosiderin and its relationship to ferritin, *J. Ultrastruct. Res.* 37:495–503.

Fletcher, J., 1971. The plasma clearance and liver uptake of iron from transferrin of low and high iron saturation, *Clin. Sci.* 41:395–402.

Frieden, F., 1973. The ferrous to ferric cycles in iron metabolism, *Nutr. Rev.* 31:41–44.

Frieden, E., and Hsieh, H. S., 1976. Ceruloplasmin: The copper transport protein with essential oxidase activity, *Adv. Enzymol. Relat. Areas Mol. Biol.* 44:187–236.

Giorgio, A. J., 1973. Primary hemochromatosis: Easily missed cause of cirrhosis, diabetes, and heart failure, *Geriatrics* 28:131–134.

Granick, S., 1946a. Ferritin: Its properties and significance for iron metabolism, *Chem. Rev.* 38:379–403.

Granick, S., 1946b. Protein apoferritin and ferritin in iron feeding and absorption, *Science* 103:107.

Granick, S., 1951. Structure and physiological functions of ferritin, *Physiol. Rev.* 31:489–511.

Grohlich, D., Morley, C. G. D., Miller, R. J., and Bezkorovainy, A., 1977. Iron incorporation into isolated rat hepatocytes, *Biochem. Biophys. Res. Commun.* 76:82–692.

Grohlich, D., Morley, C. G. D., and Bezkorovainy, A., 1979. Some aspects of iron uptake by rat hepatocytes in suspension, *Int. J. Biochem.* 10:797–802.

Halliday, J. W., Mack, U., and Powell, L. W., 1979. The kinetics of serum and tissue ferritins: Relation to carbohydrate content, *Br. J. Haematol.* 42:535–546.

Harris, D. C., 1978. Iron exchange between ferritin and transferrin *in vitro, Biochemistry* 17:3071–3078.

Harrison, P. M., 1963. The structure of apoferritin: Molecular size, shape, and symmetry from X-ray data, *J. Mol. Biol.* 6:404–422.

Harrison, P. M., 1977. Ferritin: An iron-storage molecule, *Semin. Hematol.* 14:55–70.

Harrison, P. M., and Gregory, D. W., 1968. Reassembly of apoferritin molecules from subunits, *Nature (London)* 220:578–580.

Harrison, P. M., Fischbach, F. A., Hoy, T. G., and Haggis, G. H., 1967. Ferric oxyhydroxide core of ferritin, *Nature (London)* 216:1188–1190.

Harrison, P. M., Hoy, T. G., Macara, I. G., and Hoare, R. J., 1974. Ferritin iron uptake and release. Structure–function relationships, *Biochem. J.* 143:445–451.

Harrison, P. M., Hoy, T. G., and Hoare, R. J., 1975. Towards a mechanism of iron uptake and release by ferritin molecules, *in Proteins of Iron Storage and Transport in Biochemistry and Medicine*, R. R. Crichton (ed.), North-Holland, Amsterdam, pp. 271–278.

Harrison, P. M., Banyard, S. H., and Stammers, D. K., 1979. "The structure of horse spleen apoferritin and its implications," Abstract D, Fourth International Conference on Proteins of Iron Metabolism, Davos, Switzerland, April 17–21, 1979.

Hauser, H., 1969. Fractionation of horse spleen ferritin. Relationship of Fe/N—ratio to some physio-chemical properties, *Hoppe-Seyler's Z. Physiol. Chem.* 350:1331–1339.

Hazard, J. T., and Drysdale, J. W., 1977. Ferritinaemia in cancer, *Nature (London)* 265:755–756.

Hazard, J. T., Yokota, M., Arosio, P., and Drysdale, J. W., 1977. Immunologic differences in human isoferritins: Implications for immunologic quantitation of serum ferritin, *Blood* 49:139–146.

Hegenauer, J., Ripley, L., and Saltman, P., 1977. Regulation of iron absorption by control of heme biosynthesis in the intestinal mucosa, *in Proteins of Iron Metabolism*, E. B. Brown, P. Aisen, J. Fielding, and R. R. Crichton (eds.), Grune & Stratton, New York, pp. 403–410.

Hegenauer, J., Saltman, P., and Hatlen, L., 1979. Removal of cadmium (II) from crystallized ferritin, *Biochem. J.* 177:693–695.

Hershko, C., 1977. Storage iron regulation, *Prog. Hematol.* 10:105–148.

Hershko, C., and Rachmilewitz, E. A., 1975. Non-transferrin plasma iron in patients with transfusional iron overload, *in Proteins of Iron Storage and Transport in Biochemistry and Medicine*, R. R. Crichton (ed.), North-Holland, Amsterdam, pp. 427–432.

Hoare, R. J., Harrison, P. M., and Hoy, T. G., 1975. Structure of horse-spleen apoferritin at 6 Å resolution, *Nature (London)* 255:653–654.

Hofmann, T., and Harrison, P. M., 1963. The structure of apoferritin: Degradation into and molecular weight of subunits, *J. Mol. Biol.* 6:256–267.

Hoy, T. G., and Harrison, P. M., 1976. The uptake of ferric iron by rat liver ferritin *in vivo* and *in vitro*, *Br. J. Haematol.* 33:497–504.

Hoy, T. G., Harrison, P. M., Shabbir, M., and Macara, I. G., 1974a. The release of iron from horse spleen ferritin to 1,10-phenanthroline, *Biochem. J.* 137:67–70.

Hoy, T. G., Harrison, P. M., and Shabbir, M., 1974b. Uptake and release of ferritin iron. Surface effects and exchange within crystalline core, *Biochem. J.* 139:603–607.

Huberman, A., and Barahona, E., 1978. Primary structure of rat liver apoferritin. The amino end. *Biochim. Biophys. Acta* 533:51–56.

Jacobs, A., and Worwood, M., 1975a. The biochemistry of ferritin and its clinical implications, *Prog. Hematol.* 9:1–24.

Jacobs, A., and Worwood, M., 1975b. Ferritin in serum: Clinical and biochemical implications, *N. Engl. J. Med.* 292:951–956.

Jones, T., Spencer, R., and Walsh, C., 1978. Mechanism and kinetics of iron release from ferritin by dihydroflavins and dihydroflavin analogues, *Biochemistry* 17:4012–4017.

Lavoie, D. J., Marcus, D. M., Ishikawa, K., and Listowsky, I., 1977. Ferritin and apoferritin from human liver: Aspects of heterogeneity, *in Proteins of Iron Metabolism*, E. B. Brown, P. Aisen, J. Fielding, and R. R. Crichton (eds.), Grune & Stratton, New York, pp. 71–78.

Lavoie, D. J., Ishikawa, K., and Listowsky, I., 1978. Correlations between subunit distribution, microheterogeneity, and iron content of human liver ferritin, *Biochemistry* 17:5448–5454.

Lee, S. S. C., and Richter, G. W., 1977a. Biosynthesis of ferritin in rat livers. I. Synthesis and assembly of protein subunits of ferritin, *J. Biol. Chem.* 252:2046–2053.

Lee, S. S. C., and Richter, G. W., 1977b. Biosynthesis of ferritin in rat hepatoma cells and rat livers. II. Binding of iron by ferritin protein, *J. Biol. Chem.* 252:2054–2059.

Lee, J. C. K., Lee, S. S. C., Schlesinger, K. J., and Richter, G. W., 1975. Production of ferritin by rat hepatoma cells *in vitro*: Demonstration of protein subunits and ferritin by immunofluorescence, *Am. J. Pathol.* 80:235–243.

Levine, P. H., Levine, A. J., and Weintraub, L. R., 1972. The role of transferrin in the control of iron absorption: Studies on a cellular level, *J. Lab. Clin. Med.* 80:333–341.

Linder-Horowitz, M., Ruettinger, R. T., and Munro, H. N., 1970. Iron induction of electrophoretically different ferritins in rat liver, heart, and kidney, *Biochim. Biophys. Acta* 200:442–448.

Lipschitz, D. A., Bothwell, T. H., Seftel, H. C., Wapnick, A. A., and Charlton, R. W., 1971. The role of ascorbic acid in the metabolism of storage iron, *Br. J. Haemalol.* 20:155–163.

Listowsky, I., Blauer, G., Englard, S., and Betheil, J. J., 1972. Denaturation of horse spleen ferritin in aqueous guanidinium chloride solutions, *Biochemistry* 11:2176–2182.

Loria, A., Hershko, C., and Konijn, A., 1979. Serum ferritin in an elderly population, *J. Gerontol.* 34:521–524.

Macara, I. G., Hoy, T. G., and Harrison, P. M., 1972. The formation of ferritin from apoferritin. Kinetics and mechanism of iron uptake, *Biochem. J.* 126:151–162.
Macara, I. G., Hoy, T. G., and Harrison, P. M., 1973. The formation of ferritin from apoferritin. Inhibition and metal ion-binding studies, *Biochem. J.* 135:785–789.
MacDonald, R. A., 1969. Human and experimental hemochromatosis and hemosiderosis, *in Pigments in Pathology*, M. Walman (ed.), Academic Press, New York, pp. 115–149.
Mainwaring, W. I. P., and Hofmann, T., 1968. Horse spleen apoferritin: N-terminal and C-terminal residues, *Arch. Biochem. Biophys.* 125:975–980.
Manis, J., 1973. Ferrous iron oxidation by intestinal mucosa: Possible role in mucosal iron metabolism, *Proc. Soc. Exp. Biol. Med.* 144:1025–1029.
Marston, H. R., and Allen, S. H., 1967. Function of copper in the metabolism of iron, *Nature (London)* 215:645–646.
Massover, W. H., 1978. The ultrastructure of ferritin macromolecules. III. Mineralized iron in ferritin is attached to the protein shell, *J. Mol. Biol.* 123:721–726.
May, M. E., and Fish, W. W., 1977. The isolation and properties of porcine ferritin and apoferritin, *Arch. Biochem. Biophys.* 182:396–403.
Mazur, A., Baez, S., and Shorr, E., 1955. The mechanism of iron release from ferritin as related to its biological properties, *J. Biol. Chem.* 213:147–160.
Mazur, A., Green, S., and Carleton, A., 1960. Mechanism of plasma iron incorporation into hepatic ferritin, *J. Biol. Chem.* 235:595–603.
McKay, R. H., and Fineberg, R. A., 1964. Horse spleen hemosiderin, *Arch. Biochem. Biophys.* 104:496–508.
Milsom, J. P., and Batey, R. G., 1979. The mechanism of hepatic iron uptake from native and denatured transferrin and its subcellular metabolism in the liver cell, *Biochem. J.* 182:117–125.
Munro, H. N., and Linder, M. C., 1978. Ferritin: Structure, biosynthesis, and role in iron metabolism, *Physiol. Rev.* 58:317–396.
Niitsu, Y., and Listowsky, I., 1973a. The distribution of iron in ferritin, *Arch. Biochem. Biophys.* 158:276–281.
Niitsu, Y., and Listowsky, I., 1973b. Mechanisms for the formation of ferritin oligomers, *Biochemistry* 12:4690–4695.
Niitsu, Y., Ishitani, K., and Listowsky, I., 1973. Subunit heterogeneity in ferritin, *Biochem. Biophys. Res. Commun.* 55:1134–1140.
Osaki, S., and Johnson, D. A., 1969. Mobilization of liver iron by ferroxidase (ceruloplasmin), *J. Biol. Chem.* 244:5757–5758.
Osaki, S., Johnson, D. A., and Frieden, E., 1971. The mobilization of iron from the perfused mammalian liver by a serum copper enzyme, ferroxidase I, *J. Biol. Chem.* 246:3018–3023.
Pape, L., Multani, J. S., Stitt, C., and Saltman, P., 1968a. *In vivo* reconstitution of ferritin, *Biochemistry*7:606–612.
Pape, L., Multani, J. S., Stitt, C., and Saltman, P., 1968b. The mobilization of iron from ferritin by chelating agents, *Biochemistry* 7:613–616.
Planas, J., and Frieden, E., 1973. Serum iron and ferroxidase activity in normal, copper deficient, and estrogenized roosters, *Am. J. Physiol.* 225:423–428.
Pollack, S., and Lasky, F. D., 1977. A new iron binding protein with wide tissue distribution, *in Proteins of Iron Metabolism*, E. B. Brown, P. Aisen, J. Fielding, and R. R. Crichton (eds.), Grune & Stratton, New York, pp. 393–396.
Pollycove, M., 1978. Hemochromatosis, *in The Metabolic Basis of Inherited Disease*, 4th ed., J. B. Stanbury, J. B. Wyngaarden, and D. S. Fredrickson (eds.), McGraw-Hill, New York, pp. 1127–1164.
Pootrakul, P., Christensen, A., Josephson, B., and Finch, C. A., 1977. Role of transferrin in determining internal iron distribution, *Blood* 49:957–966.

Powell, L. W., and Halliday, J. W., 1978. The detection of early hemochromatosis, *Am. J. Dig. Dis.* 23:377–379.

Powell, L. W., Alpert, E., Isselbacher, K. J., and Drysdale, J. W., 1975a. Human isoferritins: Organ specific iron and apoferritin distribution, *Br. J. Haematol.* 30:47–55.

Powell, L. W., McKeering, L., and Halliday, J. W., 1975b. Microheterogeneity of tissue ferritins in iron storage disease, *in Proteins of Iron Storage and Transport in Biochemistry and Medicine*, R. R. Crichton (ed.), North-Holland, Amsterdam, pp. 367–370.

Powell, L. W., Halliday, J. W., and Cowlishaw, J. L., 1978. Relationship between serum ferritin and total body iron stores in idiopathic hemochromatosis, *Gut* 19:538–542.

Ragan, H. A., Nacht, S., Lee, G. R., Bishop, C. R., and Cartwright, G. E., 1969. Effect of ceruloplasmin on plasma iron in copper deficient swine, *Am. J. Physiol.* 217:1320–1323.

Richter, G. W., 1965. Comparison of ferritins from neoplastic and non-neoplastic human cells, *Nature (London)* 207:616–617.

Richter, G. W., 1978. The iron-loaded cell—the cytopathology of iron storage, *Am. J. Pathol.* 91:361–404.

Richter, G. W., and Walker, G. F., 1967. Reversible association of apoferritin molecules. Comparison of light-scattering and other data, *Biochemistry* 6:2871–2881.

Russell, S. M., and Harrison, P. M., 1978. Heterogeneity in horse ferritins. A comparative study of surface charge, iron content, and kinetics of iron uptake, *Biochem. J.* 175:91–104.

Saarinen, U. M., and Siimes, M. A., 1979. Iron absorption from breast milk, cow's milk, and iron-supplemented formula: An opportunistic use of changes in total body iron determined by hemoglobin, ferritin, and body weight in 132 infants, *Pediatr. Res.* 13:143–147.

Saltman, P., Fiskin, R. D., and Bellinger, S. B., 1956a. The metabolism of iron by rat liver slices. The effect of physical environment and iron concentration, *J. Biol. Chem.* 220:741–750.

Saltman, P., Fiskin, R. D., Bellinger, S. B., and Alex, T., 1956b. The metabolism of iron by rat liver slices. The effect of chemical agents, *J. Biol. Chem.* 220:751–757.

Samarel, A., and Bern, M. M., 1978. Distribution of iron in splenic ferritin, *Lab. Invest.* 39:10–12.

Savin, M. A., and Cook, J. D., 1978. Iron transport by isolated rat intestinal mucosal cells, *Gastroenterology* 75:688–694.

Shinjyo, S., Abe, H., and Masuda, M., 1975. Carbohydrate composition of horse spleen ferritin, *Biochim. Biophys. Acta* 411:165–167.

Siimes, M. A., Addiego, J. E., and Dallman, P. R., 1974. Ferritin in serum: Diagnosis of iron deficiency and iron overload in infants and children, *Blood* 43:581–590.

Sirivech, S., Frieden, E., and Osaki, S., 1974. The release of iron from horse spleen ferritin by reduced flavins, *Biochem. J.* 143:311–315.

Smith-Johannsen, H., and Drysdale, J. W., 1969. Reversible dissociation of ferritin and its subunits *in vitro*, *Biochim. Biophys. Acta* 194:43–49.

Sturgeon, P., and Shoden, A., 1969. Hemosiderin and ferritin, *in Pigments in Pathology*, M. Wolman (ed.), Academic Press, New York, pp. 93–114.

Suran, A. A., 1966. N-terminal sequence of horse spleen apoferritin, *Arch. Biochem. Biophys.* 113:1–4.

Taylor, M. R. H., and Gatenby, P. B. B., 1966. Iron absorption in relation to transferrin saturation and other factors, *Br. J. Haematol.* 12:747–753.

Topham, R. W., 1978. Isolation of an intestinal promoter of Fe^{3+}–transferrin formation, *Biochem. Biophys. Res. Commun.* 85:1339–1345.

Torrance, J. D., Charlton, R. W., Simon, M. O., Lynch, S. R., and Bothwell, T. H., 1978. The mechanism of endotoxin-induced hypoferraemia, *Scand. J. Haematol.* 21:403–410.

Treffry, A., and Harrison, P. M., 1978. Incorporation and release of inorganic phosphate in horse spleen ferritin, *Biochem. J.* 171:313–320.

Treffry, A., Sowerby, J. M., and Harrison, P. M., 1979. Oxidant specificity in ferritin formation, *FEBS Lett.* 100:33–36.

Trump, B. F., and Berezesky, I. K., 1977. A general model of intracellular iron metabolism, *in Proteins of Iron Metabolism*, E. B. Brown, P. Aisen, J. Fielding, and R. R. Crichton (eds.), Grune & Stratton, New York, pp. 359–364.

Valberg, L. S., Birkett, N., Haist, J., Zamecnik, J., and Pelletier, O., 1979. Evaluation of the body iron status of native Canadians, *Can. Med. Assoc. J.* 120:285–290.

Van Bockxmeer, F., Hemmaplardh, D., and Morgan, E. H., 1975. Studies on the binding of transferrin to cell membrane receptors, *in Proteins of Iron Storage and Transport in Biochemistry and Medicine*, R. R. Crichton (ed.), North-Holland, Amsterdam, pp. 111–119.

Van Kreel, B. K., Van Eijk, H. G., and Leijnse, B., 1972. The isoelectric fractionation of rabbit ferritin, *Acta Haematol.* 47:59–64.

van Wyk, C. P., Linder-Horowitz, M., and Munro, H. N., 1971. Effect of iron loading on non-heme iron compounds in different liver cell populations, *J. Biol. Chem.* 246:1025–1031.

Verhoef, N. J., Kottenhagen, M. J., Mulder, H. J. M., Noordeloos, P. J., and Leijnse, B., 1978. Functional heterogeneity of transferrin-bound iron, *Acta Haematol.* 60:210–226.

Vulimiri, L., Catsimpoolas, N., Griffith, A. L., Linder, M. C., and Munro, H. N., 1975. Size and charge heterogeneity of rat tissue ferritins, *Biochim. Biophys. Acta* 412:148–156.

Wagstaff, M., Worwood, M., and Jacobs, A., 1978. Properties of human tissue isoferritins, *Biochem. J.* 173:969–977.

Wetz, K., and Crichton, R. R., 1976. Chemical modification as a probe of the topography and reactivity of horse-spleen apoferritin, *Eur. J. Biochem.* 61:545–550.

White, G. P., and Jacobs, A., 1978. Iron uptake by Chang cells from transferrin, nitriloacetate, and citrate complexes, *Biochim. Biophys. Acta* 543:217–225.

Wood, G. C., and Crichton, R. R., 1971. Optical rotatory dispersion and circular dichroism studies on ferritin and apoferritin, *Biochim. Biophys. Acta* 229:83–87.

Wyllie, J. C., 1977. Transferrin uptake by rabbit alveolar macrophages *in vitro*, *Br. J. Haematol.* 37:17–24.

Yoshino, Y., Yamakawa, S., and Hirai, Y., 1977. Iron binding compounds in particulate fraction of intestinal mucosa, *in Proteins of Iron Metabolism*, E. B. Brown, P. Aisen, J. Fielding, and R. R. Crichton (eds.), Grune & Stratton, New York, pp. 397–402.

Zähringer, J., Baliga, B. S., and Munro, H. N., 1976. Novel mechanism for translational control in regulation of ferritin synthesis by iron, *Proc. Natl. Acad. Sci. U.S.A.* 73:857–861.

Zamiri, I., and Mason, J., 1968. Electrophoresis of ferritins, *Nature (London)* 217:258–259.

Zimelman, A. P., Zimmerman, H. J., McLean, R., and Weintraub, L. R., 1977. Effect of iron saturation of transferrin on hepatic iron uptake: An *in vitro* study, *Gastroenterology* 72:129–131.

Zuyderhoudt, F. M. J., Linthorst, C., and Hengeveld, P., 1978a. On the iron content of human serum ferritin, especially in acute viral hepatitis and iron overload, *Clin. Chim. Acta* 90:93–99.

Zuyderhoudt, F. M. J., Jörning, G. G. A., and van Gool, J., 1978b. On the non-ferritin depot iron fraction in the rat liver, *Biochim. Biophys. Acta* 543:53–62.

The Interaction of Nonheme Iron with Immature Red Cells

6.1 Introduction

It was seen in Chapter 2 that circulating iron can find its way into the bone marrow cells, liver cells, and cells of other organs. Iron, in turn, can enter the circulation from the reticuloendothelial system, parenchymal storage sites, mucosal cells, and a variety of other systems. All such processes involve cell membrane transport mechanisms, and most involve serotransferrin and ferritin as well, either directly or indirectly. In this chapter we shall be concerned with the mechanisms whereby nonheme iron can find its way into the immature red cells and the pertinent intracellular iron metabolic pathways.

6.2 Early Investigations on Serotransferrin–Immature Red Cell Interactions

It was first believed that the incorporation of iron into immature red cells occurred via ferritin as an intermediate. However, Walsh *et al.*, 1949, showed that reticulocytes could accept iron directly from solution. They showed that iron incorporation into human red cells was proportional to reticulocyte counts therein. It was also shown that ferrous and ferric iron and serotransferrin-bound iron were all incorporated, whereby part of the iron was localized on the stroma and part in the heme. It was proposed that the uptake of iron by reticulocytes was a two-step process: combination with stromal *acceptors* and incorporation into heme.

The work that has served as the basis for subsequent iron–immature red cell interaction was described in a report by Jandl *et al.* (1959). These investigators incubated reticulocyte-rich blood samples with inorganic iron ($FeCl_3$ in saline) and with serotransferrin-, albumin-, and γ-globulin-bound iron. Normal red cells took up only inorganic iron and iron bound to albumin and γ-globulin, whereas reticulocytes incorporated iron from all sources. With the exception of the serotransferrin–reticulocyte experiment, most if nor all iron was localized on the red cell stroma. In the case of the serotransferrin–reticulocyte experiment, considerable amounts of iron were also found in the cell interior and hemoglobin. In contrast to inorganic iron and iron associated with albumin and γ-globulin, the incorporation of serotransferrin-bound iron increased with an increase in blood reticulocyte counts. Inclusion of EDTA into the incubation medium did not affect the uptake of iron by reticulocytes from serotransferrin, though the uptake from $FeCl_3$ was completely abolished. This indicates that free iron is not an intermediate in the process of moving iron from serotransferrin in the the reticulocytes. Iron was also taken up by the human reticulocytes from conalbumin and serotransferrins of other species. It may be mentioned, however, that some years later at least two groups were not able to observe iron uptake by rabbit reticulocytes from either human lactoferrin or hen's egg conalbumin (Zapolski and Princiotto, 1976; Brock and Esparza, 1979). Conalbumin, however, did serve to deliver iron to chick embryo red blood cells (Williams and Woodworth, 1973).

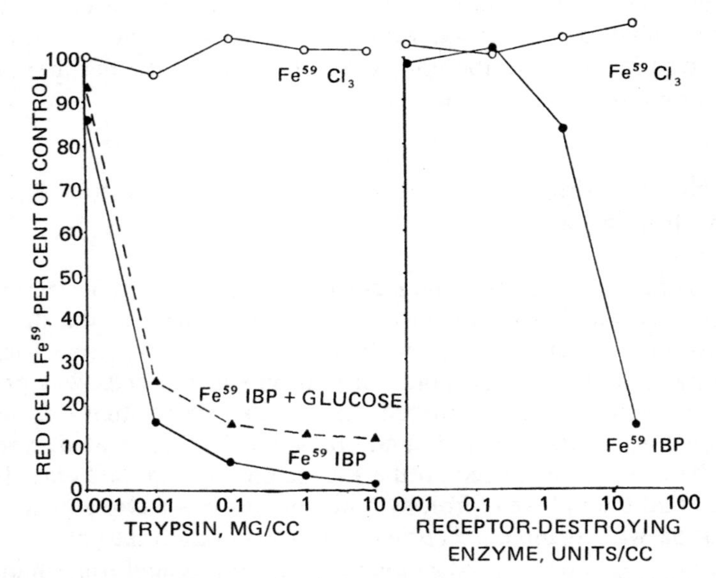

Figure 6-1. Effect of treating rabbit reticulocytes with (left) trypsin and (right) neuraminidase. Uptake of iron from $FeCl_3$ was not altered, whereas that from serotransferrin was drastically reduced. From Jandl *et al.* (1959).

The work of Jandl *et al.* (1959) also established that iron uptake by the human reticulocytes was apparently an energy-dependent process. Glucose served to increase iron uptake, whereas low O_2 tension, cyanide, fluoride, iodoacetate, azide, arsenite, and 2,4-dinitrophenol all depressed the uptake. Moreover, disrupted reticulocyte cell membranes were not able to take up iron-laden serotransferrin, indicating that the metabolic machinery of the intact cell was necessary to effect the binding of serotransferrin to the cell surface.

A very important discovery made by Jandl *et al.* (1959) was that the reticulocyte iron uptake mechanism using iron-laden serotransferrin was sensitive to trypsin and neuraminidase pretreatment of the cells. This is indicated in Figure 6-1. No effect on the uptake of iron from ferric chloride was observed following the trypsin and neuraminidase treatment. The conclusion reached was that iron is taken up by reticulocytes from iron-laden serotransferrin through the mediation of a plasma membrane receptor which may be glycoprotein in nature.

It was thus clear by the late 1950s that iron-laden serotransferrin interacted with the reticulocyte plasma membrane to deliver its iron to the heme biosynthesizing machinery of the cell.

6.3 The Mechanism of Serotransferrin–Immature Red Cell Interaction

Following the appearance of the paper by Jandl and co-workers, many additional investigators began the task of defining the exact mechanism of serotransferrin–reticulocyte interaction. Jandl and Katz (1963), using serotransferrin labeled with ^{131}I and with both ^{131}I and ^{59}Fe, found that in the case of the doubly labeled serotransferrin the ^{131}I isotope remained largely on the stroma, whereas ^{59}Fe was internalized. However, see Section 6.6 for radically different findings. Serotransferrin was not consumed during the iron abstraction process. Iron-laden serotransferrin was bound to reticulocyte stroma four to five times better than was iron-free serotransferrin. A calculation was made as to the maximum amount of serotransferrin that could be bound to 1 ml of packed reticulocytes, and the result was some 80 μg. This represents 50,000 serotransferrin molecules/cell and would cover about 2% of the cell's surface area. Further calculations revealed that at any given time, 20 mg of serotransferrin are bound to the erythroid cells in an average human organism. Since each serotransferrin molecule is on the average one-third saturated with iron and is bound to the reticulocyte cell for an average of 1–2 min before being eluted, it is normally possible to deliver 30–40 mg of iron to the hemopoietic system per day (Katz, 1965). This figure is in good agreement with that determined from ferrokinetic measurements as discussed in Chapter 2.

The quantitative aspects of serotransferrin–reticulocyte interaction was studied by Baker and Morgan using the rabbit serotransferrin–rabbit reticulocyte system (Baker and Morgan, 1969). It may be mentioned at this point that rabbit reticulocytes have become the model system for the study of serotransferrin–hemopoietic tissue interactions. Reticulocytosis is induced in rabbits by repeated bleeding, resulting in up to 40% reticulosytosis, or by phenylhydrazine injections, which gives a 90% reticulocytosis status. The behavior of reticulocytes appears to be identical in all respects to that of other immature red blood cells (Kailis and Morgan, 1974). Using Scatchard plots obtained by incubating reticulocytes with varying amounts of iron-saturated ^{131}I–serotransferrin, Baker and Morgan (1969) found that there were some 300,000 serotransferrin-binding sites on the surface of a reticulocyte and that under physiological conditions these were about 60% saturated with serotransferrin. Other estimates have given 3.3×10^5 serotransferrin binding sites in rabbit cells (Witt and Woodworth, 1978), 70,000 sites in rat reticulocytes (Black *et al.*, 1979), and 1×10^5 binding sites/cell in rabbit reticulocytes (Van Bockxmeer *et al.,* 1978). Van Bockxmeer *et al.* (1978) also estimated that there are 800 receptors/μm^2 of cell surface, assuming a mean cell volume of 120 fl. The number of serotransferrin receptors per cell may vary with the degree of the cell's maturity, where fewer receptors are present as the cell matures. Activation energy was 10.8 kcal/mole for the association reaction and 12 kcal/mole for dissociation. Free energy of the association process was -7.3 kcal/mole with negligible enthalpy effects (Baker and Morgan, 1969). Entropy was thus considered as the driving force for the serotransferrin–reticulocyte association process, indicating that hydrophobic interactions were involved.

Similar studies performed in the rat reticulocyte–serotransferrin system (Verhoef and Noordeloos, 1977) revealed that each cell had between 85,000 and 343,000 (mean 189,000 $\pm$ 96,000) serotransferrin binding sites and between 260,000 and 430,000 (mean 330,000 $\pm$ 67,000) in nucleated bone marrow cells. It is agreed that if one disregards the bone marrow cells that do not biosynthesize heme, the number of serotransferrin binding sites may be as high as 700,000 per metabolically active nucleated cell. The association constant between serotransferrin and reticulocyte was about $10\times$ higher than the 200,000 figure reported for the rabbit system (Baker and Morgan, 1969).

The interaction of iron-laden serotransferrin can also be described in the form of Michaelis–Menten kinetic concepts (Kornfeld, 1968), where serotransferrin concentration was taken as being analogous to substrate concentration in an enzyme reaction, and serotransferrin associated with reticulocyte membranes was the product. Velocity was the amount of membrane-associated serotransferrin present after 30-min incubation. The results are shown in Figure 6-2 and are indicative of the fact that there is a limited number of serotransferrin-binding sites on the reticulocyte membrane and that therefore there exists a maximum rate at which iron can be transported across the cell membrane.

In an attempt to further elucidate the mechanism of serotransferrin–reticulocyte interaction, various physical and chemical situations have been devised to interfere with the uptake of both serotransferrin and iron by the immature red cells. Morgan (1964), who used serotransferrin labeled with both ^{131}I and ^{59}Fe and was thus able to observe both the transferrin binding by and iron incorporation into the reticulocytes, used temperature variation and metabolic poisons to study this process. A biphasic uptake of serotransferrin was observed: a very rapid phase which was completely temperature independent and was not affected by arsenite and 2,4-dinitrophenol (called absorption) and a slower phase (association) which was temperature dependent and was inhibited by arsenite but not 2,4-dinitrophenol. Iron uptake was inhibited by both compounds. At 3°C, the slow phase was completely absent. Moreover, serotransferrin taken up via the *rapid* process could, to a large extent, be eluted from the reticulocytes by washing with cold saline, whereas serotransferrin taken up during the *slow* process was not removed by such washing. The elution of serotransferrin from reticulocytes following serotransferrin binding by the cells during the rapid uptake phase was also temperature dependent and sensitive to arsenite inhibition.

Iron uptake from ^{59}Fe–serotransferrin is an energy-requiring process, as was proposed on the basis of work with metabolic inhibitors (Jandl *et al.*, 1959; Morgan, 1964). This work was extended by Morgan and Baker (1969), who separated such effects on the basis of whether the inhibitor(s) affected serotransferrin binding by the cell or iron uptake only. Malonate and ouabain had no effect whatever. On the other hand, the electron transfer system inhibitors (NaCN, NaN_3, rotenone, 2,4-dinitrophenol) inhibited iron uptake but had a lesser effect on serotransferrin binding by the plasma membrane. Serotransferrin binding is also unaffected by depriving the immature red cell of energy sources such as glucose. There is, however, a negative effect on iron uptake (Martinez-Medellin and Benavides, 1979). It is unclear exactly which step(s) between the abstraction of iron from serotransferrin and its incorporation into heme represents an energy-requiring process, though it is believed that the energy requirement reflects a much more complicated situation than merely the

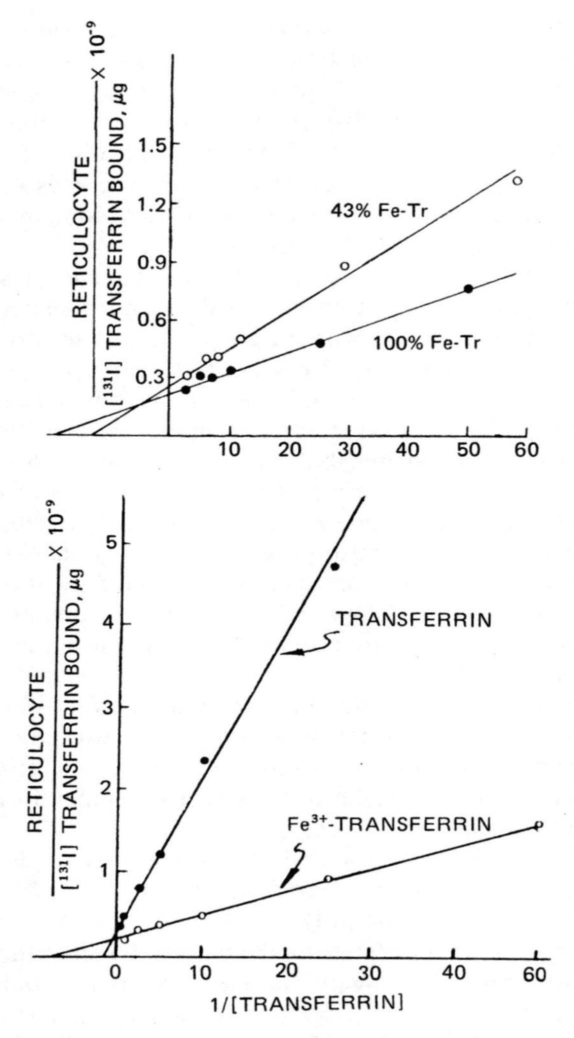

Figure 6-2. Michaelis–Menten rendition of serotransferrin uptake by rabbit reticulocytes. Graph (a) compares 43% iron-saturated serotransferrin with the fully saturated material, whereas graph (b) compares iron-free and iron-saturated serotransferrins. From Kornfeld (1969).

provision of the cell with adequate amounts of ATP (Morgan and Baker, 1969). It should also be remembered that some 90% of iron "uptake" measured in these experiments represents hemoglobin-bound ferrous iron (Martinez-Medellin and Schulman, 1972).

The binding of serotransferrin to reticulocyte membranes is inhibited by sulfhydryl reagents such as *p*-chloromercuribenzoate and iodoacetamide. Such inhibitors will, of course, also inhibit iron uptake. Morgan and Baker (1969) proposed that the sulfhydryl groups may be involved in securing serotransferrin to the plasma membrane receptor.

The role of the sulfhydryl groups was further defined by Edwards and Fielding (1971), who found that the inhibition of iron uptake but not of serotransferrin binding by reticulocytes could be observed at extremely low concentrations of *p*-chloromercuribenzoate. The effect was also extremely rapid. It was proposed then that the sulfhydryl residues of reticulocyte membranes do not function to bind iron-laden serotransferrin thereto but instead serve to abstract iron from iron-laden serotransferrin following its association with the membrane. This line of thinking has apparently not been pursued further.

We have stated the Jandl *et al.* (1959) were able to show that ferric iron chelators such as EDTA did not inhibit iron uptake by the immature red cell. This was confirmed with respect to EDTA, nitriloacetate, citrate, and *N*-β-hydroxyethyliminodiacetic acid by Morgan (1971). Morgan also discovered that 2,2′-bipyridyl and 1,10-phenanthroline inhibited iron uptake by the cells. Since these are ferrous iron chelators, it was reasoned that they entered the cell and successfully competed for iron that had been released from serotransferrin and had apparently been reduced. Since iron bound to the ferrous iron chelator cannot be utilized by the cell and must then leave the cell, this phenomenon has been developed into a very useful technique for the study of the effects of various substances on iron release from serotransferrin in the presence of reticulocytes.

A physiological inhibitor of iron uptake by the immature red cells is the heme (Ponka and Neuwirt, 1972; Karibian and London, 1965). When heme was incubated with serotransferrin in the presence of reticulocytes, it markedly inhibited the exchange of iron between serotransferrin and 2,2′-bipyridyl. However, hemin had no effect on the binding of serotransferrin to the reticulocyte membrane. It was concluded that heme (or hemin) acts as a feedback inhibitor to lower the rate of iron loss from iron-laden serotransferrin following its combination with the reticulocyte cell membrane (Ponka *et al.*, 1974). This proposal is supported by the fact that in situations where heme biosynthesis is inhibited by compounds such as isonicotinic acid hydrazide or penicillamine, iron uptake by immature red cells is enhanced, and in situations where globin synthesis is inhibited and heme accumulates, iron uptake is inhibited. It has also been suggested that sideroblastic anemia is due to a primary lesion in the heme biosynthetic pathway: in congenital sideroblastic anemia, there appears to be a molecular lesion affecting the interaction of δ-aminolevulinic acid synthetase and its cofactor, pyridoxal phosphate (Konopka and Hoffbrand, 1979). Thus the observed accumulation of nonheme iron in the red cells of such patients is due to a partial loss of inhibitory effects of heme toward the uptake of iron by the immature red cell (Ponka and Neuwirt, 1974). Moreover, in the Belgrade rat, which has a genetic block in heme biosynthesis, the most outstanding symptom is iron accumulation in the red cells (hyperchromia) and a depletion of iron stores (Garrick *et al.*, 1978).

Last, it may be pointed out that serotransferrin appears to be the only plasma iron carrier which under physiological conditions is capable of delivering iron to immature red cells. When serotransferrin is removed from blood serum by specific immunoprecipitation followed by a reconstitution of the iron in the form of $^{59}FeSO_4$, such serum fails to stimulate iron incorporation into the immature red cells (Eldor *et al.*, 1970).

6.4 Molecular Properties of Serotransferrin and its Effect on the Interaction with Immature Red Cells

Jandl *et al.* (1959) were first to show that iron-laden serotransferrin interacts much more readily with immature red cells than does iron-free serotransferrin. This has been substantiated by numerous authors. In addition, differences in reticulocyte–serotransferrin association constants have been observed among serotransferrins carrying various metal substituents such as iron, chromium, and copper (Kornfeld, 1969). It is believed that the reason metals so drastically affect the binding of serotransferrin to reticulocyte membranes is because a conformational change in the protein structure takes place when the protein binds a metal atom. This was shown by Bezkorovainy (1966) using hydrodynamic methods and has subsequently been confirmed by numerous investigators using other techniques. The conformational change presumably exposes certain areas on the surface of the protein molecule that are capable of interacting with the plasma membrane receptors of the immature red cell. More precisely, it has been proposed that serotransferrin undergoes a charge redistribution on its surface. This view has been arrived at from experiences with the chromatography of the serotransferrins on DEAE–cellulose columns (Lane, 1971, 1972). Surprisingly, in this system iron-saturated serotransferrin behaved as a weaker anion than did apotransferrin, though electrophoretically iron-saturated serotransferrin is the stronger anion (see Chapter 4). In general, behavior of the serotransferrins in the DEAE–chromatography system can be correlated with their interaction with rabbit reticulocytes: aposerotransferrin and divalent metal serotransferrin complexes (e.g., Cu, Zn, Mn) did not show a difference in DEAE–cellulose elution patterns and were very weakly bound by reticulocytes. On the other hand, the iron–serotransferrin complex, which can be separated from aposerotransferrin on DEAE–cellulose, readily interacted with reticulocyte membranes (Lane, 1971). Similar results were obtained with chromatography on Sephadex G-150, where separation occurs on the basis of Stokes' radii of the proteins: iron-saturated serotransferrin was eluted following aposerotransferrin and Cu^{2+}–serotransferrin, indicating that the iron-saturated species has a lower Stokes' radius (Kornfeld, 1969). For other differences between iron-saturated serotransferrin and aposerotransferrin, see Chapter 4.

A large volume of work in the area of serotransferrin–reticulocyte interaction has been concerned with whether or not the two iron-binding sites of serotransferrin can deliver iron to the immature red cells with equal ease. A corollary to this question is whether or not the two iron-binding sites of serotransferrin are equal with respect to the uptake of iron from the mucosal cells as well as delivery–uptake from the iron storage sites of the liver. It is well known that in regard to physical–chemical properties the two iron-binding sites of all the transferrins are not equal (see Chapter 4). The impetus for such investigations was provided by Fletcher and Huehns (1967, 1968). These workers incubated reticulocytes with plasma containing 50% iron-saturated serotransferrin labeled with ^{59}Fe until the serotransferrin became about 25% saturated. The labeled reticulocytes were removed, and the plasma as well as control plasma whose serotransferrin was about 25% saturated with iron (*in vitro*) were incubated with a fresh batch of reticulocytes. The result is depicted in Figure 6-3, which indicates that iron was taken up from the control plasma more efficiently than from plasma previously incubated with reticulocytes.

On the basis of these types of experiments, Fletcher and Huehns proposed a mechanism for the control of iron metabolism in the mammalian organism. It was postulated that the two-binding sites are biologically unequal: Site A (arbitrarily designated) prefers to give up its iron to erythropoietic tissue and accepts iron readily from iron storage sites in the liver and mucosal cells of the small intestine. The degree to which this site is occupied by iron determines how iron is to be absorbed from the gastrointestinal tract. It is thus supposed to act as the *messenger* described in Chapters 3 and 5. Site B was believed also to be able to release its iron to the erythropoietic tissue, except that it allegedly acts more sluggishly. It was proposed that site B is " . . . concerned with the storage of iron," presumably releasing its iron to the liver parenchymal cells in preference to immature red cells.

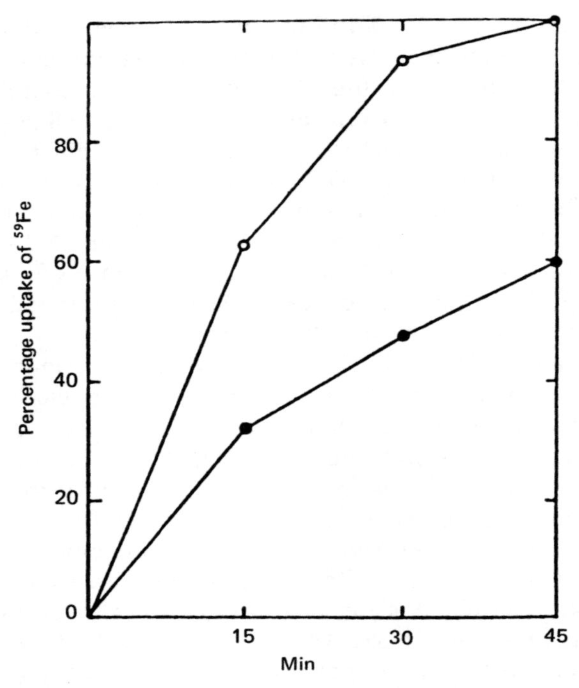

Figure 6-3. Comparison of iron uptake by reticulocytes from control plasma sample labeled with ^{59}Fe and from plasma previously incubated with reticulocytes. The latter shows a decidedly lower rate of iron uptake. From Fletcher and Huehns (1967).

The Fletcher–Huehns hypothesis has received confirmation from several groups of workers. Most if not all of them have utilized a variation of the methodology used by Fletcher and Huehns: Serotransferrin or iron-free plasma is labeled with ^{59}Fe to the desired iron saturation level (this may be low or high) and then the protein or plasma is incubated with reticulocytes until about half the iron originally present has been removed by the cells. The cells are then removed, and the serotransferrin is reconstituted to its original iron content with either unlabeled iron or, more commonly, ^{55}Fe (Figure 6-4). This doubly labeled preparation either may be used *in vivo* (see below), or a fresh batch of reticulocytes can be incubated with it, and the results are compared with a control labeled randomly with ^{59}Fe and ^{55}Fe. Using essentially this procedure, Verhoef *et al.* (1978), using 80% saturated rat serotransferrin and rat plasma with rat immature red cells, were able to confirm the predictions of Fletcher and Huehns: There was a site (site A) on the serotransferrin molecule which was specific for erythropoietic cells, fibroblasts, and lymphoblasts and another site (termed site B) which gives up its iron preferentially to liver parenchymal cells. The same conclusions were reached by Fletcher (1969) using 50% iron-saturated rabbit and human serum transferrin with rabbit reticulocytes.

In vivo confirmation of the Fletcher-Huehns hypothesis was apparently provided by Hahn (1973) and in a series of papers by E. Brown's group (Awai *et al.*, 1975a, 1975b, Okada *et al.*, 1977). By using the doubly labeled rat serotransferrin, both preincubated and not preincubated with the reticulocytes as already described, it was shown that following injection of such preparation into rats, tissue distribution of label was unequal, whereas randomly labeled serotransferrin (i.e., labeled with both ^{55}Fe and ^{59}Fe) showed a ^{55}Fe–^{59}Fe ratio of nealy 1 in all tissues. The ^{59}Fe-labeled serotransferrin preincubated with reticulocytes and then reconstituted with ^{55}Fe showed ratios higher than 1 in red blood cells, bone marrow, placenta, and spleen. It was less than 1 in liver, especially in the parenchymal cells. The results, which were statistically highly significant, are depicted in Figure 6-5 (Awai *et al.*, 1975a). In another very clever experiment, Okada *et al.* (1977) permitted the rat to absorb ^{59}Fe from the intestine into the circulation and then cannulated the portal vein. According to the Fletcher–Huehns prediction, iron absorbed from the intestine into the bloodstream is positioned into the red-cell-oriented site A of serotransferrin. The prediction indeed proved to be valid: When

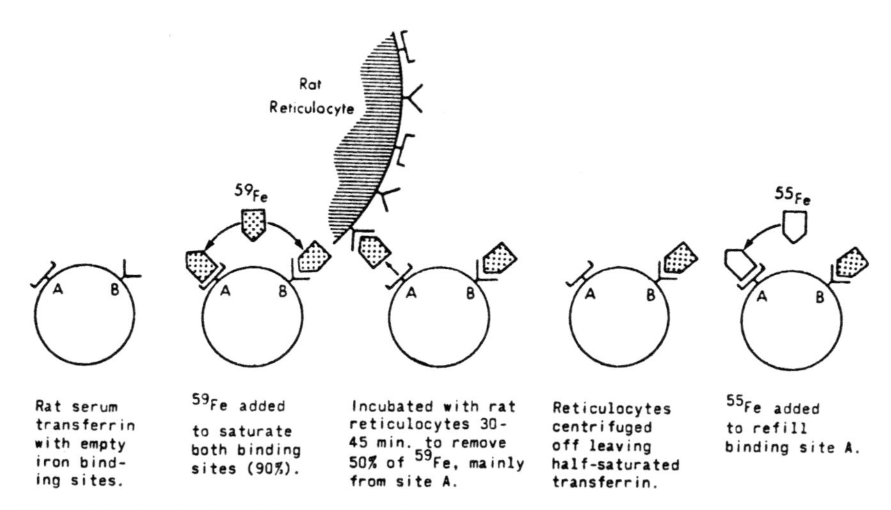

Figure 6-4. Mechanics of doubly labeled serotransferrin preparation. Reprinted by permission from Awai, M., Chipman, B., and Brown, E. B., 1975, In vivo evidence for the functional heterogeneity of transferrin-bound iron. I. Studies in normal rats, *J. Lab. Clin. Med.* 85:769–784. Copyright by C. V. Mosby Co.

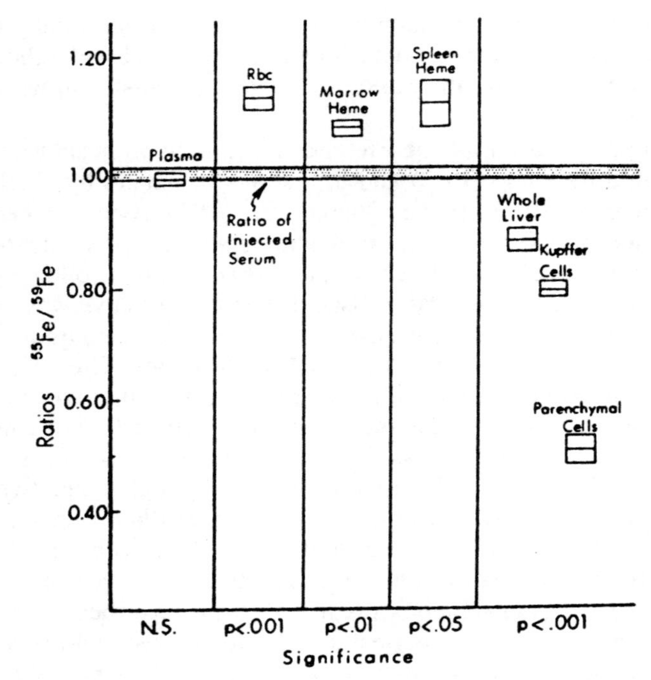

Figure 6-5. Ratios of ^{55}Fe–^{59}Fe in various tissues of the rat following injection of serotransferrin selectively labeled at site A (^{55}Fe, erythropoietic tissue preference) and site B (^{59}Fe, parenchymal tissue preference). Reprinted with permission from Awai, M., Chipman, B., and Brown, E. B., 1975, In vivo evidence for the functional heterogeneity of transferrin-bound iron. I. Studies in normal rats *J. Lab. Clin. Med.* 85:769–784. Copyright by C. V. Mosby Co.

portal vein plasma containing the labeled serotransferrin was incubated with reticulocytes, it proved to be a better iron donor than ^{59}Fe-labeled plasma from peripheral circulation where the iron label is supposedly randomized. This is illustrated in Figure 6-6.

There are several investigations, however, which are not consistent with the Fletcher–Huehns hypothesis and purport to show that the iron-binding sites act equally well in regard to all tissues. Thus Harris and Aisen (1975a, 1975b) proceeded to selectively double-label (see above) both human and rabbit serotransferrins with ^{55}Fe and ^{59}Fe and incubated these with both human and rabbit reticulocytes. When human serotransferrin was used with human reticulocytes and rabbit serotransferrin with rabbit reticulocytes, no site selectivity was noted. However, when a heterologous system involving human serotransferrin and rabbit reticulocytes was tried, selectivity predicted by the Fletcher–Huehns hypothesis was observed. These results were confirmed by using human serotransferrin labeled at site A (pH 6) on the one hand and at site B (pH 7.5) on the other with human reticulocytes (Harris, 1977; Morgan *et al.*, 1978). The same results were observed by Huebers *et al.* (1978), who prepared rat serotransferrins labeled at either the *acid-labile* site (^{59}Fe) or the *acid-stable* site (^{55}Fe) or doubly labeled at both the sites, one with ^{59}Fe and the other with ^{55}Fe. These were injected into rats and isotope distribution in various tissues measured. No differences were found, and the two serotransferrin variants (Tf_s and Tf_f) found in rat serum were also found to be functionally identical. Table 6-1 shows the ^{59}Fe–^{55}Fe ratios in various tissues, which are seen to be close to 1.

Brown's group, which provided much of the evidence in favor of the Fletcher–Huehns hypothesis, has recently reexamined their experiments using isolated isotransferrins (Tf_s and Tf_f) (Okada *et al.*, 1979). Their observations now indicate that the functional heterogeneity previously ascribed to sites A and B (Awai *et al.*, 1975a, 1975b; Okada *et al.*, 1977) was really the result of the functional difference between Tf_s and Tf_f. Thus, Tf_s was a better iron donor with respect to reticulocytes and a better iron acceptor from intestinal mucosal cells (portal circulation) than Tf_f. On the other hand, Tf_f was a better iron donor in regard to liver cells than Tf_s. The two iron-binding sites, A and B, were functionally equivalent in

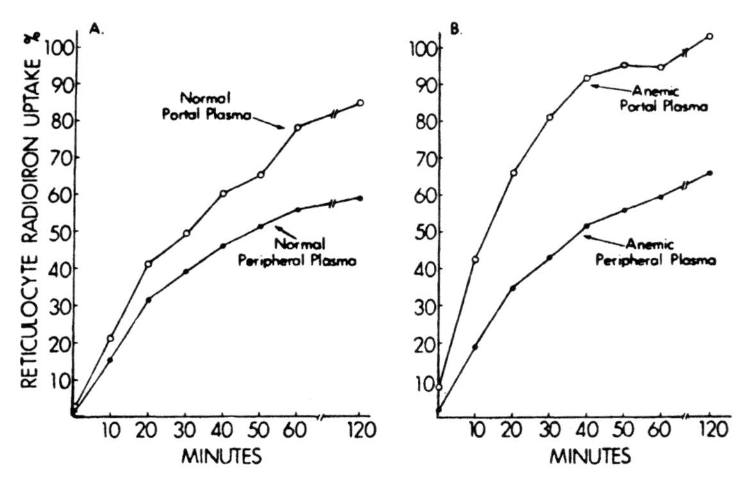

Figure 6-6. Uptake by rat reticulocytes of iron from rat peripheral plasma and portal vein plasma following the feeding of radioiron to the animals. Reprinted by permission from Okada, S., Chipman, B., and Brown, E. B., 1977, In vivo evidence for the functional heterogeneity of transferrin-bound iron. IV. Selective uptake by erythroid precursors of radioiron from portal vein plasma transferrin during intestinal iron absorption, *J. Lab. Clin. Med.* 89:51–64. Copyright by C. V. Mosby Co.

Table 6-1. Isotope Uptake by Various Rat Tissues Following the Administration of Diferric Serotransferrin Labeled with ^{55}Fe at the Acid-Labile Site and with ^{59}Fe at the Acid-Stable Site[a]

Tissue	^{55}Fe–^{59}Fe ratios: From normal rats	From iron-deficient rats
Red cells		
Skeleton	0.93 ± 0.01	0.99 ± 0.03
Circulatory	0.96 ± 0.01	0.95 ± 0.03
Liver	0.93 ± 0.01	0.99 ± 0.01
Spleen	0.95 ± 0.01	0.99 ± 0.03

[a] Adapted from Huebers *et al.* (1978).

both Tf_s and Tf_f. These results are summarized in Table 6-2. It is unclear at present why Brown's group has observed functional differences between Tf_s and Tf_f (Okada *et al.*, 1979), whereas Finch's group has not observed any such heterogeneity (Huebers *et al.*, 1978).

Much work has been done on the uptake of iron by immature red

Table 6-2. Radioiron Distribution in Rat Tissues Following the Injection of Isotransferrin Tf_s labeled with ^{55}Fe and Isotransferrin Tf_f labeled with ^{59}Fe in a Ratio of 1:1[a]

Tissue	Mean ^{55}Fe–^{59}Fe ratios	Significance[b]
Plasma	0.93 ± 0.01	$p < 0.025$
Reticulocytes	1.30 ± 0.15	—
Bone marrow		
Heme	1.25 ± 0.08	$p < 0.025$
Ferritin	0.96 ± 0.13	Not significant
Liver		
Total	0.99 ± 0.10	Not significant
Ferritin	0.83 ± 0.06	$p < 0.05$
Spleen		
Total	0.92 ± 0.06	Not significant
Heme	1.79 ± 0.19	$p < 0.01$

[a] Adapted from Okada *et al.* (1979).
[b] In regard to the uptake of the two isotopes of iron, tissues with preference for Tf_s show ratios higher than 1, and those showing preference for Tf_f show ratios of less than 1.

cells from serotransferrins saturated with iron to varying degrees. With some notable exceptions, it was found that iron uptake from iron-saturated serotransferrin by reticulocytes *in vitro* as well as its clearance *in vivo* is greater than that of incompletely iron-saturated serotransferrin. This fact has been used both for and against the validity of the Fletcher–Huehns hypothesis.

Thus Brown *et al.* (1975) prepared rat serotransferrins saturated with iron to 90% and 10% and followed the clearance of both in rats *in vivo*. The iron-saturated material was more efficient in donating its iron to red cells, bone marrow, spleen, and liver. The half-lives were 36 min for the iron-poor serotransferrin and 27 min for the iron-saturated species. Essentially the same results were obtained in human beings, where half-lives of iron were 92 ± 33 min for 4% iron-saturated serotransferrin and 54 ± 15 min for the 80% iron-saturated species (Hahn *et al.*, 1975). Brown *et al.* (1975) interpreted their results as being supportive of the Fletcher–Huehns hypothesis.

Finch and his co-workers have carried out a number of both *in vivo* and *in vitro* experiments; the aim was to evaluate the relative effectiveness of monoferric and diferric serotransferrins to distribute iron throughout the organism. They invariably found that the diferric species is a more efficient iron donor than the monoferric serotransferrin. This is most likely due to differences in the three-dimensional structures of the proteins, as pointed out in Chapter 4. Thus Skarberg *et al.* (1978) determined iron clearance rates with primarily diferric and monoferric serotransferrins in human subjects and found half-lives to be 74 ± 6 and 93 ± 7 min, respectively. Tissue distribution of isotope from mono- and diferric serotransferrins was, however, the same. Christensen *et al.* (1978) prepared diferric and monoferric serotransferrins by isoelectric focusing and found that tissue uptake was much greater from the diferric species: Iron used from the diferric serotransferrin was taken up 1.56–2.1 times faster by immature red cells and 2.38 to 2.65 times faster by the liver than iron bound to the monoferric species. And finally, in an *in vitro* experiment, Huebers *et al.* (1978) incubated diferric serotransferrin with rat reticulocytes until about half of the iron was incorporated into reticulocytes. The residual plasma was again subjected to an isoelectric focusing experiment to assess the presence of monoferric serotransferrin species. None were found. Huebers *et al.* (1978) concluded that iron is removed from the diferric serotransferrin species from both sites before the apotransferrin is eluted. Their conclusion was that the two iron-binding sites of serotransferrin are biologically equivalent and that differences in tissue iron uptake are due to the relative abundance of the mono- and diferric serotransferrin species in the medium.

One may inquire at this point as to the current viability of the Fletcher–Huehns hypothesis. If the hypothesis is defined in terms of the biological heterogeneity of sites A and B on the same serotransferrin molecule, it must be admitted that the bulk of available evidence points to its invalidity. One is, however, left with a nagging suspicion that the last word on the biologic identity or nonidentity of sites A and B has not been spoken. There is no doubt that chemically and physically the two sites differ from each other. Why would nature do such a thing without a purpose? And then, when one speaks of "monoferric" serotransferrin, one really does not distinguish between the site A and site B monoferric serotransferrins. Are these two serotransferrin species biologically different? Further *in vivo* studies utilizing these compounds should be most instructive and feasible, since iron does not appear to randomize from one site to another during short-term experiments (Okada *et al.*, 1978). On the other hand, if the Fletcher–Huehns hypothesis is to include differences in the behavior of monoferric vs. diferric serotransferrins, then the hypothesis is certainly valid. In either case, the work of Fletcher and Huehns has stimulated a tremendous amount of activity in the area of serotransferrin–tissue interaction, and much has been learned in the area of iron metabolism as a result.

6.5 Transferrin Receptors in the Immature Red Cells

When the transferrins assume the appropriate three-dimensional conformation, they can interact with various tissues, the purpose being to dispense iron into those tissues. A receptor for human serotransferrin in human reticulocytes was detected as early as 1959 (Jandl *et al.*, 1959), and many attempts have been made since then to isolate and characterize it. The immature red cells are, of course, not the only tissue(s) that possesses serotransferrin receptors on its surface. Others may include placenta, liver, spleen, kidney (Van Bockxmeer *et al.*, 1975), and both lymphoblastoid B and T cells (Larrick and Cresswell, 1979). Placental trophoblasts are especially rich in such receptors. Their function may be to transport iron from the mother to fetus or, more likely, to impede the proliferation of white cells by sequestering the serotransferrin. The latter stimulates the transition of leukocytes from the G_1 to the S phase during proliferation (Faulk and Galbraith, 1979).

Garrett *et al.* (1973) were one of the first groups to demonstrate directly the fact that serotransferrin is bound to reticulocytes. They incubated doubly labeled serotransferrin (human?) (^{125}I–^{59}Fe = 2.9) with rabbit reticulocytes and then isolated the stroma. The ^{125}I–^{59}Fe ratio there was 0.7, indicating loss of iron and binding of serotransferrin. The latter could be extracted from the stroma, bound to a macromolecular fraction, by a variety of detergents.

The first purification of the serotransferrin receptor was achieved from human reticulocytes by Speyer and Fielding (1974). These workers used the nonionic detergent Triton X-100 to extract the receptor–serotransferrin complex and then proceeded to purify it by gel filtration of Sepharose 2B and 6B. The final product had a molecular weight of 230,000, and if allowance is made for the 80,000 daltons attributable to serotransferrin, the "receptor" itself was deemed to have a molecular weight of 150,000. Using essentially the same techniques, Sly *et al.* (1975a, 1975b) purified the serotransferrin receptor material from rabbit reticulocytes, where reticulocytosis was induced by phenylhydrazine. The complex had a molecular weight of near 200,000 and the receptor material itself, 120,000. There was evidence to suggest the existence of subunits in the receptor material.

These initial studies were followed by numerous other attempts at purification and characterization of serotransferrin receptors from a number of tissues, especially reticulocytes. It appears, however, that in spite of a large volume of such work, we still have precious little information on the properties of this molecule or molecules. It is possible to mention at least three groups that have used essentially the approach of Speyer and Fielding, which assumes a rather tight bond between the serotransferrin and its receptor. Leibman and Aisen (1977) chromatographed their ^{125}I–serotransferrin-labeled Triton X-100 extracts of rabbit reticulocyte membranes on Ultragel AcA22 columns and isolated a receptor-serotransferrin complex with a molecular weight of 445,000. Upon SDS–polyacrylamide gel electrophoresis, serotransferrin (m.w. 80,000) and two other components, m.w. 95,000 and 176,000, were identified. These workers speculated that the receptor molecule may consist of either two subunits, m.w. 176,000 each, or three subunits, one with a m.w. of 176,000 and the other two with 95,000 each. Van Bockxmeer and Morgan (1977) used the anionic detergent Teric 12A9 to extract ^{125}I–serotransferrin-labeled rabbit reticulocyte membranes and, following chromatography on Sephadex G-200 media, isolated a serotransferrin complex with a m.w. of 350,000. The receptor itself therefore had a m.w. of near 275,000. Unlabeled serotransferrin could displace the label from the serotransferrin–receptor complex. Rat reticulocyte serotransferrin receptor molecule was prepared by VanderHeul *et al.* (1978) and was shown to have a m.w. of less than 230,000. Moreover, these workers were able to prepare antibodies against the receptor and to show that it inhibited the uptake of serotransferrin by normal reticulocytes.

Another approach to the identification of the serotransferrin receptor was provided by Witt and Woodworth (1978), who reasoned that serotransferrin could protect the receptor against chemical modification. They labeled reticulocytes, either serotransferrin depleted or serotransferrin saturated, with iodine and then subjected the membranes to SDS–polyacrylamide gel electrophoresis. The membrane component, which was protected against iodination, had a m.w. of 190,000.

Chemical modification in the form of cross-linking the receptor with ^{125}I–serotransferrin followed by electrophoresis on SDS gels was used by Nunez *et al.* (1977) to identify serotransferrin-binding components of rabbit reticulocyte membranes. They were able to identify two such receptor substances: one with a m.w. of 60,000 and the other with a m.w. of 145,000. They speculated that the 60,000-dalton particle may be a subunit of the 145,000-dalton molecule, which would correspond to the results of Sly *et al.* (1975a, 1975b).

The most promising method for the purification of the serotransferrin receptor of reticulocytes is by affinity chromatography involving insolubilized serotransferrin and extracts of reticulocyte membranes. Light (1977) reported on the chromatography of Triton X-100 extracts of rabbit reticulocyte membranes on such columns whereby receptor molecules with a m.w. of 30,000–35,000 were identified. These had a tendency to form dimers with m.w. of 60,000–70,000. Sullivan and Weintraub (1978) have prepared rat reticulocyte serotransferrin receptors by affinity chromatography and have identified particles with m.w. of 95,000 and 145,000 as having the ability to bind serotransferrin. These were precipitated in the presence of serotransferrin with serotransferrin antibodies.

There has been considerable speculation in regard to what happens to the serotransferrin receptor molecules as the immature red cell matures. Leibman and Aisen (1977) have presented evidence that the receptor molecule loses its carbohydrate moiety and thus loses the ability to combine with serotransferrin. Others have maintained that there is an actual loss (degradation) of the receptor as the cell matures (Van Bockxmeer and Morgan, 1977; Witt and Woodworth, 1978). Light has reported the isolation of a glycoprotein from rabbit bone marrow erythroid cells which had a m.w. of only 18,000 but could interact with serotransferrin (Light, 1978). He speculated that the serotransferrin receptors in the very immature, and therefore most active in regard to transferrin binding, erythroid cells are of a very small size and that they lose their activity in regard to serotransferrin binding as the cell matures because of polymerization. This fact may thus explain the discrepancy in molecular weights reported by the various authors (see Table 6-3 for a summary). Light's work would imply that such polymerized receptors might

Table 6-3. Serotransferrin Receptors from Various Tissues and Their Properties

Tissue	Species	Molecular weight	Subunits	Notes	Reference
Reticulocytes	Human	150,000	—	—	Speyer and Fielding, 1974
	Rabbit	120,000	Two, m.w. 60,000 each	—	Sly *et al.*, 1975a, 1978
	Rabbit	350,000–400,000	Two types, m.w. 95,000 and 176,000	Contains carbohydrate, which disappears upon cell maturation	Leibman and Aisen, 1977
	Rabbit	275,000	—	Disappears upon cell maturation	Van Bockxmeer and Morgan, 1977
	Rabbit	190,000	—	Disappears upon cell maturation	Witt and Woodworth, 1978
	Rabbit	145,000 (?)	Two, m.w. 60,000 each	—	Nunez *et al.*, 1977
	Rabbit	30,000–35,000	—	Dimerizes to form particles with m.w. of 60,000–70,000	Light, 1977
	Rabbit	110,000	Two types, m.w. 17,000 and 48,000	Contains carbohydrate and lipid	Sly *et al.*, 1978
	Rabbit	175,000	Two, m.w. 95,000 each	—	Hu and Aisen, 1978
	Rat	230,000	—	$S^{\circ}_{20} = 7$, antibodies prepared in rabbits	VanderHeul *et al.*, 1978
	Rat	95,000 and 145,000	—	—	Sullivan and Weintraub, 1978
Chick embryo red cells	Hen	35,000	—	—	Witt and Woodworth, 1975
Bone marrow	Rabbit	18,000	—	Polymerizes as cell matures	Light, 1978

bind serotransferrin with lesser and lesser affinity as the reticulocyte matures. This, however, has not proved to be the case, since Van Bockxmeer and Morgan (1977) have shown that with increasing maturity rabbit erythroid cells lose serotransferrin receptors but that the affinity for serotransferrin remains the same throughout.

Muller and Shinitzky (1979) have recently proposed that the loss of transferrin receptors by the erythropoietic cells may be a two-stage process, consisting first of a removal of the receptor from plasma membrane surface followed by its degradation step. The initial step was termed *passive modulation,* which is based on the observation that when cell plasma membrane fluidity is decreased (by cholesterol depletion), membrane proteins, including the serotransferrin receptor, become more accessible to the aqueous environment, both extracellular and intracellular. A increase in membrane microviscosity (fluidity), as occurs during red cell maturation, leads to a stronger interaction between lipids and the hydrophobic side chains of membrane proteins, thus making the latter less accessible to the aqueous environment(s) (Borochov and Shinitzky, 1976). Thus, as the red cell matures, its membrane becomes more fluid and *masks* the serotransferrin receptor protein. Following such masking of the receptor by the membrane lipid layer(s), the protein is presumably degraded.

There are only two reports dealing at some length with the physical–chemical properties of serotransferrin receptors. Both are concerned with receptor molecules from rabbit reticulocytes isolated by extraction with nonionic detergents and purified by gel filtration. In one case (Sly *et al.,* 1978), the m.w. of the receptor was near 110,000, and upon gel electrophoresis in the presence of SDS, urea, and mercaptoethanol, subunits with m.w. of 17,000 and 48,000 could be identified in proportions consistent with a model containing three of the smaller subunits and one larger one. The receptor contained both carbohydrate and lipid, the latter being consistent with the observation that uptake of iron by reticulocytes inhibited by phospholipase A (Hemmaplardh and Morgan, 1977). Its composition is given in Table 6-4. The other report (Hu and Aisen, 1978) has determined the m.w. of the serotransferrin–receptor complex to be near 250,000 and that of the receptor molecule to be 175,000. The latter probably consisted of two subunits, m.w. 95,000 each. Physical parameters for the serotransferrin–receptor complex were 0.735 cm^3/g for partial specific volume, 9.49 S for the sedimentation constant, 3.5×10^{-7} cm^2/sec for the diffusion constant, 61 Å for the Stokes' radius, and 1.46 for frictional ratio. The receptor itself was not thoroughly characterized.

The detergent-extracted receptor material from rabbit reticulocytes

Table 6-4. Composition of Serotransferrin Receptor from Rabbit Reticulocyte Membranes[a]

Component	g/100 g
Lysine	2.4
Histidine	1.8
Arginine	4.4
Aspartic acid–asparagine	11.9
Threonine	3.0
Serine	8.1
Glutamic acid–glutamine	13.2
Proline	5.1
Glycine	5.2
Alanine	6.2
Valine	3.6
Methionine	0.6
Isoleucine	1.8
Leucine	11.8
Tyrosine	3.8
Phenylalanine	6.1
Hexose	4.6
Hexosamine	4.4
Sialic acid	1.0
Fucose	1.0
"Lipid"	5.0

[a] Adapted from Sly *et al.* (1978).

was used to estimate certain binding parameters (Van Bockxmeer *et al.*, 1978). Since Scatchard-type plots using serotransferrin and the purified receptor material gave straight lines (Figure 6-7), it was concluded that the system obeyed the simple equation,

$$\text{serotransferrin} + \text{receptor} \underset{K_2}{\overset{K_1}{\rightleftharpoons}} \text{serotransferrin–receptor complex} \qquad (6\text{-}1)$$

and that therefore there exists a single type of serotransferrin receptor. The association constant ($K_a = K_1/K_2$) was $4.1 \pm 0.3 \times 10^7 \ M^{-1}$, which at 300°K gives a $\Delta G° = -10.4 \pm 0.7$ kcal/mole. By using the van't Hoff equation and K_a values obtained at different temperatures, $\Delta H°$ and $\Delta S°$ were -106 cal/mole and $+34.5$ e.u., respectively. Because of the low $\Delta H°$ and the hefty $\Delta S°$ values, it may be surmised that the serotransferrin–receptor interaction involves hydrophobic binding. This is in agreement with the system involving intact reticulocytes (Baker and Morgan, 1969; see above). Ca ions were also required for the interaction, as they are in the intact reticulocyte–serotransferrin reaction.

6.6 Iron Removal from Serotransferrin

There is now no doubt that the first step in the departure of iron from serotransferrin into the immature red cell is the combination of serotransferrin with a specific receptor or receptor complex on the cell's plasma membrane. Subsequent steps are, however, under intense debate. There is a school of thought maintaining that iron-laden serotransferrin is internalized by the cell through the process of endocytosis, that iron is abstracted therefrom, and that iron-poor serotransferrin is then refluxed back into the extracellular medium. Another and more populous school of thought holds the view that the internalization of serotransferrin is at best a minor pathway in iron uptake by the immature red cell and that most of the iron is lost from the serotransferrin during its sojourn as the serotransferrin–receptor complex. There exists evidence supporting both views.

The nonpenetration theory is supported first of all by the apparent existence of Fe^{3+}–receptors in the plasma membrane of immature red cells. This was first reported by Garrett *et al.* (1973), who were able to show that rabbit reticulocytes labeled with ^{59}Fe–and ^{125}I–serotransferrin upon extraction with detergent solutions yielded a certain fraction which contained ^{59}Fe but little if any ^{125}I. These results were amplified by Fielding and Speyer (1974), who made an extensive analysis of Triton X-100

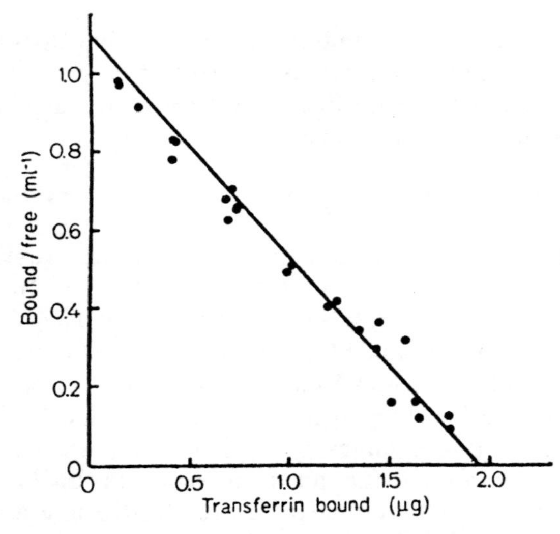

Figure 6-7. Scatchard plot relating the binding of serotransferrin to its specific receptor isolated from reticulocyte membranes. From Van Bockxmeer *et al.* (1978).

extracts of human reticulocyte membranes labeled with ^{59}Fe- and ^{125}I-containing serotransferrin, the latter having been absorbed by the intact reticulocyte before membrane penetration. In addition to the serotransferrin receptor (termed B-2), they were able to purify two membrane components and one cytosol component that bound iron but not serotransferrin. One membrane component, B_1, had a m.w. of near 10^6, whereas the other, component A, was very large and sedimented at 95,000 $\times$ g for 60 min. The cytosol component C was also very large and was later identified as ferritin (Nunez *et al.*, 1978; Speyer and Fielding, 1979). Kinetic experiments indicated the relationship among these components in regard to iron transport as shown in Figure 6-8. It would be interesting to investigate whether component A could differentiate between iron located at the acid-stable and the acid-labile sites of serotransferrin.

Serotransferrin attached to Sepharose and therefore unable to penetrate the cell membrane was able to donate iron to reticulocytes quite efficiently (Loh *et al.*, 1977).

There are several reports showing that iron can be mobilized from the reticulocyte membrane by the reticulocyte cytosol. Thus, rabbit reticulocyte membranes labeled with ^{59}Fe–serotransferrin can donate iron to unlabeled reticulocyte cytosol and utilize such iron for the biosynthesis of hemoglobin and incorporation into ferritin. An iron-binding protein with a m.w. near 5000 and specific for ferrous iron has been implicated in the mobilization process (Workman and Bates, 1974). In similar system obtained from mouse reticulocytes, ferritin and a cytoplasmic protein with a m.w. of 17,000 were apparently capable of mobilizing iron from reticulocytes. Ferritin iron was also available for hemoglobin synthesis (Nunez *et al.*, 1978; Speyer and Fielding, 1979), and it has in fact been argued that ferritin is the "obligatory" vehicle for iron incorporation into protoporphyrin IX (Speyer and Fielding, 1979). Ferritin, in the presence of FMN or FAD, succinate, and mitochondria can donate iron to protoporphyrin IX (Ulvik and Romslo, 1978). This is not too surprising as iron is readily mobilized from ferritin by the preceding reducing agents in model systems in the absence of any subcellular particles (see Chapter 5).

Additionally, in support of the nonpenetration view of iron loss by serotransferrin, one may also mention the older series of experiments (Jandl and Katz, 1963) where following the binding of serotransferrin labeled with both ^{131}I and ^{59}Fe to reticulocytes, ^{59}Fe was found largely inside the cell, whereas ^{131}I remained on the stroma.

The penetration hypothesis was first proposed by Morgan and Baker (1969) and followed up by radioautographic evidence (Morgan and Appleton, 1969), where following incubation of reticulocytes with

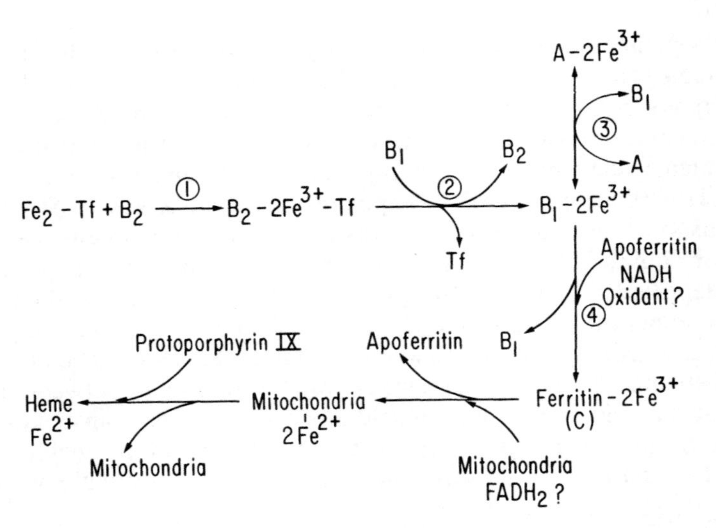

Figure 6-8. Interaction of iron-laden serotransferrin with the reticulocyte membrane according to Speyer and Fielding (1974); Fielding and Speyer (1974); and Speyer and Fielding (1979). Fe_2–Tf is the iron-laden serotransferrin, B_2 is the membrane serotransferrin receptor, A and B_1 are iron-binding macromolecules in the membrane, and C is ferritin. Reactions 1, 2, and 3 take place in the membrane, whereas reaction 4 serves to transfer iron from the membrane into the cytosol. Entry of iron into ferritin requires reduction (see Chapter 5) and subsequent reoxidation. These steps are combined in step 4. Ferritin subsequently interacts with mitochondria.

^{125}I–serotransferrin, the latter could be visualized in the cell interior by electron microscope by autoradiography. This led a number of workers to investigate this proposal. Martinez-Medellin and Schulman (1972) reported that, contrary to the work of Fielding and Speyer (1974), serotransferrin was the only iron-containing stromal macromolecule. Moreover, after 2.5 hr of incubation, most of the serotransferrin taken up by the cell was located in the cytosol. This labeled serotransferrin could be chased out by cold serotransferrin. Labeled serotransferrin previously incubated with reticulocytes was also found in the reticulocyte cytosol after cell lysis by Sly *et al.* (1975b), and moreover, it appeared to be combined with a small-molecular-weight carrier protein. Microtubule disrupting agents, colchisine, vinblastine, vincristine, strychnine, and D_2O, inhibited serotransferrin and iron uptake by both the reticulocytes and bone marrow cells. A representative experiment is shown in Figure 6-9. Microfilament inhibitors had no effect. This was interpreted to indicate that serotransferrin is internalized by the immature red cell through the process of endocytosis involving microtubules (Hemmaplardh *et al.*, 1974). In fact, if serotransferrin is conjugated with ferritin and incubated with rat reticulocytes or normoblasts, the conjugate can be visualized in the micropinocytic vesicles by electron microscopy. Ferritin itself, or ferritin conjugated with IgG, does not penetrate the cell (Sullivan *et al.*, 1976).

It will be recalled that the interaction of serotransferrin with an immature red cell involves two steps: *adsorption*, which is temperature independent, and the temperature-dependent *association*. It has been proposed that the adsorption involves the combination of the serotransferrin with its specific membrane receptor, whereas association is the endocytosis step. It is the latter phase of serotransferrin (and iron) uptake by the immature red cells that is inhibited by the microtubule disrupting agents, which have no effect on the combination of serotransferrin with its specific receptor (Hemmaplardh and Morgan, 1977). This concept received support from the work of Martinez-Medellin *et al.* (1977), who found that only a small amount of serotransferrin present on a reticulocyte

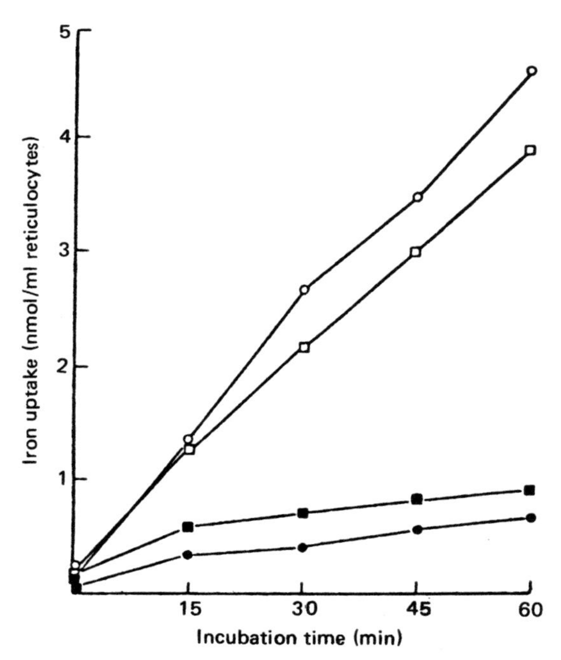

Figure 6-9. Effect of vinblastine on the uptake of iron by reticulocytes from iron-laden serotransferrin and citrate. Open circles and open squares are control cells taking up iron from serotransferrin and citrate respectively, whereas the closed circles and squares represent the same using vinblastine-treated cells. From Hemmaplardh, D., Kailis, S. G., and Morgan, E. H., Brit. *J. Haematol.* 28:53–65, 1974. By permission of the Blackwell Scientific Publications, Ltd.

is available for iodination with lactoperoxidase, the greater part of serotransferrin being presumably in the cytosol or being buried deeply in the membrane. As expected, the reaction of the isolated serotransferrin receptor with serotransferrin was only moderately temperature dependent compared to that of iron uptake by the cells (Van Bockxmeer *et al.*, 1978).

It is interesting to note that reticulocyte cytosol is not capable of abstracting iron from serotransferrin. The intact cell is apparently necessary for this to occur (Morgan, 1976).

Regardless of whether or not serotransferrin penetrates the immature red cell membrane during the process of iron assimilation by the latter, there must exist a mechanism whereby iron is removed from serotransferrin. The ultimate destination of iron is the mitochondrion, where it is incorporated into protoporphyrin IX, and, moreover, somewhere during this process iron must be reduced from the ferric to the ferrous state. We shall now discuss the mechanisms that have been proposed to be operative in these processes and which could be applied to iron removal from serotransferrin both in the cell membrane and the cytosol.

There is a view which states that iron is released from its serotransferrin complex following an enzymatic removal of carbonate, the synergistic anion required for the ternary complex to form (Aisen and Leibman, 1973; Egyed, 1973). This conclusion was reached because the reticulocyte is unable to abstract iron from serotransferrin if carbonate is substituted by oxalate or malonate. Metabolic inhibitors implicated CO_3^{2-} in the iron release process, albeit circumstantially: The sulfhydryl inhibitors preventing both the uptake of iron and serotransferrin by reticulocytes (Morgan and Baker, 1969) also prevented the loss of carbonate; the group of inhibitors, which had no effect on serotransferrin uptake by reticulocytes but did inhibit iron uptake, inhibited carbonate release to exactly the same extent as the uptake of iron, and compounds that stimulated the uptake of iron (isonicotinic acid hydrazide, a pyridoxal antagonist) also stimulated the loss of carbonate (Schulman *et al.*, 1974). Carbonate, incidentally, does not readily exchange with the medium: the half-life of serotransferrin-bound carbonate is close to 20 days (Aisen *et al.*, 1973).

A number of small-molecular-weight substances have the capability of removing iron from serotransferrin *in vitro* and could thus function as the means whereby serotransferrin loses its iron to the immature red cell. Citrate has been implicated as an effective iron-exchange medium between various types of transferrins (Aisen and Leibman, 1968), and 2,3-diphosphoglycerate and various nucleotides are also very effective (Morgan, 1977). Of special interest are the nucleotides, since it has been shown that a large proportion of nonferritin nonhemoglobin iron in the human red cells is coordinated with ATP, ADP, AMP, and NADP (Konopka and Szotor, 1972). Carver and Frieden (1978) made a careful study of the ability of ATP to abstract iron from diferric serotransferrin at slightly acid pH (pH 6.1). Such a pH may locally occur in various intracellular compartments, and diferric serotransferrin, according to Huebers *et al.* (1978), may be the actual iron-donating serotransferrin species. Taking into consideration the proposal by Morgan and Baker (1969) that serotransferrin may have to be protonated before losing its iron (apparently at the histidyl residues), Carver and Frieden proposed the following scheme for the loss of iron by serotransferrin:

$$\begin{array}{l} (CO_{3\,+}^{2\,-})_2\text{–Tf–}(Fe^{3+})_2 \xrightarrow{H^+,\ pH\ 6.1} (CO_3^{2-})_2\cdots\text{Tf}\cdots(Fe^{3+})_2 \xrightarrow{\text{ATP}\ \ 2\,CO_3^{2-}} \\ \text{ATP–Tf–}(Fe^{3+})_2 \xrightarrow[\text{Tf}]{} \text{ATP–}(Fe^{3+})_2 \xrightarrow{\text{ascorbate}} \text{ATP–}Fe^{2+} + \text{ascorbate–}Fe^{2+} \end{array} \qquad (6\text{-}2)$$

It should be noted that reducing agents such as ascorbate or glutathione are inactive in reducing iron to the ferrous state as long as the iron is coordinated with serotransferrin. On the other hand, in the presence of ATP, such iron is readily reduced. The ascorbate– and ATP–iron complexes then presumably interact with mitochondria or ferritin.

That mitochondria are a milestone in the path of iron incorporation into protoporphyrin IX may be surmised from the location of the ferrochelatase enzyme, which is present in mitochondria, and from reticulocytes poisoned by isonicotinic acid hydrazide (INH), an antagonist of pyridoxal phosphate in this system. INH inhibits the incorporation of iron into heme. In reticulocytes treated with INH and subsequently with ^{59}Fe–serotransferrin, some 40–50% of the iron is located in the mitochondria, 25–30% in a low-molecular-weight fraction, 10–15% in ferritin, and the rest in hemoglobin. In the normal reticulocyte under these conditions most of the iron is found in hemoglobin (Borova *et al.*, 1973). The iron can be mobilized from the mitochondria by pyridoxal and pyridoxal-5-phosphate and especially effectively by pyridoxal isonicotinoylhydrazone (PIH), which is formed from pyridoxal-5-phosphate and INH when INH-treated reticulocytes loaded with iron are incubated with pyridoxal-5-phosphate (Ponka *et al.*, 1979a, 1979b). These chelators are more effective in removing iron from mitochondria than is 2,2′-bipyridyl, which, while preventing iron incorporation into protoporphyrin IX, will not remove iron from mitochondria (Ponka *et al.*, 1979a, 1979b).

There is some evidence which has been interpreted to indicate that transferrin entering the immature red cell interacts directly with mitochondria to deposit its iron there. Thus, reticulocyte mitochondria both free and stroma associated were shown to be able to bind serotransferrin and to remove iron therefrom to biosynthesize heme. Hemin inhibited the removal of iron from serotransferrin but enhanced the actual binding of serotransferrin to mitochondria (Neuwirt *et al.*, 1975; Ponka *et al.*, 1977). It should be pointed out that the latter hypothesis may not necessarily require that iron-laden serotransferrin penetrate the immature red cell membrane, since mitochondria were shown to be closely associated with the red cell stroma of reticulocytes. This hypothesis would, however, imply that the actual removal of iron from the iron-laden serotransferrin may occur through its reduction from the ferric to the ferrous state using the redox machinery of the mitochondria. This would also be consistent with the action of metabolic inhibitors such as CN^-, 2,4-dinitrophenol, rotenone, etc., which inhibit the uptake of iron by reticulocytes but do not inhibit the uptake of serotransferrin. The proposal by Ponka and Neuwirt's group (Neuwirt *et al.*, 1975; Ponka *et al.*, 1977) may be diagrammatically represented as in Figure 6-10.

In closing this section, it would be appropriate to speculate in regard to the scenario one might encounter when iron is removed from serotransferrin and is incorporated into hemoglobin. The weight of current evidence seems to favor the idea that iron is mostly removed from serotransferrin by the membrane-bound machinery to be transported via ferritin to mitochondria. A minor amount of iron may enter the cell bound to serotransferrin via a pinocytotic or endocytotic mechanism. Both processes may involve the same well-characterized plasma membrane serotransferrin receptor. The former mechanism, assuming the participation of ferritin, must involve at least three redox steps: The ferric iron abstracted from transferrin via the removal of the synergistic anion, pro-

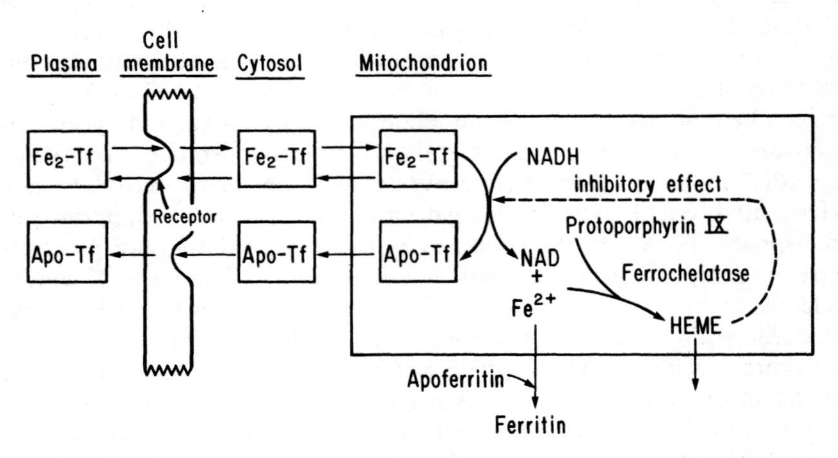

Figure 6-10. Uptake of iron via the penetration hypothesis as viewed by Ponka *et al.* (1977) and Neuwirt *et al.* (1975).

tonation of the ligand groups, or chelation by small nucleotides (e.g., ATP) must be reduced to the ferrous state before it will be taken up by the ferritin (or apoferritin) molecule (see Chapter 5). Once ferrous iron enters ferritin, it is reoxidized to the ferric state. To be removed from ferritin, the iron must once again be reduced to the ferrous state, after which it may, as a small-molecular-weight ferrous chelate, interact with the mitochondria and become incorporated into protoporphyrin IX in the ferrochelatase reaction. The smaller amount of iron entering the immature red cell via pinocytosis of the serotransferrin–iron complex may then lose its iron directly to the mitochondria through the reducing action of the mitochondrial redox machinery.

Summary

The delivery of iron to the immature red cells is accomplished by serotransferrin, which under normal conditions appears to be the only substance capable of doing this. The process of iron transfer from serotransferrin to the reticulocytes (reticulocytes are used routinely as model cells for the bone marrow immature red cell system) involves first the interaction of serotransferrin with the cell followed by the abstraction of iron. Iron-laden serotransferrin interacts with immature red cells much more effectively than does aposerotransferrin, and this is apparently due to conformational differences between the two types of proteins. The interaction of iron-laden serotransferrin with the reticulocyte cell surface in turn occurs in two steps: a loose *adsorption* followed by a firm *association*.

It has been proposed that the two iron-binding sites of serotransferrin do not have an equal propensity to donate iron to immature red cells; however, most recent evidence indicates that the two iron-binding sites of serotransferrin are biologically identical. On the other hand, diferric serotransferrin is a much more efficient iron donor in regard to immature red cells than is monoferric transferrin.

Serotransferrin interacts with a specific receptor on the plasma membrane of the immature red cell, and it has been proposed that there are some 300,000 such receptors for each rabbit reticulocyte cell. The receptor has been isolated and shown to be a glycoprotein with a molecular weight variously reported to be between 17,000 and 400,000 daltons. There is evidence to indicate that following the combination of serotransferrin with its receptor, some of the iron-laden protein is internalized by the cell through the process of pinocytosis, where the iron-laden serotransferrin may interact directly with the immature red cell mitochondria and iron incorporated into the protoporphyrin molecule. The major portion of iron seems to enter the immature red cell by being removed from serotransferrin by an appropriate plasma membrane machinery. It then apparently interacts with ferritin, which then serves as an iron donor to the mitochondrial hemoglobin synthesizing system.

References

Aisen, P., and Leibman, A., 1968. Citrate-mediated exchange of Fe^{3+} among transferrin molecules, *Biochem. Biophys. Res. Commun.* 32:220–226.

Aisen, P., and Leibman, A., 1973. The role of the anion-binding site of transferrin in its interaction with the reticulocyte, *Biochim. Biophys. Acta* 304:797–804.

Aisen, P., Leibman, A., Pinkowitz, R. A., and Pollack, S., 1973. Exchangeability of bicarbonate specifically bound to transferrin, *Biochemistry* 12:3679–3684.

Awai, M., Chipman, B., and Brown, E. B., 1975a. *In vivo* evidence for the functional heterogeneity of transferrin-bound iron. I. Studies in normal rats, *J. Lab. Clin. Med.* 85:769–784.

Awai, M., Chipman, B., and Brown, E. B., 1975b. *In vivo* evidence for the functional heterogeneity of transferrin-bound iron. II. Studies in pregnant rats, *J. Lab. Clin. Med.* 85:785–796.

Baker, E., and Morgan, E. H., 1969. The kinetics of the interaction between rabbit transferrin and reticulocytes, *Biochemistry* 8:1133–1141.

Bezkorovainy, A., 1966. Comparative study of metal-free, iron-saturated, and sialic acid-free transferrins, *Biochim. Biophys. Acta* 127:535–537.

Black, C., Glass, J., Nunez, M. T., and Robinson, S. H., 1979. Transferrin binding and iron transport in iron-deficient and iron replete rat reticulocytes, *J. Lab. Clin. Med.* 93:645–651.

Borochov, H., and Shinitzky, M., 1976. Vertical displacement of membrane proteins mediated by changes in microviscosity, *Proc. Natl. Acad. Sci. U.S.A.* 73:4526–4530.

Borova, J., Ponka, P., and Neuwirt, J., 1973. Study of intracellular iron distribution in rabbit reticulocytes with normal and inhibited heme synthesis, *Biochim. Biophys. Acta* 320:143–156.

Brock, J. H., and Esparza, I., 1979. Failure of reticulocytes to take up iron from lactoferrin saturated by various methods, *Br. J. Haematol.* 42:481–483.

Brown, E. B., Okada, S., Awai, M., and Chipman, B., 1975. *In vivo* evidence for the functional heterogeneity of transferrin-bound iron. III. Studies of transferrin at high and low iron saturation, *J. Lab. Clin. Med.*, 85:576–585.

Carver, F. J., and Frieden, E., 1978. Factors affecting the adenosine triphosphate induced release of iron from transferrin. *Biochemistry* 17:167–172.

Christensen, A. C., Huebers, H., and Finch, C., 1978. Effect of transferrin saturation on iron delivery in rats, *Am. J. Physiol.* 235:R18–R22.

Edwards, S. A., and Fielding, J., 1971. Studies of the effect of sulfhydryl and other inhibitors on reticulocyte uptake of doubly-labelled transferrin, *Br. J. Haematol.* 20:405–416.

Egyed, A., 1973. The significance of transferrin-bound bicarbonate in the uptake of iron by reticulocytes, *Biochim. Biophys. Acta* 304:805–813.

Eldor, A., Manny, N., and Izak, G., 1970. The effect of transferrin-free serum on the utilization of iron by rabbit reticulocytes, *Blood* 36:233–238.

Faulk, W. P., and Galbraith, G. M. P., 1979. Trophoblast transferrin and transferrin receptors in the host–parasite relationship in human pregnancy, *Proc. R. Soc. London Ser. B* 204:83–97.

Fielding, J., and Speyer, B. E., 1974. Iron transport intermediates in human reticulocytes and membrane binding site of iron–transferrin, *Biochim. Biophys. Acta* 363:387–396.

Fletcher, J., 1969. Variation in the availability of transferrin-bound iron for uptake by immature red cells, *Clin. Sci.* 37:273–297.

Fletcher, J., and Huehns, E. R., 1967. Significance of the binding of iron by transferrin, *Nature (London)* 215:584–586.

Fletcher, J., and Huehns, E. R., 1968. Function of transferrin, *Nature (London)* 218:1211–1214.

Garrett, N. E., Garrett, R. J. B., and Archdeacon, J. W., 1973. Solubilization and chromatography of iron-binding compounds from reticulocyte stroma, *Biochem. Biophys. Res. Commun.* 52:466–474.

Garrick, L. M., Edwards, J. A., and Hoke, J. E., 1978. The effect of hemin on globin synthesis and iron uptake by reticulocytes of the Belgrade rat, *FEBS Lett.* 93:109–114.

Hahn, D., 1973. Functional behaviour of transferrin, *Eur. J. Biochem.* 34:311–316.

Hahn, D., Baviera, B., and Ganzoni, A. M., 1975. Functional heterogeneity of the transport iron compartment, *Acta Haematol.* 53:285–291.

Harris, D. C., 1977. Functional equivalence of iron bound to human transferrin at low pH or high pH, *Biochim. Biophys. Acta* 496:563–565.

Harris, D. C., and Aisen, P., 1975a. Functional equivalence of the two iron-binding sites of human transferrin, *Nature (London)* 257:821–823.

Harris, D. C., and Aisen, P., 1975b. Iron-donating properties of transferrin, *Biochemistry* 14:262–268.

Hemmaplardh, D., and Morgan, E. H., 1977. The role of endocytosis in transferrin uptake by reticulocyte and bone marrow cells, *Br. J. Haematol.* 36:85–96.

Hemmaplardh, D., Kailis, S. G., and Morgan, E. H., 1974. The effects of inhibitors of microtubule and microfilament function on transferrin and iron uptake by rabbit reticulocytes and bone marrow, *Br. J. Haematol.* 28:53–65.

Hu, H.-Y. Y., and Aisen, P., 1978. Molecular characteristics of the transferrin–receptor complex of the rabbit reticulocyte, *J. Supramol. Struct.* 8:349–360.

Huebers, H., Huebers, E., Csiba, E., and Finch, C. A., 1978. Iron uptake from rat plasma transferrin by rat reticulocytes, *J. Clin. Invest.* 62:944–951.

Jandl, J. H., and Katz, J. H., 1963. The plasma-to-cell cycle of transferrin, *J. Clin. Invest.* 42:314–326.

Jandl, J. H., Inman, J. H., Simmons, R. L., and Allen, D. W., 1959. Transfer of iron from serum iron-binding protein to human reticulocytes, *J. Clin. Invest.* 38:161–185.

Kailis, S. G., and Morgan, E. H., 1974. Transferrin and iron uptake by rabbit bone marrow cells *in vitro*, *Br. J. Haematol.* 28:37–52.

Karibian, D., and London, I. M., 1965. Control of heme synthesis by feedback inhibition, *Biochem. Biophys. Res. Commun.* 18:243–249.

Katz, J. H., 1965. The delivery of iron into the immature red cell: A critical review, *Ser. Haematol.* 6:15–29.

Konopka, L., and Hoffbrand, A. V., 1979. Haem synthesis in sideroblastic anemia, *Br. J. Haematol.* 42:73–83.

Konopka, K., and Szotor, M., 1972. Determination of iron in the acid-soluble fraction of human erythrocytes, *Acta Haematol.* 47:157–163.

Kornfeld, S., 1968. The effects of structural modifications on the biologic activity of human transferrin, *Biochemistry* 7:945–954.

Kornfeld, S., 1969. The effect of metal attachment to human apotransferrin on its binding to reticulocytes, *Biochim. Biophys. Acta* 194:25–33.

Lane, R. S., 1971. Binding of transferrin and metal ions by suspensions of reticulocyte-rich rabbit blood, *Biochim. Biophys. Acta* 243:193–197.

Lane, R. S., 1972. Transferrin–reticulocyte binding: Evidence for the functional importance of transferrin conformation, *Br. J. Haematol.* 22:309–317.

Larrick, J. W., and Cresswell, P., 1979. Transferrin receptors on human B and T lymphoblastoid cell lines, *Biochim. Biophys. Acta* 583:483–490.

Leibman, A., and Aisen, P., 1977. Transferrin receptor of the rabbit reticulocyte, *Biochemistry* 16:1268–1272.

Light, N. D., 1977. The isolation and partial characterization of transferrin binding components of the rabbit reticulocyte plasma membrane, *Biochim. Biophys. Acta* 495:46–57.

Light, N. D., 1978. Further studies on the rabbit erythroid cell plasma membrane transferrin receptor, *Biochem. Biophys. Res. Commun.* 81:261–267.

Loh, T. T., Yeung, Y. G., and Yeung, D., 1977. Transferrin and iron uptake by rabbit reticulocytes, *Biochim. Biophys. Acta* 471:118–124.

Martinez-Medellin, J., and Benavides, L., 1979. The rate-limiting step in the reticulocyte uptake of transferrin and transferrin iron, *Biochim. Biophys. Acta* 584:84–93.

Martinez-Medellin, J., and Schulman, H. M., 1972. The kinetics of iron and transferrin incorporation into rabbit erythroid cells and the nature of stromal-bound iron, *Biochim. Biophys. Acta* 264:272–284.

Martinez-Medellin, J., Schulman, H. M., DeMiguel, E., and Benavides, L., 1977. New evidence for the internalization of functional transferrin in rabbit reticulocytes, *in Proteins of Iron Metabolism,* E. B. Brown, P. Aisen, J. Fielding, and R. R. Crichton (eds.), Grune & Stratton, New York, pp. 305–310.

Morgan, E. H., 1964. The interaction between rabbit, human, and rat transferrin and reticulocytes, *Br. J. Haematol.* 10:442–452.

Morgan, E. H., 1971. A study of iron transfer from rabbit transferrin to reticulocytes using synthetic chelating agents, *Biochim. Biophys. Acta* 244:103–116.

Morgan, E. H., 1976. Failure of a cell-free system from rabbit reticulocytes to remove iron from transferrin, *Biochem. J.* 158:489–491.

Morgan, E. H., 1977. Iron exchange between transferrin molecules mediated by phosphate compounds and other cell metabolites, *Biochim. Biophys. Acta* 499:169–177.

Morgan, E. H., and Appleton, T. C., 1969. Autoradiographic localization of ^{125}I-labelled transferrin in rabbit reticulocytes, *Nature (London)* 223:1371–1372.

Morgan, E. H., and Baker, E., 1969. The effect of metabolic inhibitors on transferrin and iron uptake and transferrin release from reticulocytes, *Biochim. Biophys. Acta* 184:442–454.

Morgan, E. H., Huebers, H., and Finch, C. A., 1978. Differences between the binding sites for iron binding and release in human and rat transferrin, *Blood* 52:1219–1228.

Muller, C., and Shinitzky, M., 1979. Modulation of transferrin receptors in bone marrow cells by changes in lipid fluidity, *Br. J. Haematol.* 42:355–362.

Neuwirt, J., Borova, J., and Ponka, P., 1975. Intracellular iron kinetics in erythroid cells, *in Proteins of Iron Storage and Transport in Biochemistry and Medicine,* R. R. Crichton (ed.), North-Holland, Amsterdam, pp. 161–166.

Nunez, M., Fischer, S., Glass, J., and Lavidor, L., 1977. The crosslinking of ^{125}I-labelled transferrin to rabbit reticulocytes, *Biochim. Biophys. Acta* 490:87–93.

Nunez, M. T., Glass, J., and Robinson, S. H., 1978. Mobilization of iron from the plasma membrane of the murine reticulocytes. The role of ferritin, *Biochim. Biophys. Acta* 509:170–180.

Okada, S., Chipman, B., and Brown, E. B., 1977. *In vivo* evidence for the functional heterogeneity of transferrin-bound iron. IV. Selective uptake by erythroid precursors of radioiron from portal vein plasma transferrin during intestinal iron absorption, *J. Lab. Clin. Med.* 89:51–64.

Okada, S., Rossmann, M. D., and Brown, E. B., 1978. The effect of acid pH and citrate on the release and exchange of iron on rat transferrin, *Biochim. Biophys. Acta* 543:72–81.

Okada, S., Jarvis, B., and Brown, E. B., 1979. *In vivo* evidence for the functional heterogeneity of transferrin-bound iron. V. Isotransferrin: An explanation of the Fletcher–Huehns phenomenon in the rat, *J. Lab. Clin. Med.* 93:189–198.

Ponka, P., and Neuwirt, J., 1972. The effect of plasma and transferrin on the hemin inhibition of iron uptake by reticulocytes, *Experientia* 28:189–190.

Ponka, P., and Neuwirt, J., 1974. Annotation: Haem synthesis and iron uptake by reticulocytes, *Br. J. Haematol.* 28:1–5.

Ponka, P., Neuwirt, J., and Borova, J., 1974. The role of heme in the release of iron from transferrin in reticulocytes, *Enzyme* 17:91–99.

Ponka, P., Neuwirt, J., Borova, J., and Fuchs, O., 1977. The role of mitochondria in the control of iron delivery to hemoglobin molecules, *in Proteins of Iron Metabolism*, E. B. Brown, P. Aisen, J. Fielding, and R. R. Crichton (eds.), Grune & Stratton, New York, pp. 319–326.

Ponka, P., Borova, J., Neuwirt, J., and Fuchs, O., 1979a. Mobilization of iron from reticulocytes, *FEBS Lett.* 97:317–321.

Ponka, P., Borova, J., Neuwirt, J., Fuchs, O., and Necas, E., 1979b. A study of intracellular iron metabolism using pyridoxal isonicotinyl hydrazone and other synthetic chelating agents, *Biochim. Biophys. Acta* 586:278–297.

Schulman, H. M., Martinez-Medellin, J., and Sidloi, R., 1974. The reticulocyte-mediated release of iron and bicarbonate from transferrin: Effect of metabolic inhibitors, *Biochim. Biophys. Acta* 343:529–534.

Skarberg, K., Eng, M., Huebers, H., Marsaglia, G., and Finch, C., 1978. Plasma radioiron kinetics in man: Explanation for the effect of plasma iron concentration, *Proc. Natl. Acad. Sci. U.S.A.* 75:1559–1561.

Sly, D. A., Grohlich, D., and Bezkorovainy, A., 1975a. Transferrin receptors from reticulocyte membranes and cytosol, *in Proteins of Iron Storage and Transport in Biochemistry and Medicine*, R. R. Crichton (ed.), North-Holland, Amsterdam, pp. 141–145.

Sly, D. A., Grohlich, D., and Bezkorovainy, A., 1975b. Transferrin in the reticulocyte cytosol, *Biochim. Biophys. Acta* 385:36–40.

Sly, D. A., Grohlich, D., and Bezkorovainy, A., 1978. Transferrin receptor from rabbit reticulocyte membranes, *in Cell Surface Carbohydrate Chemistry*, R. E. Harmon (ed.), Academic Press, New York, pp. 255–268.

Speyer, B. E. and Fielding, J., 1974. Chromatographic fractionation of human reticulocytes after uptake of doubly labelled (^{59}Fe, ^{125}I) transferrin, *Biochim. Biophys. Acta* 332:192–200.

Speyer, B. E., and Fielding, J., 1979. Ferritin as a cytosol iron transport intermediate in human reticulocytes, *Br. J. Haematol.* 42:255–267.

Sullivan, A. L., and Weintraub, L. R., 1978. Identification of ^{125}I-labeled rat reticulocyte membrane proteins with affinity for transferrin, *Blood* 52:436–446.

Sullivan, A. L., Grasso, J. A., and Weintraub, L. R., 1976. Micropinocytosis of transferrin by developing red cells: An electron microscopic study utilizing ferritin-conjugated transferrin and ferritin-conjugated antibodies to transferrin, *Blood* 47:133–143.

Ulvik, R., and Romslo, I., 1978. Studies on the utilization of ferritin iron in the ferrochelatase

reaction of isolated rat liver mitochondria, *Biochim. Biophys. Acta* 541:251–262.
Van Bockxmeer, F. M., and Morgan, E. H., 1977. Identification of transferrin receptors in reticulocytes, *Biochim. Biophys. Acta* 468:437–450.
Van Bockxmeer, F., Hemmaplardh, D., and Morgan, E. H., 1975. Studies on the binding of transferrin to cell membrane receptors, *in Proteins of Iron Storage and Transport in Biochemistry and Medicine,* R. R. Crichton (ed.), North-Holland, Amsterdam, pp. 111–119.
Van Bockxmeer, F. M., Yates, G. K., and Morgan, E. H., 1978. Interaction of transferrin with solubilized receptors from reticulocytes, *Eur. J. Biochem.* 92:147–154.
VanderHeul, C., Kroos, M. J., and Van Eijk, H. G., 1978. Binding sites of iron transferrin on rat reticulocytes. Inhibition by specific antibodies, *Biochim. Biophys. Acta* 511:430–441.
Verhoef, N. J., and Noordeloos, P. J., 1977. Binding of transferrin and uptake of iron by rat erythroid cells *in vitro, Clin. Sci. Mol. Med.* 52:87–96.
Verhoef, N. J., Kottenhagen, M. J., Mulder, H. J. M., Noordeloos, P. J., and Leijnse, B., 1978. Functional heterogeneity of transferrin-bound iron, *Acta Haematol.* 60:210–226.
Walsh, R. J., Thomas, E. D., Chow, S. K., Fluharty, R. G., and Finch, C. A., 1949. Iron metabolism. Heme synthesis *in vitro* by immature erythrocytes, *Science* 110:396–398.
Williams, S. C., and Woodworth, R. C., 1973. The interaction of iron–conalbumin (anion) complexes with chick embryo red blood cells, *J. Biol. Chem.* 248:5848–5853.
Witt, D. P., and Woodworth, R. C., 1975. Interaction of affinity labeled conalbumin with reticulocyte membranes, *in Proteins of Iron Storage and Transport in Biochemistry and Medicine,* R. R. Crichton (ed.), North-Holland, Amsterdam, pp. 133–140.
Witt, D. P., and Woodworth, R. C., 1978. Identification of the transferrin receptor of the rabbit reticulocyte, *Biochemistry* 17:3913–3917.
Workman, E. F., and Bates, G. W., 1974. Mobilization of iron from reticulocyte ghosts by cytoplasmic agents, *Biochem. Biophys. Res. Commun.* 58:787–794.
Zapolski, E. J., and Princiotto, J. V., 1976. Failure of rabbit reticulocytes to incorporate conalbumin or lactoferrin iron, *Biochim. Biophys. Acta* 421:80–86.

Microbial Iron Uptake and the Antimicrobial Properties of the Transferrins

7.1 Introduction

Most if not all microorganisms including the fungi require ferric iron for growth, where iron concentration in the medium must be between 0.4 and 4×10^{-6} *M* (Weinberg, 1978). Yet, because the solubility product of ferric hydroxide, which is present at pH values of about 6 and above, is near 4×10^{-36}, ferric iron concentrations in the growth medium are near 10^{-15} *M*. Moreover, many biological fluids contain one or several proteins of the transferrin class, which have association constants of around 10^{36} with respect to iron. Indeed, it has been said that the frequency of iron dissociation from human serum transferrin in the absence of small chelating agents or a specific biological system designed to abstract iron from this protein is one iron atom per 10,000 years (see Chapter 4). To overcome such obstacles, microorganisms have evolved the ability to synthesize small-molecular-weight chelators, which seek out iron in the environment and return it to the microorganism. Such chelators are termed siderochromes or, alternately, siderophores. It has been argued that the latter is the preferred term (Neilands, 1977), and we shall use it in this chapter.

The chemistry and biology of the siderophores are probably the best understood and most intensively investigated aspects of microbial iron metabolism. A number of recent reviews have been written on the subject (Neilands, 1973, 1974, 1977; Keller-Schierlein, 1977; Llinas, 1973; Snow, 1970; Raymond and Carrano, 1979).

It should be mentioned that there are numerous small-molecular-weight compounds with iron-chelating activity that have been isolated from microorganisms. A number of these have been shown to function as siderophores under physiological conditions, whereas the exact physiological function of several others as yet remains uncertain. As might be expected, microorganisms prefer to incorporate iron from siderophores produced by the homologous strain of the organism; nevertheless, siderophores produced by one strain or species may serve as iron donors for another strain or species as well (e.g., Wiebe and Winkelmann, 1975).

7.2 Structure of Siderophores

Iron in the siderophores is generally chelated in the ferric state by oxygen ligands, forming a *hexadentate* octahedral complex. From the structural point of view, two types of siderophores exist: the catechol-like compounds, which are colored red to purple in the iron-saturated state, and the reddish-brown hydroxamate-like compounds. There are, however, numerous siderophores that do not neatly fit into either of these classifications. It is said that with few exceptions the hydroxamate-like siderophores occur in fungi, yeasts, and bacteria, whereas the catechol-like compounds occur in bacteria only.

7.2.1 Catechol-like Siderophores

There is a limited number of siderophores that can be classified as being strictly of the catechol type. The best known member of this group is enterochelin (enterobactin) isolated from *Aerobacter aerogenes, Salmonella typhimurium LT2*, and *E. coli K-12* cultures. Its yield from *E. coli K-12* amounts to some 15 mg/liter of culture medium (O'Brien and Gibson, 1970). Enterochelin may be viewed as being the siderophore of the entire enteric bacterial class, though it is not restricted to these microorganisms.

Upon hydrolysis with HCl, enterochelin yields serine and 2,3-dihydroxybenzoic acid. The serine residues are connected with each other through ester linkages, as shown in Figure 7-1. Electron paramagnetic resonance studies of the iron–enterochelin complex have shown the presence of a g value of 4.3, which is characteristic of a high-spin ferric iron with a rhombic symmetry. Upon the binding of iron, each catechol–OH group loses two protons, so that a ferri enterochelin molecule has a net charge of -3. Enterochelin is active as a growth promoter in the cultures of various mutant strains of *Salmonella typhimurium* that have lost their ability to biosynthesize the siderophore, so that there is no question as to the *bona fide* nature of enterochelin as an iron carrier (Pollack and Neilands, 1970).

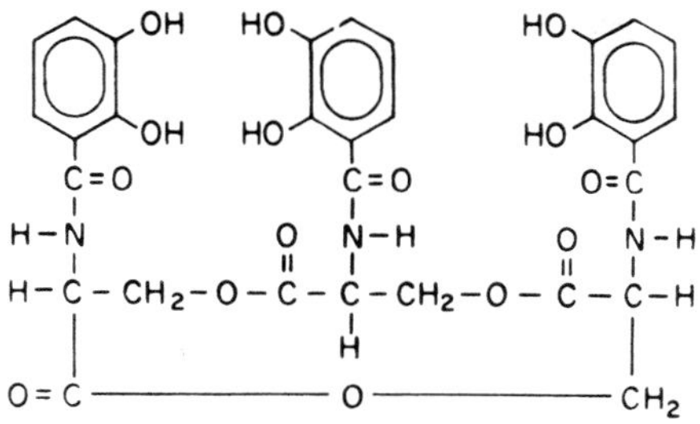

Figure 7-1. Structure of enterochelin. From Pollack and Neilands (1970).

Other compounds of the catechol type that have been isolated from various microorganisms are 2,3-dihydroxybenzoyl glycine (itoic acid), which happens to be the first catechol-like siderophore to be isolated (in 1958); 2,3-dihydroxy-*N*-benzoyl-L-serine; bis-(2,3-dihydroxybenzoyl)-L-lysine; and the siderophores containing spermidine. Their structures and other pertinent information are shown in Table 7-1.

Of the various catechol-like siderophores described, only enterochelin and agrobactin (Table 7-1) can bind one atom of iron for each siderophore molecule (coordination number of iron in these cases is 6). Each iron atom will thus require three 2,3-dihydroxybenzoyl glycine molecules, whereas three bis-(2,3-dihydroxybenzoyl)-L-lysine moleules would be required to sequester two iron atoms.

Enterochelin is capable of binding ferrous iron. At pH 7, the enterochelin-bound iron exists in the ferric state and possesses a wine color in both water and methanol. Its association constant is believed to be in the range of 10^{35}–10^{45}. If the pH is lowered to 4, the ferric iron is reduced by the catechol residues, and the complex acquires a blue color if in methanol. Upon the addition of water, the blue color changes to pale purple, indicating that water competes with the catechol residues for iron and that the ferrous iron–enterochelin complex is much less stable than that involving ferric iron. Further, titration studies indicate that four protons become associated with the iron-binding site at pH 4 and are probably attached to the cathechol groups. This results in an electrically neutral complex, where the six negative charges on the oxygen atoms are exactly balanced out by the four protons and the two positive charges on the ferrous iron atom. It is believed that the ferrous iron–enterochelin complex exists in the vicinity of the bacterial cell membranes, whose environment, because of the various proton fluxes, may be quite acidic. Being neutral, the complex is much easier to transport across the membrane than would be the triply negatively charged ferric–enterochelin complex. The ferrous iron–enterochelin complex oxidizes spontaneously if the pH is raised to 7 (Hider *et al.*, 1979).

Table 7-1. Properties of Some Catechol-like Siderophores[a]

Structure	Name of compound	Components	Source	Reference
OH; OH; C=O; HN—CH_2—COOH	Dihydroxybenzoyl glycine (DBG); also itoic acid	Dihydroxybenzoic acid, glycine	*B. subtilis* (50 mg/liter of culture medium)	Neilands, 1973
OH; OH; C=O; CH_2OH; N—C—COOH; H	2,3-Dihydroxybenzoyl serine (DBS)	Dihydroxybenzoic acid, serine	*E. coli K-12, Aerobacter aerogenes, Salmonella typhimurium*	Neilands, 1973
OH; OH; C=O; H; H—N—C—COOH; $HOCHCH_3$	2,3-Dihydroxybenzoyl threonine	Dihydroxybenzoic acid, threonine	*Klebsiella oxytoca*	Neilands, 1973

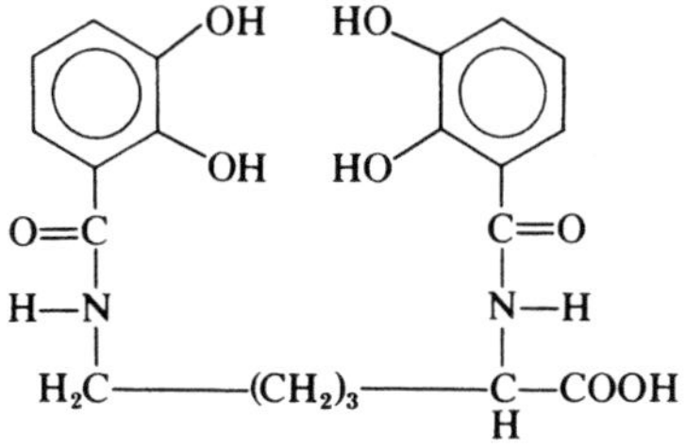 OH HO OH HO O=C C=O H—N N—H H_2C—$(CH_2)_3$—C—COOH H	Bis-(2,3-dihydroxybenzoyl)-L-lysine	Dihydroxybenzoic acid (2 moles), lysine	*Azobacter vinlandii type 0* (90 mg/liter of culture medium)	Neilands, 1973
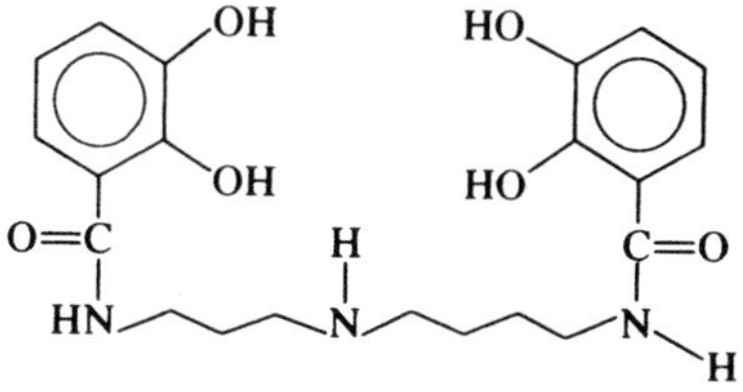 OH HO OH HO O=C H C=O HN N N H	*N′N*[8]-bis-(2,3-dihydroxybenzoyl) spermidine	Dihydroxybenzoic acid (2 moles), spermidine	*Micrococcus denitrificans*	Tait, 1975
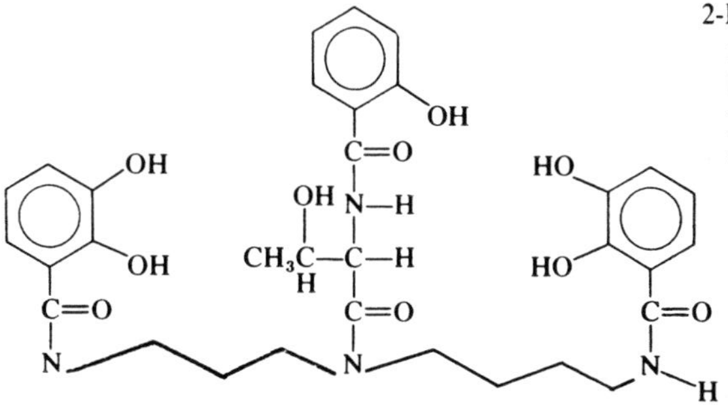 OH C=O OH HO OH N—H CH_3C—C—H H OH HO C=O C=O C=O N N N H	2-Hydroxybenzoyl-*N*-L-threonyl-*N*[4]-[*N′N*[8]-bis-(dihydroxybenzoyl)] spermidine	Dihydroxybenzoic acid (2 moles), salicylic acid, threonine, spermidine	*Micrococcus denitrificans*	Tait, 1975

continued overleaf

Table 7-1. (Continued)

Structure	Name of compound	Components	Source	Reference
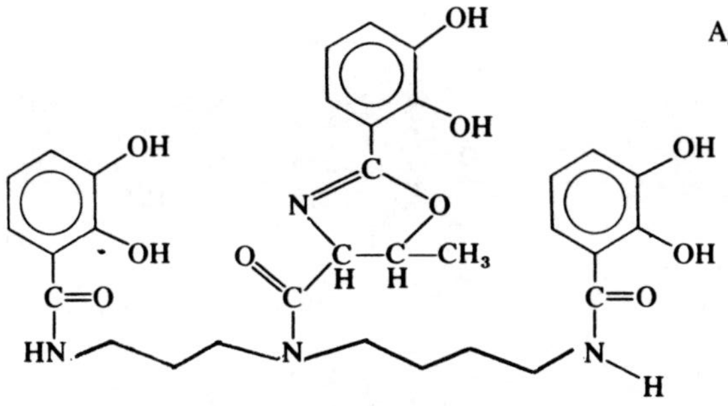 OH, OH, C=O, HN, N, OH, OH, C, N, O, O, C, C—C—CH_3, H H, OH, OH, C=O, N, H	Agrobactin	Dihydroxybenzoic acid (3 moles), threonine, spermidine	*Agrobacterium tumefaciens* (5–10 mg/liter of culture medium)	Ong *et al.*, 1979
Structure unknown but is similar to that of enterochelin	—	—	*Vibrio cholerae*	Payne and Finkelstein, 1978a

[a] Structures drawn without regard to stereochemistry.

7.2.2 Hydroxamate-like Siderophores

Siderophores derived from various types of hydroxamic acids have been known to exist since the beginning of the twentieth century, though their true physiological function was not established until the late 1940s and the 1950s. Structurally, these compounds are characterized by the presence of hydroxyl groups on the N atoms, as, e.g., in

$$\overset{|}{\text{—N—OH}}$$

where the —OH group serves as an iron ligand and loses its proton upon the binding of iron. The hydroxamate-like siderophores are found in various gram-positive and acid-fast bacilli as well as in fungi, yeasts, and other microorganisms. Several have been used in iron overload therapy.

Several diverse kinds of hydroxamate-like siderophores are known to exist. These may be classified as follows.

The Ferrioxamine Family. These compounds, characterized by the presence of α-amino-ω-hydroxyaminoalkane, may be either linear or cyclic and are found in the cultures of all actinomycetes. Desferal (deferroxamine B, desferrioxamine B) (*Streptomyces pylosus*) was the first compound of this type whose structure was elucidated (1960). Its structure in the iron-saturated state (termed ferrioxamine B) is shown in Figure 7-2, as is its cyclic analogue, ferrioxamine A. The antibiotic ferrimycin A is related to the ferrioxamine group of compounds (see Figure 7-3).

The Ferrichrome Family. The ferrichromes contain amino acids and other substituents. The amino acids, unlike those of enterochelin, are linked in peptide linkages. An amino acid common to all ferrichromes is

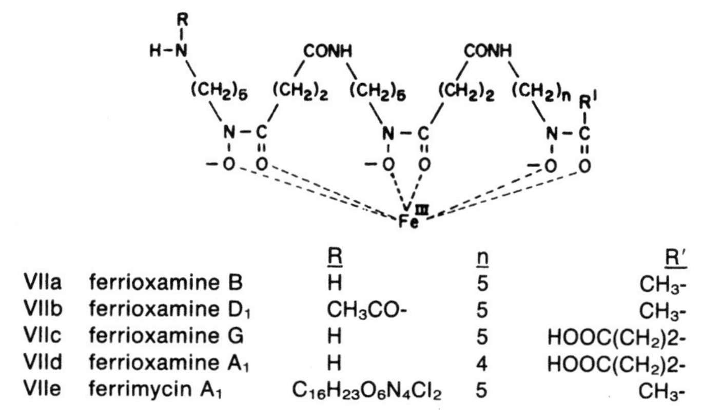

		R	n	R′
VIIa	ferrioxamine B	H	5	CH_3-
VIIb	ferrioxamine D_1	CH_3CO-	5	CH_3-
VIIc	ferrioxamine G	H	5	$HOOC(CH_2)2$-
VIId	ferrioxamine A_1	H	4	$HOOC(CH_2)2$-
VIIe	ferrimycin A_1	$C_{16}H_{23}O_6N_4Cl_2$	5	CH_3-

Figure 7-2. Structure of the ferrioxamine group of siderophores. From Neilands (1977).

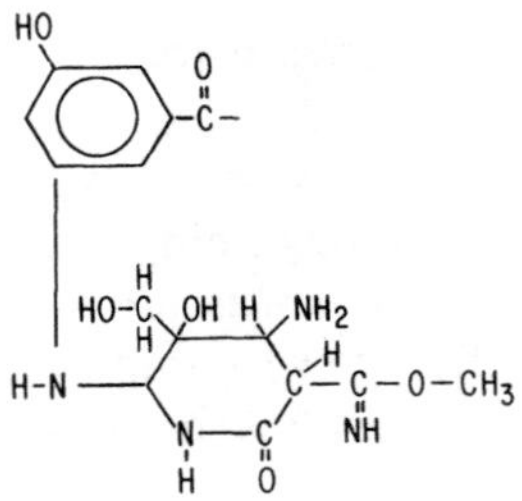

Figure 7-3. Structure of the R substituent (see Figure 7-2) found in the antibiotic ferrimycin A_1. R′ is —CH_3 and n is 5.

L-ornithine. Other substituents include amino acids such as glycine, alanine, and serine; acetic acid; or *trans*-β-methylglutaconic acid. The general structure of a ferrichrome is shown in Figure 7-4, which represents the structure both schematically and in three dimensions as determined by X-ray diffraction studies. The amino acids are arranged in an antiparallel β-pleated-sheet fashion (Winkelmann, 1979). The ferrichromes occur in the ascomycetes, basidiomycetes, and *Fungi imperfecti*, such as the *Aspergillus, Penicillium, Neurospora*, and *Ustilago* species.

An interesting derivative of the ferrichromes is a group of antibiotics called the albomycins (griseins), which have pyrimidine-containing substituents in the R position of Figure 7-4. For instance, albomycin E has the following substituent attached to a seryl residue:

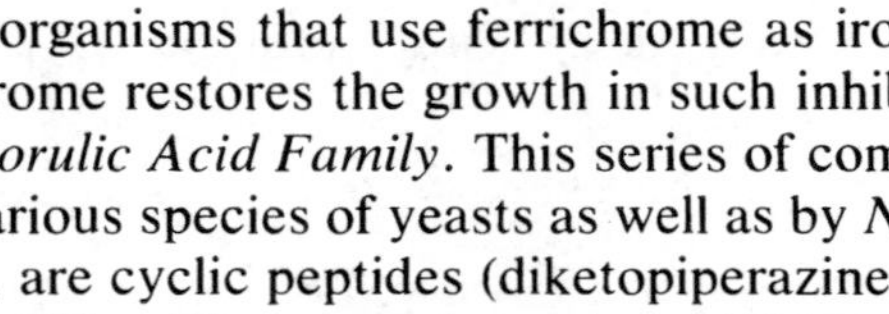

$$H{-}N{=}\,(\text{ring: } N{-}CH_3,\ C{=}O)\,N{-}SO_2{-}O{-}\text{serine} \quad (7\text{-}1)$$

Albomycin is produced in *Actinomyces griseus* cultures and inhibits the growth of microorganisms that use ferrichrome as iron scavengers. Addition of ferrichrome restores the growth in such inhibited cultures.

The Rhodotorulic Acid Family. This series of compounds, produced principally by various species of yeasts as well as by *Neurospora* and the penicillin molds, are cyclic peptides (diketopiperazines) of substituted δ-*N*-hydroxy-L-ornithine. The general structure of this group of compounds is given in Figure 7-5, where R is acetyl in rhodotorulic acid. The yeast *Rhodotorula pilimanae* can produce as much as 10 g of rhodotorulic acid per liter of culture medium. Rhodotorulic acid possesses four iron ligands per molecule, so that three rhodotorulic acid molecules would be required to bind two iron atoms. Since two ligands per rhodotorulic acid molecule would lose their protons upon the binding of iron, such a complex would

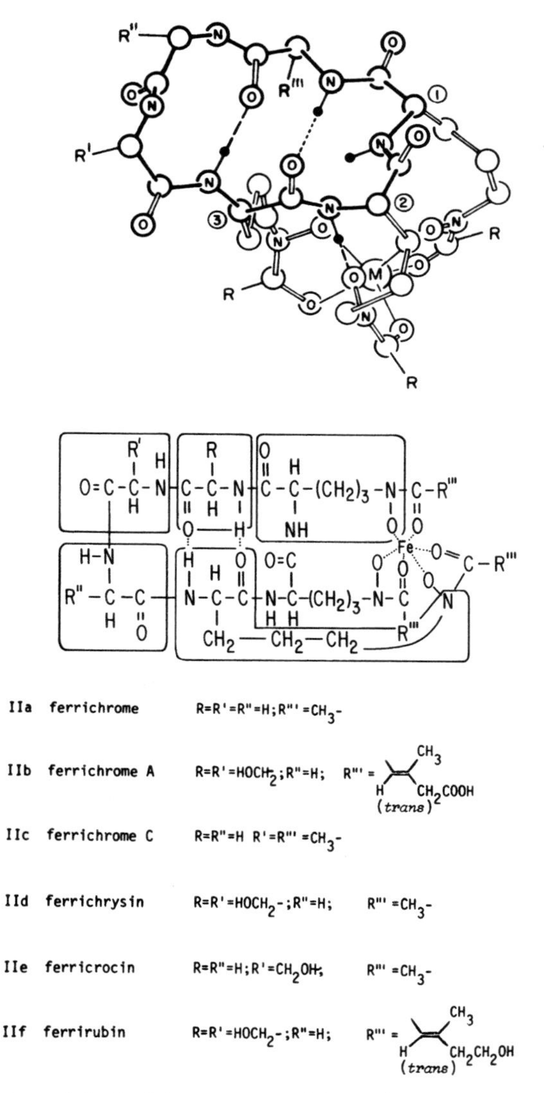

Figure 7-4. Structure of the ferrichrome group of siderophores. (a) Three-dimensional view of the molecule (from Neilands, 1975). (b) Two-dimensional representation of the molecule(s) without regard to stereochemistry or bond lengths. Individual amino acid residues are boxed in.

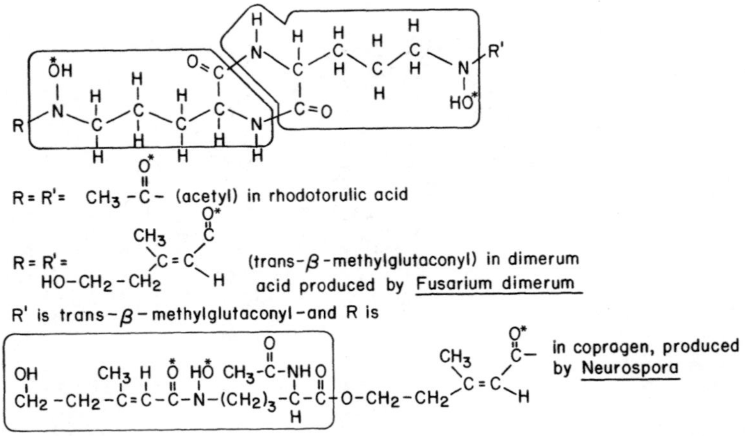

R = R' = $CH_3-\overset{O^*}{\overset{\|}{C}}-$ (acetyl) in rhodotorulic acid

R = R' = $HO-CH_2-CH_2(CH_3)C=CH-C(=O^*)-$ (trans-β-methylglutaconyl) in dimerum acid produced by Fusarium dimerum

R' is trans-β-methylglutaconyl-and R is $HO-CH_2-CH_2-C(CH_3)=CH-C(=O^*)-N(O^*H)-(CH_2)_3-CH(NH-C(=O)-CH_3)-C(=O)-O-CH_2-CH_2-C(CH_3)=CH-C(=O^*)-$ in coprogen, produced by Neurospora

Figure 7-5. Structures of the rhodotorulic acid group of siderophores. The iron-binding ligands are marked with an asterisk.

be expected to have a net charge of 0. The same situation would be present in dimerum acid. In coprogen, on the other hand, one molecule of the siderophore would bind one atom of iron, and the net charge of the complex would still remain at 0.

The Aerobactin Family. This group of substances can be characterized as containing citrate. Its general structure is given in Figure 7-6. The best characterized compounds of this group are aerobactin and schizokinen. The former is produced by *Aerobacter aerogenes* in yields approximating 1 g/liter of culture medium. Upon hydrolysis it yields ϵ-*N*-hydroxy-L-lysine, citrate, and acetate. Schizokinen, on the other hand, is produced by *Bacillus megaterium*, and it consists of citrate, acetate, and 2 moles of 1-amino-3-acethydroxamidopropane. Both aerobactin and schizokinen bind one iron atom per molecule of siderophore, where the iron-binding ligands are provided by the —COOH and —OH of citrate, the carboxyl groups of acetate, and the —OH groups of the two hydroxamate residues.

The Mycobactins. Mycobactins are fat-soluble siderophores associated with various species of the mycobacteria. They were recognized as growth factors for *Mycobacterium paratuberculosis* (*M. johnei*) as early as 1911. This organism does not have the capacity to synthesize mycobactins and depends on other microorganisms to provide this growth factor. The organism used as a source for such a *M. paratuberculosis* growth promoter has traditionally been *M. phlei*. The yields of mycobactins have ranged from 0.045% of dry cell weight in *M. Kansasii* to 2% in *M. fortuitum* (Snow, 1970).

Structurally, the mycobactins encompass both the elements of the hydroxamate-like and the phenol-like siderophores and must have fatty acid substituents. They have potentially five asymmetric centers. The general structure of the mycobactins is given in Figure 7-7, which gives the positions of the various substituents (R^1–R^5) and centers of asymmetry (a–f). Structure proofing of a mycobactin involves various hydrolytic degradative procedures. The best characterized mycobactin, called mycobactin P (from *M. phlei*), upon such degradation gives 6-methylsalicylic acid (or its degradation products, CO_2 and 5-hydroxytoluene or *m*-cresol), serine (from the oxazoline ring), 2 moles of ε-*N*-hydroxy-L-lysine, a β-hydroxy acid called 3-hydroxy-2-methylpentanoic acid, and *cis*-octadec-2-enoic acid ($HOOC—CH{=}CH—(CH_2)_{14}—CH_3$). In terms of the structure in Figure 7-7, R^1 is the *cis*-octadec-2-enoyl group, R^2 and R^5 are $—CH_3$, R^3 is —H, and R^4 is $—C_2H_5$. Mycobactin S, isolated from *M. smegmii* (another well-studied mycobacterium) differs from mycobactin P only in that its R^5 is —H and R^4 is $—CH_3$. Mycobactins P and S belong to the so-called P-type class of mycobactins, where the fatty acid side chain is in the R^1 position. In the M-type class of mycobactins, exemplified by mycobactins M and N, the fatty acid side chain is in position R^4, and R^1 is either $—CH_3$ (mycobactin M) or $—C_2H_5$ (mycobactin N).

Each mycobactin molecule can sequester one ferric iron atom, and since three protons are lost thereby, no increase or decrease of net charge takes place in the complex.

The Exochelins. A group of small-molecular-weight siderophores have recently been discovered in cultures of *Mycobacterium smegmatis*. They have an important function in the acquisition of iron by the mycobacteria and may interact with the mycobactins that are restricted to the cell membranes or the waxy coat associated with them. Their structures have not yet been elucidated, though they are believed to be cyclic

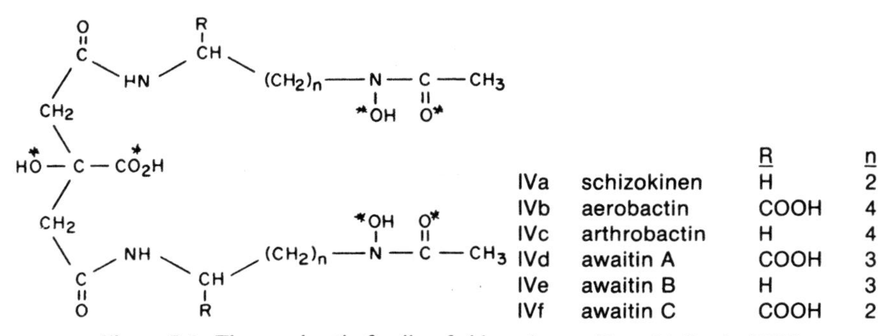

Figure 7-6. The aerobactin family of siderophores. From Neilands (1975).

MYCOBACTIN FAMILY

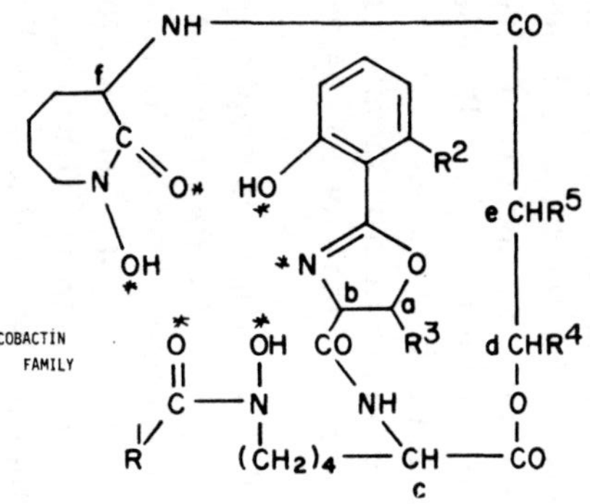

Structure	R^1	R^2	R^3	R^4	R^5	a	b	c	d	e	f
Va mycobactin A	13Δ	CH_3	H	CH_3	H						
Vb mycobactin F	17,15,13,11,9Δ	H	CH_3	CH_3	H	*threo*			S	( )	L
Vc mycobactin H	19,17Δ	CH_3	CH_3	CH_3	H	R	L	L	S	( )	L
Vd mycobactin M	1	H	CH_3	18,17,16,15	CH_3				RS	S	
Ve mycobactin N	2	H	CH_3	18,17,16,15	CH_3						
Vf mycobactin P	19,17,15*cis*Δ[1]n	CH_3	H	C_2H_5	CH_3	( )	L	L	S	R	L
Vg mycobactin R	19Δ	H	H	C_2H_5	CH_3	( )	L	L	R	S	L
Vh mycobactin S	19,17,15,13*cis*Δ	H	H	CH_3	H	( )	L	L	S	( )	L
Vi mycobactin T	20,19,18,17 20,19,18,17Δ	H	H	CH_3	H	( )			R	( )	L

Vj nocobactin. From *Nocardia ceteroides*. Resembles mycobactin M, except the oxazoline ring is replaced by an oxazole ring and the hydrocarbon side chain is shorter.

Figure 7-7. Structure of the mycobactin group of siderophores. The iron-binding ligands are marked with an asterisk. The numbers in the above table show the main types of side chains. Those of greatest abundance are underlined. Asymmetric centers are labeled a–f; lack of an asymmetric center is indicated by the symbol ( ). Blank spaces indicate that the configuration has not been determined. From Neilands (1975).

hexapeptides whose main component is ϵ-acetyl-ϵ-*N*-hydroxylysine (3 moles/mole) (Macham and Ratledge, 1975).

The Fusarinine Family. These compounds occur in the *Fusarium* fungal cultures and consist of fusarinine units (Figure 7-8). Fusarinine, in turn, is a complex of δ-*N*-hydroxy-L-ornithine and *cis*-β-methylglutaconic acid. The individual fusarinine units are linked via ester linkages, and the oligomer may be either linear or cyclic. Each fusarinine unit provides two iron ligands, so that each mole of fusarinine B ($n = 3$) binds 1 gram atom of iron, as does the cyclic fusarinine C (in the iron-saturated state it is called fusigen and is produced by *Fusarium cubense*).

There are numerous other hydroxamate-like substances produced by microorganisms whose role in iron metabolism is uncertain. Notable among them are the aspergillic acids, which were the first hydroxamate-like substances from microbial culture media to be structure-proofed (1947); thioformin (fluopsin) from *Pseudomones fluorescens*; actinonin; lipoxamycin; hadacidin; mycelianamide; and others. Most of these have moderate to potent antibiotic properties.

7.3 Chemical and Physical Properties of the Siderophores

The solubilities of the siderophores are variable. The catechol-like compounds are very sparingly soluble in aqueous solutions at neutrality, though the solubility increases upon the binding of iron. They are freely soluble in solvents with moderately high dielectric constants such as acetone and ethanol, and at pH 2 or below (iron is lost at such low pH values), they become soluble in typical organic solvents such as ethylacetate.

The hydroxamate-like siderophores are freely soluble in aqueous media both in the iron-free and iron-saturated states. An exception is presented by the mycobactins, which are insoluble in water but are soluble in organic solvents. This property is perhaps of importance in the process of transporting iron through the waxy coat possessed by most mycobacteria.

The iron–siderophore stability constants are extremely high, as they

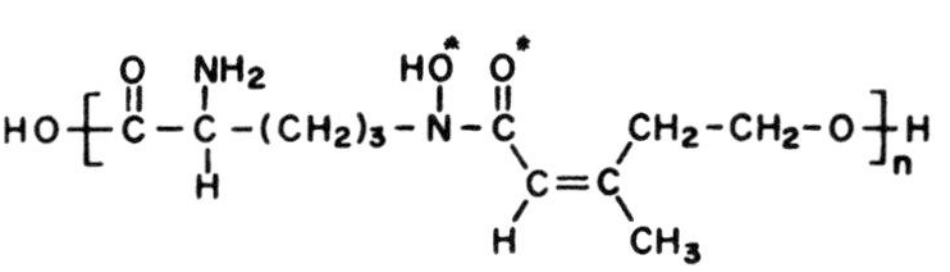

$HO-[C(=O)-CH(NH_2)-(CH_2)_3-N(OH^*)-C(=O^*)-CH=C(CH_3)-CH_2-CH_2-O]_n-H$

VIa	fusarinine	n = 1
VIb	fusarinine A	n = 2
VIc	fusarinine B	n = 3
VId	fusarinine C (=deferrifusigen), n = 3, cyclo[10]	

Figure 7-8. Structure of the fusarinine group of siderophores. The iron-binding ligands are marked with an asterisk. From Neilands (1975).

Table 7-2. Logarithms of Stability Constants (K_f) of Various Ferrisiderophores and of Free Ferric Iron Concentrations in the Presence of the Siderophores

Siderophore	log K_f	$-\log[Fe^{3+}]^a$	$-\log[Fe^{3+}]^b$
Enterochelin	52	33.3	31.4
Ferrioxamine B	30.6	28.6	26.6
Ferrichrome	29.1	27.2	25.2
Aerobactin	22.9	25.4	23.3
Rhodotorulic acid	31.2[c]	25.0	21.9
Transferrin	36	25.6	23.6

[a] When the total iron concentration is 10^{-6} *M* and the siderophore concentration is 10^{-3} *M* at pH 7.4.
[b] When the total iron concentration is 10^{-6} *M* and the siderophore concentration is 10^{-5} *M* at pH 7.4.
[c] For the dimer, per each iron atom.

must be to both maintain ferric iron in solution at neutral pH in view of the low K_s of $Fe(OH)_3$ and to successfully compete for iron with the transferrin class of proteins. These constants are summarized in Table 7-2, from which it is seen that enterochelin has the highest association constant yet discovered with respect to iron.

All siderophores have absorption maxima in the ultraviolet light range and in the visible range when coordinated with iron. The ferrichromes have absorption maxima in the range of 420–440 mm, whereas the mycobactins absorb in the region of 450 mm. In fact, the absorption of hydroxamate-like siderophores, especially aspergillic acid, in this range of the spectrum prompted the early workers in transferrin chemistry to propose the hydroxamic acid residues were the iron ligands in the transferrins.

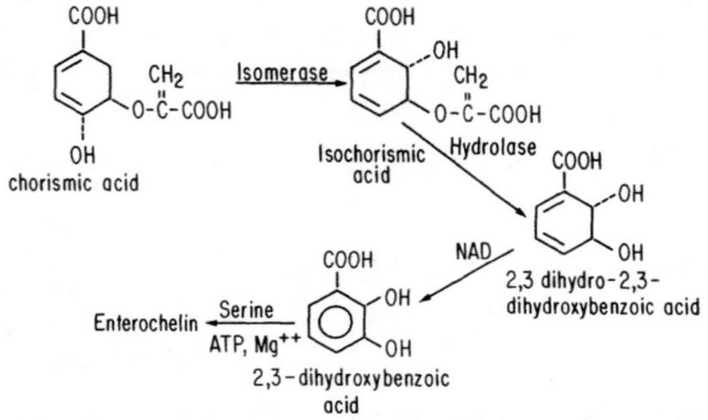

Figure 7-9. Pathway of enterochelin biosynthesis.

7.4 Metabolism and Biological Properties of the Siderophores

7.4.1 Biosynthesis of Siderophores

Biosynthesis of the catechol-like siderophores proceeds from chorismic acid, an intermediate in the biosynthesis of phenylalanine and tyrosine in microorganisms, as shown in Figure 7-9.

Biosynthetic pathways for the hydroxamate-like siderophores must be taken on a case-by-case basis due to the great variety of siderophores known. However, very little information on the pertinent pathways is available. It is believed that hydroxylation of the appropriate amino groups (e.g., the δ-amino group of ornithine) takes place enzymatically using the free amino acid or other compound and O_2 as substrates. The *N*-hydroxyamino acid or other hydroxamate thus biosynthesized can then form peptide linkages with other amino acids by a mechanism differing from that of protein biosynthesis and not involving the genetic code (Emery, 1974).

7.4.2 Mode of Iron Delivery to Microbial Cells by Siderophores

An important area of interest in siderophore physiology is the mode of iron transport into the bacterial cell. There are essentially three mechanisms of siderophore-mediated iron uptake, and they are illustrated in Figure 7-10. In the first case, exemplified by enterochelin and fusarinine, termed the *American approach* (Raymond and Carrano, 1979), the siderophore bearing the iron is internalized by the cell, and the iron is abstracted from the siderophore by hydrolyzing the latter. In case of enterochelin, a very specific esterase breaks up the seryl–serine linkages to release the iron. The siderophore is thus used once and discarded. The reason for this approach is apparently because in order to release the iron from the siderophore without destroying the latter, iron should be reduced to the ferrous state, as does happen with the second group of siderophores (Leong and Neilands, 1976). In the case of enterochelin-bound iron, however, the redox potential of the iron is extremely low (-0.7 and -0.986 V at pH 7 and 10, respectively), which is well beyond the capabilities of any biological redox system. The enterochelin hydrolysis product, dihydroxybenzoyl serine, is excreted into the medium. Although it is capable of binding iron in its own right, dihydroxybenzoyl serine does so much less efficiently than does enterochelin and becomes important as an iron carrier only in the absence of the latter.

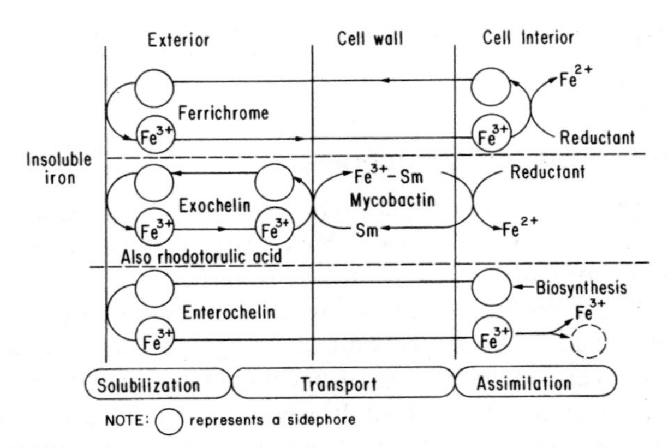

Figure 7-10. Mechanisms for the siderophore-bound iron transport into bacterial cells. Adapted from Raymond and Carrano (1979).

Before iron-laden enterochelin is internalized by the microbial cell, it must combine with a specific outer membrane receptor. This receptor is apparently identical to that of Colicin B-D in *E. coli K-12* (Pugsley and Reeves, 1977b) and is the product of the feu *B* gene. The receptor is a protein with a molecular weight of 75,000–90,000, and its biosynthesis is derepressed in iron deficiency states. It is interesting to note that enterobacilli that do not normally produce hydroxamate-like siderophores can nevertheless acquire iron from compounds such as ferrichrome. The outer membrane of the enterobacilli therefore possesses receptors for such hydroxamate-like siderophores (this is a product of the ton *A* gene) as well as for citrate (Neilands and Wayne, 1977). The ferrichrome receptor is apparently partially identical to that of bacteriophage T_1, T_5, and 80, since ferrichrome was able to protect the otherwise susceptible *E. Coli* cells against phage infection (Wayne and Neilands, 1975). Alternately, mutants that could not accept iron from ferrichromes and that were albomycin resistant were also resistant to bacteriophage ϕ80 infection.

The second mode of iron uptake from siderophores is exemplified by the ferrichromes, where the iron-laden siderophore is internalized, iron is removed therefrom by reduction to the ferrous state, and the siderophore is reexcreted into the medium. In the case of ferrichromes, the redox potential of the iron is near -0.4 V, which is within the range of biological redox systems. Bacterial cells are apparently sensitive to siderophore conformation, since *Neurospora crassa* and the *Aspergillus* refused to utilize iron bound to ferrichrome synthesized with D-ornithine instead of L-ornithine (Winkelmann, 1979).

The third mechanism of acquiring iron operates in mycobacteria and organisms utilizing rhodotorulic acid. In the mycobacteria, the water-soluble compounds used for iron sequestration are salicylates and the exochelins. The structures of exochelins have not yet been elucidated. The exochelins, when present in high concentrations in the medium, exchange iron with the mycobactins, which are located in the lipophilic portions of the cell surface. The mycobactins, in turn, translocate the iron into the cell interior, where it is reduced by an NAD-dependent oxidoreductase system and is thus removed from the siderophore (Macham *et al.*, 1975). A similar situation is believed to exist in case of *Rhodotorula pilimanae*, where the water-soluble siderophore rhodotorulic acid complexed with chromium was found not to enter the cell. It is postulated that the microorganism must possess a membrane-bound siderophore that would acquire the iron from rhodotorulic acid and would translocate the iron into the cell interior (Carrano and Raymond, 1978). And finally, the *Pseudomonas aeruginosa* organism has been shown to produce both water-soluble and water-insoluble siderophores, the latter being firmly associated with the cell but extractable with ethyl acetate. These have been called pyochelins B and pyochelins A, respectively (Liu and Shokrani, 1978). It is entirely possible that pyochelins B act in the same capacity as the exochelins, while pyochelins A function like the mycobactins.

A pyochelin from *P. aeruginosa* medium acted as a growth stimulant in culture media deficient in iron. It was extractable from the medium by ethyl acetate and had a minimum molecular weight of 400–600. It behaved as a phenol rather than a catechol and contained no hydroxamate groupings (Cox and Graham, 1979). The structures of any of the pyochelins are yet to be determined.

It should be noted that there are some microorganisms whose growth in an iron-deficient medium cannot be stimulated by the addition of any of the siderophores. One example is the *Yersinia pestis*, which will grow only if iron is provided in the form of hemin or inorganic iron with protoporphyrin IX. Other *Yersiniae*, however, have not lost their ability to utilize siderophore-bound iron (Perry and Brubaker, 1979).

7.4.3 Energy Requirements during Iron Uptake by Microorganisms

The uptake of iron by the microorganisms from siderophores is an energy-requiring process. This can be demonstrated using the usual metabolic inhibitors such as the thiol reagents, cyanide, dinitrophenol, azide, arsenite, and anaerobic conditions. This has been shown in regard to enterochelin and *E. coli* on the one hand (Pugsley and Reeves, 1977a) and exochelins and *Mycobacterium smegmatis* (Stephenson and Ratledge, 1979) on the other, though in the latter case the situation is somewhat

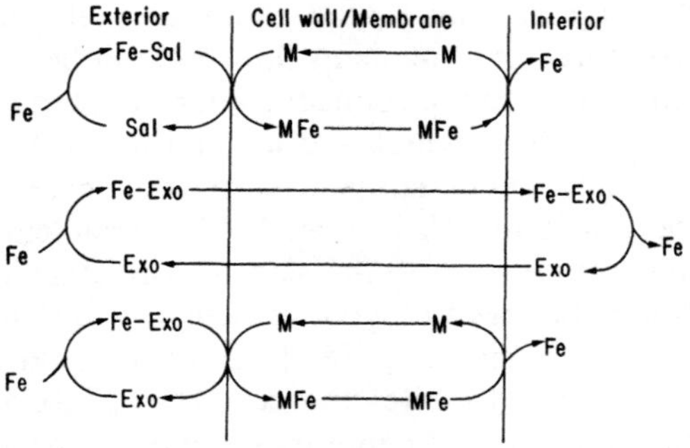

Figure 7-11. Transport of iron into *Mycobacterium smegmii*. Adapted from Stephenson and Ratledge (1979). Sal is salicylic acid, Exo is exochelin, and M is mycobactin.

complicated: At low external exochelin concentration, the metabolic inhibitors were highly effective in inhibiting iron uptake by the cells, whereas at high concentrations, the inhibitors were relatively inactive. It was proposed on the basis of these and other experiments that the first system involves the internalization of the iron-laden exochelin and does not involve the mycobactins. Such an internalization would then require a highly "energized" membrane and would be analogous to either the enterochelin or the ferrichrome situations. The second system does not involve the internalization of exochelin but does depend on the exchange of iron between iron-laden exochelin and mycobactin. This dual system of iron transport, present in *Mycobacterium smegmii*, is illustrated in Figure 7-11.

7.4.4 Control of Siderophore Biosynthesis

Large amounts of enterochelin or other siderophores are usually obtained only by growing the appropriate microorganism in media deficient in iron. This means that iron deficiency induces the biosynthesis of an array of enzyme systems and membrane receptor proteins whose function is to assure an adequate iron supply for the organism's growth. Under conditions where iron supply is adequate, such enzyme biosynthesis is repressed. Schematically, the control mechanism regulating siderophore and receptor protein production and utilization is shown in Figure 7-12. Expression of such a control mechanism, at least in enterobacteria which can utilize iron from both enterochelin and the ferrichromes, may be delineated as follows (Ernst *et al.*, 1978):

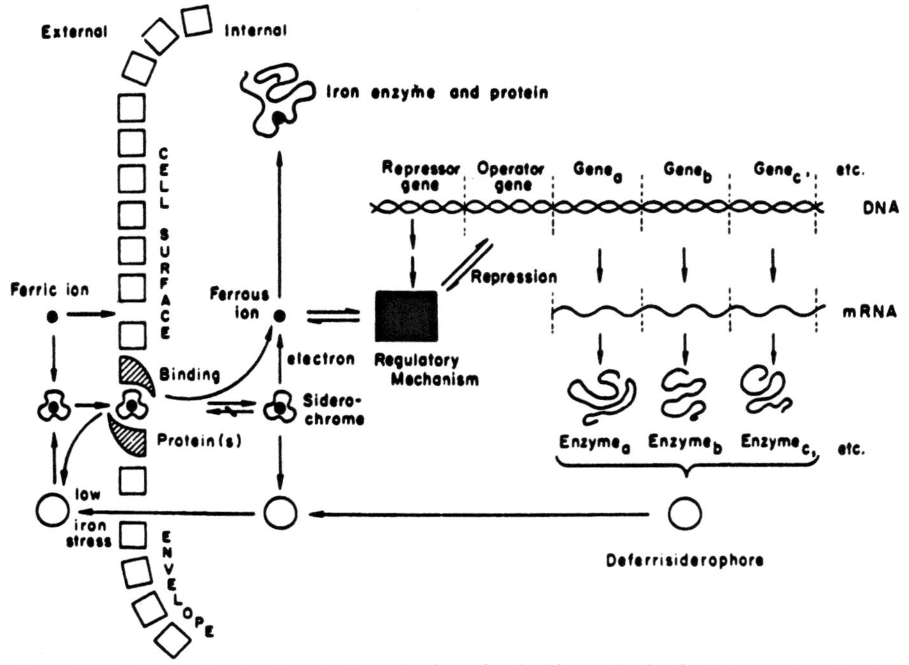

Figure 7-12. Molecular-level control mechanisms for the biosynthesis of substances involved in bacterial iron transport. From Neilands (1975).

1. Enzyme levels involved in the biosynthesis of enterochelin are greatly elevated.
2. Enzyme levels responsible for the intracellular degradation of ferrienterochelin are greatly elevated.
3. Components responsible for the cellular uptake of ferrienterochelin and ferrichrome become greatly elevated. This includes membrane receptors for the two types of siderophores.
4. There is an increase in the levels of certain outer membrane proteins whose function is as yet unclear.

The specific nature enzyme or enzymes whose levels are increased in iron deficiency states have not yet been defined.

7.5 Siderophores of Mammalian Origin

The growth of certain tranformed cells in tissue culture (e.g., the BALB/3T3 cells transformed by SV 40 virus) can be arrested by picolinic acid, which apparently interferes with their iron metabolism. Iron is required for the proper DNA replication processes. It is possible to isolate from such transformed cells certain mutants that have overcome the inhibitory effect of picolinic acid. In the culture media of such mutants it is possible to demonstrate the presence of a siderophore–ionophore-like substance which is capable of binding iron. It is apparently a peptide, heat stable, and insensitive to trypsin digestion (Fernandez-Pol, 1978). A provocative possibility is that cancer cells secrete siderophore-like substances *in vivo* which satisfy the metal requirements of the neoplasms. Presumably, should the production of such siderophore-like substances cease, the growth of the cancer cells would also cease. This offers a variety of approaches that one might devise to control neoplastic growth. Normal cells in tissue culture (e.g., normal rat kidney cells) are also sensitive to iron deprivation brought about by the treatment with picolinic acid. They are arrested in the G_1 (G_0) phase of replication, and growth is not restored unless the offending agent is removed and transferrin-bound iron is made available to the culture (Fernandez-Pol *et al.*, 1978). Normal mammalian cells therefore do not seem to be able to produce siderophore-like substances in response to iron deficiency.

7.6 Iron Chelators in Clinical Practice

Tissue iron overload is a serious consequence of a variety of hematological disorders such as the thalassemias, various types of hemolytic anemias, and, in general, conditions calling for frequent blood transfusions. These as well as the primary iron overload diseases such as hemochromatosis and hemosiderosis and the accidental iron poisoning cases require the removal of iron from the organism, especially the liver. Accumulation of iron in the liver will destroy the tissue, resulting in fibrosis and loss of function. Less serious is the accumulation of iron in the reticuloendothelial tissue (Fairbanks, 1978). In fact, liver iron overload is a major cause of death in homozygous β-thalassemia. The removal of iron from the organism cannot be accomplished through physiologic means, since the mammalian organism is geared to iron conservation rather than iron loss, and spontaneous iron excretion can rarely exceed 5 mg of iron per day in patients with iron overload.

Iron chelators utilized clinically have been the naturally occurring siderophores as well as chemicals made in the laboratory using microbial siderophores as model substances. Figure 7-13 illustrates the structure of a synthetic iron chelator with potential clinical application which was

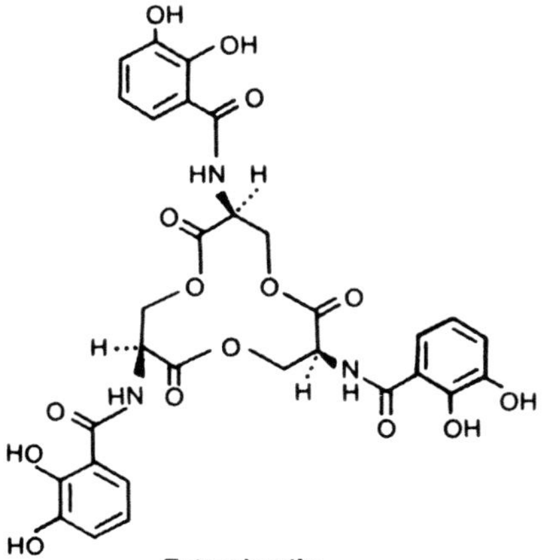

Enterobactin

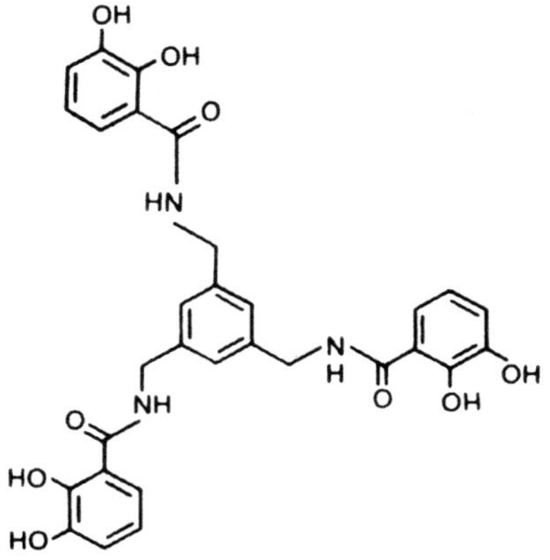

1,3,5-tris(*N,N′,N″*-2,3-dihydroxybenzoyl)-aminomethylbenzene

Figure 7-13. Structure of a synthetic enterochelin-like iron chelator, a potential agent for iron overload therapy. Reprinted with permission from *Chem. Eng. News*, Feb. 19, 1979. Copyright by the American Chemical Society.

synthesized on the basis of the enterochelin structure. The compound that has been used to treat iron overload, both acute and chronic, in the human being has been almost exclusively desferal (desferrioxamine) (Waxman and Brown, 1969). Recently, however, it has been recognized that better chelating agents are necessary to treat the great variety of the relatively commonplace iron overload syndromes, and a number of laboratories are currently involved in the systematic testing and evaluation of iron-chelating agents for their clinical use. A "useful" iron chelator must satisfy several criteria:

1. It must have a high association constant with respect to iron.
2. It must be nontoxic, nonmutagenic, and noncarcinogenic.
3. It must be effective in removing iron from parenchymal storage sites (principally liver).

The transfused mouse and rat have been used to evaluate the effectiveness of a number of iron chelators as well as to study the fundamental mechanism of iron mobilization from the iron-overloaded organism. Table 7-3 shows some rat organ iron values under normal, iron-deficient, and iron-overloaded conditions. Although results obtained with such model animals have been extremely instructive, certain differences in regard to the mode of action of the iron chelators between the rodent and human

Table 7-3. Rat Tissue Iron Content (in μmoles of iron/g of dry tissue) in Normal, Iron-Deficient, and Iron-Overloaded Conditions[a]

Tissue	Normal (n = 15)	Iron deficiency anemia[b] (n = 5)	Iron overload[c] (n = 10)
Liver	7.6	2.6	22.1
Spleen	34.8	7.3	37.8
Heart	5.3	5.1	8.0
Kidney	7.6	5.5	7.1
Brain	2.8	1.7	2.8
Muscle	2.5	3.0	10.7
Urine (μmoles/24 hr)	15×10^{-3}	13×10^{-3}	30×10^{-3}

[a] From Bobeck-Rutsaert *et al.* (1974).
[b] Via bleeding.
[c] Via injection of 40 mg of iron–dextran.

organisms have been noted (Hershko, 1978). It is therefore prudent to view the human situation as similar but not identical to that of the mouse and rat.

Chaim Hershko and his co-workers are largely responsible for elucidating the mechanism of action of the clinically tested iron chelators in the rat (Hershko *et al.*, 1973; Hershko, 1978; Hershko *et al.*, 1978). This was possible by devising iron probes that would specifically label either the reticuloendothelial or the parenchymal systems. Thus, when heat-damaged red blood cells containing ^{59}Fe are injected into the animal, practically all label appears in the reticuloendothelial system. On the other hand, when ^{59}Fe–ferritin or ^{59}Fe–transferrins are injected, only liver iron stores become labeled. With the different tissues thus labeled, it becomes possible to inject into such animals various types of iron chelators and to study their effect on the removal of iron from the organism.

There are apparently two operational mechanisms for the removal of iron from the rat's organism by exogenous iron chelators. The first mechanism, called *intracellular* and exemplified by desferrioxamine (desferal), cholylhydroxamic acid (CHA), and rhodotorulic acid, involves the penetration of the parenchymal cells by the chelators, sequestration of iron therein, and excretion of the iron–chelator complex through the bile and into the feces. Thus, in rats whose liver iron stores are labeled with ^{59}Fe, desferrioxamine and rhodotorulic acid will cause the excretion of iron label through the feces, and little or no label will be found in the urine. The second mechanism whereby iron can be removed from hypertansfused rats involves the extracellular compartment, and the source of such iron is the reticuloendothelial system. The chelators most effective within the framework of the extracellular mechanism is diethylenetriaminepentaacetic acid (DTPA), a compound similar in structure to ethylenediaminetetraacetic acid (EDTA). DTPA cannot penetrate the reticuloendothelial system cells (or the hepatocytes); however, it will sequester iron if transferrin in the bloodstream becomes saturated with iron. Transferrin is the principal if not the only recipient of the reticuloendothelial system iron. Iron complexed with DTPA is excreted via urine. It should be noted that neither iron-laden DTPA on the one hand nor iron-laden desferrioxamine, CHA, and rhodotorulic acid on the other can penetrate any of the cells where iron is stored.

When rats with labeled reticuloendothelial system iron are treated with desferrioxamine or rhodotorulic acid, labeled iron is found both in feces (one-third) and in the urine (two-thirds). Hence, under these circumstances, these two substances, but not CHA, can participate in both the intra- and extracellular mechanisms of iron removal. Again, desferrioxamine and rhodotorulic acid, like DTPA, will not bind iron released from the reticuloendothelial system unless transferrin is iron saturated, as it usually is in hypertransfused animals. Apparently these chelators cannot effectively compete for iron with transferrin and can only pick up the slack. The source of fecal desferrioxamine and rhodotorulic-acid-bound iron in these circumstances is transferrin-bound iron that is normally taken up by the hepatocytes. It should be noted that iron very rapidly reappears in plasma following destruction of red cells by the reticuloendothelial system, whereas the exchange between hepatocyte iron and plasma amounts to only 4% of the hepatocyte iron per day. The excretion of iron from rats with hepatocyte and reticuloendothelial iron lable is shown in Table 7-4.

A large number of potential iron-chelating compounds have been tested for their effectiveness of removing iron from various organs of the hypertransfused mouse. The results are indicated in Table 7-5 and show that ethylenediamine-N,N^1-bis-(2-hydroxyphenylacetic acid) dihydrochloride (EDHPA–HCl) was the most effective agent tested. It apparently removes iron primarily from the liver, and much of it apparently appears in the urine. No fecal iron data were given for this compound.

In human beings, experiments similar to those for rats and mice have not been performed. However, in patients with β-thalassemia major treated with desferrioxamine, the main route of excretion of a tracer ^{59}Fe–transferrin dose was not the feces as in the rat but via urine ($\frac{1}{3}:\frac{2}{3}$, respectively). The source of this urinary iron was shown to be an extremely *labile* iron pool of the reticuloendothelial system (Hershko and Rachmilevitz, 1979). Moreover, the degree of transferrin saturation with

Table 7-4. Excretion of ^{59}Fe from Rats (in % of injected dose) Following Chelation Therapy[a]

Chelator used	Route of excretion	From hepatocytes				From reticuloendothelial cells			
		Day 1	Day 2	Day 3	Day 4	Day 1	Day 2	Day 3	Day 4
Desferrioxamine	Urine	0.21	0.24	0.26	0.27	5.02	5.25	5.28	5.31
	Feces	6.26	12.52	16.09	17.19	3.24	3.31	2.90	3.24
Rhodotorulic acid	Urine	0.65	0.72	0.74	0.75	2.15	2.19	2.20	2.20
	Feces	16.93	20.28	20.98	20.81	1.95	1.25	0.30	0.74
CHA[b]	Urine	0.08	0.10	0.15	0.15	0	0	0	0
	Feces	0.76	1.11	1.19	1.61	0	0	0	0
DTPA[c]	Urine	0.57	0.71	0.72	0.79	6.67	6.87	6.96	7.02
	Feces	0.52	0	0.26	0.40	0	0	0	0

[a] Hepatocytes were labeled by the injection of ^{59}Fe–ferritin, whereas the reticuloendothelial system was labeled by the injection of ^{59}Fe red cells. From Hershko *et al.* (1978).
[b] Cholylhydroxamic acid.
[c] Diethylenetriaminepentaacetic acid.

iron does not have much of an influence on the urinary iron excretion during desferrioxamine therapy. The nature of the labile iron pool is yet to be elucidated.

7.7 Antimicrobial Properties of the Transferrins

The transferrins have been associated with bacteriostatic activity since their discovery in the 1940s and 1950s. Briefly, the basis for such activity is that because of the high association constant characterizing the

Table 7-5. Effectiveness of Various Chelating Agents in the Removal of Iron from Liver and Spleen of Hypertransfused Mice[a]

Compound tested	Dose (mg/kg)	% Change as per controls		
		Spleen	Liver	Urine
Desferrioxamine	250	0	−32	+400
Rhodotorulic acid	300	+19	0	+229
Schizokinen	300	0	0	+42
Triacetylfusarinine C	300	0	−18	+247
DTPA[b]	200	0	0	+131
EDHPA[c]	25	+14	−33	+285
EDHPA–HCl	50	0	−54	+1439

[a] Adapted from Pitt *et al.* (1979).
[b] Diethylenetriaminepentaacetic acid.
[c] Ethylenediamine-*N*,*N*′-bis(2-hydroxyphenylacetic acid).

transferrin–iron interaction, the transferrins can make iron unavailable to the microorganisms and thus arrest their growth. Certain microorganisms apparently can overcome such inhibition by producing potent siderophores that are capable of combining with the transferrins. Lactoferrin appears to be the most potent antibacterial agent among the transferrins. There is a growing suspicion that the iron-binding proteins of the transferrin class play a major role in host defense mechanisms, both in adults and in newborn infants. Consequently a number of review articles discussing the role of transferrins in the normal defense mechanisms of animals have been published in recent years (Reiter, 1978; Bullen *et al.*, 1972a,b; 1974, 1978; Bezkorovainy, 1977; Pearson and Robinson, 1976; Goldman, 1973).

The most forceful writer and protagonist in this area is probably Eugene Weinberg (Weinberg, 1974, 1977, 1978). He has pointed out, by quoting a passage from Shakespeare's *King Lear*,

> I'll fetch some flax and whites of eggs (conalbumin)
> To apply to his bleeding face

that the antimicrobial properties of the transferrins were known in Elizabethan England. But more importantly, he is probably responsible for delineating what exactly is involved in the so-called *nutritional immunity* mechanism, a term applied to the process whereby the iron is withheld from invading microorganisms by the hosts' transferrins. He summarizes these criteria as follows (Weinberg, 1977):

> During actual episodes of microbial invasion, hosts intensify their efforts to withhold iron from the invaders. Methods whereby this is accomplished include: (a) iron is shunted away from plasma...and into storage in hepatocytes..; (b) intestinal absorption of iron is supressed...; (c) leucocytes migrate to the invaded area and secrete lactoferrin...; hyperthermia develops....

Each aspect of nutritional immunity will be discussed in this chapter.

7.7.1 *In Vitro* Effects of Transferrins on Microbial Growth

The growth of numerous microorganisms is inhibited by the presence of the transferrins (Marcelis *et al.*, 1978). This can be demonstrated by using the purified proteins with a suitable bacterial growth media or more often by using serum or milk. In many cases microorganisms fail to grow in such biological fluids unless iron is added to a level required to saturate transferrin with iron to from 60 to 80% of its iron-binding capacity. One notable exception is *Staph. aureus*, which is inhibited neither by serotransferrin nor by lactoferrin (Masson *et al.*, 1966; Marcelis *et al.*, 1978).

A typical growth curve for *E. coli* in human milk with and without added iron is shown in Figure 7-14.

Several investigators have recognized that though the transferrins may exert bacteriostatic effects in certain microorganisms, such effects may be magnified severalfold by additional factors. Thus, the inhibition of growth of *Cl. welchii* by serotransferrin can be increased by the addition of a β_2- or γ-globulin fraction of serum (Rogers *et al.*, 1970). Likewise, lactoferrin by itself has a moderate bacteriostatic effect with respect to *E. coli 0111/B_4*. However, in concert with a specific IgA-type antibody from human milk, lactoferrin has a much greater effect on this microorganism in the presence of bicarbonate. The antibody alone has no bacteriostatic effect (see Figure 7-15) (Bullen *et al.*, 1972b; Spik *et al.*, 1978; Griffiths and Humphreys, 1977). It has been suggested that whereas the lactoferrin effectively sequesters the iron of the medium, the antibody prevents either the synthesis or secretion of siderophores to counteract the effect of the lactoferrin (Griffiths and Humphreys, 1978; Bullen *et al.*, 1978). Human serum also has a powerful inhibitory activity with respect to the growth of *Nisseria gonorrheae*. Such an effect can be abolished

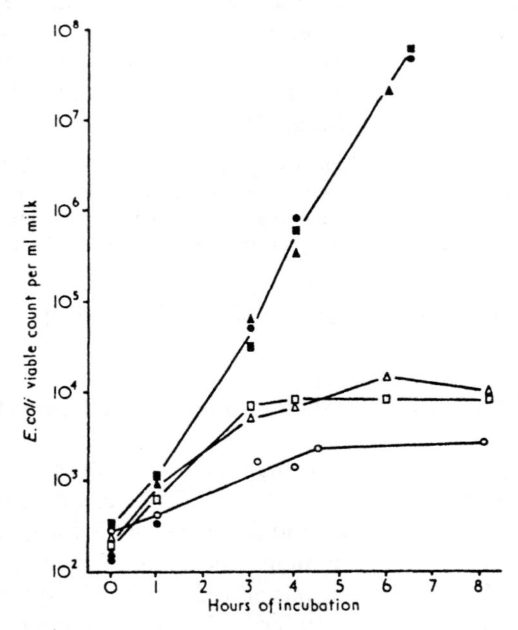

Figure 7-14. The growth of *Escherichia coli* in human milk in the presence and absence of added iron. Open markers are milk samples without exogenous iron, closed markers are milk samples with exogenously added iron. From Bullen *et al.* (1972b).

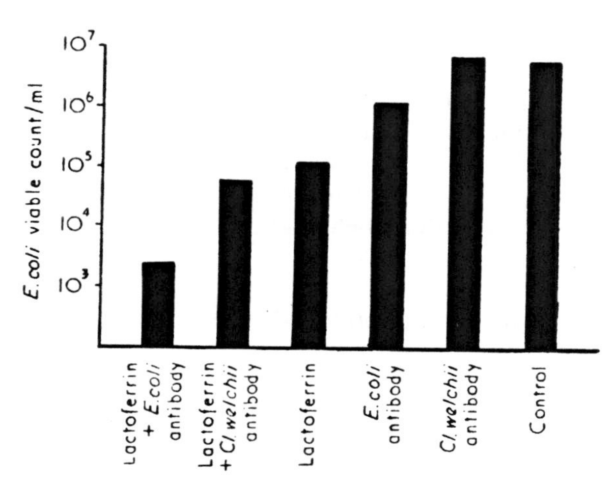

Figure 7-15. Growth of *Escherichia coli 0111* in the presence of lactoferrin and *E. coli* antibody. From Bullen *et al.* (1972b).

only by adding iron to the serum but in amounts far exceeding those required to saturate the serotransferrin present. These data were used to argue that serotransferrin is not the only iron-sensitive antimicrobial agent present in serum (Norrod and Williams, 1978). The nature of the non-transferrin factors remains to be determined.

Serotransferrin and presumably other transferrins as well have powerful antifungal activities, especially if a *stimulatory* factor (complement?, antibody?) from serum is present (Shiraishi and Arai, 1979). The inhibitory activity was directly related to the UIBC (see Chapter 1) of the transferrin.

It has been stated that the antimicrobial properties of the transferrins can be abolished by saturating or nearly saturating them with iron. In addition to iron, citrate has been shown to "inactivate" the transferrins (Griffiths and Humphreys, 1977; Law and Reiter, 1977), presumably because citrate can function as an iron shuttle between the transferrin and the microorganism. Another series of compounds that can abolish the antimicrobial effects of the transferrins in specific instances is not unexpectedly the siderophores. It had been presumed that certain siderophores are capable of abstracting iron from the transferrins and thus make it available to the microorganisms (e.g., Griffiths and Humphreys, 1978). It has been claimed, however, that the mechanism of siderophore action in the presence of the transferrins is not to compete with the iron-laden transferrins for iron but instead to combine with the former. What then follows is the binding of the siderophore–transferrin complex to the microorganism surface via specific outer membrane receptors of perhaps a lipopolysaccharide nature. Membrane-associated receptors would then proceed to separate the apotransferrin from the iron-laden transferrin–siderophore complex, and the siderophore–iron complex thus formed would be internalized by the bacterial cell (Kvach *et al.*, 1977; Kochan *et al.*, 1978). This system was investigated in detail with the enterochelin–*Salmonella typhimurium* system, though the chemical nature of the transferrin–enterochelin complex remains to be elucidated.

7.7.2 Iron Status and Infection in Whole Organisms

A relationship between iron and microbial infection can be observed *in vivo* in a number of animals. Such observations can be divided into the well-controlled experiments using mice, rats, or other laboratory animals and clinical studies on human beings where the incidence of systemic infection is correlated with changes in iron status due to some pathological condition. Regardless of the conditions under which the systemic sepsis is observed, a certain physiological response to the bacterial invasion is observed, namely total plasma iron (TI) levels go down, and unsaturated iron-binding capacity (UIBC) goes up. Total iron-binding capacity (TIBC) generally remains normal. This effect has been traced to a lowered intestinal iron absorption and the movement of iron from the circulation into the liver and reticuloendothelial system storage sites. There may also be a lowered volume of iron entry into the circulation from the reticuloendothelial system (see Chapter 5). As an example, in the mouse infected with *Escherichia coli* the TI dropped from 353 ± 65 to 38 ± 62 μg/dl (Ganzoni and Puschmann, 1977), and in rabbits infected with *Pasteurella multocida* the TI dropped from a normal of 250–270 to 70–120 μg/dl, with the TIBC remaining constant (Kluger and Rothenburg, 1979). It should be noted that when iron saturation of serotransferrin drops to below 16% of saturation, erythropoiesis is impeded and anemia may develop. In fact, persons with chronic infections are sometimes severely anemic (Kumar, 1979).

Endotoxin and other agents simulating inflammation will bring about effects similar to those obtained with bacterial infection. Endotoxin seems to work through the release of the so-called leukocyte endogenous mediator (LEM) from the white cells, which appears to be the factor that directly affects iron uptake by the liver cells (Weinberg, 1978). Endotoxin per se had no effect upon iron uptake by isolated rat hepatocytes (Grohlich *et al.*, 1979).

Inflammation, whether brought about by bacterial infection or by endotoxin brings about a rise in temperature (fever), however, the fever per se has been excluded as a cause for the lowering of the TI in the affected animal (Grieger and Kluger, 1978). Nevertheless, high body temperatures enhance the bacteristatic effects of the transferrins (Kluger and Rothenburg, 1979). It has been theorized that elevated temperatures suppress siderophore production by the invading microorganisms. Inflammation also causes a proliferation of various types of white blood cells, including the phagocytic polymorphonuclear leukocytes (neutrophils). Lactoferrin has been shown to be an important component of these cells (Masson *et al.*, 1969; Bennett and Kokocinski, 1978). It apparently participates in the destruction of the invading bacterium by withholding iron from it. This was cleverly shown by Bullen and Armstrong (1979), who incubated neutrophils with ferritin–antibody and with apoferritin–antibody complexes. Both were phagocytosed, but only those cells that received the ferritin–antibody complex showed decresed bactericidal activity. Moreover, disrupted leukocytes release lactoferrin into the plasma, which removes iron from serotransferrin because of its higher affinity for iron (see Chapter 4). The iron-laden lactoferrin is easily cleared by the reticuloendothelial system. It is thus argued by some that white cell lactoferrin is in part responsible for the hypoferremia observed in systemic bacterial infections by moving serotransferrin-bound iron into the reticuloendothelial system (Van Snick *et al.*, 1974).

Given the preceeding information, it may be expected that experimental animals as well as human beings with increased iron status should be more susceptible to systemic infection, the other side of the coin being that iron-deficient animals and/or human beings should be less susceptible to such infections. Several studies have shown that in certain situations where TI levels are high and UIBC values are low, the patients are much more likely to develop systemic bacterial or fungal infections than are normal human beings. Thus patients with acute leukemia, whose UIBC levels are extremely low, are prone to develop candidiasis, and their sera readily support the growth of *Candida albicans*. If aposerotransferrin is added to such sera, the yeast does not grow any better than in the normal serum (Figure 7-16) (Caroline *et al.*, 1969). Further, patients with various hemolytic anemias that are accompanied by high iron levels in the bloodstream had a much greater incidence of systemic infections than did patients with iron deficiency or dimorphic anemia (Masawe *et al.*, 1974). On the other hand, it is reported that in patients with hemochromatosis, sickle cell anemia, and thalassemia major, where TI levels are also high and UIBC levels low, the incidence of bacterial sepsis is no greater than that of normal populations (Pearson and Robinson, 1976).

If iron-overloaded subjects generally show an increased susceptibility to infection, it may be expected that iron deficient persons would show

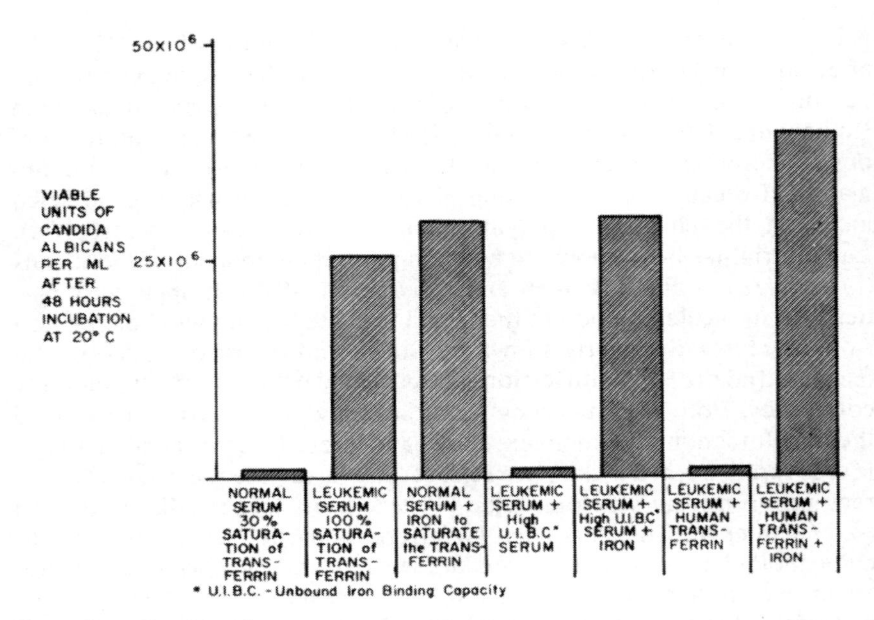

Figure 7-16. Growth of fungus in normal and leukemic sera. Reprinted by permission from Caroline, L., Rosner, F., and Kozina, P. J., 1969, Elevated serum iron, low unbound transferrin, and candidiasis in acute leukemia, *Blood* 34:441–451. Copyright by Grune and Stratton, Publ.

a greater than normal resistance to systemic infection. This is not the case, as iron is apparently required for the normal development of lymphoid tissue, so that in the presence of chronic nutritional iron deficiency, the affected individuals in fact have a greater incidence of infectious disease. The incidence of infectious disease can also be lowered in children receiving oral iron supplements (Gross and Newberne, 1980). It should be remembered that oral iron supplementation would not be expected to lead to iron overload because of the "mucosal block" control system.

There is no question, however, that the parenteral or intramuscular administration of exogenous iron, either to human beings or to experimental animals, will increase the likelihood of septicemia. Thus, the intramuscular administration of iron–dextran to newborn infants increases the likelihood of sepsis, so that this practice is no longer recommended (Barry and Reeve, 1977; Becroft *et al.*, 1977). Iron–dextran is not available to saturate circulating transferrin even though TI serum levels may reach 1000 μg/dl; however, such iron is apparently available to the microorganisms. Finally, in kwashiorkor it has been routine practice to administer iron parenterally at the beginning of the treatment. A number of youngsters so treated succumb, presumably because of overwhelming infection after the initiation of such treatment. It has been noted that the serotransferrin levels of children who die are significantly lower than in those who survive, 0.33 and 1.30 mg/ml, respectively, after 2 weeks of treatment. The iron apparently serves to diminish the bacteriostatic activity of an already low serotransferrin level (McFarlane *et al.*, 1970).

In iron-deficient individuals the situation in regard to susceptibility to infection is unclear. There are reports claiming a decline in immunocompetence of iron-deficient individuals (Macdougall *et al.*, 1975), whereas others are less certain (Strauss, 1978).

In experimental animals, the administration of iron to infected animals may or may not develop lethal sepsis. The so-called *virulent* organisms are said to be able to take greater advantage of high iron status of the host. Thus, among approximately 120 microorganisms tested, only those known to be extremely virulent in animals could be induced to develop infections in guinea pigs by high iron status. Those shown to be virulent were BCG, *Corynbacterium ovis, C. murium, Listeria monocytogenes, Erysipelothrix rhusiopathiae, Cl. perfringens, Cl. septicum, Cl. edematiens,* and some strains of Klebsiella, Proteus, and *Aeromonas hydrophila* (Miles *et al.*, 1979). Other workers have shown similar effects of iron on *E. coli, K. pneumoniae, P. aeruginosa, S. aureus, B. anthracis, Y. pestis, M. tuberculosis,* and *N. gonorrheae*. Virulence of these organisms can, of course, also be increased by the injection of specific siderophores (along with the iron) (Jones *et al.*, 1977; Kochan *et al.*, 1978). Payne and Finkelstein (1978b) have divided various microorganisms into four classes according to how they respond to iron:

I. Highly virulent microorganisms that can successfully compete for iron even at very low iron levels in the medium. These include *Pseudomonas aeruginosa, E. coli Kl, Salmonella typhimurium, Haemophilus influenzae type B, Neisseria gonorrheae DGI,* and *Neisseria meningitidis*. These microorganisms generally cause disseminated infections.

II. Pathogens whose virulence is largely proportional to the iron available to them. Iron-poor transferrins are successful in inhibiting their growth. They are normally associated with localized infections. This class includes *Shigella flexnerii* and *dysenteriae* and *Vibrio cholerae*.

III. Avirulent organisms that cannot compete for iron but will grow if sufficient and available iron is present in the medium. It is speculated that though such organisms do produce siderophores, they cannot utilize siderophore–iron complexes. Members of this group are certain mutants of *S. flexnerii, E. coli,* and *V. cholerae*.

IV. Organisms that do not become virulent even if sufficient available quantities of iron are present.

7.7.3 Lactoferrin and Protection of the Breast-Fed Infant against Digestive and Systemic Disease

Numerous clinical studies since the 1920s have demonstrated that breast-fed human infants are more resistant to various infections, especially gastrointestinal disorders. This host resistance has been ascribed to a number of factors present in human milk but not in cow's milk formula. These factors include a relatively high white blood cell count including polymorphonuclear leukocytes and lymphocytes, IgA-type antibodies (especially anticoliforms), lysozyme, the *Lactobacillus bifidus* growth factors, and, last but not least, lactoferrin (Welsh and May, 1979; Goldman, 1973). Thus, when human milk and a widely used artificial formula were inoculated with a culture of *E. coli*, the human milk showed inhibitory activity, whereas the microorganism grew unimpeded in the formula. It grew at maximal rates regardless of any added iron, indicating no transferrin-related bacteriostasis whatever (Baltimore *et al.*, 1978).

Lactoferrin is present in human milk in rather large quantities (see Chapter 4), and the degree of its iron saturation is only 9% of capacity (Weinberg, 1978); hence it should exert a powerful bacteriostatic effect upon a number of microorganisms. Moreover, it has been claimed that milk proteins remain intact in an infant's stomach for up to 90 min after feeding and are apparently passed unchanged into the small intestine (Bullen *et al.*, 1974). In guinea pigs, gastrointestinal disorders are also much less common when the the pups are suckled than in animals artificially fed. The protective effect of the homologous raw milk can be abolished in the guinea pig pups if they are fed hematin along with the mother's milk (Bullen *et al.*, 1972a,b). These and other results prompted many authorities to question the widsom of supplementing artificial infant formulas with iron. However, the Committee on Nutrition (1978) has ruled that the benefits of iron supplementation outweigh the possibility of higher incidence of gastrointestional distress in artificially fed human infants.

The function of lactoferrin in human milk does not appear to be restricted to its antimicrobial activity. It has been determined that nearly 50% of all iron present in human milk is available for absorption, whereas only 10% is so available from bovine milk (Saarinen *et al.*, 1977). Since most if not all of the iron present in human milk must be bound to lactoferrin, the latter clearly must be involved in iron absorption by the infant's small intestine. Recent studies have demonstrated that lactoferrin but not serotransferrin or conalbumin can donate iron to sections of human duodenal mucosa. Since lactoferrin itself was not internalized by the mucosal cells, it was presumed that the brush borders have specific receptors that will combine with iron-laden lactoferrin and abstract iron from it (Cox *et al.*, 1979).

In the area of veterinary medicine, the function of the transferrins as antimicrobial agents was not appreciated until very recently. Thus serotransferrin was not even considered when bactericidal effects of bovine sera were evaluated (Carroll, 1971), nor was lactoferrin mentioned when immunity to mastitis in cattle was reviewed (Norcròss and Stark, 1970). Mastitis, as is well known, is a great economic burden upon the American cattle industry (Janzen, 1970). It has been determined that lactoferrin levels in milks of mastitic udders increase some 30-fold, so that lactoferrin becomes a major whey protein (in one milk sample it reached 6.2 mg/ml). Polymorphonuclear leukocytes were excluded as a major source of the lactoferrin, and the epithelial cells were implicated as the site of lactoferrin synthesis. It thus appears that the biosynthesis of lactoferrin by the udder is a major physiological defense mechanism in mastitis (Harmon *et al.*, 1976). It is likely that in the future the transferrins in general will prove to be of considerable interest to researchers in the area of veterinary medicine.

Summary

Most microorganisms require iron for survival and growth. Since free ferric iron is present in the medium in extremely low amounts and is often sequestered by one of the transferrins, microorganisms, under iron deficiency conditions, can biosynthesize small-molecular-weight compounds called siderophores that serve as iron scavengers in the medium. Chemically, siderophores are of two types: the catechol-like, exemplified by enterochelin, and the hydroxamic-acid-like, exemplified by the siderochromes. Siderophores can deliver iron to the microorganisms by essentially three mechanisms: Some can penetrate the cell membrane and release iron intracellularly by being destroyed; some can penetrate the cell, release the iron, and be percolated back into the medium; and others can deliver iron to an outer cell membrane receptor without penetrating the cell. Those siderophores that can acquire iron in the presence of the transferrins combine with the latter, and the complex is bound to specific receptors on the outer cell membrane, whence iron is internalized.

Iron overload in the human being can be treated by the adminstration of natural or synthetic siderophores. To be clinically "useful," the siderophore must be able to mobilize iron from the liver. Some siderophores such as cholylhydroxamic acid act only on the parenchymal cells and are excreted bound up with the iron through the bile duct into the small intestine. Other chelating agents, such as desferal and rhodotorulic acid, acquire iron from both the reticuloendothelial and the parenchymal systems and are excreted via both the urine and feces. A third type of chelators mobilize iron only from the reticuloendothelial system.

The transferrins are bacteriostatic agents because they can withhold iron from the microorganisms. *In vivo*, this effect is called *nutritional immunity*. The antimicrobial effect of the transferrins can be abolished by saturating them with iron. The virulence of microorganisms can be related to the effectiveness with which they can acquire iron from the medium containing the transferrins. The most virulent organisms can proliferate in the presence of transferrins even with a very low degree of iron saturation, whereas the less virulent organisms can be effectively inhibited by iron-free transferrins. Physiologically, the organism responds to systemic bacterial infection by lowering plasma TI and increasing its UIBC. This is accomplished by moving iron into the parenchymal storage sites and decreasing iron absorption by the small intestine. It is believed that lactoferrin plays an important role in the resistance of the breast-fed infant against gastrointestinal disease.

References

Baltimore, R. S., Vecchitto, J. S., and Pearson, H. A., 1978. Growth of *Escherichia coli* and concentration of iron in an infant feeding formula, *Pediatrics* 62:1072–1073.

Barry, D. M. J., and Reeve, A. W., 1977. Increased incidence of gram-negative neonatal sepsis with intramuscular iron administration, *Pediatrics* 60:908–912.

Becroft, D. M. O., Dix, M. R., and Former, K., 1977. Intramuscular iron–dextran and susceptibility of neonates to bacterial infections, *Arch. Dis. Child.* 52:778–781.

Bennett, R. M., and Kokocinski, T., 1978. Lactoferrin content of peripheral blood cells, *Br. J. Haematol.* 39:509–521.

Bezkorovainy, A., 1977. Human milk and colostrum proteins: A review, *J. Dairy Sci.* 60:1023–1037.

Bobeck-Rutsaert, M. M. J. C., op den Kelder, A. M., Wiltink, W. F., Van Eijk, H. G., and Leijnse, B., 1974. Site of action of desferrioxamine in removing iron in normal and pathological conditions, *Acta Haematol.* 51:151–158.

Bullen, J. J., and Armstrong, J. A., 1979. The role of lactoferrin in the bactericidal function of polymorphonuclear leukocytes, *Immunology* 36:781–791.

Bullen, J. J., Rogers, H. J., and Griffiths, E., 1972a. Iron binding proteins and infection, *Br. J. Haematol.* 23:389–392.

Bullen, J. J., Rogers, H. J., and Leigh, L., 1972b. Iron-binding proteins in milk and resistance to *Escherichia coli* infection in infants, *Br. Med. J.* 1:69–75.

Bullen, J. J., Rogers, H. J., and Griffiths, E., 1974. Bacterial iron metabolism in infection and immunity, *in Microbial Iron Metabolism*, J. B. Neilands (ed.), Academic Press, New York, pp. 517–551.

Bullen, J. J., Rogers, H. J., and Griffiths, E., 1978. Role of iron in bacterial infection, *Curr. Top. Microbiol. Immunol.* 80:1–35.

Caroline, L., Rosner, F., and Kozina, P. J., 1969. Elevated serum iron, low unbound transferrin, and candidiasis in acute leukemia, *Blood* 34:441–451.

Carrano, C. J., and Raymond, K. N., 1978. Coordination chemistry of microbial iron transport compounds: Rhodotorulic acid and iron uptake in *Rhodotorula pilimanae*, *J. Bacteriol.* 136:69–74.
Carroll, E. J., 1971. Bactericidal acitivity of bovine serums against coliform organisms isolated from milk of mastitic udders, udder skin, and environment, *Am. J. Vet. Res.* 32:689–701.
Committee on Nutrition, 1978. Relationship between iron status and incidence of infection in infancy, *Pediatrics* 62:246–250.
Cox, C. D., and Graham, R., 1979. Isolation of an iron-binding compound from *Pseudomonas aeruginosa*, *J. Bacteriol.* 137:357–364.
Cox, T. M., Mazurier, J., Spik, G., Montreuil, J., and Peters, T., 1979. Iron binding proteins and influx of iron across the duodenal brush border: Evidence for specific lactoferrin receptors in the human intestine, *Biochim. Biophys. Acta* 588:120–128.
Emery, T., 1974. Biosynthesis and mechanism of action of hydroxamate-type siderochromes, *in Microbial Iron Metabolism*, J. B. Neilands (ed.), Academic Press, New York, pp. 107–123.
Ernst, J. F., Bennett, R. L., and Rothfield, L. I., 1978. Constitutive expression of the iron–enterochelin and ferrichrome uptake systems in a mutant strain of *Salmonella typhimurium*, *J. Bacteriol.* 135:928–934.
Fairbanks, V. F., 1978. Chronic iron overload: New chelators and new strategies, *J. Lab. Clin. Med.* 92:141–143.
Fernandez-Pol, J. A. 1978. Isolation and characterization of a siderophore-like growth factor from mutants of SV40-transformed cells adapted to picolinic acid, *Cell* 14:489–499.
Fernandez-Pol, J. A., Klos, D., and Donati, R. M., 1978. Iron transport in NRK cells synchronized in G_1 by picolinic acid, *Cell. Biol. Int. Rep.* 2:433–439.
Ganzoni, A. M., and Puschmann, M., 1977. Iron status and host defense, *in Proteins of Iron Metabolism*, E. B. Brown, P. Aisen, J. Fielding, and R. R. Crichton (eds.), Grune & Stratton, New York, pp. 427–432.
Goldman, A. S., 1973. Host resistance factors in human milk, *J. Pediatr.* 82:1082–1090.
Grieger, T. A., and Kluger, M. J., 1978. Fever and survival: The role of serum iron, *J. Physiol.* 279:187–196.
Griffiths, E., and Humphreys, J., 1977. Bacteriostatic effect of human milk and bovine colostrum on *Escherichia coli*: The importance of bicarbonate, *Infect. Immun.* 15:396–401.
Griffiths, E., and Humphreys, J., 1978. Alterations in *t*RNAs containing 2-methylthio-N^6-(Δ^2-isopentenyl)-adenosine during growth of enteropathogenic *Escherichia coli* in the presence of iron-binding proteins, *Eur. J. Biochem.* 82:503–513.
Grohlich, D., Morley, C. G. D., and Bezkorovainy, A., 1979. Unpublished observations.
Gross, R. L., and Newberne, P. M., 1980. Role of nutrition in immunologic function, *Physiol. Rev.* 60:188–290.
Harmon, R. J., Schanbacher, F. L., Ferguson, L. C., and Smith, K. L., 1976. Changes in lactoferrin, immunoglobulin G, bovine serum albumin, and α-lactalbumin during acute experimental and natural coliform mastitis in cows, *Infect. and Immun.* 13:533–542.
Hershko, C., 1978. Determinants of fecal and urinary iron excretion in desferrioxamine-treated rats, *Blood* 51:415–423.
Hershko, C., and Rachmilevitz, E. A., 1979. Mechanism of desferrioxamine-induced iron excretion in thalassemia, *Br. J. Haematol.* 42:125–132.
Hershko, C., Cook, J. D., and Finch, C. A., 1973. Storage iron kinetics. III. Study of desferrioxamine by selective radioiron labels of the RE and parenchymal cells, *J. Lab. Clin. Med.* 81:876–886.

Hershko, C., Grady, R. W., and Cerami, A., 1978. Mechanism of iron chelation in the hypertransfused rat: Definition of two alternative pathways of iron mobilization, *J. Lab. Clin. Med.* 92:144–151.

Hider, R. C., Silver, J., Neilands, J. B., Morrison, I. E. G., and Rees, L. V. C., 1979. Identification of iron(II) enterobactin and its possible role in *Escherichia coli* iron transport, *FEBS Lett.* 102:325–328.

Janzen, J. J., 1970. Economic losses resulting from mastitis. A review, *J. Dairy Sci.* 53:1151–1161.

Jones, R. L., Peterson, C. M. Grady, R. W., Kumbaraci, T., and Cerami, A., 1977. Effects of iron chelators and iron overload on salmonella infection, *Nature (London)* 267:63–65.

Keller-Schierlein, W., 1977. Chemistry of iron-chelating agents from microorganisms; development and characterization of desferrioxamine B, *in Proceedings of Symposium on Development of Iron Chelators for Clinical Use,* W. F. Anderson, and M. C. Hiller (eds.), Department of Health, Education, and Welfare Publication No. (NIH) 77-994, Washington, D.C., pp. 53–82.

Kluger, M. J., and Rothenburg, B. A., 1979. Fever and reduced iron: Their interaction as a host defense response to bacterial infection, *Science* 203:374–376.

Kochan, I., Wsynczuk, J., and McCabe, M. A., 1978. Effects of injected iron and siderophores on infections in normal and immune mice, *Infect. Immun.* 22:560–567.

Kumar, R., 1979. Mechanism of aneaemia of chronic infection—estimation of labile iron pool and interpretation of ferrokinetic data, *Indian J. Med. Res.* 70:455–462.

Kvach, J. T., Wiles, T. I., Mellencamp, M. W., and Kochan, I., 1977. Use of transferrin–iron–enterobactin complexes as the source of iron by serum exposed bacteria, *Infect. Immun.* 18:439–445.

Law, B. A., and Reiter, B., 1977. The isolation and bacteriostatic properties of lactoferrin from bovine milk whey, *J. Dairy Res.* 44:595–599.

Leong, J., and Neilands, J., 1976. Mechanisms of siderophore iron transport in enteric bacteria, *J. Bacteriol.* 126:823–830.

Liu, P. V., and Shokrani, F., 1978. Biological activities of pyochelins: Iron-chelating agents of *Pseudomones aeruginosa, Infect. Immun.* 22:878–890.

Llinas, M., 1973. Metal–polypeptide interactions: The conformational state of iron proteins. II. The siderochromes, *Struct. Bonding (Berlin)* 17:139–156.

Macdougall, L. G., Anderson, R., McNab, G. M., and Katz, J., 1975. The immune response in iron-deficient children: Impaired cellular defense mechanisms with altered humoral components, *J. Radiol.* 86:833–843.

Macham, L. P., and Ratledge, C., 1975. A new group of water-soluble iron-binding compounds from mycobacteria: The exochelins, *J. Gen. Microbiol.* 89:379–382.

Macham, L. P., Ratledge, C., and Nocton, J. C., 1975. Extracellular iron acquisition by mycobacteria: Role of the exochelins and evidence against the participation of mycobactin, *Infect. Immun.* 12:1242–1251.

Marcelis, J. H., den Daas-Slagt, H. J., and Hoogkamp-Korstanje, J. A. A., 1978. Iron requirement and chelator production of staphlycocci. *Streptococcus faecalis* and *enterobacteriacea, Antonie van Leeuwenhoek J. Microbiol. Serol.* 44:257–267.

Masawe, A. E. J., Muindi, J. M., and Swai, G. B. R., 1974. Infections in iron deficiency and other types of anaemia in the tropics, *Lancet* 2:314–317.

Masson, P. L., Heremans, J. F., Prognot, J. J., and Wanters, G., 1966. Immunochemical localization and bacteriostatic properties of an iron-binding protein from bronchial mucus, *Thorax* 21:538–544.

Masson, P. L., Heremans, J. F., and Schonne, E., 1969. Lactoferrin, and iron-binding protein in neutrophilic leucocytes, *J. Exp. Med.* 130:643–658.

McFarlane, H., Reddy, S., Adcock, K. J., Adeshina, H., Cooke, A. R., and Akene, J., 1970. Immunity, transferrin, and survival in kwashiorkor, *Br. Med. J.* 4:268–270.

Miles, A. A., Khimji, P. L., and Maskell, J., 1979. The variable response of bacteria to excess ferric iron in host tissues, *J. Med. Microbiol.* 12:17–28.

Neilands, J. B., 1973. Microbial iron transport compounds (siderochromes), *in Inorganic Biochemistry,* Vol. I, G. L. Eichhorn (ed.), Elsevier, Amsterdam, pp. 167–202.

Neilands, J. B., 1974. Iron and its role in microbial physiology, *in Microbial Iron Metabolism,* J. B. Neilands (ed.), Academic Press, New York, pp. 4–34.

Neilands, J. B., 1977. Microbial iron transport compounds (siderophores), *in Proceedings of Symposium on Development of Iron Chelators for Clinical Use,* W. F. Anderson and M. C. Hiller (eds.), Department of Health, Education, and Welfare Publication No. (NIH) 77-994, Washington, D.C., pp. 5–44.

Neilands, J. B., and Wayne, R. R., 1977. Membrane receptors for microbial iron transport compounds (siderophores), *in Proteins of Iron Metabolism,* E. B. Brown, P. Aisen, J. Fielding, and R. R. Crichton (eds.), Grune & Stratton, New York, pp. 365–369.

Norcross, N. L., and Stark, D. M., 1970. Immunity to mastitis, a review, *J. Dairy Sci.* 53:387–393.

Norrod, P., and Williams, R. P., 1978. Effects of iron and culture filtrates on killing of *Neisseria gonorrhoeae* by normal human serum, *Infect. Immun.* 21:918–924.

O'Brien, I. G., and Gibson, F., 1970. The structure of enterochelin and related 2,3-dihydroxy-*N*-benzoyl serine conjugates from *Escherichia coli, Biochim. Biophys. Acta* 215:393–402.

Ong, S. A., Peterson, T., and Neilands, J. B., 1979. Agrobactin, a siderophore from *Agrobacterium aumefaciens, J. Biol. Chem.* 254:1860–1865.

Payne, S. M., and Finkelstein, R. A., 1978a. Siderophore production by *Vibrio Cholerae, Infect. Immun.* 20:310–311.

Payne, S. M., and Finkelstein, R. A., 1978b. The critical role of iron in host bacterial interactions, *J. Clin. Invest.* 61:1428–1440.

Pearson, H. A., and Robinson, J. E., 1976. The role of iron in host resistance, *Adv. Pediatr.* 23:1–33.

Perry, R. D., and Brubaker, R. R., 1979. Accumulation of iron by *Yersiniae, J. Bacteriol.* 137:1290–1298.

Pitt, C. G., Gupta, G., Estes, W. E., Rosenkrantz, H., Metterville, J. J., Crumbliss, A. L., Palmer, R. A., Nordquest, K. W., Sprinkle-Hardy, K. A., Whitcomb, D. R., Byers, B. R., Arceneaux, J. E. L., Gaines, C. G., and Sciortino, C. V., 1979. The selection and evaluation of new chelating agents for the treatment of iron overload, *J. Pharmacol. Exp. Ther.* 208:12–18.

Pollack, J. R., and Neilands, J. B., 1970. Enterobactin, an iron transport compound, *Biochem. Biophys. Res. Commun.* 38:989–992.

Pugsley, A. P., and Reeves, P., 1977a. Uptake of ferrienterochelin by *Escherichia coli*: Energy-dependent state of uptake, *J. Bacteriol.* 130:26–36.

Pugsley, A. P., and Reeves, P., 1977b. The role of colicin receptors in the uptake of ferroenterochelin by *Escherichia coli K-12, Biochem. Biophys. Res. Commun.* 74:903–911.

Raymond, K. N., and Carrano, C. J., 1979. Coordination chemistry and microbial iron transport, *Acc. Chem. Res.* 12:183–190.

Reiter, B., 1978. Review of the progress of dairy science: Antimicrobial systems in milk, *J. Dairy Res.* 45:131–147.

Rogers, H. J., Bullen, J. J., and Cushnie, G. H., 1970. Iron compounds and resistance to infection. Further experiments with *Clostridium welchii type A in vivo* and *in vitro, Immunology* 19:521–538.

Saarinen, U. M., Siimes, M. A., and Dallman, P. R., 1977. Iron absorption in infants: High bioavailability of breast milk iron as indicated by the extrinsic tag method of iron absorption and by the concentration of serum ferritin, *J. Pediatr.* 91:36–39.
Shiraishi, A., and Arai, T., 1979. Antifungal activity of transferrin, *Sabouraudia* 17:79–83.
Snow, G. A., 1970. Mycobactins: Iron-chelating growth factors from mycobacteria, *Bacteriol. Rev.* 34:99–125.
Spik, G., Cheron, A., Montreuil, J., and Dolby, J. M., 1978. Bacteriostasis of a milk sensitive strain of *Escherichia coli* by immunoglobulins and iron-binding proteins in association, *Immunology* 35:663–671.
Stephenson, M., and Ratledge, C., 1979. Iron transport in *Mycobacterium smegmatis*: Uptake of iron from ferriexochelin, *J. Gen. Microbiol.* 110:193–202.
Strauss, R. G., 1978. Iron deficiency, infections, and immune function: A reassessment, *Am. J. Clin. Nutr.* 31:660–666.
Tait, G. H., 1975. The identification and biosynthesis of siderochromes formed by *Micrococcus denitrificans*, *Biochem. J.* 146:191–204.
Van Snick, J. L., Masson, P. L., and Heremans, J. F., 1974. The involvement of lactoferrin in the hyposideremia of acute inflammation, *J. Exptl. Med.* 140:1068–1084.
Waxman, H. S., and Brown, E. B., 1969. Clinical usefulness of iron chelating agents, *Prog. Hematol.* 6:338–373.
Wayne, R., and Neilands, J. B., 1975. Evidence for common binding sites for ferrichrome compounds and bacteriophage ϕ 80 in the cell envelope of *Escherichia coli*, *J. Bacteriol.* 121:497–503.
Weinberg, E. D., 1974. Iron and susceptibility to infectious disease, *Science* 184:952–956.
Weinberg, E. D., 1977. Infection and iron metabolism, *Am. J. Clin. Nutr.* 30:1485–1490.
Weinberg, E. D., 1978, Iron and infection, *Microbiol. Rev.* 42:45–66.
Welsh, J. K., and May, J. T., 1979. Anti-infective properties of breast milk, *J. of Pediatr.* 94:1–9.
Wiebe, C., and Winkelmann, G., 1975. Kinetic studies on the specificity of chelate–iron uptake in *Aspergillus*, *J. Bacteriol.* 123:837–842.
Winkelmann, G., 1979. Evidence for stereospecific uptake of iron chelates in fungi, *FEBS Lett.* 97:43–46.

The Iron–Sulfur Proteins

8.1 Introduction and Classification

The field of iron–sulfur proteins, the proteins containing iron complexed with sulfhydryl residues and in most cases with inorganic sulfur, can trace its beginnings to the 1950s. In fact, amazement has been expressed at the lateness of recognition of these ubiquitous components of various electron transport systems with microorganisms, plants, and animals (Beinert, 1973). At the time this chapter was written, however, numerous review articles, chapters in books, and books were available on the subject (e.g., Lovenberg, 1973, 1974, 1977; Hall *et al.*, 1975; Orme-Johnson, 1973; Orme-Johnson and Orme-Johnson, 1978; Lippard, 1973, Yasunobu *et al.*, 1976).

The classification and nomenclature of the iron–sulfur proteins have been the subject of much debate, and the Nomenclature Committee of the International Union of Biochemistry (1979) has finally recommended the classification method depicted in the Scheme 1. According to this recommendation, iron–sulfur proteins are placed into a major class of iron proteins, the other two classes being the heme-containing proteins on the one hand and various other nonheme iron proteins such as the transferrins, ferritin, and oxygenases on the other. The iron–sulfur proteins are further divided into the *simple* type, which contain only iron, amino acids, and in most cases inorganic sulfur, and the *complex* iron–sulfur proteins that may also contain molybdenum, flavenoids, or heme. Simple iron–sulfur proteins include the rubredoxins and ferredoxins.

The rubredoxins (rubredoxin is often abbreviated Rd) are relatively small proteins and are red in color when in the oxidized state. The iron is bonded to sulfhydryl groups of cysteine residues. They contain no in-

Scheme 1. Classification of Iron Proteins with Special Reference to Iron–Sulfur Proteins

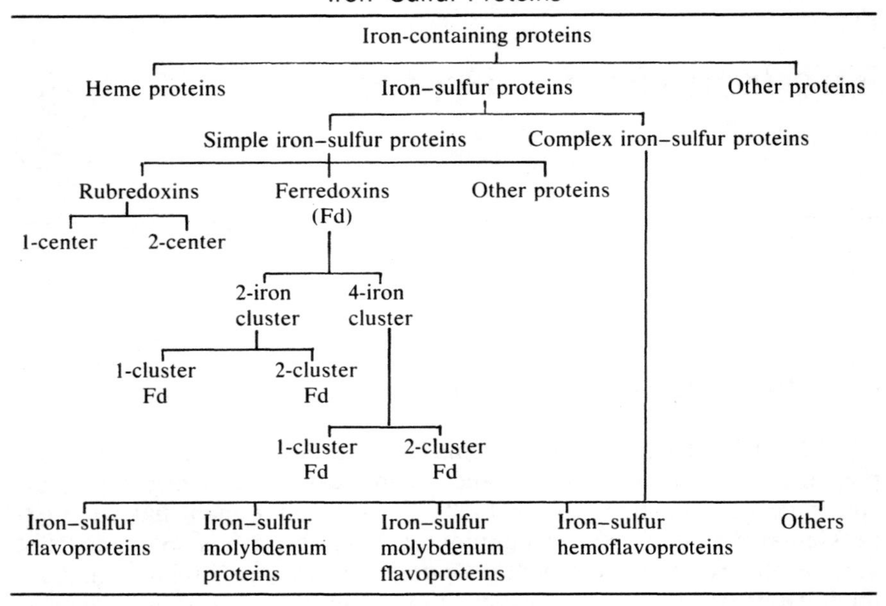

organic sulfur, and hence none is released from the rubredoxins by acid. A rubredoxin molecule may contain one or two iron atoms. Although they may participate in redox reactions, their specific physiological function, with some notable exceptions, is still unknown.

The much larger group of simple iron–sulfur proteins known as ferredoxins (ferredoxin is often abbreviated Fd) are brown in color and are specifically required in various redox systems. They are generally larger in molecular weight than are the rubredoxins and contain *clusters* of two or four iron atoms. The iron atoms are sequestered by sulfhydryl groups of the cysteinyl residues and by inorganic sulfur. The latter can be released as H_2S with acid. It was previously required that in order to be classified a ferredoxin, an iron–sulfur protein had to have a negative redox potential and be able to reduce pyridine nucleotides. These requirements have now been abandoned, and classification should now rest solely on the structure of the iron-containing centers of the protein.

The ferredoxins may be subdivided on the basis of their iron–sulfur centers into the plant-type and the bacterial-type ferredoxins. The former contain iron clusters consisting of two iron atoms, two atoms of inorganic sulfur (S^{2-}), and the cysteine ligands. The bacterial-type ferredoxins contain clusters of four iron atoms, four atoms of inorganic sulfur, and the sulfhydryl ligands. It is notable that the plant-type ferredoxins may occur in bacteria and of course in algae, but bacterial ferrodoxins do not occur in plants. The structure of chelated iron as it appears in rubredoxins and the two types of ferredoxins is shown in Figure 8-1.

Ferredoxins are usually involved in the transfer of but a single electron. Thus, even in clusters containing more than one iron atom, only one iron atom will undergo a redox reaction. The charge that ferredoxin iron residues can carry may be +2 or +3. Iron in the +3 redox state is paramagnetic and gives electron paramagnetic resonance (EPR) signals, whereas state +2 is, in most cases, diamagnetic, and no EPR signal is given. The number of iron and sulfur atoms as well as the oxidation state of the iron present in an iron–sulfur protein may be indicated by a shorthand system as follows:

- $[2\ Fe–2\ S]^{2+}$: one plant-type cluster with a +2 oxidation state in the iron atoms
- $[4\ Fe–4\ S]^{2+}$: one bacterial-type cluster with a +2 oxidation state in the iron atoms
- $2[4\ Fe–4\ S]^{2+}$: two bacterial-type clusters with a +2 oxidation state in the iron atoms

In some cases, one may wish to indicate the involvement of the

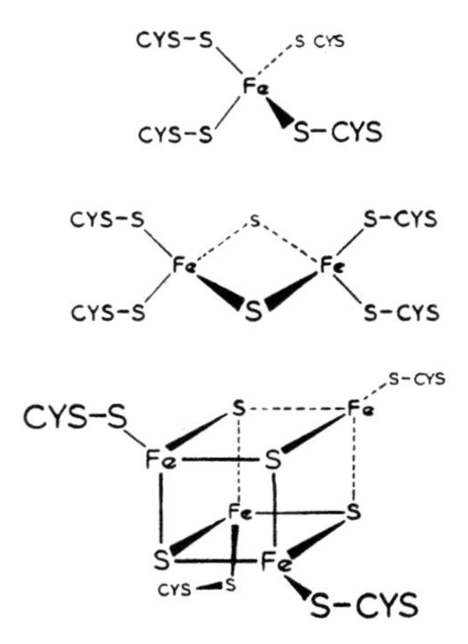

Figure 8-1. Structures of iron clusters in rubredoxin and the ferredoxins. Only ferredoxin clusters contain inorganic sulfur. (top) Rubredoxin cluster, (middle) iron clusters found in higher plant-type ferredoxins, and (bottom) iron clusters found in bacterial ferredoxins. Reproduced with permission from the *Annu. Rev. Biochem.*, Vol. 42. Copyright 1973 by Annual Reviews Inc.

sulfhydryl residues (RS) in iron clusters of rubredoxins and ferredoxins. Remembering that each sulfhydryl group loses a proton upon reacting with iron and indicating inorganic sulfur (sulfide, S^{2-}) by S*, we may write the following with the superscripts indicating the net charge of the cluster:

$[Fe(RS)_4]^{2-}$:	reduced rubredoxin (Fe^{2+} + 4 RS^-)
$[Fe(RS)_4]^{1-}$:	oxidized rubredoxin (Fe^{3+} + 4 RS^-)
$[Fe_2S_2^*(RS)_4]^{2-}$:	oxidized plant-type ferredoxin with a single two-iron cluster (2 Fe^{3+} + 2 S^{2-} + 4 RS^-)
$[Fe_2S_2^*(RS)_4]^{3-}$:	reduced plant-type ferredoxin in a single two-iron cluster (Fe^{3+} + Fe^{2+} + 2 S^{2-} + 4 RS^-)
$[Fe_4S_4^*(RS)_4]^{1-}$:	oxidized HIPIP* bacterial-type ferredoxin in a single four-iron cluster (3 Fe^{3+} + Fe^{2+} + 4 S^{2-} + 4 RS^-)
$[Fe_4S_4^*(RS)_4]^{2-}$:	reduced HIPIP bacterial-type ferredoxin or oxidized non-HIPIP bacterial-type ferredoxin with a single four-iron cluster (2 Fe^{3+} + 2 Fe^{2+} + 4 S^{2-} + 4 RS^-)
$[Fe_4S_4^*(RS)_4]^{3-}$:	reduced non-HIPIP bacterial-type ferredoxin with a single four-iron cluster (Fe^{3+} + 3 Fe^{2+} + 4 S^{2-} + 4 RS^-)

It is clear from this preliminary discussion that in the iron–sulfur proteins, iron is coordinated with four ligands. Since the 3*d*, 4*s*, and 4*p* orbitals of iron are the valence (bonding) orbitals, the eight electrons accounting for the coordinated covalent bonds between iron and sulfur may occupy any of these orbitals. Spectroscopic studies, however, have shown that in most if not all cases the 3*d* orbitals contain unpaired electrons. The 4*s* and $4p_{x,y,z}$ orbitals of iron can therefore be viewed as being responsible for the iron–sulfur bonding.

8.2 Rubredoxins

Only a few rubredoxins have been studied in great detail. These can be divided into the anaerobic and aerobic bacterial proteins. The former are usually of a low molecular weight and contain but a single iron atom. Those isolated from aerobic bacteria are larger and may carry two iron atoms per protein molecule. The function of the anaerobic bacterial rubredoxins, by far the more numerous class, is unknown, whereas the aerobic bacterial rubredoxins apparently participate in a system that hydrox-

* See Section 8.3 for definition.

ylates the ω position of alkanes and fatty acids as follows:

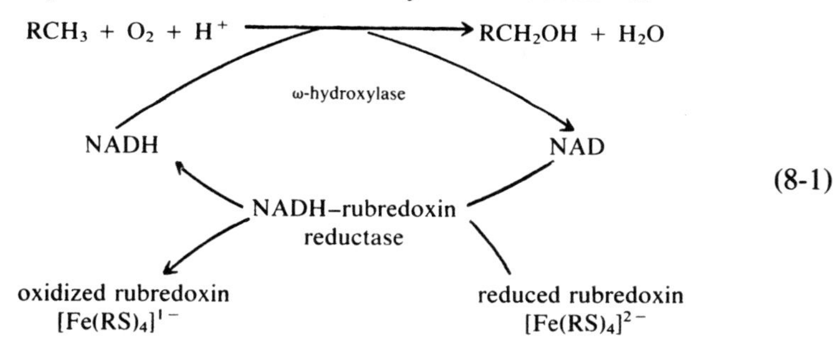

(8-1)

The most thoroughly investigated aerobic bacterial rubredoxin was isolated from *Pseudomonas oleovorans*. It has a molecular weight of around 20,000 and, as isolated, contains one iron atom sequestered to four cysteinyl residues. It can bind an additional iron atom when one is added to the monoferric preparation. In the visible range of the spectrum, it has absorption maxima at 380 and 497 nm in the oxidized form and has a redox potential of -0.037 V/mole.

Both the monoferric and diferric rubredoxins (oxidized forms) from *P. oleovorans* give signals at $g = 4.3$ and $g = 9.4$, indicating iron in a high-spin state in a rhombic field. No signals are given in the reduced state. The signals obtained are similar if not identical to those of the rubredoxins from anaerobic bacteria, indicating that the iron ligands are essentially the same for the two types of compounds. The signals obtained are illustrated in Figure 8-2.

The amino acid composition of the *Pseudomonas oleovorans* is given in Table 8-1, from which it is seen that the protein has 174 amino acid residues. Its N-terminal and C-terminal amino acids are alanine and lysine, respectively. Its primary structure is (Yasunobu and Tanaka, 1973a)

```
 1                                       10                                      20
H2N—Ala-Ser-Tyr-Lys-Cys-Pro-Asp-Cys-Asn-Tyr-Val-Tyr-Asp-Glu-Ser-Ala-Gly-Asn-Val-His-

                                         30                                      40
Glu-Gly-Phe-Ser-Pro-Gly-Thr-Pro-Trp-His-Leu-Ile-Pro-Glu-Asp-Trp-Asp-Cys-Pro-Cys-

                                         50                                      60
Cys-Ala-Val-Arg-Asp-Lys-Leu-Asp-Phe-Met-Leu-Ile-Glu-Ser-Gly-Val-Gly-Glu-Lys-Gly-

                                         70                                      80
Val-Thr-Ser-Thr-His-Thr-Ser-Pro-Asn-Leu-Ser-Glu-Val-Ser-Gly-Thr-Ser-Leu-Thr-Ala-
                                         90                                     100
Glu-Ala-Val-Val-Ala-Pro-Thr-Ser-Leu-Glu-Lys-Leu-Pro-Ser-Ala-Asp-Val-Lys-Gly-Gln-

                                        110                                     120
Asp-Leu-Tyr-Lys-Thr-Glu-Pro-Pro-Arg-Ser-Asp-Ala-Glu-Gly-Gly-Lys-Ala-Tyr-Leu-Lys-

                                        130                                     140
Trp-Ile-Cys-Ile-Thr-Cys-Gly-His-Ile-Tyr-Asp-Trp-Glu-Ala-Leu-Gly-Asp-Glu-Ala-Glu-
                                        150                                     160
Gly-Phe-Thr-Pro-Gly-Thr-Arg-Phe-Glu-Asp-Ile-Pro-Asp-Trp-Asp-Cys-Cys-Trp-Cys(Asx,

                                        170             174
Pro)Gly-Ala-Thr-Lys-Glu-Asn-Tyr-Val-Leu-Tyr-Glu-Glu-Lys—COOH
```

(8-2)

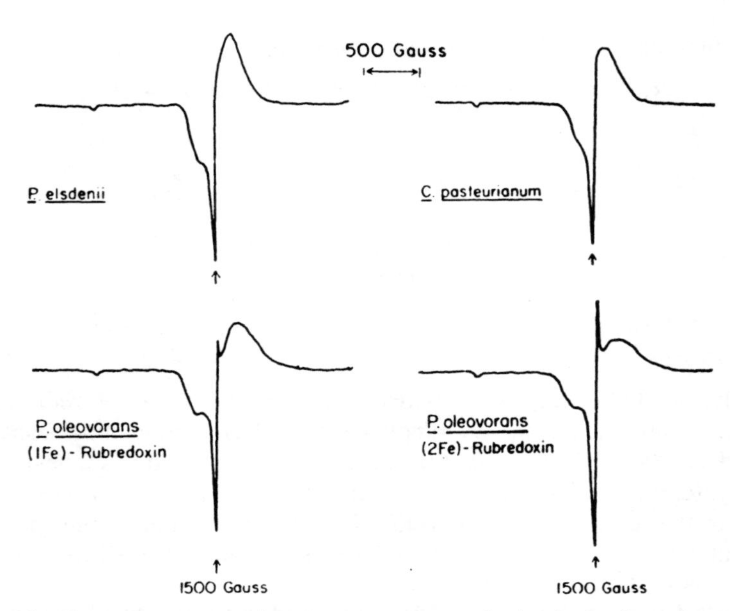

Figure 8-2. Electron paramagnetic resonance spectra of rubredoxins from four microorganisms. From Lode and Coon (1971).

which indicates that this rubredoxin, similar to the others, is an acidic protein. From the amino acid sequence given herein it is possible to surmise that two centers of iron sequestration may exist in the aerobic bacterial rubredoxins: one at the N-terminal and the other at the C-terminal end. Studies with cyanogen bromide fragments of this rubredoxin have shown that the iron is located at the C-terminus when the protein is isolated and that this iron center is involved in the catalytic activity of the protein in the hydroxylation reaction. If iron is also reconstituted with the N-terminal end of the protein, it takes no part in the hydroxylation reaction electron transfer system. The iron–protein complexes in aerobic bacterial rubredoxins are illustrated in Figure 8-3.

The best characterized rubredoxins are those of the anaerobic bacteria, especially the rubredoxin from *Clostridium pasteurianum*. This and other anaerobic bacterial rubredoxins have molecular weights of near 6000 and contain a single iron center per each protein molecule. The redox potential of the *C. pasteurianum* rubredoxin is -0.057 V/mole, and in the oxidized state it absorbs visible light at 490 and 380 nm. When reduced, it absorbs at 333 and 311 nm.

Spectroscopic data employing EPR, Mössbauer, and magnetic susceptibility methodology have indicated that the iron atom in *C. pasteurianum* rubredoxin is ligated in a weak tetrahedral arrangement and that

Table 8-1. Amino Acid Composition of Various Rubredoxins (in moles/mole of protein)[a]

Amino acid	*P. oleovorans* (aerobic)[b]	*C. pasteurianum* (anaerobic)	*P. elsdenii* (anaerobic)	*M. aerogenes* (anaerobic)
Lys	10	4	4	2
His	4	0	0	0
Arg	3	0	0	0
Asx	19	11	10	8
Thr	12	3	2	2
Ser	12	0	1	1
Glx	18 (19)	6	3	8
Pro	13	5	2	4
Gly	15	6	5	5
Ala	12	0	7	3
Cys	10	4	4	4
Val	10	5	3	4
Met	1	1	2	1
Ile	6	2	2	1
Leu	11	1	1	3
Tyr	8	3	3	3
Phe	4	2	2	3
Trp	6 (4)	1	1	1
Total	174	54	52	53

[a] Adapted from Yasunobu and Tanaka (1973a).
[b] Values in parentheses are those of Lode and Coon (1971).

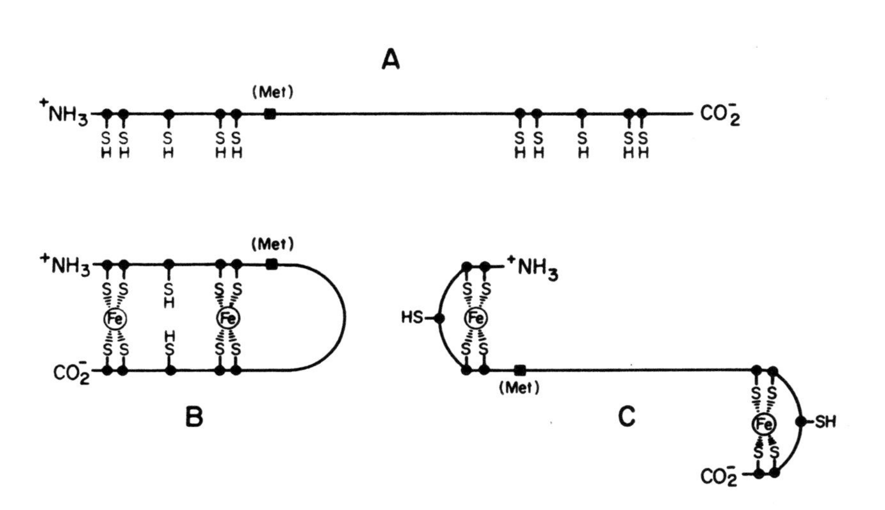

Figure 8-3. Schematic representation of the structure of the two-iron rubredoxin from *P. oleovorans*. A is the extended structure; B and C are possible structures, which depend on the positions of the disulfide bridges. Available evidence indicates that C is the correct representation. From Lode and Coon (1971).

iron is in the high-spin state both in the reduced and oxidized states (5.05 ± 0.20 and 5.85 ± 0.20 Bohr magnetons, respectively), corresponding to four unpaired electrons in the reduced state and five unpaired electrons in the oxidized state. These reside in the $3d$ orbitals of the iron atom. As is the case of the rubredoxin from *P. oleovorans*, the anaerobic bacterial rubredoxin gives EPR signals at $g = 4.3$ and $g = 9.4$ at low temperatures (1.5–5.0°K) (Figure 8-1).

The amino acid composition of several anaerobic bacterial rubredoxins is shown in Table 8-1. These proteins contain only 52–54 amino acids and are quite acidic. The isoelectric point of the *C. pasteurianum* rubredoxin is at pH 2.93 (Dutton and Rogers, 1978). Following are the primary structures of three anaerobic bacterial rubredoxins:

```
                          5                   10                  15
C.P. f-Met-Lys-Lys-Tyr-Thr-Cys-Thr-Val-Cys-Gly-Tyr-Ile-Tyr-Asp-Pro-Glu-Asp-
P.E.   Met-Asp-Lys-Tyr-Glu-Cys-Ser-Ile-Cys-Gly-Tyr-Ile-Tyr-Asp-Glu-Ala-Glu-
M.A.   Met-Gln-Lys-Phe-Glu-Cys-Thr-Leu-Cys-Gly-Tyr-Ile-Tyr-Asp-Pro-Ala-Leu-

               20                  25                  30
       Gly-Asp-Pro-Asp-Asp-Gly-Val-Asn-Pro-Gly-Thr-Asp-Phe-Lys-Asp-Ile-Pro-
       Gly-Asp- - -Asp-Gly-Asn-Val-Ala-Ala-Gly-Thr-Lys-Phe-Ala-Asp-Leu-Pro-
       Val-Gly-Pro-Asp-Thr-Pro-Asp-Gln-Asp-Gly- - -Ala-Phe-Glu-Asp-Val-Ser-

        35                  40                  45                  50
       Asp-Asp-Trp-Val-Cys-Pro-Leu-Cys-Gly-Val-Gly-Lys-Asp-Glu-Phe-Glu-Glu-     (8-3)
       Ala-Asp-Trp-Val-Cys-Pro-Thr-Cys-Gly-Ala-Asp-Lys-Asp-Ala-Phe-Val-Lys-
       Glu-Asn-Trp-Val-Cys-Pro-Leu-Cys-Gly-Ala-Gly-Lys-Glu-Asp-Phe-Glu-Val-

               54
       Val-Glu-Glu—COOH
       Met-Asp—COOH
       Tyr-Glu-Asp—COOH
```

It will be noted that methionine is the N-terminal amino acid in all cases and that, interestingly, the methionine is formylated in the *C. pasteu-*

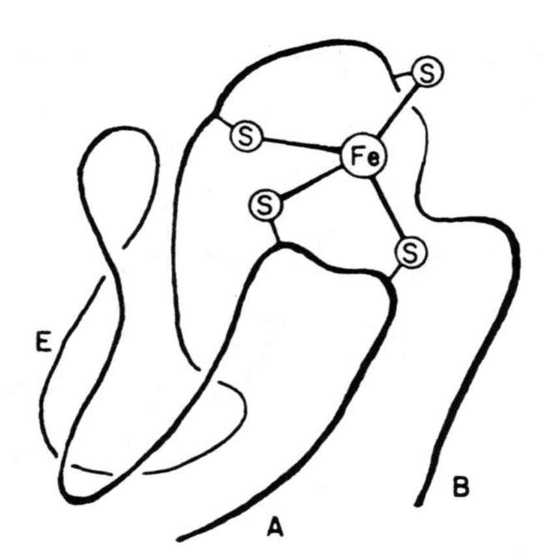

Figure 8-4. Three-dimensional representation of the structure of a one-iron rubredoxin from *C. pasteurianum*. From Watenpaugh *et al.* (1972). Copyright 1972 by Cold Spring Harbor Laboratory.

rianum rubredoxin. All four cysteinyl residues of this protein are involved in iron chelation (Cys appears in positions 6, 9, 39, and 42). From X-ray diffraction data, which have resolved the *C. pasteurianum* rubredoxin down to 1.5 Å (Watenpaugh *et al.*, 1972), the structure shown in Figure 8-4 has been constructed.

The primary structures of the anaerobic bacterial rubredoxins shown in equation (8-3) demonstrate extensive homologies among these proteins. Surprisingly, rubredoxin from *P. oleovorans* also shows a rather well-defined homology with the anaerobic bacterial rubredoxins in the 1–59 and the 119–174 sections of the protein molecule (line up Cys 6 of anaerobic rubredoxins with Cys 5 of the 1–59 fragment on the one hand and Cys 123 of the 119–174 fragment on the other). Such studies also indicate internal homology within the *P. oleovorans* rubredoxin, suggesting that this protein might have originated through a glue duplication process in the course of evolution.

8.3 Ferredoxins

The ferredoxins can be broadly defined as those simple iron–sulfur proteins which contain *labile* inorganic sulfur in addition to the iron. In this section, we shall focus our attention on higher plant and microbial ferredoxins, which, for the most part, are involved in electron transport systems in these organisms. Such ferredoxins can be divided into the higher plant type (clusters containing two iron atoms; see Figure 8-1), the bacterial high-potential iron protein (HIPIP) type (molecules containing a single four-iron cluster), and the clostridial type (molecules containing two four-iron clusters for a total of eight iron atoms/molecule).

8.3.1 Plant-Type Ferredoxins (Two-Iron Clusters)

Ferredoxins of the higher plant type are present in chloroplasts and participate in various aspects of the photosynthetic phenomenon. They are reddish in color when in the oxidized form. Originally, the chloroplast ferredoxins were observed as factors promoting the photoreduction of methemoglobin, where the latter acted as a terminal electron acceptor.

The isolation of a plant ferredoxin is exemplified by the preparation procedure of the alfalfa ferredoxin (Keresztes-Nagy and Margoliash, 1966): Freshly picked alfalfa is stored frozen; when required, some 10 kg of the frozen material is crushed and homogenized in a blender in batches of 0.5 kg with 1.3 liter of water containing 1 g of Tris. This maintains the pH of the homogenate at 7.5–8.0, which is necessary to prevent loss of H_2S. The homogenate is filtered, and enough NaCl is added to give an ionic strength of 0.15. The solution is then passed through a DEAE–cellulose column, which retains the ferredoxin. The column is washed repeatedly with 0.15 *M* NaCl, and then the ferredoxin is eluted with 0.8 *M* NaCl and concentrated. It is then passed through a Sephadex G-25 column, where it emerges in the void volume. To this crude ferredoxin solution is added 0.6 g of $(NH_4)_2SO_4$ per ml of solution, whereby ferredoxin remains in the supernatant. It is diluted 50-fold with H_2O and is chromatographed on the DEAE–cellulose column as before. The yield at this point is 350 mg of ferredoxin. Final purification of the ferredoxin is achieved by 3-fold chromatography on DEAE-cellulose using an ionic strength gradient system of 0.2 to 0.4 *M* NaCl buffered with 0.01 *M* Tris–HCl at pH 7.5. The final product is deaerated and stored under nitrogen. The procedure should be carried out as rapidly as possible in order to avoid loss of the organic sulfur and the oxidation of any ferrous iron present.

The iron clusters, depicted in Figue 8-1(b) and (c), are radically different from those encountered in the rubredoxins. EPR spectroscopy reveals signals in the g = 1.8–2.1 range, and EPR patterns given by the various types of ferredoxins differ from each other as well (Figure 8-5). However, all share a signal at g = 1.94 when in the reduced state and at low temperatures (about 80°K). Plant-type as well as most other ferredoxins do not give EPR signals when in the oxidized state. This is explained by viewing the two iron atoms of the two-iron cluster as being in the high-spin Fe^{3+} state ($S = \frac{5}{2}$) and being coupled antiferromagnetically. The spins should thus cancel each other out. On the other hand, when one of the iron atoms gets reduced, it becomes a high spin Fe^{2+} ($S = \frac{4}{2}$) and the net spin then becomes $+\frac{1}{2}$ inducing an EPR signal (Figure 8-6).

In the oxidized state, then, both iron atoms would be expected to exist in the $\frac{5}{2}$ spin state and thus have five unpaired electrons in the 3*d* orbitals. Although there is some uncertainty about this, it is believed that the ligand field characterizing iron in the plant-type ferredoxins is tetrahedral in nature, and a recent X-ray diffraction analysis of the *Spirulina platensis* ferredoxin at 2.8-Å resolution seems to bear this out. The X-ray study also demonstrated the involvement of cysteine residues 41, 46, 49, and 79 in the sequestration of iron (Tsukihara *et al.*, 1978). The active center of the *S. platensis* ferredoxin is shown in Figure 8-7.

Ferredoxins of the higher plant type have been isolated from higher plants, algae, and some bacteria. With some exceptions, these participate in various aspects of the photosynthetic phenomenon. Traditionally, soluble ferredoxins of the higher plant type have been viewed as electron

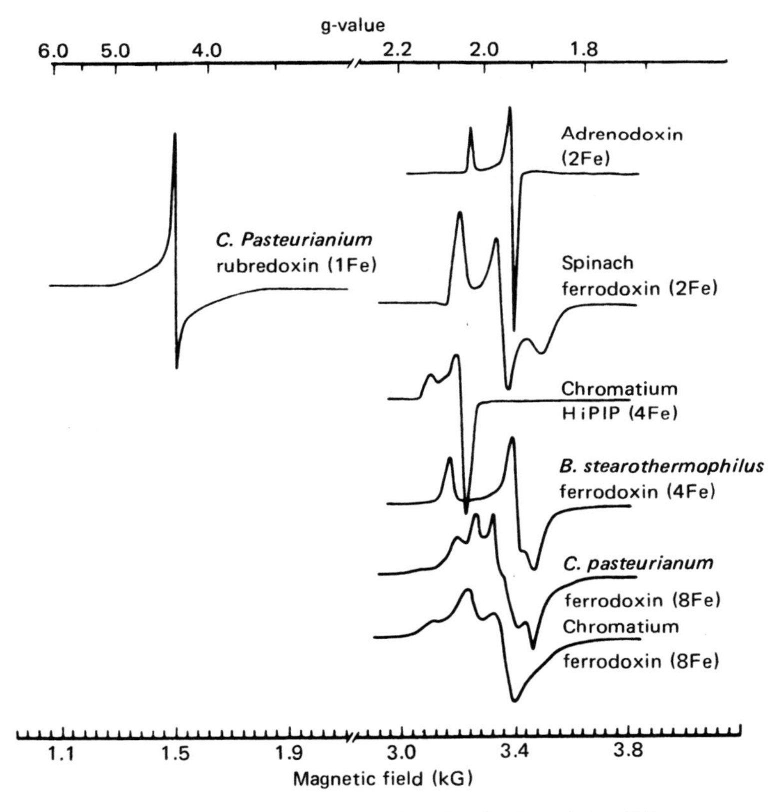

Figure 8-5. Electron paramagnetic spectra of a rubredoxin and the different types of ferredoxins. Rubredoxin and adrenodoxin were analyzed at 77°K, the other ferredoxins at 20°K. From Hall, D. O., Rao, K. K., and Cammack, R., *Sci. Prog. Oxf.* 62:285–317, 1975, by permission of the Blackwell Scientific Publications, Ltd.

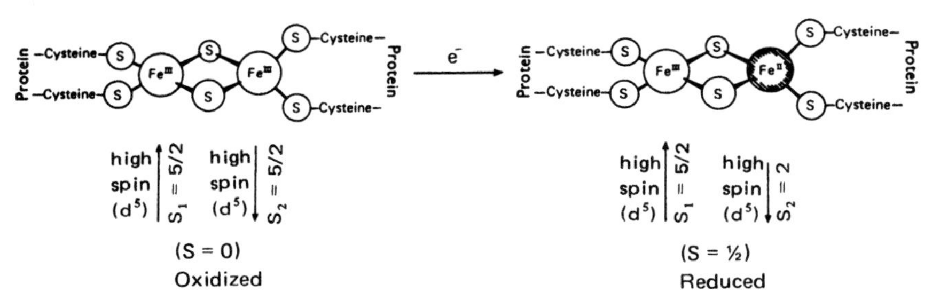

Figure 8-6. Iron–sulfur clusters in the plant-type ferredoxin, indicating antiferromagnetic coupling in the oxidized state. From Rao *et al.* (1971).

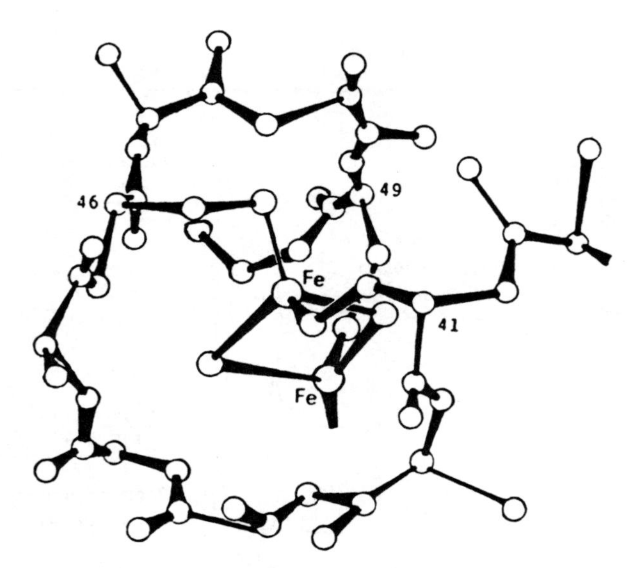

Figure 8-7. Three-dimensional representation of the environment of a plant-type ferredoxin iron cluster. From Tsukihara *et al.* (1978).

donors to NADP and therefore must possess electrode potentials lower (more negative) than that of the NADP system ($\Delta E_0^{\prime} = -0.32$ V). These and other physical parameters of some plant-type ferredoxins are given in Table 8-2.

The primary structures of numerous ferredoxins have been deter-

Table 8-2. Some Physical–Chemical Parameters of Higher Plant-Type Ferredoxins

	Ferredoxin source			
	Nostoc MAC[a]			*Pseudomonas putida*
Parameter	Fd I	Fd II	Spinach[b]	(putidaredoxin)[b]
Absorption maxima (oxidized, nm)	330, 423, 465	330, 423, 465	330, 420, 463	328, 412, 455
Molecular weight	10,850	10,500	12,000	12,000
No. of amino acids	98	96	97	106
Isoelectric point	3.2	3.1	—	—
Redox potential (V)	−0.350	−0.455	−0.420	−0.235
Electrons transferred	1	1	1	1
EPR *g* values	—	—	1.94	1.94, 2.01

[a] A blue-green alga; from Hutson *et al.* (1978).
[b] From Yasunobu and Tanaka (1973a).

Table 8-3. Amino Acid Composition of Various Plant-Type (two iron atoms/cluster) Ferredoxins (in moles of amino acid/mole of protein)

Amino Acid	Organism source[a]							
	1	2	3	4	5	6	7	8
Lysine	5	4	5	4	4	5	3	3
Histidine	2	1	2	2	2	1	1	2
Arginine	1	1	1	1	1	1	3	5
Aspartic Acid	10	13	11 (10)	13	10	13	15	13
Threonine	6	8	5	6	8	10	6	5
Serine	8	7	7 (8)	7	7	7	7	7
Glutamic acid	17	13	18 (17)	15	16	13	12	10
Proline	3	4	5 (4)	3	2	3	3	4
Glycine	7–8	6	6	6	6	6	7	8
Alanine	10	9	7	9	7	6	10	9
½-Cystine	6	5	5 (4)	5	6	5	4	6
Valine	9	7	7 (8)	5	6	8	6	14
Methionine	0	0	1	0	1	0	0	3
Isoleucine	4	4	4 (5)	5	6	6	6	6
Leucine	6	8	7 (8)	10	8	8	7	6
Tyrosine	4	4	4	5	6	5	3	3
Phenylalanine	2	2	1	2	2	1	3	1
Tryptophan	1	1	1	0	0	0	0	1
Total	111–112	97	97 (97)	98	98	98	96	106

[a] Organisms:
1. *Medicago sativa* (alfalfa); Keresztes-Nagy and Margoliash (1966).
2. *Spinacea oleracea* (spinach); Shin *et al.* (1979).
3. *Triticum aestivum* (wheat); Shin *et al.* (1979). Numbers in parentheses are values obtained by Takruri and Boulter (1979).
4. *Cyanidium caldorium*, a eukaryotic red alga; Hase *et al.* (1978b).
5. *Porphyria umbilicalis*, a red alga; Takruri *et al.* (1978).
6. *Nostoc MAC*, a blue-green alga, ferredoxin I; Hutson *et al.* (1978).
7. *Nostoc MAC*, ferredoxin II.
8. *Pseudomonas putida* (putidaredoxin); Dayhoff (1976).

mined. They are of considerable interest as indicators of pathways taken by organisms during the course of their evolution from the prokaryotic to the eukaryotic states and as aids in the taxonomy of plants. The amino acid composition for representative higher-plant-type ferredoxins is given in Table 8-3, and amino acid sequences for a number of such ferredoxins are given in Figure 8-8. It can be surmised that all ferredoxins show homologies with each other, though the extent to which this is true varies enormously. It is possible to construct a matrix, shown in Table 8-4, which would indicate differences among the various sequences. It is seen that the ferredoxins from wheat, spinach, alfalfa, taro, and kao differ by only some 20 amino acid residues and are thus quite close from an evolutionary point of view. It is thus possible to classify plants of doubtful taxonomy on the basis of their ferredoxin sequence, as was done with, for instance,

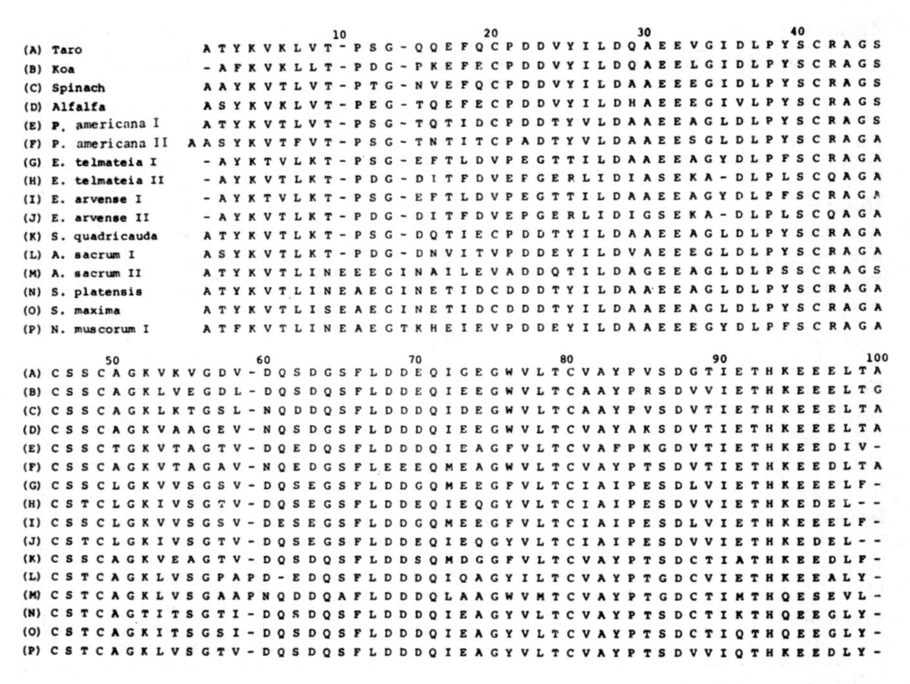

Figure 8-8. Amino acid sequences of several plant-type ferredoxins. Abbreviations for amino acid residues are standard, and a biochemistry text should be consulted for further information. From Wakabayashi *et al.* (1978).

Halobacterium halobium. The ferredoxin from this organism had a higher-plant-type iron cluster but contained 128 amino acid residues/mole (m.w. 14,330). Nevertheless, its amino acid sequence, when properly aligned, was quite similar to the ferredoxin of *Nostoc muscorum*, a blue-green alga. *H. halobium* therefore shares a kinship with *N. muscorum* from an evolutionary point of view (Hase *et al.*, 1978a). Phylogenetic trees have also been constructed on the basis of such matrices showing amino acid sequence differences (Figure 8-9).

As indicated previously, the function of most of the higher-plant-type ferredoxins is to participate in the photosynthetic reaction, i.e., fixation of CO_2, as indicated in

$$CO_2 + 2\,H_2O + \text{light} \rightarrow (CH_2O) + H_2O + O_2 + 112 \text{ kcal/gram atom of C} \quad (8\text{-}4)$$

From this equation it is evident that the *fixation* of each molecule of CO_2 is accompanied by the transfer of four electrons and the production of O_2. In this reaction, CO_2 is reduced, and water, i.e., oxygen of the water molecule, is oxidized.

Table 8-4. Matrix of Amino Acid-Sequence Differences for Representative Plant-Type Ferredoxins.[a] I and II refer to two different ferredoxin types

Organism																
Type	Common name	Latin name	*Triticum aestivum*	*Spinacea oleracea*	*Medicago sativa*	*Colocasia esculenta*	*Leucaena glauca*	*Equisetum arvense* I	*Equisetum telmateia* I	*E. arvense* II	*E. telmateia* II	*Scenedesmus*	*Aphanothece*	*Spirulina maxima*	*Nostoc muscorum*	*Porphyra umbilicalis*
Higher plants	Wheat	*Triticum aestivum*	0													
	Spinach	*Spinacea oleracea*	21	0												
	Alfalfa	*Medicago sativa*	19	19	0											
	Taro	*Colocasia esculenta*	20	18	16	0										
	Koa	*Leucaena glauca*	20	20	23	20	0									
Horsetails	—	*Equisetum arvense* I	34	40	40	37	37	0								
	—	*Equisetum telmateia* I	32	39	39	36	36	1	0							
	—	*E. arvense* II	44	46	46	44	40	31	40	0						
	—	*E. telmateia* II	42	45	45	43	39	30	29	1	0					
Green alga	—	*Scenedesmus*	30	30	29	26	33	31	30	43	42	0				
Blue-green algae	—	*Aphanothece sacrum* I	30	33	34	37	34	41	41	41	40	29	0			
	—	*Spirulina maxima*	33	34	37	36	39	41	40	45	44	25	30	0		
	—	*Nostoc muscorum*	27	34	34	36	32	39	38	41	40	28	26	21	0	
Red alga		*Porphyra umbilicalis*	37	37	37	37	40	39	38	41	43	31	41	28	37	0

[a] From Takruri and Boulter, 1979.

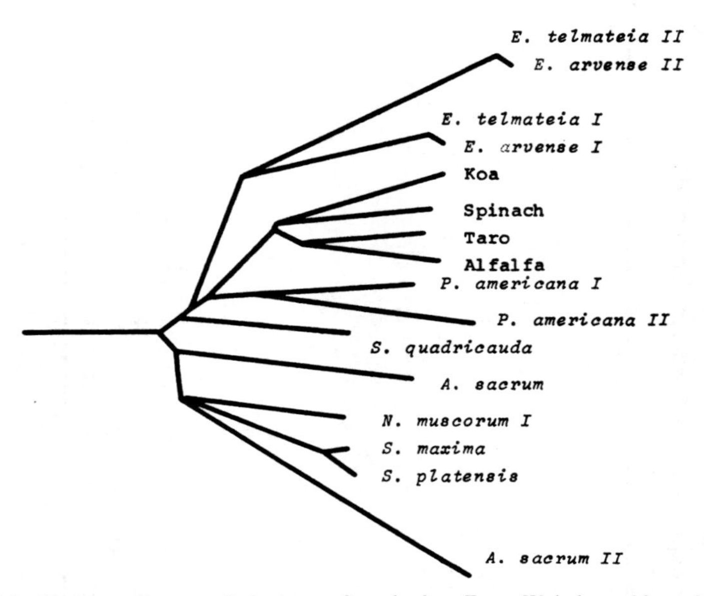

Figure 8-9. Phylogenetic tree of plant-type ferredoxins. From Wakabayashi *et al.* (1978).

This process is of course not as simple as depicted in equation (8-4), and numerous intermediates, including NADP and ferredoxins, are involved in the process. The most remarkable aspect of the photosynthetic process is that electrons are pumped "uphill" from the O_2–H_2O system with an E_0^1 of $+0.82$ compared to that of the Fd_{ox}–Fd_{red} system with an E_0^1 of -0.42 (spinach). This, of course, requires an input of an enormous amount of energy, and the latter is supplied by light, which is harnessed through a system of chloroplast pigments. In terms of energy equivalents, we are speaking of a total E_0^1 of 1.24 V, which would correspond to about 57,000 cal/two electrons transferred if the system were 100% efficient. In actuality, some 170,000 cal are required to accomplish the oxidation of one mole of water.

The photosynthetic electron transport machinery of higher plants is illustrated in Figure 8-10. It is sometimes referred to as a Z diagram to emphasize its schematic appearance (Zelitch, 1979). There are two light-sensitive centers in this system, termed PS (photosystem) I and PS II, responding to light at 700 and 670 nm, respectively. Both systems apparently contain chlorophyll. They may be considered to represent "pumping" stations along the way electrons take from water to CO_2: The initial reaction, involving PS II, transfers electrons from water to a *strong oxidant* (compound Z in Figure 8-10). It apparently has a ΔE_0^1 of greater

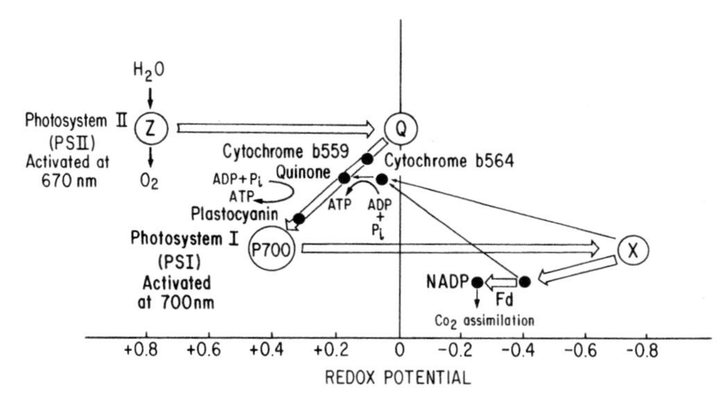

Figure 8-10. Photosynthetic process showing the function of iron–sulfur proteins therein. This representation is often referred to as the Z-scheme. See text for details.

than +0.8 V, and it is believed to be chlorphyll in nature. The next step, boosted by light energy, involves the uphill transfer of electrons from Z_{red} to a *weak reductant* termed Q in Figure 8-10. This compound is believed to be a quinone in nature. The electrons can now be transferred spontaneously (without energy input) from Q_{red} to a weak oxidant of PS I. In Figure 8-10 it is designated as P-700 and is probably chlorophyll in nature. Next, in a light-dependent event, the electrons are once again pumped uphill to a *strong reductant* of PS I, termed X in Figure 8-10. It is believed to consist of iron–sulfur protein centers. Associated with PS I are two additional iron–sulfur centers, chloroplast-bound and therefore insoluble in water. They are apparently of the four-iron type and are termed centers A and B. Their EPR g values are 2.05, 1.94, and 1.86 for center A and 2.05, 1.92, and 1.89 for center B. Their $\Delta E_0^{\prime}$ values are −0.0540 V for center A and −0.0590 for center B (Malkin and Bearden, 1978). The exact position or function of iron–sulfur centers in PS I is not certain.

Following reduction of X, electrons are transferred to the soluble ferredoxins, which, under the influence of ferredoxin–NADP reductase (a flavoprotein), transfers its electrons to NADP. The NADPH + H^+ can then catalyze the reduction of CO_2, depicted in general terms as

$$3\,CO_2 + 9\,ATP + 6\,NADPH + 6\,H^+ + 5\,H_2O \rightarrow \text{triose-P} + 9\,ADP + 8\,P_i + 6\,NADP \quad (8\text{-}5)$$

It is obvious from equation (8-5) that the *fixing* of CO_2 requires ATP in addition to NADPH + H^+. The ATP must be generated either by respiration or photosynthesis. As indicated in Figure 8-10, the latter process is feasible, where for each NADP reduced, one ATP is generated in the process of electron transfer from Q_{red} to P-700_{ox}. However, the fixation of each CO_2 molecule requires three ATP molecules, so that the additional ATPs must be supplied from elsewhere. This is accomplished by short-circuiting the system in what is a termed cyclic phosphorylation process, where reduced X or ferredoxin loses its electrons to cytochrome b_{564}, and in the process one ATP molecule is generated. Cyclic phosphorylation thus provides a total of two ATP molecules but does not result in the reduction of NADP. On the other hand, noncyclic phosphorylation results in the production of one ATP molecule and the reduction of one molecule of NADP. One could therefore supply the needed ingredients for the reduction of CO_2 by appropriately balancing the extent to which cyclic and noncyclic phosphorylations occur in the chloroplast [e.g., four passes of the noncyclic phosphorylation pathway to yield 4(NADPH + H^+) and 4 ATP, plus one pass of the cyclic phosphorylation pathway to yield two ATP molecules to reduce two CO_2 molecules] (Yocum *et al.*, 1973).

A ferredoxin (putidaredoxin) containing the higher-plant-type iron center has been isolated from *Pseudomonas putida*; however, its function and structure are unlike those of the two-iron ferredoxins. The amino acid composition of putidaredoxin is given in Table 8-3, and its physical–chemical properties are listed in Table 8-2.

Putidaredoxin is a component of an electron transport system which hydroxylates camphor; hence, functionally this ferredoxin may be classified as a component of an oxygenase system. The other components are cytochrome P-450 (m.w. 45,000), NAD, and a flavoprotein. The sequence of events is represented as follows:

$$\underset{(E_0^1 = -0.38\ \text{V})}{\text{P-450–Fe}^{3+}} + \text{camphor} \rightarrow \underset{(E_0^1 = -0.18\ \text{V})}{\text{P-450–Fe}^{3+}\text{–camphor complex}} \quad (8\text{-}6a)$$

$$\text{P-450–Fe}^{3+}\text{–camphor} + \underset{(E_0^1 = -0.235\ \text{V})}{\text{Fe}^{2+}\text{–putidaredoxin}} \rightarrow \text{P-450–Fe}^{2+}\text{–camphor} + \text{Fe}^{3+}\text{–putidaredoxin} \quad (8\text{-}6b)$$

$$\text{P-450–Fe}^{2+}\text{–camphor} + O_2 \rightarrow \text{P-450–Fe}^{2+}\text{–}O_2\text{–camphor} \quad (8\text{-}6c)$$

$$\text{P-450–Fe}^{2+}\text{–}O_2\text{–camphor} \rightleftharpoons \text{P-450–Fe}^{3+}\text{–}O_2^-\text{–camphor} \quad (8\text{-}6d)$$

$$\text{P-450–Fe}^{3+}\text{–}O_2^-\text{–camphor} + \text{Fe}^{2+}\text{–putidaredoxin} \rightarrow \text{P-450–Fe}^{2+}\text{–}O_2^-\text{–camphor} + \text{Fe}^{3+}\text{–putidaredoxin} \quad (8\text{-}6e)$$

$$\text{P-450–Fe}^{2+}\text{–}O_2^-\text{–camphor} \rightarrow \underset{(\text{position 5})}{\text{camphor–OH}} + 2\,H_2O + \text{P-450–Fe}^{3+} \quad (8\text{-}6f)$$

The putidaredoxin is reduced by a flavoprotein ($E_0^1 = -0.285$ V), which in turn is regenerated by the NAD–NADH system ($E_0^1 = -0.32$ V). The crucial event in this series of reactions appears to be the binding between P-450–Fe^{3+} and camphor, whereby the ΔE_0 of P-450–Fe^{3+} of -0.38 V is increased to −0.18 V, thus making it possible to be reduced by putidaredoxin. This bacterial system has been of great help in unraveling the mechanisms of mammalian hydroxylation reactions, especially those catalyzed by a protein functionally similar to putidaredoxin, namely adrenodoxin (see Sectin 8.5).

8.3.2 Bacterial-Type Ferredoxins Containing Two Four-Iron Clusters

The eight-iron ferredoxins have been found largely in the anaerobic organisms (Hall *et al.*, 1975), and in fact clostridial ferredoxins have been the subject of most such investigations. These include ferredoxins from *Clostridium pasteurianum, Clostridium butyricum, Peptococcus aerogenes, Clostridium acidiurici, Clostridium tartarivorum, Clostridium thermosaccharolyticum,* and *Peptostreptococcus elsdenii.* Since they contain so much inorganic sulfide (eight atoms/molecule) and iron, they are almost black in color when in the crystalline state. Isolation procedures are similar to those of plant ferredoxins, where advantage is taken of the acidic nature of this protein: The microorganisms are broken down (disrupted) by sonication, and ferredoxin is absorbed from the soluble portion of the cells by DEAE–cellulose. After eluting the ferredoxin, the preparation is chromatographed on Sephadex G-25 and then again on DEAE–cellulose at about pH 7.5 and is finally crystallized from ammonium sulfate solution. This procedure was used by Lovenberg *et al.* (1963) to prepare five ferredoxins from various anaerobic microorganisms.

Each eight-iron ferredoxin molecule contains two equivalent iron clusters, each cluster carrying four iron and four inorganic sulfur atoms, though there are reports on the existence of ferredoxin-like proteins with two unequal iron centers. One such substance was postulated to have one four-iron and one two-iron center (Stout, 1979a). The two centers are about 12 Å apart, at least in the well-characterized *Peptococcus aerogenes* ferredoxin (Adman *et al.*, 1973). The appearance of each center is illustrated in Figure 8-1(c), so that each iron atom has a tetrahedral ligand field, and the entire cluster looks like a cube. Each iron cluster in the clostridial type of ferredoxins is associated with an aromatic residue, most likely tyrosine. In three dimensions, the *P. aerogenes* ferredoxin appears as shown in Figure 8-11. The eight-iron ferredoxins give no EPR spectrum when in the oxidized state, though in the reduced state, signals with *g* values of 1.89, 1.96, and 2.01 can be observed. The *oxidized* form of such ferredoxins is considered to be $2[4\ Fe–4\ S^*–4\ RS]^{2-}$, i.e., where both clusters contain 2 Fe^{3+} and 2 Fe^{2+}, whereas the *reduced state* is defined

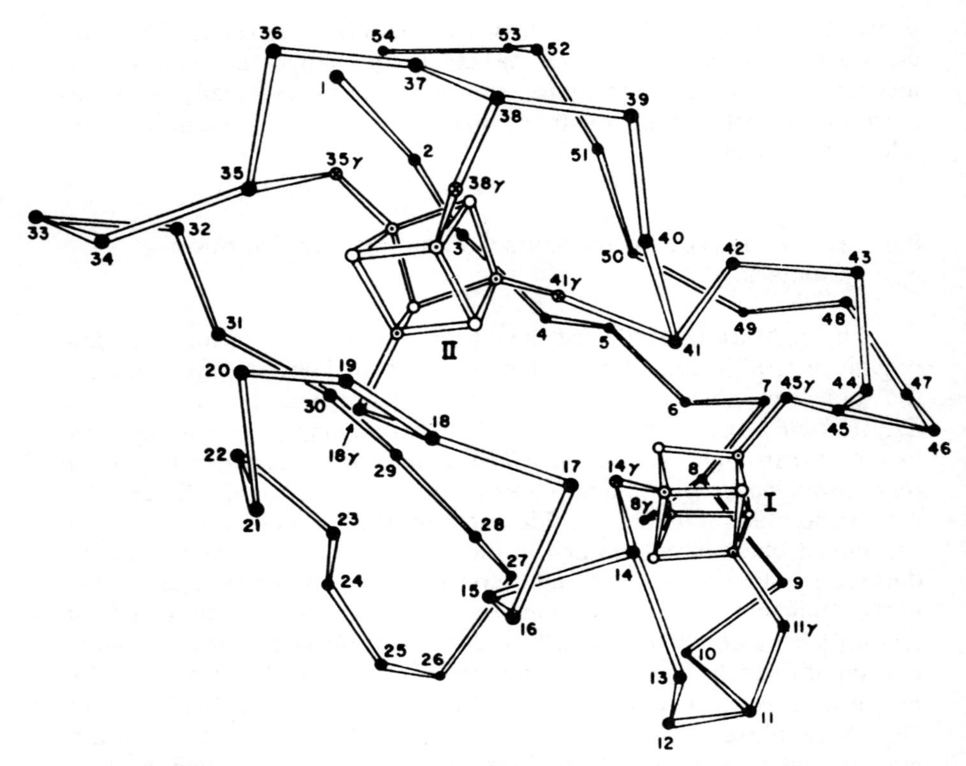

Figure 8-11. Three-dimensional structure of an eight-iron ferredoxin from *Peptococcus aerogenes*. Fe is ⊙; inorganic sulfur is ○; cysteine sulfur is ⊗; α-carbon atom is ●. From Adman *et al.* (1973).

as the state where each cluster contains 3 Fe^{2+} and 1 Fe^{3+}, i.e., 2[4 Fe–4 S*–4 RS]$^{3-}$. The transition between each state is characterized by a ΔE_0^1 of -0.410 V. There is also a *super oxidized* state, seldom if ever occurring in nature, where we have 1 Fe^{2+} and 3 Fe^{3+} in each cluster. A molecular species where one cluster is in the reduced and the other in the oxidized state has been observed, and this gives an EPR pattern similar to that of ferredoxins containing a single four-iron cluster (see Section 8.3.3). The ΔE_0^1 of both clusters in most of the eight-iron ferredoxins are almost identical, regardless of whether or not the companion cluster is in the oxidized or reduced state. This fact has been used to argue that the two clusters do not interact with each other (Sweeney and McIntosh, 1979). A ferredoxin molecule, in spite of its two redox centers, exchanges only one electron at a time. It is believed that the reason oxidized forms of bacterial ferredoxins do not give EPR signals is the nature of the 2 Fe^{3+}–2 Fe^{2+} center. As one iron is reduced to give the 1 Fe^{3+}–3 Fe^{2+} centers, the iron atoms can couple antiferromagnetically to give a net spin of $\frac{1}{2}$ and therefore an EPR signal.

There are ferredoxins that give widely differing redox potentials for the two iron clusters found therein. These ferredoxins (termed the azobacter type) have been isolated from *Azobacter vinelandii, Rhodospirillum rubrum* (Fd IV), *Mycobacterium flavum*, and *Mycobacterium smegmatis* (Hase *et al.*, 1979). The difference in redox potentials of the two centers is generally 0.7 V, where one center generally has a ΔE_0^1 of ~ −0.45 and the other +0.25 V. These ferredoxins operate, in contrast to the clostridial types, between 2[4 Fe–4 S*–4 RS]$^{2-}$ (reduced) and 2[4 Fe–4 S*–4 RS]$^{-1}$ (oxidized) states. In such cases, the reduced form will be diamagnetic and will give no EPR signal, whereas the oxidized form will give a signal with a g value of 2.01. It is speculated that the arrangement of the cysteine side-chain ligands in the protein molecule is in part responsible for the differences between the clostridial and azobacter types of ferredoxins. The azobacter-type ferredoxins have higher molecular weights than those of the clostridial type. Physical properties of various eight-iron ferredoxins are given in Table 8-5.

In addition to the clostridial and azobacter types of eight-iron ferredoxins, we can distinguish ferredoxins derived from photosynthetic bacteria. These include *Chlorobium limicola* (Fd I and Fd II), *Chromatium vinosum*, and *Chlorobium thiosulfatophilum* (Hase *et al*, 1978c). It would appear that the molecular weights of the azobacter-type ferredoxins are the largest, followed by those of the photosynthetic bacteria, and finally by the clostridial ferredoxins. Differences can also be observed from amino acid composition, shown in Table 8-6.

Amino acid sequences of numerous eight-iron ferredoxins have been determined. All are, of course, homologous proteins, though the greatest degrees of identity are seen among the clostridial ferrodoxins (Hall *et al.*, 1975), the azobacter-type ferredoxins (Hase *et al.*, 1979), and the photosynthetic bacterial ferredoxins (Hase *et al.*, 1978c). An interesting feature in the clostridial ferredoxin area is their internal homology. It has been proposed that such proteins originated from the gene duplication and gene fusion processes during the course of evolution, a situation not dissimilar to the apparent origin of the transferrin group of proteins. However, this proposal has been challenged (Yasunobu and Tanaka, 1973a) because clostridial ferredoxins could not be considered as consisting of two independent halves with respect to their active sites: One iron cluster involves Cys residues 8, 11, 14, and 45, and the other involves residues

Table 8-5. Physical–Chemical Properties of Eight-Iron Microbial Ferredoxins[a]

	Microorganism source			
Parameter	*Cl. pasteurianum*	*Azobacter vinelandii* Fd I[b]	*Peptococcus aerogenes*	*Chromatium vinosum*
Absorption maxima (oxidized, nm)	390	—	—	390
Molecular weight	5600	14,500	5600	10,000
Isoelectric point	2.75	—	—	—
Redox potential (V)	−0.410	−0.424; +0.320	—	−0.480
Electrons transferred	1	1	—	1
EPR *g* values	1.89, 1.96, 2.01 (red.)	2.01 (ox.)	—	1.94 (red.)
Number of amino acids	56	—	54	81
Number of iron atoms	8	8	8	8
Number of iron clusters	2	2	2	2

[a] From Yasunobu and Tanaka (1973a) and Rabinowitz (1971), unless otherwise noted.
[b] From Stout (1979b).

18, 35, 38, and 41 in the *Peptococcus aerogenes* Fd. Thus both halves of the protein are intimately interconnected. Amino acid sequences of a number of clostridial ferredoxins are given in Figure 8-12, and the phylogenetic tree constructed therefrom is given in Figure 8-13.

The functions of the eight-iron ferredoxins are many and varied. The clostridial ferredoxins, because of their low ΔE_0^1, can substitute for the higher-plant-type ferredoxins in the NADP–reductase system of chloroplasts. This, of course, is an artifical situation. However, in photosynthetic bacteria, ferredoxins participate in the reduction of NADP and in ATP production by the cyclic phosphorylation process. Ferredoxins are an essential component of the nitrogen-fixing mechanism in a number of bacteria. The details of this reaction are dealt with in Section 8.4. But perhaps the best studied bacterial reactions involving ferredoxins are the CO_2 assimilation reactions and other processes involving carbohydrate and lipid metabolism. These and other functions of bacterial ferredoxins will be discussed in greater detail in Section 8.3.4.

8.3.3 Bacterial Ferredoxins with One Four-Iron Cluster

Several microorganisms produce ferredoxins with single four-iron clusters. The structures of the iron-containing active centers are exactly identical to those of the eight iron-atom ferredoxins; i.e., they contain, in addition to the iron, four inorganic sulfur atoms and are associated with four sulfhydryl residues. They have the appearance of a cube, as illustrated in Figure 8-1(c). Detailed X-ray diffraction studies have been performed on at least one such protein (Carter *et al.*, 1972). Microorganisms that biosynthesize ferredoxins of the four-iron type are *Desulfovibrio gigas* (a nonphotosynthetic anaerobe), *Bacillus polymyxa* (facultative nitrogen-fixing anaerobe), *Chromatium vinosum, Rhodopseudomonas palustrus, Rhodopseudomonas gelatinosa, Rhodospirillum spheroides, Rhodospirillum capulatus, Paracoccus denitrificans* (a denitrifying organism), and *Rhodospirillum tenue*.

Table 8-6. Amino Acid Composition of Various Eight-Iron Microbial Ferredoxins (in moles of amino acid/mole of protein)

	Microorganism source[a]						
Amino acid	1	2	3	4	5	6	7
Lysine	4	6	7	1	1	0	2
Histidine	1	2	2	0	0	0	2
Arginine	1	1	3	0	0	0	2
Aspartic acid	15	15	10	8	8	3	8
Threonine	2	4	6	1	0	4	6
Serine	4	2	5	5	5	2	4
Glutamic acid	15	20	8	4	4	8	16
Proline	10	9	7	3	5	4	5
Glycine	7	4	6	4	4	5	5
Alanine	11	8	12	8	7	12	3
½-Cystine	8	8	8	8	8	9	9
Valine	9	10	11	6	4	4	6
Methionine	1	1	1	0	0	0	1
Isoleucine	6	7	6	5	6	5	6
Leucine	3	8	5	0	0	1	3
Tyrosine	6	2	1	1	2	3	3
Phenylalanine	2	5	4	1	0	1	0
Tryptophan	1	1	1	0	0	0	0
Total	106	113	102	55	54	61	81

[a] Microorganisms:
1. *Mycobacterium smegmatis*; Hase *et al.* (1979).
2. *Azobacter vinelandii*; Yates *et al.* (1978).
3. *Mycobacterium flavum*; Yates *et al.* (1978).
4. *Clostridium pasteurianum*; Yates *et al.* (1978).
5. *Peptococcus aerogenes*; Adman *et al.* (1973).
6. *Chlorobium thiosulfatophilum*, a green photosynthetic bacterium; Hase *et al.* (1978c).
7. *Chromatium vinosum*, a purple photosynthetic bacterium; Yasunobu and Tanaka (1973a).

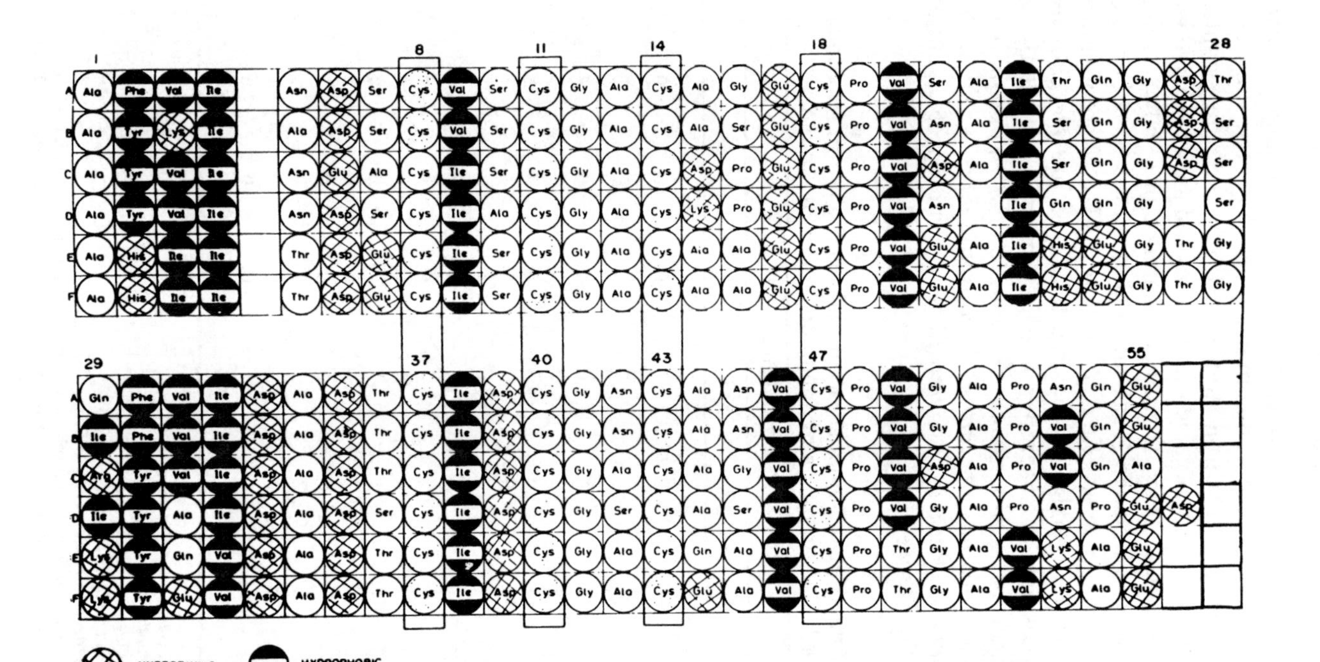

Figure 8-12. Amino acid sequences of a number of clostridial-type (eight-iron) ferredoxins. A, *Cl. butyricum*; B, *Cl. pasteurianum*; C, *Cl. acidiurici*; D, *Peptococcus aerogenes*; E, *Cl. tartarivorum*; F, *C. thermosaccharolyticum*. From Hall, D. O., Rao, K. K., and Cammack, R., *Sci. Prog. Oxf.* 62:285–317, 1975, by permission of the Blackwell Scientific Publications, Ltd.

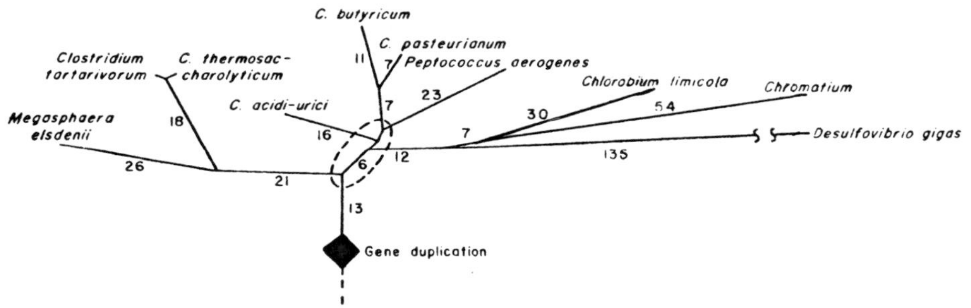

Figure 8-13. Phylogenetic tree of clostridium-type (eight-iron) ferredoxins. From Dayhoff (1976).

Ferredoxins from the last seven microorganisms (most are photosynthetic) are of the so-called high-potential iron protein (HIPIP) type. They are so called because their redox potentials are near +0.35, in contrast to most other ferredoxins that have negative redox potentials. The high potential is explained by the fact that the normal *reduced state* for HIPIPs is [4 Fe–4 S*–4 RS]$^{-2}$ (2 Fe^{2+} and 2 Fe^{3+}), whereas the normal *oxidized* state is [4 Fe–4 S*–4 RS]$^{-1}$ (1 Fe^{2+} and 3 Fe^{3+}). There also exists a *superreduced* state, artificially produced, where we have [4 Fe–4 S*–4 RS]$^{-3}$ (3 Fe^{2+} and 1 Fe^{3+}). The latter is, of course, the normal reduced state for other ferredoxins, including those from *Desulfovibrio gigas* and *Bacillus polymyxa*. It is also noteworthy that the HIPIPs do not give EPR signals in the reduced state and do give signals in the oxidized state, as expected. Some physical properties of the four-iron atom ferredoxins are given in Table 8-7.

Amino acid compositions of several four-iron atom ferredoxins are given in Table 8-8. It is seen that these proteins are indeed a diverse lot, some containing a high proportion of basic amino acids, others containing no tryptophan, and so on. From an evolutionary point of view, comparing amino acid sequences, *Desulfovibrio gigas* is closest to the clostridial type of ferredoxins (Yasunobu and Tanaka, 1973a), though only 19% homology is observed when whole proteins are compared. However, a 45% homology is obtained when the N-terminal halves of the two proteins are compared:

```
      1                                           10
C.P.  Ala-Tyr-lys-Ile-Ala-Asp-Ser-Cys-Val-Ser-Cys-Gly-Ala-Cys-Ala-
D.G.      Pro-Ile-Glu-Val-Asp-Asn-Cys-Met-Ala-Cys-Glu-Ala-Cys-Ile-
      N-terminus

                     20                                       30
C.P.  -Ser-Glu-Cys-Pro-Val-Asn-Ala-Ile-Ser-Gln-Gly-Asp-Ser-Ile-Phe-
D.G.  -Asn-Glu-Cys-Pro-Val-Asp-Val-Phe-Gln-Met-Asp-Glu-Gln-Gly-Asp-      (8-7)

                                              40
C.P.  -Val-Ile-Asp-Ala-Asp-Thr-Cys-Ile-Asp-Cys-Gly-Asn-Cys-Ala-
D.G.  -Lys-Ala-Val-Asn-Ile-Pro-Asn-Ser-Asn-Leu-Asp-Asp-Gln-Cys-

                          50
C.P.  -Asn-Val-Cys-Pro-Val-Gly-Ala-Pro-Val-Gln-Glu—COOH
D.G.  -Val-Glu-Ala-Ile-Gln-Ser-Cys-Pro-Ala-Ala-Ile-Arg-Ser—COOH
```

Table 8-7. Physical Parameters of Some Four-Iron Ferredoxins

Parameter	Microorganism source		
	Desulfovibrio gigas[a]	*Bacillus polymyxa*, Fd I[b]	*Chromatium vinosum* D_1 HIPIP[c]
Absorption maxima (oxidized, nm)	305, 390	395	325, 375, 450, 388 (red.)
Molecular weight	6570	9500	9650
Number of amino acids	56	76	86
Atoms of Fe/mole	4	4	4
EPR *g* values	—	2.06, 1.93, 1.91, 1.88 (red.)	2.04, 2.12 (ox.)
Redox potential (V)	−0.33	−0.38	+0.35
Electrons transferred	1	1	1
Isoelectric point	—	3.9	—

[a] Yasunobu and Tanaka (1973a).
[b] Stombaugh *et al.* (1973).
[c] Hall *et al.* (1975).

It has been argued on the basis of these results that the *D. gigas* Fd is a more ancient Fd from an evolutionary point of view than are the clostridial ferredoxins. The HIPIPs are undoubtedly homologous proteins in spite of their rather wide differences in amino acid composition and molecular weights. These homologies are illustrated by the amino acid sequences presented in Table 8-9.

8.3.4 Removal of Iron Clusters from Ferredoxins

The iron–sulfur cores may be quantitatively removed (*extruded*) from ferredoxins by suitable procedures. It is desirable to do this in order to avoid the interference of amino acid or other prosthetic groups when spectroscopic investigations of the iron–sulfur cluster are performed. The strategy involved is given as (Orme-Johnson and Holm, 1978; Wong *et al.*, 1979)

ferredoxin −(2 Fe–3 S*–2 RS or 4 Fe–4 S*–4 RS) + denaturing solvent (e.g., DMSO) → unfolded ferredoxin

unfolded ferredoxin + R^1SH (reducing agent) → ferredoxin apoprotein + $[Fe_2S^*_2(SR^1)_4]$ or $[Fe_4S^*_4(SR^1)_4]$ (8-8)

The extruded cluster may carry any oxidation state; i.e., the oxidation state of the native ferredoxin may be preserved in the extruded clusters. This technique has been applied to numerous ferredoxins and other iron–sulfur proteins containing labile inorganic sulfur (Wong *et al.*, 1979; Maskiewicz and Bruice, 1977; Gillum *et al.*, 1977). Synthetic $Fe_4S_4^*(SR^1)_4$ compounds have been prepared [e.g., $Fe_4S_4^*$(phenylthiol)$_4$] so that the properties of the extruded product can be compared to those of the synthetic materials and the former thus identified (e.g., Hill *et al.*, 1977a).

8.3.5 Functions of Bacterial Ferredoxins

As pointed out earlier, bacterial ferredoxins participate in numerous redox reactions. We shall now endeavor to look at a few of these. The

Table 8-8. Amino Acid Composition of Some Four-Iron Ferredoxins (in moles of amino acid/mole of protein)

	Microorganism source[a]					
Amino acid	1	2	3	4	5	6
Lysine	1	4	5	7	6	1
Histidine	0	0	1	2	0	2
Arginine	1	1	2	2	1	0
Aspartic acid	11	15	10	18	6	10
Threonine	0	6	4	8	3	2
Serine	3	2	3	4	5	4
Glutamic acid	9	7	9	23	5	16
Proline	4	4	5	8	3	6
Glycine	1	6	6	8	6	3
Alanine	6	9	20	14	11	10
½-Cystine	6	4	4	5	4	4
Valine	5	2	3	10	3	2
Methionine	2	0	1	0	1	1
Isoleucine	5	8	2	6	4	0
Leucine	1	3	5	7	0	4
Phenylalanine	1	1	1	4	2	3
Tyrosine	0	4	2	2	3	1
Tryptophan	0	0	3	2	0	2
Total	56	76	86	130	63	71
Mol. weight	6000	9000	10,000	14,000	—	—

[a] Microorganisms:
1. *Desulfovibrio gigas*; Yasunobu and Tanaka (1973a)
2. *Bacillus polymyxa*; Yasunobu and Tanaka (1973a)
3. *Chromatium vinosum* HIPIP; Matsumoto *et al.* (1976)
4. *Pseudomonas* fd; Matsumoto *et al.* (1976)
5. *Rhodospirillum tenue* HIPIP; Tedro *et al.* (1979)
6. *Paracoccus denitrificans* HIPIP; Tedro *et al.* (1977)

Table 8-9. Sequence Alignment of High Potential Iron–Sulfur Proteins[a]

Organism	Residues 1–43 (numbered 10, 20, 30, 40)	Residues 44–85 (numbered 50, 60, 70, 80)
Chromatium vinosum	S A P A N A V A A N D A T A I A L K Y N Q D A T K S E R V A A A R P G L P P E E Q H C	A N C Q F M Q A D A A G A T D E W K G C Q L F P G K L – I N V N G W C A S W T L K A G
Thiocapsa pfennigii	E D L P H V D A A T N P I A Q S L H Y I E D A N A S E R N P V T K T E L P G S E Q F C	H N C S F I Q A D – – – – S G A W R P C T L Y P G Y T – V S E D G W C L S W A H K T A
Rhodopseudomonas gelatinosa	A P V D – E K N P Q A V A L G Y V S D A A K A D K – A K Y K Q F V A G S – – H C	G N C A L F Q G K – – – A T D A V G G C P L F A G K Q – V A N K G W C S A W A K K A
Paracoccus sp.	Q D L P P L D – P S A E Q A Q A L N Y V K D T A E A A D H P A H Q E G – – – – E Q – C	D N C M F F Q A D – – – – – – – S Q G C Q L F P Q N S – V E P A G W C Q S W T A Q N
Rhodospirillum tenue	G T N A S M R K A F N Y Q E – – – – – – – – – V S K T A – – – – K N – C	A N C A Q F I P G A S – – A S A A G A C K V I P G D S Q I Q P T G Y C D A Y I V K K

[a] From Tedro *et al.*, 1979.

nitrogen fixation process will be dealt with in Section 8.4. No definite function has been assigned to the HIPIPs.

The Phosphoroclastic Reaction. Anaerobic organisms have various ways of disposing of the end product of glucose fermentation, pyruvate. Clostridial-type microorganisms generally degrade pyruvate to acetyl CoA, which can then proceed to form butyryl CoA or acetyl phosphate. Ferredoxins are involved in the reaction giving rise to acetyl CoA, as depicted in (Gottschalk, 1979, p. 183)

a. pyruvate + thiamine pyrophosphate–enzyme $\rightleftharpoons CO_2$
(TPP–E)
+ hydroxyethylthiamine pyrophosphate–enzyme (HETPP)
b. HETPP + ferredoxin_{ox} + CoA $\rightleftharpoons$ TPPE + acetyl CoA + ferredoxin_{red}
c. $\text{ferredoxin}_{red} + 2\ H^+ \rightleftharpoons \text{ferredoxin}_{ox} + H_2$ (8-9)
d. acetyl CoA + $P_i \rightleftharpoons$ acetyl phosphate + CoA
e. acetyl phosphate + ADP $\rightleftharpoons$ acetate + ATP

Steps a and b of reaction (8-9) are catalyzed by pyruvate–ferredoxin oxidoreductase (pyruvate dehydrogenase; see Section 8.6); step c is catalyzed by hydrogenase, an iron–sulfur protein in its own right (Section 8.6); and reaction d is catalyzed by phosphotransacetylase. Acetyl phosphate can generate ATP from ADP.

Reduction of Sulfate and Sulfite. Certain microorganisms of the anaerobic variety can use sulfate, sulfite, and thiosulfate as terminal electron acceptors, and ferredoxins are involved in such reactions. Of the organisms already discussed, *Desulfovibrio gigas* is a good example, though others are capable of performing such reactions as well. The immediate electron donor in the reduction of sulfite or thiosulfate to S^{2-} is apparently hydrogen,

$$3\ SO_3^{2-} \xrightarrow[\text{Fd?}]{H_2} S_3O_6^{2-} \xrightarrow{H_2} S_2O_3^{2-} \text{ (thiosulfate)} \xrightarrow{H_2} H_2S \quad (8\text{-}10)$$

(each H_2 step after $S_3O_6^{2-}$ releases SO_3^{2-}, which is returned to sulfite)

Using an extract of *Desulfovibrio gigas* and molecular hydrogen (H_2), it was shown that the production of sulfide from sulfite and thiosulfate was markedly stimulated by ferredoxin. Exactly what role ferredoxin plays in the very complex mechanism of the sulfite and thiosulfate reduction process is as yet not well understood (Hatchikian *et al.*, 1972).

Nitrate Reduction. A number of microorganisms and higher plants can assimilate nitrate and convert it to ammonia. The reactions leading to ammonia require ferredoxins:

$$\begin{array}{ccccc} & 2\,H^+ \quad H_2 & \leftarrow \text{hydrogenase} \rightarrow & 6\,H^+ \quad 3\,H_2 & \\ & 2\,Fd_{red} \quad 2\,Fd_{ox} & & 6\,Fd_{red} \quad 6\,Fd_{ox} & \\ NO_3^- & \xrightarrow[\text{nitrate reductase}]{} & NO_2^- & \xrightarrow[\text{nitrite reductase}]{} & NH_3 \\ \text{nitrate} & & \text{nitrite} & & \text{ammonia} \end{array} \tag{8-11}$$

Nitrate and nitrite reductases are themselves iron–sulfur proteins, and their properties are summarized in Table 8-12.

Conversion of CO_2 to Formate. Ferredoxin operates as an electron donor in the *Cl. pasteurianum* conversion of carbon dioxide to formate. The formate so produced enters the metabolic machinery of this strict anaerobe by being bound to tetrahydrofolate. The enzyme involved, ferredoxin–CO_2 oxidoreductase, is an iron–sulfur protein, and its properties are listed in Table 8-12.

Bacterial Photosynthesis. There are numerous bacteria that can perform photosynthesis, i.e., derive chemical energy from light, which in turn promotes the assimilation of CO_2. However, the bacterial photosynthetic electron transport mechanism is somewhat simpler than that of algae and higher plants and may be equated in general terms to the cyclic phosphorylation system of photosystem II in plants. It may be described briefly as follows: Light strikes special receptors containing bacterial chlorophylls called antennae. The energy is transducted to an electron donor called the *reaction center* or P-890. P-890 is a complex of several proteins, bacterial chlorophylls, caratenoids, and probably iron–sulfur protein(s). P-890 transmits the electron to an acceptor, termed X. This is most likely an iron–sulfur protein complex, since in the reduced state there are EPR signals detectable with g values at 1.68, 1.82, and 1.91 (Parson, 1974). The electron acceptor X then transfers the electron to a quinone, and the latter transfers it to cytochrome c_2. This reaction is coupled with a phosphorylation step, whereby one molecule of ATP is generated from ADP and phospate. Cytochrome c_2 passes the electron back to P-890 to complete the cycle.

Note that this system does not provide reducing equivalents to the bacterial cells. NADH may be generated by a system using sulfide or hydrogen as electron donors in various bacteria utilizing ATP (derived from the photosynthetic process) to drive these reactions (Gottschalk, 1979, pp. 242–245). However, none of these NAD(P)H generating systems require iron–sulfur proteins, and we shall not dwell on them.

The function of ferredoxins in the overall bacterial photosynthetic process is best understood as regards the so-called reductive carboxylic acid cycle, where CO_2 is fixed unto succinyl and acetyl CoA and α-keto acids (Buchanan, 1973). This series of reactions takes place parallel to the classical Calvin CO_2 fixation process. These two processes are, of course, responsible for the growth of many microorganisms on CO_2 as their sole carbon source. The cycle in question is represented in Figure 8-14. It is seen that a net of 4 CO_2 molecules are *fixed* with the expenditure of four NADH and three ATP equivalents. In two of these reactions, fixation of CO_2 by succinyl CoA and by acetyl CoA, ferredoxin is required. The product of the cycle is oxaloacetate. The system may be short-circuited by the fixation of two CO_2 molecules only, the product being acetyl CoA.

There are other CO_2-fixing reactions in bacteria that require reduced ferredoxins. Several of these give rise to precursors of amino acids that are essential in man. Some such reactions are the following:

$$\text{propionyl CoA} + CO_2 \xrightarrow[\text{Fd}_{red}\ \text{Fd}_{ox}]{\alpha\text{-ketobutyrate synthase}} \alpha\text{-ketobutyrate} \rightarrow \text{isoleucine} \quad (8\text{-}12)$$

$$\text{isobutyryl CoA} + CO_2 \xrightarrow[\text{Fd}_{red}\ \text{Fd}_{ox}]{\alpha\text{-ketoisovalerate synthase}} \alpha\text{-ketoisovalerate} \rightarrow \text{valine} \quad (8\text{-}13)$$

$$\text{phenylacetyl CoA} + CO_2 \xrightarrow[\text{Fd}_{red}\ \text{Fd}_{ox}]{\text{phenylpyruvate synthase}} \text{phenylpyruvate} \rightarrow \text{phenylalanine} \quad (8\text{-}14)$$

8.3.6 Evolution of Ferredoxins

Throughout our discussion of ferredoxins, references have been made to the evolution of the various groups of these proteins. In this section we shall briefly summarize the data on ferredoxins as a whole.

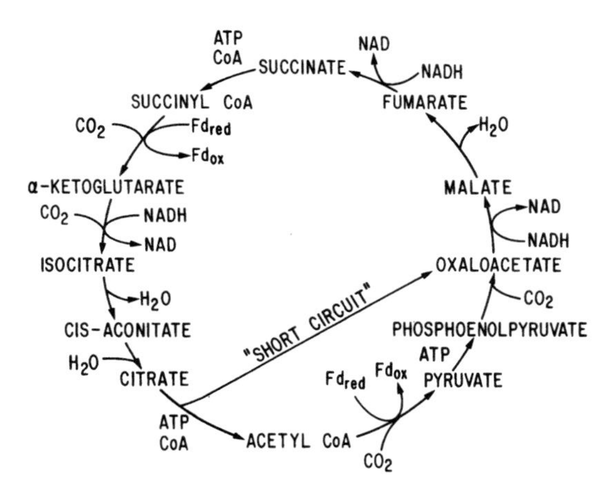

Figure 8-14. Reductive carboxylic acid cycle which results in the biosynthesis of gluconeogenesis intermediates.

The ferredoxins are believed to be representative of the most ancient proteins yet discovered. Presumably, the uptake of iron and inorganic sulfur by the polypeptide chain is not enzymatic, and since iron and sulfide were plentiful in the reducing atmosphere of the primeval earth, a suitably constructed polypeptide chain could acquire these elements spontaneously. Moreover, ferredoxins contain an abundance of amino acids found in the Murchison meteorite and which can be produced under conditions simulating the atmosphere and events of the primitive earth (Yasunobu and Tanaka, 1973b). *Protoferredoxins* may in fact have appeared on earth as long ago as 3 billion years ago.

The oldest ferredoxins are believed to be those found in the anaerobic *Clostridia* species. These organisms are representative of life as it existed in the oxygenless atmosphere of this earth, which ceased to exist about a billion years ago and was replaced by the present-day oxidizing atmosphere. Some have argued that the *Desulfovibrio gigas* ferredoxin is even older than those of the *Clostridia* (see Section 8.3.3). Hypothetical ancestors for plant and bacterial ferredoxins have been inferred from amino acid sequences (Dayhoff, 1972; Dayhoff, 1976), and from these, evolutionary trees for ferredoxins have been constructed. That for plant ferredoxins is shown in Figure 8-9, and that for the clostridial ferredoxins is shown in Figure 8-13. Their relationship is indicated in Figure 8-15.

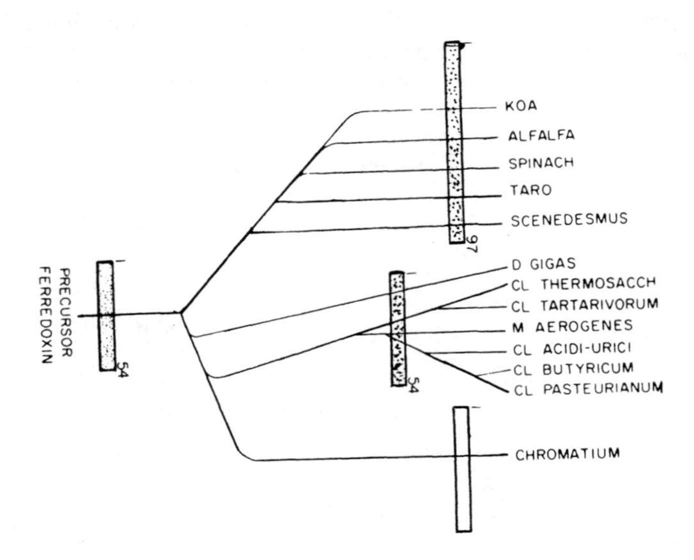

Figure 8-15. Phylogenetic tree of all ferredoxins. From Yasunobu and Tanaka (1973b).

8.4 Iron–Sulfur Proteins and Nitrogen Fixation

It has been appreciated for many years that certain prokaryotic organisms are able to convert molecular nitrogen into ammonia, which can be utilized by plants to make protein. Iron–sulfur proteins are intimately involved in this process, which is crucial to the maintenance of life on earth. Several review articles dealing with this process are available (Hardy and Burns, 1973; Burris and Orme-Johnson, 1974; Gottschalk, 1979, pp. 250–257; Mortenson and Thorneley, 1979; Ljones, 1979). Nitrogen fixation requires both reducing equivalents and ATP; in the process of converting one nitrogen molecule to 2 NH_3, six electrons are utilized (it is hence a reduction reaction), and at least 12 ATP molecules are hydrolyzed to ADP and phosphate. The basic nitrogen-fixing unit is called nitrogenase. It receives its electrons from reduced ferredoxins. We have dealt with the process by which microorganisms can reduce oxidized ferredoxins (see pp. 371 and 372). The ferredoxins required by the nitrogenase reaction have low redox potentials.

The nitrogenase system consists of two iron–sulfur proteins. The first is simply called the Fe–protein, and the other is termed the MoFe–protein because it also contains molybdenum. These proteins can be separated, and neither will catalyze nitrogen reduction by itself. The Fe–protein has a molecular weight of around 58,000 (*Cl. pasteurianum*) and consists of two identical subunits. It contains a single 4 Fe–4 S* cluster, and it has been proposed that the iron cluster joins the two subunits by being sequestered through the cysteine groups of both subunits. It appears that in its redox capactiy, the Fe–protein operates between the $[Fe_4S_4^*(SR)_4]^{-3}$ (reduced) and the $[Fe_4S_4^*(SR)_4]^{-2}$ (oxidized) states and transmits one electron at a time. The Fe–protein has the ability to bind Mg–ATP in both the reduced and oxidized states, which event changes its three-dimensional structure and the redox potential. The amino acid sequence of the Fe–protein subunits has been determined, but no homologies to any known protein have been discovered. The physical properties of the Fe–protein are presented in Table 8-10, and the amino acid composition of the *Klebsiella pneumoniae* material is given in Table 8-11.

The MoFe–protein is larger than the Fe–protein and consists of two pairs of subunits with unequal molecular weights and amino acid compositions (Table 8-11) (Kennedy *et al.*, 1976). There are two types of iorn–sulfur clusters in the MoFe–protein. The so-called M clusters include 12 iron atoms and most likely the Mo atoms. Each cluster apparently contains 6 iron atoms. The other 16 iron atoms (out of a total of 30 found in the *Azobacter vinelandii* MoFe–protein) are present in the classical

Table 8-10. Physical–Chemical Properties of the Nitrogenase Components

Parameter	Fe–protein[a]	MoFe–protein
Molecular weight	58,000	220,000
Number of subunits	2 (identical)	4 (nonidentical)
Number of amino acid residues/ subunit	273	—
Number of iron atoms	4	24–32 (+2 Mo atoms)
EPR *g* values	2.05, 1.94, 1.89 (reduced)	4.3, 3.6, 2.01 (oxidized)
Isoelectric point	4.4	4.95
α-Helix content	35%	—
Pleated-sheet content	34%	—
Redox potential (V)	−0.250 to −0.295[b]	−0.02 to −0.07
Absorption maximum (nm)	420 (reduced)	—

[a] From *Cl. pasteurianum*; Fe–proteins from other microorganisms may have different physical–chemical properties.
[b] This changes to −0.400 V upon binding of Mg–ATP.

Table 8-11. Amino Acid Composition of the Nitrogenase Proteins from *Klebsiella pneumoniae* (in moles of amino acid/mole of protein)

	MoFe–protein[a]		
Amino acid	Fast subunit	Slow subunit	Fe–protein[b]
Lysine	25	26	36
Histidine	12	13	6
Arginine	27	23	24
Aspartic acid	54	59	60
Threonine	25	38	36
Serine	22	22	24
Glutamic acid	60	67	84
Proline	24	28	18
Glycine	49	44	60
Alanine	38	42	60
Valine	29	31	42
Methionine	17	18	36
Isoleucine	27	18	48
Leucine	42	57	42
Tyrosine	19	16	18
Phenylalanine	21	27	12
Tryptophan	—	—	0
½-Cystine	—	—	18

[a] From Kennedy *et al.* (1976). The fast subunit has a molecular weight of 54,000; the slow subunit has a molecular weight of 57,000.
[b] From Hardy and Burns (1973). The molecular weight assumed is 68,000. Divide by 2 to get the composition of the monomer subunit.

$Fe_4S_4^*$ clusters, which are termed the P clusters. It is believed that the M clusters in reality comprise a cofactor tightly bound to the MoFe–protein and are crucial for its biological activity. The M cluster may assume three oxidation states: the *native* state, which is responsible for the EPR signals at g values of 3.6, 4.3, and 2.0 (Orme-Johnson *et al.*, 1972); the *oxidized* state, artificially produced with potent oxidizing agents; and the *super-reduced* state, found during the nitrogen fixation process and arising via reduction by the reduced Fe–protein (Zimmermann *et al.*, 1978; Huynh *et al.*, 1979). Other workers have claimed that the M clusters have Fe–Mo ratios of 8 and S*–Mo ratios of 6 and have, in fact, postualted a plausible structure for the M cluster (Figure 8-16). It provides for a total of 6 iron atoms in a tetrahedral ligand field with antiferromagnetic coupling and a spin of $S = \frac{3}{2}$ and for two iron atoms in the octahedral ligand field associated with a molybdenum atom (Teo and Averill, 1979).

The mechanism of the nitrogenase reaction is fairly well worked out. It is a reductive process, and the system is not necessarily specific for nitrogen. Other substrates that can be reduced are hydrogen ions (to hydrogen gas), acetylene (to ethylene), N_2O (to N_2 and H_2O), N_3^- (to N_2 and NH_3), and HCN (to CH_4 and NH_3). Microorganisms that can carry out the nitrogenase reaction include *Anabaena cylindrica* and *Gloeocapsa* (blue-green bacteria), *Rhodospirillum rubrum* (photosynthetic bacterium), *Clostridium pasteurianum* and *Desulfovibrio gigas* (strict anaerobes), and *Klebsiella pneumoniae, Bacillus polymyxa, Azobacter vinelandii,* and *Mycobacterium flavum* (aerobes and facultative anaerobes). The mechanism of nitrogen reduction is summarized in Figure 8-17. Briefly, it can be described as follows: Reduced ferredoxin reduces the Fe–protein, at which time the latter binds 2 Mg–ATP molecules/molecule

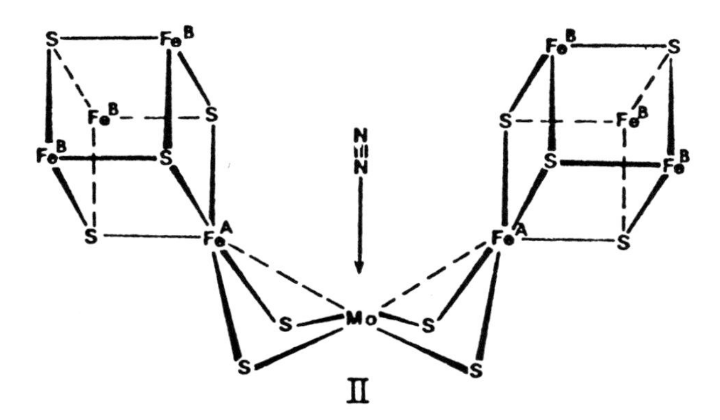

Figure 8-16. Structure of the M cluster from a nitrogenase. From Teo and Averill (1979).

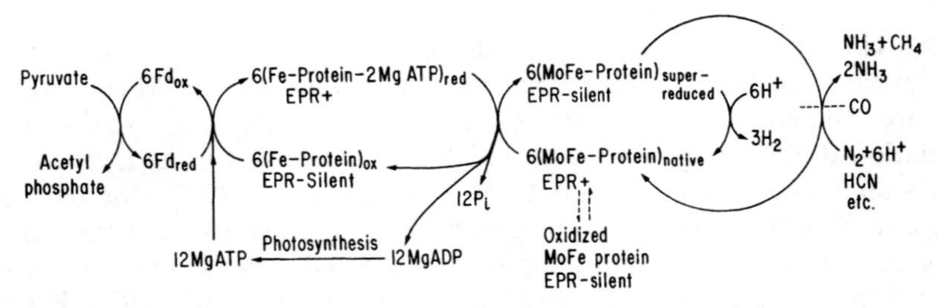

Figure 8-17. Mechanism of action of nitrogenase.

of protein. It then becomes associated with the MoFe–protein in a ratio of 1 : 1 or 1 : 2 and transfers its electron to the latter. In the process, ATP is hydrolyzed. MoFe–protein then proceeds to reduce the substrate (e.g., N_2) after which the complex dissociates. The cycle begins anew when reduced ferredoxin reduces the Fe–protein. Note that in the oxidized state the Fe–protein gives no EPR signals, whereas in the reduced state it gives signals with g values of 2.05, 1.94, and 1.89. On the other hand, the MoFe–protein gives the typical M-cluster signals in the native state but is silent when the latter is reduced. In fact, EPR signal differences have been used to follow the transfer of electrons from the Fe–protein to the MoFe–protein. Figure 8-17 also implies that different mechanisms exist for the reduction of H^+ and the other substrates. This has been inferred because CO is capable of inhibiting the reduction of all substrates other than H^+. H_2 itself inhibits the reduction of only N_2. The nature of intermediates between N_2 and NH_3 is not clearly understood.

8.5 Iron–Sulfur Proteins of Mammalian Electron Transport Mechanisms

Iron–sulfur proteins are components of several mammalian redox systems. We shall concern ourselves with essentially three such topics: cleavage of cholesterol side chains and hydroxylation of derivatives of cholesterol during the biosynthesis of various steroid hormones, hydroxylation of 25-hydroxycholecalciferol to the active form of vitamin D, and components of the classical oxidative phosphorylation pathway.

8.5.1 Adrenodoxin and the Biosynthesis of Steroid Hormones

The conversion of cholesterol into the various steroid hormones in the cortex of the adrenal glands is illustrated in Figure 8-18. The figure shows that the first step in such conversions is the oxidative cleavage of the side chain by a desmolase followed by a series of hydroxylation steps. Some of these steps occur in the mitochondria and some in microsomes (Gunsalus and Sligar, 1978). The enzymes performing the hydroxylation reactions are called monoxygenases, since only one atom of the molecular oxygen molecule is incorporated into the compound being modified. The oxygenases are discussed in more detail in Chapter 9.

A central figure in the conversion of cholesterol to corticosterone is an adrenal mitochondrial ferredoxin called adrenodoxin. Functionally, its

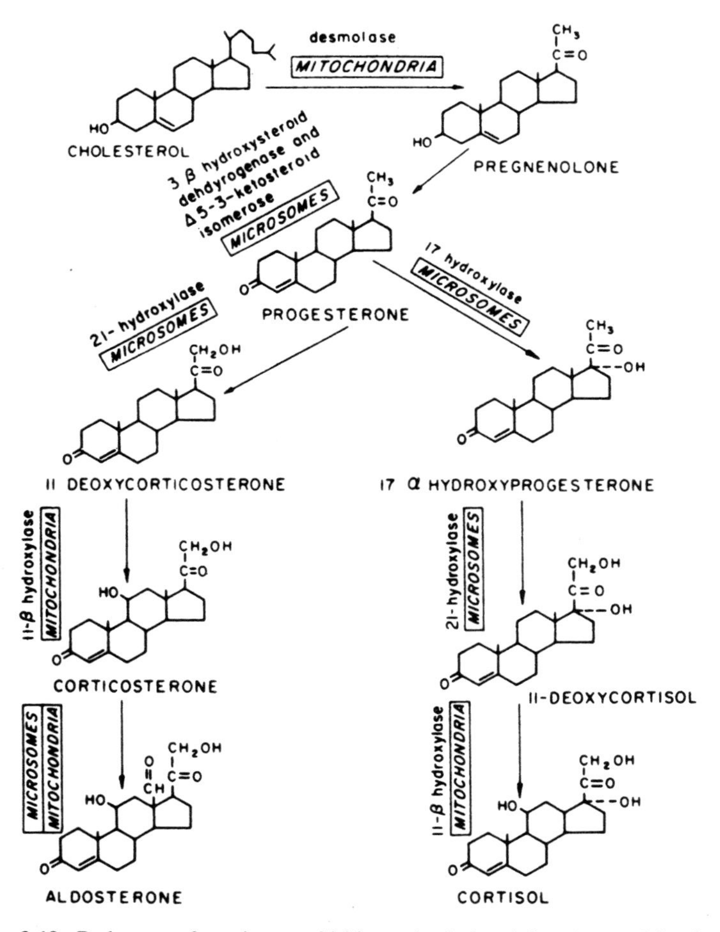

Figure 8-18. Pathways of corticosteroid biosynthesis involving the participation of mammalian ferredoxins. From Gunsalus, I. C., and Sligar, S. G., 1978. Oxygen reduction by the P450 monoxygenase systems, *Adv. Enzymol.* 47:1–44. Reprinted by permission of John Wiley & Sons. Copyright 1978 by John Wiley & Sons, Inc.

closest counterpart in the plant kingdom is putidaredoxin. However, the amino acid sequence of adrenodoxin bears no evolutionary resemblance to any plant ferredoxins (Tanaka *et al.*, 1973). Adrenodoxin appears to be a typical higher-plant-type ferredoxin with a single 2 Fe–2 S* center per molecule. Its physical properties are summarized in Table 8-12, and its amino acid composition is given in Table 8-13. The amino acid sequence is

10 20
Ser-Ser-Ser-Gln-Asp-Lys-Ile-Thr-Val-His-Phe-Ile-Asn-Arg-Asp-Gly-Glu-Thr-Leu-Thr-

30 40
Thr-Lys-Gly-Lys-Ile-Gly-Asp-Ser-Leu-Leu-Asp-Val-Val-Val-Gln-Asn-Asn-Leu-Asp-Ile-

50 60
Asp-Gly-Phe-Gly-Ala-Cys-Gly-Gly-Thr-Leu-Ala-Cys-Ser-Thr-Cys-His-Leu-Ile-Phe-Glu- (8-15)

70 80
Gln-His-Ile-Phe-Gly-Lys-Leu-Glu-Ala-Ile-Thr-Asn-Glu-Glu-Asn-Asn-Met-Leu-Asp-Leu-

90 100
Ala-Tyr-Gly-Leu-Thr-Asp-Arg-Ser-Arg-Leu-Gly-Cys-Gln-Ile-Cys-Leu-Thr-Lys-Ala-Met-

110
Asp-Asn-Met-Thr-Val-Arg-Val-Pro-Asp-Ala-Val-Ser-Asp-Ala—COOH

where the iron–sulfur cluster is sequestered by cysteinyl residues at positions 52, 55, 92, and 95. Both ends of the molecule are thus involved in binding the iron–sulfur cluster (Tanaka *et al.*, 1973).

Adrenodoxin participates in a typical hydroxylation reaction by reducing cytochrome P-450, and the reduced adrenodoxin is then regenerated by an NADP-linked flavoprotein (adrenodoxin reductase). The sequence of events taking place in such reactions is outlined in Figure 8-19. The rate-limiting reaction is apparently the transfer of electrons from adrenodoxin reductase to adrenodoxin (Lambeth and Kamin, 1979).

The rate-limiting reaction in the formation of hormones from cholesterol is the mitochondrial conversion of the latter to pregnenolone. This involves first the hydroxylation of carbons 20 and 22 of cholesterol, fol-

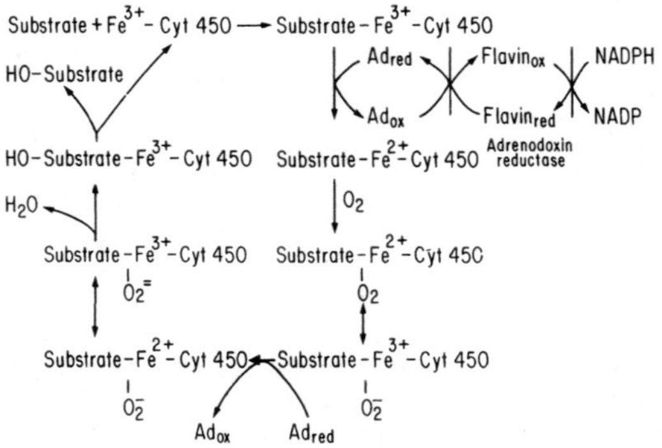

Figure 8-19. Typical hydroxylation reaction involving adrenodoxin (Ad). The substrate is a steroid biosynthetic intermediate.

Table 8-12. Physical Properties of Mammalian Ferredoxins

Parameter	Bovine adrenal gland adrenodoxin[a]	Chick liver ferredoxin[b]	Rieske's protein[c]
Molecular weight	12,638	53,000	24,500
Moles of iron/mole of protein	2	2	2
Absorption maxima (oxidized, nm)	518, 455, 414, 320	411	—
EPR *g* values (red.)	2.02, 1.94	2.03, 1.96	2.02, 1.90, 1.81, 1.78
Redox potential (V)	−0.36[d], −0.27[e]	—	+0.280

[a] From Yasunobu and Tanaka (1973a).
[b] From Kulkoski and Ghazarian (1979).
[c] From Trumpower and Edwards (1979).
[d] From Estabrook *et al.* (1973).
[e] From Huang and Kimura (1973).

lowed by a cleavage of the bond between the two carbon atoms to produce pregnenolone and isocaproic aldehyde. Adrenodoxin is involved in both hydroxylations. A specific cytochrome P-450 and adrenodoxin reductase are also necessary to complete the electron transport mechanism (Baron, 1976a). The other well-studied reaction involving adrenodoxin is the mitochondrial hydroxylation of 11-deoxycorticosterone to corticosterone.

Table 8-13. Amino Acid Composition of Mammalian Ferredoxins (in moles/mole of protein)

Amino acid	Bovine adrenal gland adrenodoxin[a]	Chick liver[b] ferredoxin
Lysine	5	35
Histidine	3	7
Arginine	4	14
Aspartic acid	18	68
Threonine	10	26
Serine	7	30
Glutamic acid	11	70
Proline	1	31
Glycine	8	76
Alanine	7	41
Cysteine	5	3
Valine	7	24
Methionine	3	8
Isoleucine	8	18
Leucine	12	25
Tyrosine	1	9
Phenylalanine	4	13
Tryptophan	0	4
Total	114	502

[a] Yasunobu and Tanaka (1973a).
[b] Kulkoski and Ghazarian (1979).

It is interesting that this enzyme system requires a specific cytochrome P-450 (Seybert *et al.*, 1978). The 11-β-hydroxylase is apparently present in very low amounts in the placenta, whose mitochondria are, however, extremely active in the cholesterol side-chain cleavage reaction. It is said that human placenta can biosynthesize as much as 300 mg of progesterone per day (Simpson and Miller, 1978). A different sort of cholesterol side-chain hydroxylation occurs in the liver as a prelude to cleavage in the conversion of cholesterol to cholic acid. The hydroxylation takes place at position 26 and also involves an adrenodoxin-like iron protein (Atsuta and Okada, 1978). Adrenodoxin-like proteins also occur in other steroidogenic tissues such as testicles and the ovary (Baron, 1976b) as well as in brain (Oftebro *et al.*, 1979). Hydroxylations taking place in the microsomes apparently do not require iron–sulfur proteins.

8.5.2 Hydroxylation of 25-Hydroxycholecalciferol

Cholecalciferol is hydroxylated in the 25 position by the liver microsomes and then in the 1 position by the kidney mitochondria. As already pointed out, microsomal hydroxylations do not require ferredoxins, whereas mitochondrial ones do, and the hydroxylation of 25-cholecalciferol by the 1-hydroxylase follows this rule. Cytochrome P-450 is again the terminal oxidase, which in turn is reduced by the kidney ferredoxin. The latter is regenerated by NADP-linked ferredoxin reductase containing flavinoids. The chicken kidney ferredoxin has been isolated and partially characterized (Kulkoski and Ghazarian, 1979). Its properties are listed in Table 8-12 and 8-13. Surprisingly, this ferredoxin showed a partial identity pattern with adrenodoxin in the Ouchterlony immunodiffusion system (Kulkoski *et al.*, 1979).

8.5.3 Iron–Sulfur Centers in the Oxidative Phosphorylation Pathway

The mitochondria of nonsteroidogenic tissues such as beef heart contain iron–sulfur proteins which take part in the oxidative phosphorylation process. In fact, mitochondria contain more ferredoxin-bound iron than heme iron. Many of these iron–sulfur proteins have been detected through EPR spectroscopy of mitochondria and in a few instances have been purified. With such scanty information at hand, it has nevertheless been possible to put together a scheme, illustrated in Figure 8-20, which indicates the location of the iron–sulfur centers. Thus complex I, often referred to

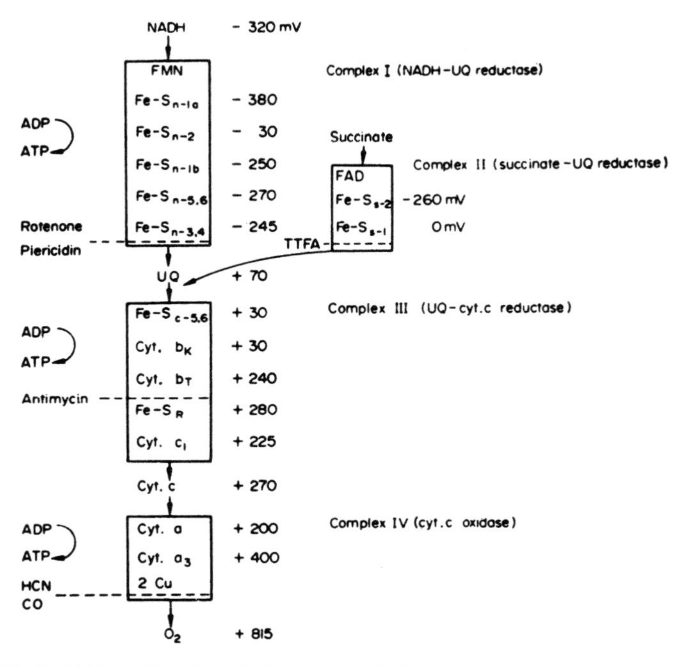

Figure 8-20. Oxidative phosphorylation system indicating the participation of iron–sulfur centers. The relative positions of such centers within complexes I, II, and III are largely tentative. From Hall, D. O., Rao, K. K., and Cammack, R., *Sci. Prog. Oxf.* 62:285–317, 1975, by permission of Blackwell Scientific Publications, Ltd.

as NADH dehydrogenase and located on the inner membrane of mitochondria, contains FMN (1 mole/mole of protein) and 28 iron–sulfur units/molecule of protein. The most notable center, probably center S_n-1*a*, exhibits an EPR signal at $g = 1.94$ when NADP dehydrogenase is reduced with NADH. Other signals observed in complex I are at *g* values of 2.022 and 1.923 (center 1); 2.054 and 1.922 (center 2); 2.101, 1.886, and 1.864 (center 3); 2.103 and 1.861 (center 4); and 2.11, 2.06, 2.03, 1.93, 1.90, and 1.88 (center 5) (Singer *et al.*, 1973; Beinert, 1977). The molecular weight of NADH dehydrogenase is as high as 800,000 and probably consists of several component proteins.

Iron–sulfur centers are also found in the succinic dehydrogenase complex, called simply succinic dehydrogenase. It contains a total of eight iron atoms and one FAD molecule, its molecular weight being close to 150,000 (Singer *et al.*, 1973). It appears that there are two 2 Fe–2 S* clusters and one 4 Fe–4 S* cluster in succinic dehydrogenase. Centers S-1 and S-2 both contain the 2 Fe–2 S* clusters, and the 4 Fe–4 S* cluster may be present in an as yet poorly characterized S-3 center of the enzyme (Salerno *et al.*, 1979).

An iron–sulfur component of complex III was also isolated and originally termed Rieske's protein ($Fe–S_R$ in Figure 8-20). It has recently been purified and shown to be required in the reduction of cytochrome c_1 by ubiquinone and succinate (Trumpower and Edwards, 1979). Its properties are shown in Table 8-12. Because it has a redox potential of near 0.28 V, this component may be considered to be a high-potential (HIPIP) type of ferredoxin.

8.6 Miscellaneous Iron–Sulfur Proteins

A number of enzymes exist whose prosthetic groups are iron–sulfur centers. In addition, many also contain other components such as flavins, molybdenum, or heme and are thus classified as *complex* iron–sulfur proteins. Many of these enzymes are involved in catalyzing electron transport involving ferredoxins. The properties of these enzymes are summarized in Table 8-14.

Many enzymes listed in Table 8-14 have been previously mentioned and their functions described (e.g., hydrogenase and pyruvate dehydrogenase). We shall now briefly mention the properties of some of the others.

Xanthine Oxidase. This enzyme catalyzes the hydroxylation of purines, and is present in liver, milk, and even microorganisms.

hypoxanthine $\xrightarrow[O_2]{H_2O^*,\ 2e^-}$ xanthine $\xrightarrow[O_2]{H_2O^*,\ 2e^-}$ uric acid (8-16)

Purines other than hypoxanthine or its analogues can serve as substrates for xanthine oxidase, especially drugs such as allopurinol. Internally, the

Table 8-14. Physical–Chemical Properties of a Number of Iron–Sulfur Enzymes

Enzyme	Source	Molecular weight	Moles of Fe/mole of protein	Type of cluster	Other cofactors	EPR *g* values	Redox potential (V)	Reference
Pyruvate dehydrogenase	*Cl. acidi-urici*	240,000	6	—	Thiamine	—	—	Orme-Johnson, 1973
Xanthine oxidase	Bovine milk	300,000	8	4 Fe–4 S* (2)	2 FAD, 2 Mo	2.12, 2.01, 1.91	—	Massey, 1973
Dihydroorotate dehydrogenase	*Zymobacterium oroticum*	115,000	4	2 Fe–2 S* (2)	2 FAD, 2 FMN	2.01, 1.94, 1.92	−0.252	Singer *et al.*, 1973
Aldehyde oxidase	Liver (mammalian)	280,000	8	—	2 FAD, 2 Mo	2.01, 1.94, 1.92	—	Massey, 1973; Orme-Johnson, 1973
Hydrogenase	*Alcaligenes lutropus*	186,000	12	4 Fe–4 S* (2), 2 Fe–2 S* (2)	2 FMN	2.04, 2.0, 1.95, 1.93, 1.86	−0.445 & −0.325	Schneider *et al.*, 1979
Aconitase	Beef heart	83,000	2–3	2 Fe–2 S* (1)	None	2.01 (ox.)	—	Kurtz *et al.*, 1979
Hydrogenase	*Cl. pasteurianum*	60,000	4	2 Fe–2 S* (?)	None	2.03, 1.86	—	Orme-Johnson, 1973

continued overleaf

Table 8-14. (*Continued*)

Enzyme	Source	Molecular weight	Moles of Fe/mole of protein	Type of cluster	Other cofactors	EPR *g* values	Redox potential (V)	Reference
Nitrite reductase	Spinach	60,000	4	4 Fe–4 S* (1)	Siroheme	2.04, 2.00, 1.93	−0.57	Lancaster *et al.*, 1979; Cammack, 1978
Nitrate reductase	*Micrococcus denitrificans*	160,000	8	—	Mo, heme (plants)	2.06, 1.95, 1.88	—	Orme-Johnson, 1973
Sulfite reductase	*Escherichia coli*	670,000	16	—	4 FAD, 4 FMN, 4 heme	2.04, 1.93, 1.90	—	Orme-Johnson, 1973
Ferredoxin–CO_2 oxidoreductase	*Cl. pasteurianum*	118,000 & aggregates	24	—	1 Mo	—	−0.40	Scherer and Thauer, 1978
Trimethylamine dehydrogenase	*Bacterium W3A1*	147,000	4	4 Fe–4 S*	FAD (?)	—	—	Hill *et al.*, 1977b
Amidophosphoribosyl transferase	*Bacillus subtilis*	100,000 & 200,000	3	2 Fe–2 S* (?)	None	—	—	Wong *et al.*, 1977

electron chain apparently follows the scheme in

$$\text{substrate} \xrightarrow{e^-} \underset{Mo^{6+} \rightarrow Mo^{5+}}{Mo} \rightarrow \underset{Fe^{3+} \rightarrow Fe^{2+}}{Fe} \rightarrow FAD \rightarrow \text{oxygen} \qquad (8\text{-}17)$$

Aldehyde Oxidase. Like xanthine oxidase, aldehyde oxidase serves to oxidize various substrates including aldehydes; however, it does not use hypoxanthine as a substrate. This property distinguishes it from xanthine oxidase. It can utilize oxygen as the ultimate electron acceptor; however, the hydroxyl group incorporated into the substrate comes from water:

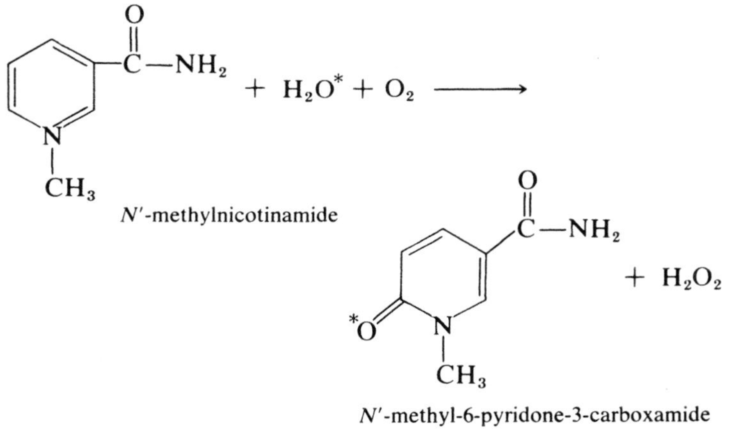

$$N'\text{-methylnicotinamide} + H_2O^* + O_2 \longrightarrow N'\text{-methyl-6-pyridone-3-carboxamide} + H_2O_2 \qquad (8\text{-}18)$$

N′-methylnicotinamide

N′-methyl-6-pyridone-3-carboxamide

The electron chain is apparently the same as that in the xanthine oxidase reaction, with $FADH_2$ being the immediate electron donor of oxygen.

Dihydroorotate Dehydrogenase. This bacterial enzyme catalyzes the following sequence of reactions:

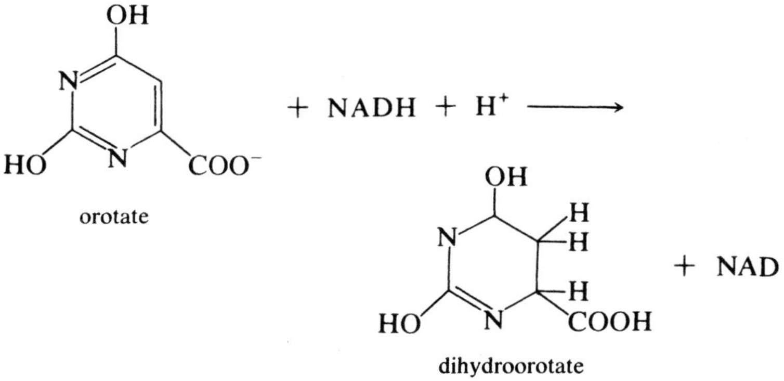

$$\text{orotate} + NADH + H^+ \longrightarrow \text{dihydroorotate} + NAD \qquad (8\text{-}19)$$

orotate

dihydroorotate

$$\text{dihydroorotate} + O_2 \rightarrow \text{orotate} + H_2O_2 \quad (8\text{-}20)$$

$$\text{NADH} + H^+ + O_2 \rightarrow \text{NAD} + H_2O_2 \quad (8\text{-}21)$$

Aconitase. Mitochondrial aconitase (a Krebs cycle enzyme) has been isolated from a number of sources. The most remarkable characteristic of this enzyme is that the EPR signal at $g = 2.01$ in the oxidized state is typical of the high-potential iron–sulfur proteins with 4 Fe–4 S* clusters, yet aconitase has a single 2 Fe–2 S* cluster similar to those of the higher-plant ferredoxins. Another noteworthy point is that aconitase is one of the few iron–sulfur proteins not involved in a redox reaction.

Trimethylamine Dehydrogenase. Bacterial trimethylamine dehydrogenase catalyzes the demethylation of trimethylamine to dimethylamine and formaldehyde as shown:

$$(CH_3)_3N + H_2O + \text{electron acceptor} \rightarrow (CH_3)_2NH + CH_2O + \text{reduced electron acceptor} \quad (8\text{-}22)$$

Amidophosphoribosyl Transferase (glutamine phosphoribosyl pyrophosphate amidotransferase). This enzyme catalyzes the first step in the *de novo* biosynthesis of purines. It is an iron–sulfur protein and is present both in bacteria and mammalian tissues (Itakura and Holmes, 1979). It is rapidly inactivated by O_2, presumably because its iron–sulfur cluster is entirely in the ferrous state, and O_2 converts this to the ferric state. It is also one of the few iron–sulfur proteins not normally involved in a redox reaction.

Summary

Iron–sulfur proteins comprise a major group of iron-containing proteins in nature. They may be subdivided into the rubredoxins, the ferredoxins, and the iron–sulfur enzymes, which are usually complexed with other cofactors as well. Rubredoxins are present in microorganisms only and are characterized by a tetrahedral ligand field involving iron and sulfhydryl groups of cysteine residues. The ferredoxins and the iron–sulfur enzymes contain iron *clusters*, where iron is associated with inorganic sulfur and sulfhydryl residues of cysteine. The higher-plant-type ferredoxins are characterized by clusters containing only two iron and two inorganic sulfur atoms each, whereas the bacterial (clostridial) types of ferredoxins contain clusters with four iron and four inorganic sulfur atoms each. Whereas microbial ferredoxins often contain the higher-plant-type iron clusters, the higher-plant ferredoxins do not contain bacterial-type iron clusters. Ferredoxin-like proteins are also present in animals.

The function of ferredoxins is to transmit electrons in a variety of redox systems. Ferredoxins are essential components of plant and bacterial photosynthetic mechanisms, nitrogen fixation processes, nitrate reduction, and others. In animals, ferredoxins, one of which is called adrenodoxin, participate in various hydroxylation reactions, including hydroxylation of steroids and provitamin D. Iron–sulfur centers participate in animal oxidative phosphorylation pathways. Most notable of these are succinic dehydrogenase and NADH dehydrogenase, which also contain flavins as cofactors. A number of enzymes that may or may not have a function in redox reactions are iron–sulfur proteins.

References

Adman, E. T., Sieker, L. C., and Jensen, L. H., 1973. The structure of a bacterial ferredoxin, *J. Biol. Chem.* 248:3987–3996.

Atsuta, Y., and Okada, K., 1978. Isolation of rat liver mitochondrial ferredoxin and its reductase active in the 5β-cholestane-3α, 7α, 12α-triol 26-hydroxylase, *J. Biol. Chem.* 253:4653–4658.

Baron, J., 1976a. Immunochemical studies on adrenal ferredoxin: Involvement of adrenal ferredoxin in the cholesterol side-chain cleavage reaction of mammalian adrenals, *Arch. Biochem. Biophys.* 174:226–238.

Baron, J., 1976b. Immunochemical studies on adrenal ferredoxin: Involvement of adrenal ferredoxin-like iron–sulfur proteins in the cholesterol side-chain cleavage reaction of mammalian steroidogenic tissues, *Arch. Biochem. Biophys.* 174:239–248.

Beinert, H., 1973. Development of the field and nomenclature, *in Iron–Sulfur Proteins*, Vol. I, W. Lovenberg, (ed.), Academic Press, New York, pp. 1–36.

Beinert, H., 1977. Iron–sulfur centers of the mitochondrial electron transfer system—recent developments, *in Iron–Sulfur Proteins*, Vol. III, W. Lovenberg (ed.), Academic Press, New York, pp. 61–110.

Buchanan, B. B., 1973. Ferredoxin and carbon assimilation, *in Iron–Sulfur Proteins*, Vol. I, W. Lovenberg (ed.), Academic Press, New York, pp. 129–150.

Burris, R. H., and Orme-Johnson, W. H., 1974. Survey of nitrogenase and its EPR properties, *in Microbial Iron Metabolism*, J. B. Neilands (ed.), Academic Press, New York, pp. 187–209.

Cammack, R., Hucklesby, D. P., and Hewitt, E. J., 1978. Electron-paramagnetic-resonance studies of the mechanism of leaf nitrite reductase, *Biochem. J* 171:519–526.

Carter, C. W., Freer, S. T., Xuong, N. H., Alden, R. A., and Kraut, J., 1972. Structure of the iron–sulfur cluster in the *Chromatium* iron protein at 2.25 Å resolution, *Cold Spring Harbor Symp. Quant. Biol.* 36:381–385.

Dayhoff, M. O., 1972. *Atlas of Protein Sequence and Structure*, Vol. 5, National Biomedical Research Foundation, Washington, D.C., pp. D35–D46.

Dayhoff, M. O., 1976. *Atlas of Protein Sequence and Structure*, Vol. 5, Suppl. 2, National Biomedical Research Foundation, Washington, D.C., pp. 51–64.

Dutton, J. E., and Rogers. L. J., 1978. Isoelectric focusing of ferredoxins, flavodoxins, and a rubredoxin, *Biochim. Biophys. Acta* 537:501–506.

Estabrook, R. W., Suzuki, K., Mason, J. I., Baron, J., Taylor, W. E., Simpson, E. R., Purvis, J., and McCarthy, J., 1973. Adrenodoxin: An iron–sulfur protein of adrenal cortex mitochondria, *in Iron–Sulfur Proteins*, Vol. I., W. Lovenberg (ed.), Academic Press, New York, pp. 193–223.

Gillum, W. O., Mortenson, L. E., Chen, J.-S., and Holm, R. H., 1977. Quantitative extrusions of the $Fe_4S_4^*$ cores of the active sites of ferredoxins and the hydrogenase of *Clostridium pasteurianum*, *J. Am. Chem. Soc.* 99:584–595.

Gottschalk, G., 1979. *Bacterial Metabolism*, Springer, Berlin.

Gunsalus, I. C., and Sligar, S. G., 1978. Oxygen reduction by the P450 monoxygenase system, *Adv. Enzymol. Relat. Areas Mol. Biol.* 47:1–44.

Hall, D. O., Rao, K. K., and Cammack, R., 1975. The iron–sulfur proteins: Structure, function and evolution of a ubiquitous group of proteins, *Sci. Prog. Oxford* 62:285–317.

Hardy, R. W. F., and Burns, R. C., 1973. Comparative biochemistry of iron–sulfur proteins and dinitrogen fixation, *in Iron–Sulfur Proteins*, Vol. I, W. Lovenberg (ed.), Academic Press, New York, pp. 65–110.

Hase, T., Wakabayashi, S., Matsubara, H., Kerscher, L., Oesterhelt, D., Rao, K. K., and Hall, D. O., 1978a. Complete amino acid sequence of *Halobacterium halobium* ferredoxin containing an N^{ϵ}-acetyllysine residue, *J. Biochem.* 83:1657–1670.

Hase, T., Wakabayashi, S., Wada, K., Matsubara, H., Jüttner, F., Rao, K. K., Fry, I., and Hall, D. O., 1978b. *Cyanidium caldarium* ferredoxin: A red algal type?, *FEBS Lett.* 96:41–44.

Hase, T., Wakabayashi, S., Matsubara, H., Evans, M. C. W., and Jennings, J. V., 1978c. Amino acid sequence of a ferredoxin from *Chlorobium thiosulfatophilum* strain *Tassajara*, a photosynthetic green sulfur bacterium, *J. Biochem.* 83:1321–1325.

Hase, T., Wakabayashi, S., Matsubara, H., Imai, T., Matsumoto, T., and Tobari, J., 1979. *Mycobacterium smegmatis* ferredoxin: A unique distribution of cysteine residues constructing iron–sulfur centers, *FEBS Lett.* 103:224–228.

Hatchikian, E. C., Le Gall, J., Buishi, M., and Dubourdieu, M., 1972. Regulation of the reduction of sulfite and thiosulfate by ferredoxin, flavodoxin, and cytochrome cc_3 in extracts of the sulfate reducer *Desulfovibrio gigas*, *Biochim. Biophys. Acta* 258:701–708.

Hill, C. L., Renaud, J., Holm, R. H., and Mortenson, L. E., 1977a. Synthetic analogues of the active sites of iron–sulfur proteins. 15. Comparative polarographic potentials of the $[Fe_4S_4(SR)_4]^{2-,3-}$ and *Clostridium pasteurianum* ferredoxin redox couples, *J. Am. Chem. Soc.* 99:2549–2557.

Hill, C. L., Steenkamp, D. J., Holm, R. H., and Singer, T. P., 1977b. Identification of the iron–sulfur center in trimethylamine dehydrogenase, *Proc. Natl. Acad. Sci. U.S.A.* 74:547–551.

Huang, J. J., and Kimura, T., 1973. Studies on adrenal steroid hydroxylases. Oxidation–reduction properties of adrenal iron sulfur protein (adrenodoxin), *Biochemistry* 12:406–409.

Hutson, K. G., Rogers, L. J., Haslett, B. G., Boulter, D., and Cammack, R., 1978. Comparative studies on two ferredoxins from the cyanobacterium *Nostoc* strain MAC, *Biochem. J.* 172:465–477.

Huynh, B. H., Münck, E., and Orme-Johnson, W. H., 1979. Nitrogenase XI: Mössbauer studies on the cofactor centers of the MoFe protein from *Azobacter vinelandii OP*, *Biochim. Biophys. Acta* 527:192–203.

Itakura, M., and Holmes, E. W., 1979. Human amidophosphoribosyltransferase, *J. Biol. Chem.* 254:333–338.

Kennedy, C., Eady, R. R., Kondorosi, E., and Rekosh, D. K., 1976. The molybdenum–iron protein of *Klebsiella pneumoniae* nitrogenase, *Biochem. J.* 155:383–389.

Keresztes-Nagy, S., and Margoliash, E., 1966. Preparation and characterization of alfalfa ferredoxin, *J. Biol. Chem.* 241:5955–5966.

Kulkoski, J. A., and Ghazarian, J. G., 1979. Purification and characterization of the ferredoxin component of 25-hydroxycholecalciferol 1α-hydroxylase, *Biochem. J.* 177:673–678.

Kulkoski, J. A., Peterson, B. L., Elcombe, B., Winkelhake, J. L., and Ghazarian, J. G., 1979. Ferredoxin of 25-hydroxyvitamin D_3-1α-hydroxylase, *FEBS Lett.* 99:183–188.

Kurtz, D. M., Holm, R. H., Ruzicka, F. J., Beinert, H., Coles, C. J., and Singer, T. P., 1979. The high potential iron–sulfur cluster of aconitase in a binuclear iron–sulfur cluster, *J. Biol. Chem.* 254:4967–4969.

Lambeth, J. D., and Kamin, H., 1979. Adrenodoxin reductase–adrenodoxin complex, *J. Biol. Chem.* 254:2766–2774.

Lancaster, J. R., Vega, J. M., Kamin, H., Orme-Johnson, N. R., Orme-Johnson, W. H., Krueger, R. J., and Siegel, L. M., 1979. Identification of the iron–sulfur center of spinach ferredoxin–nitrite reductase as a tetranuclear center, and preliminary EPR studies of mechanism, *J. Biol. Chem.* 254:1268–1272.

Lippard, S., 1973. Iron–sulfur coordination compounds, *Acc. Chem. Res.* 6:282–288.

Ljones, T., 1979. Nitrogen fixation and bioenergetics: The role of ATP in nitrogenase catalysis, *FEBS Lett.* 98:1–8.

Lode, E. T., and Coon, M. J., 1971. Enzymatic ω-oxidation, V. Forms of *Pseudomonas oleovorans* rubredoxin containing one or two iron atoms: Structure and function in ω-hydroxylation, *J. Biol. Chem.* 246:791–802.

Lovenberg, W. (ed.), 1977. *Iron–Sulfur Proteins,* Vol. III, Academic Press, New York.

Lovenberg, W., 1974. Ferredoxin and rubredoxin, *in Microbial Iron Metabolism,* J. B. Neilands (ed.), Academic Press, New York, pp. 161–185.

Lovenberg, W. (ed.), 1977. *Iron–sulfur Proteins,* Vol. III, Academic Press, New York.

Lovenberg, W., Buchanan, B., and Rabinowitz, J. C., 1963. Studies on the chemical nature of clostridial ferredoxin, *J. Biol. Chem.* 238:3899–3913.

Malkin, R., and Bearden, A. J., 1978. Membrane-bound iron–sulfur centers in photosynthetic systems, *Biochim. Biophys. Acta* 505:147–181.

Maskiewicz, R., and Bruice, T. C., 1977. Kinetic study of the dissolution of $Fe_4S_4{}^{2-}$—cluster core ions of ferredoxins and high potential iron protein, *Biochemistry* 16:3024–3029.

Massey, V., 1973. Iron–sulfur flavoprotein hydroxylases, *in Iron–Sulfur Proteins,* Vol. I, W. Lovenberg (ed.), Academic Press, New York, pp. 302–348.

Matsumoto, T., Tobari, J., Suzuki, K., Kimura, T., and Tchen, T. T., 1976. Purification and properties of a four iron–four sulfur protein from a *Pseudomonas* species, *J. Biochem.* 79:937–943.

Mortenson, L. E., and Thorneley, R. N. F., 1979. Structure and function of nitrogenase, *Annu. Rev. Biochem.* 48:387–418.

Nomenclature Committee of the International Union of Biochemistry, 1979. Nomenclature of iron–sulfur proteins. Recommendations, 1978, *Eur. J. Biochem.* 93:427–430.

Oftebro, H., Stormer, F. C., and Pedersen, J. I., 1979. The presence of an adrenodoxin-like ferredoxin and cytochrome P-450 in brain mitochondria, *J. Biol. Chem.* 254:4331–4334.

Orme-Johnson, W. H., 1973. Iron–sulfur proteins: Structure and function, *Annu. Rev. Biochem.* 42:159–204.

Orme-Johnson, W. H., and Holm, R. H., 1978. Identification of iron–sulfur clusters in proteins, *Methods Enzymol.* 58:268–274.

Orme-Johnson, W. H., and Orme-Johnson, N. R., 1978. Overview of iron–sulfur proteins, *Methods Enzymol.* 58:259–268.

Orme-Johnson, W. H., Hamilton, W. D., Jones, T. L., Tso, M.-Y. W., Burris, R. H., Shah, V. K., and Brill, W. J., 1972. Electron paramagnetic resonance of nitrogenase and nitrogenase components from *Clostridium pasteurianum W5* and *Azobacter vinelandii OP, Proc. Natl. Acad. Sci. U.S.A.* 69:3142–3145.

Parson, W. W., 1974. Bacterial photosynthesis, *Annu. Rev. Microbiol.* 28:41–59.

Rabinowitz, J. C., 1971. Clostricial ferredoxin: An iron–sulfur protein, *in Bioinorganic Chemistry*, R. F. Gould (ed.), American Chemical Society, Washington, D.C., pp. 322–345.

Rao, K. K., Cammack, R., Hall, D. O., and Johnson, C. E., 1971. Mössbauer effect in *Scenedesmus* and spinach ferredoxins, *Biochem. J.* 122:257–265.

Salerno, J. C., Lim, J., King, T. E., Blum, H., and Ohnishi, T., 1979. The spatial relationships and structure of the binuclear iron–sulfur clusters in succinate dehydrogenase, *J. Biol. Chem.* 254:4828–4835.

Scherer, P. A., and Thauer, R. K., 1978. Purification and properties of reduced ferredoxin: CO_2 oxidoreductase from *Clostridium pasteurianum*, a molybdenum iron–sulfur protein, *Eur. J. Biochem.* 85:125–135.

Schneider, K., Cammack, R., Schlegel, H. G., and Hall, D. O., 1979. The iron–sulfur centres of soluble hydrogenase from *Alcaligenes eutrophus, Biochim. Biophys. Acta* 578:445–461.

Seybert, D. W., Lambeth, J. D., and Kamin, H., 1978. The participation of a second molecule of adrenodoxin in cytochrome P-450-catalyzed 11-β-hydroxylation, *J. Biol. Chem.* 253:8355–8358.

Shin, M., Yokoyama, Z., Abe, A., and Fukasawa, H., 1979. Properties of common wheat ferredoxin, and a comparison with ferredoxins from related species of *Triticum* and *Aegilops, J. Biochem.* 85:1075–1081.

Simpson, E. R., and Miller, D. A., 1978. Cholesterol side-chain cleavage, cytochrome P-450, and iron–sulfur protein in human placental mitochondria, *Arch. Biochem. Biophys.* 190:800–808.

Singer, T. P., Gutman, M., and Massey, V., 1973. Iron–sulfur flavoprotein dehydrogenases, in *Iron–Sulfur Proteins*, Vol. I, W. Lovenberg (ed.), Academic Press, New York, pp. 285–294.

Stombaugh, N. A., Burris, R. H., and Orme-Johnson, W. H., 1973. Ferredoxins from *Bacillus polymyxa, J. Biol. Chem.* 248:7951–7956.

Stout, C. D., 1979a. Structure of the iron–sulfur clusters in *Azobacter* ferredoxin at 4.0 Å resolution, *Nature (London)* 279:83–84.

Stout, C. D., 1979b. Two crystal forms of *Azobacter* ferredoxin, *J. Biol. Chem.* 254:3598–3599.

Sweeney, W. V., and McIntosh, B. A., 1979. Determination of cooperative interaction between clusters in *Clostridium pasteurianum* 2(4Fe–4S) ferredoxin, *J. Biol. Chem.* 254:4499–4501.

Takruri, I., and Boulter, D., 1979. The amino acid sequence of ferredoxin from *Triticum aestivum* (wheat), *Biochem. J.* 179:373–378.

Takruri, I., Haslett, B. G., Boulter, D., Andrew, P. W., and Rogers, L. J., 1978. The amino acid sequence of ferredoxin from the red alga *Porphyra umbilicalis, Biochem. J.* 173:459–466.

Tanaka, M., Haniu, M., and Yasunobu, K. T., 1973. The amino acid sequence of bovine adrenodoxin, *J. Biol. Chem.* 248:1141–1157.

Tedro, S. M., Meyer, T. E., and Kamen, M. D., 1977. Primary structure of a high potential iron–sulfur protein from a moderately halophilic denitrifying coccus, *J. Biol. Chem.* 252:7826–7833.

Tedro, S. M., Meyer, T. E., and Kamen, M. D., 1979. Primary structure of a high potential four-iron–sulfur ferredoxin from the photosynthetic bacterium *Rhodospirillum tenue*, *J. Biol. Chem.* 254:1495–1500.

Teo, B.-K., and Averill, B. A., 1979. A new cluster model for the FeMo-cofactor of nitrogenase, *Biochem. Biophys. Res. Comm.* 88:1454–1461.

Trumpower, B. L., and Edwards, C. A., 1979. Identification of oxidation factor as a reconstitutively active form of the cytochrome b-c_1 segment of the respiratory chain, *FEBS Lett.* 100:13–16.

Tsukihara, T., Fukuyama, K., Tahara, H., Katsube, Y., Matsuura, Y., Tanaka, N., Kakudo, M., Wada, K., and Matsubara, H., 1978. X-Ray analysis of ferredoxin from *Spirulina platensis*. II. Chelate structure of active center, *J. Biochem.* 84:1645–1647.

Wakabayashi, S., Hase, T., Wada, K., Matsubara, H., Suzuki, K., and Takaichi, S., 1978. Amino acid sequences of two ferredoxins from pokeweed, *Phytolacca americana*, *J. Biochem.* 83:1305–1319.

Watenpaugh, K. D., Sieker, L. C., Herriott, J. R., and Jensen, L. H., 1972. The structure of a non-heme iron protein: Rubredoxin at 1.5 A resolution, *Cold Spring Harbor Symp. Quant. Biol.* 36:359–367.

Wong, J. Y., Meyer, E., and Switzer, R. L., 1977. Glutamine phosphoribosylpyrophosphate amidotransferase from *Bacillus subtilis*, *J. Biol. Chem.* 252:7424–7426.

Wong, G. B., Kurtz, D. M., Holm, R. H., Mortenson, L. E., and Upchurch, R. G., 1979. A ^{19}F NMR method for identification of iron–sulfur cores extruded from active centers of proteins, with applications to milk xanthine oxidase, and the iron–molybdenum proteins of nitrogenase, *J. Am. Chem. Soc.* 101:3078–3090.

Yasunobu, K. T., and Tanaka, M., 1973a. The types, distribution in nature, structure–function, and evolutionary data of the iron–sulfur proteins, *in Iron–Sulfur Proteins*, Vol. II, W. Lovenberg (ed.), Academic Press, New York, pp. 27–130.

Yasunobu, K. T., and Tanaka, M., 1973b. The evolution of iron–sulfur protein containing organisms, *Syst. Zool.* 22:570–589.

Yasunobu, K. T., Mower, H. F., and Hayaishi, O. (eds.), 1976. *Iron and Copper Proteins*, Plenum, New York.

Yates, M. G., O'Donnell, M. J., Lowe, D. J., and Bothe, H., 1978. Ferredoxins from nitrogen fixing bacteria. Physical and chemical characterization of two ferredoxins from *Mycobacterium flavum 301*, *Eur. J. Biochem.* 85:291–299.

Yocum, C. F., Siedow, J. N., and San Pietro, A., 1973. Iron–sulfur proteins in photosynthesis, *in Iron–Sulfur Proteins*, Vol. I, W. Lovenberg (ed.), Academic Press, New York, pp. 111–127.

Zelitch, I., 1979. Photosynthesis and plant productivity, *Chem. Eng. News*, Feb. 5, pp. 28–48.

Zimmermann, R., Münck, E., Brill, W. J., Shah, V. K., Henzl, M. T., Rawlings, J., and Orme-Johnson, W. H., 1978. Nitrogenase X: Mössbauer and EPR studies on reversibly oxidized MoFe protein from *Azobacter vinelandii OP*. Nature of the iron centers, *Biochim. Biophys. Acta* 537:185–207.

Miscellaneous Aspects of Iron Metabolism

9.1 Introduction

The preceding chapters have amply demonstrated the crucial role that iron plays in the process of life. These chapters have dealt with a number of topics which have received much attention from biomedical researchers and are especially important in human health and disease. A number of subjects dealing with nonheme iron have not been mentioned in this volume, and this chapter is designed to survey such areas, where a considerable body of knowledge is available.

9.2 Phosvitin, an Egg Iron-Binding Phosphoprotein

It has been said that the iron-binding protein of egg white, conalbumin (see Chapter 4), acts to protect the egg against microbial invasion. However, the iron required for the developing embryo is presumably provided by the egg yolk protein called phosvitin. It is found in the eggs of many species, including birds, reptiles, amphibians, and fish (Taborsky, 1974). In the hen's egg, phosvitin accounts for some 7% of all yolk protein and about 60% of all yolk phosphoprotein. It is localized in the granular fraction of yolk material, which makes it possible to concentrate phosvitin by simple centrifugation of the yolks. The protein must then be extracted from the sediment by a variety of methods. Phosvitin preparations generally consist of a major and a minor component. These can be detected electrophoretically and by gel chromatography. They differ from each other in regard to molecular weight and amino acid composition.

The remarkable thing about phosvitins is their high phosphorus and serine content. The phosphorus is of course linked to serine, though some threonine residues are also involved in the binding of phosphate (Allerton and Perlmann, 1965). Phosvitins contain as much as 10% phosphorus, and over half of their amino acid residues are serine. Moreover, the proteins have no cystine; hence they are not covalently cross-linked. Phosvitins also contain carbohydrate. The amino acid and phosphorus contents of some phosvitins are given in Table 9-1.

There has been considerable uncertainty in regard to the true molecular weights of phosvitins, which are summarized in Table 9-2. It will be noted that rather large differences exist among phosvitins isolated from the eggs of different species. The generally accepted molecular weight of hen's egg phosvitin is 36,000 (Taborsky, 1974). Partial specific volumes of these proteins are very low, ranging from 0.540 to 0.570 cc/g (Clark, 1972). This indicates that there is a considerable degree of electrostriction, which is in turn due to the polyelectrolyte nature of these proteins.

The primary structure of phosvitin has been investigated to a limited extent because of the uncertainty in regard to the purity of phsovitin

Table 9-1. Amino Acid Contents of a Number of Iron-free Phosvitins (in moles of amino acid residue/mole of protein)[a]

	Species			
Amino acid	Hen	Frog	Rainbow trout	Pacific herring
Lys	17	15.6	5	1
His	11	10.7	0	0
Arg	12	12.4	16	1
Gly	6	11.8	4	0
Ala	8	4.8	3	0
Val	3	1.3	0	0
Leu	3	2.0	0	3
Ile	2	1.8	2	0
Pro	3	3.6	4	0
Phe	2	0.4	0	1
Tyr	1	2.9	1	0
Trp	1	0.2	0	0
Ser	122	93.7	65	23–25
Thr	5	4.2	2	0
½-Cys	0	0.4	0	0
Met	1	0.2	0	0
Asx	14	12.6	9	1
Glx	13	19.7	4	0
Phosphate	120	114.1	9.8%	10.6%
Total (except for phosphate)	224	198.3	115	30–32
Molecular weight	35,500	32,000	19,350	4620

[a] Adapted from Taborsky (1974).

Table 9-2. Physical Properties of Phosvitins

Species	Molecular weight	Method	Reference
Hen's egg	40,000–45,000	Trp, Tyr, Met content	Allerton and Perlmann, 1965
	34,000 (major)	Ultracentrifugation	Clark, 1970
	28,000 (minor)	Ultracentrifugation	Clark, 1970
	21,000	Osmotic pressure	Mecham and Olcott, 1949
	38,000–39,000 with Mg^{2+}	Osmotic pressure	Mecham and Olcott, 1949
	34,400 (major)	Ultracentrifugation	Clark, 1972
Duck S_{1b}	36,600	Ultracentrifugation	Clark, 1972
Duck S_{2b}	28,400	Ultracentrifugation	Clark, 1972
Turkey S_1	29,400	Ultracentrifugation	Clark, 1972
Turkey S_5	30,500	Ultracentrifugation	Clark, 1972
Ostrich S_2	52,500	Ultracentrifugation	Clark, 1972
Ostrich S_4	44,700	Ultracentrifugation	Clark, 1972
Rainbow trout	19,350	Ultracentrifugation	Taborsky, 1974
Dog salmon	19,000	Ultracentrifugation	Taborsky, 1974
Pacific herring	4,620	Composition, ultracentrifuge	Taborsky, 1974
Frog	32,000	Ultracentrifugation	Taborsky, 1974

preparations. Its N-terminal group is alanine (major component) and lysine (minor component) (Clark, 1972). Some investigators have also detected N-terminal serine in the whole protein preparations. The C-terminus appears to be tyrosine. The presence of a single methionine residue permits the splitting of phosvitin into two halves by cyanogen bromide (Taborsky, 1974). The protein has also been subjected to proteolytic digestion for the purpose of isolating its carbohydrate moiety, and there appears to be but a single carbohydrate chain attached to an asparagine residue. The sequence of amino acids in the glycopeptide isolated is (Shainkin and Perlmann, 1971a)

```
        carbohydrate
             |
–Ser–Asn–Ser–Gly- (phosphoserine)8–Arg–Ser–Val–Ser–His–His-        (9-1)
```

Equation (9-1) also indicates that there are clusters of phosphoserine, eight residues each, present throughout the phosvitin molecule. The structure of the oligosaccharide chain is (Shainkin and Perlmann, 1971b)

```
NANA–Gal–GlcNac–Man
                   \                  |
                    Man–Man–GlcNac–Asn            (9-2)
                   /                  |
NANA–Gal–GlcNac–GlcNac
```

The secondary structure of phosvitin has been investigated via the optical rotatory dispersion and circular dichroism techniques. At neutral pH values, the protein exists in an extended form because of its contiguous charged phosphate residues (Grizzuti and Perlmann, 1970). However, if the charge is suppressed, as at low pH values and in the presence of organic solvents, the molecule assumes a structured conformation involving the β pleated sheet and possible α helix (Perlmann and Grizzuti, 1971).

Phosvitin binds ferric iron, but if ferrous iron is offered, phosvitin apparently catalyzes a rapid oxidation thereof with the result that ferric iron thus formed becomes sequestered (Taborsky, 1963). The nature of the bound iron is polynuclear, and the stoichiometry is one iron atom for each two phosphorus atoms (Gray, 1975; Webb *et al.*, 1973). It is assumed that the oxygen atoms of the phosphate groups are the actual iron ligands. Iron can account for as much as 7% of the total weight of phosvitin, which translates into some 45 iron atoms/molecule of protein.

There are apparently two possible phosvitin complexes that can be formed. The so-called *green* phosvitin is obtained when iron is added to native phosvitin. The result is a series of polynuclear iron clusters, where the iron ligand field is tetrahedral (Webb *et al.*, 1973). Such clusters apparently correspond to areas on the protein molecule with contiguous phosphoserine residues. Equilibrium dialysis studies have shown that the association constant between iron and phosvitin is in the vicinity of 10^{18}, which is considerably below that of serotransferrin (Hegenauer *et al.*, 1979). The binding of iron by phosvitin serves to neutralize or suppress the negative charges carried by the phosphate residues, so that the iron–phosvitin molecule can assume an ordered conformation (α helix or pleated sheet) even at neutrality.

Brown phosvitin is obtained when heated apophosvitin is reconstituted with iron. The latter is again present in a polynuclear (clustered) state with a P–Fe ratio of 2; however, the iron is present in the octahedral ligand field. It has been proposed that the heating step modifies the structure of the polypeptide chain in such a way that iron can easily become hydrolyzed while sequestered by the phosphate residues (Webb *et al.*, 1973).

It has been proposed that the physiological function of phosvitin is to act as a ferroxidase and to carry iron from iron storage sites (liver) to egg yolk. Osaki *et al.* (1975) have carried out a number of model experiments on iron oxidation in the presence of phosvitin by O_2 and the exchange of iron between the iron–phosvitin complex and serotransferrin. They point out that about 1 mg of iron is transferred by the egg-laying hen to egg yolk in the form of iron–phosvitin complex and that the transfer of iron to serotransferrin assures an iron supply for hemopoietic purposes. Phosvitin is biosynthesized in the liver in response to estrogens (Planas and Frieden, 1973) and is transported to the site of yolk assembly via the bloodstream. The role of phosvitin in the iron metabolism of the egg-laying hen is summarized in Figure 9-1.

9.3 The Oxygenases

Enzymes that catalyze the incorporation of elemental oxygen into organic molecules are called oxygenases and can be divided into the dioxygenases and the monoxygenases (also called mixed-function oxidases). The former class catalyzes the introduction of both atoms of molecular oxygen into organic substances, whereas the latter reaction results in the fixation of one oxygen atom, with the other appearing as water (Gunsalus *et al.*, 1975). Enzymes of the latter class are often called hydroxylases. Cofactors that are associated with the oxygenases are NAD, NADP, the flavinoids, the iron–sulfur clusters (described in Chapter 8), heme, and metal ions sequestered by amino acid side chains. Among the latter we

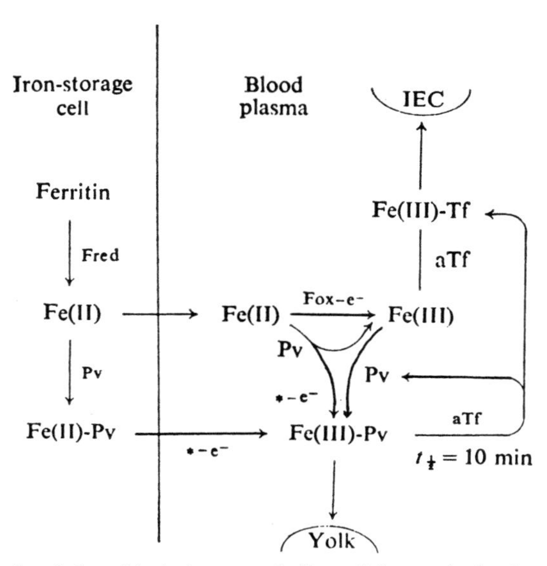

Figure 9-1. The role of phosvitin in iron metabolism of the egg-laying hen. Abbreviations: Pv, phosvitin; F_{ox}, ferroxidase; F_{red}, ferritin reducing system; aTf, aposerotransferrin. An asterisk indicates reaction at a relatively high pO_2. From Osaki *et al.* (1975).

find iron, and in this section we shall be concerned with some selected iron-containing oxygenases. Table 9-3 lists a number of oxygenases where the participation of iron has been demonstrated. In many such cases, the involvement of iron is surmised by the fact that added Fe^{2+} stimulated enzyme activity. A monograph is available on the chemistry and biology of oxygenases (Hayaishi, 1974).

9.3.1 Phenylalanine Hydroxylase

Phenylalanine hydroxylase is the enzyme responsible for the conversion of the essential amino acid phenylalanine to tyrosine; it is absent in the classical phenylketonuria. In addition to iron, which appears to be tightly bound to the enzyme, phenylalanine hydroxylase requires tetrahydrobiopterin as a cofactor, as shown:

$$\text{phenylalanine} + O_2 \xrightarrow[\text{tetrahydrobiopterin} \to \text{dihydrobiopterin}]{\text{phenylalanine hydroxylase}} \text{tyrosine} + H_2O \qquad (9\text{-}3)$$

$$\text{dihydrobiopterin} \xrightarrow[\text{NADPH} + H^+ \to \text{NADP}]{\text{dihydrobiopterin reductase}} \text{tetrahydrobiopterin}$$

The oxidized dihydrobiopterin is regenerated by reduced NADP, so that hydroxylation of phenylalanine is tightly coupled with NADPH oxidation.

Phenylalanine hydroxylase contains a maximum of two iron atoms per protein molecule, i.e., one iron atom per subunit. It has an EPR signal at $g' = 4.23$, indicating a high-spin ferric state. This signal disappears when substrate is added to the enzyme, most likely because iron becomes reduced and changes the nature of its ligand field. It is believed that the iron operates as a *link* in the flow of electrons from tetrahydrobiopterin to oxygen. Iron-deficient rats do not seem to have a decreased activity in phenylalanine hydroxylase even though blood phenylalanine levels in such animals are extremely high (Mackler *et al.*, 1979). Table 9-4 lists some physical–chemical properties of phenylalanine hydroxylase as well as those of other iron-containing oxygenases.

9.3.2 Tyrosine and Tryptophan Hydroxylases

There are two other enzymes of aromatic amino acid metabolism that require tetrahydrobiopterin and contain iron. Tyrosine hydroxylase converts tyrosine into dihydroxyphenylalanine (DOPA), which, in turn, is an

Table 9-3. Some Oxygenases Containing Nonheme Iron[a]

Enzyme	Source	Type[b]	Reaction catalyzed
Phenylalanine metabolism			
Phenylalanine hydroxylase	Liver	M	Converts phenylalaine to tyrosine
Tyrosine hydroxylase	Adrenals & brain	M	Converts tyrosine to DOPA
4-Hydroxyphenylpyruvate hydroxylase	Liver & *Pseudomonas*	D	Converts 4-hydroxyphenylpyruvate to homogentisic acid
Homogentisic acid oxygenase	Liver, *Pseudomonas fluorescens*	D	Converts homogentisic acid to maleylacetoacetic acid
Tryptophan metabolism			
Tryptophan hydroxylase	Liver, brain, intestine	M	Converts tryptophan to 5-OH-tryptophan
3-Hydroxyanthranilate oxygenase	Liver, kidney	D	Converts anthranilate to α-amino-β-carboxymuconic semialdehyde[c]
Lipoxygenase[d]	Soybeans	D	Forms hydroperoxides from polyunsaturated fatty acids
Catechol oxygenases			
Pyrocatechase	*Acinetobacter calcoaceticus*, *Pseudomonas arvilla*	ID	Converts catechol to *cis,cis*-muconic acid
Protocatechuate-3,4-dioxygenase	*Pseudomonas aeruginosa*	ID	Converts protocatechuate to β-carboxymuconic acid
Metapyrocatechase	*Pseudomonas arvilla*	ED	Converts catechol to α-hydroxymuconic semialdehyde

continued overleaf

Table 9-3. (Continued)

Enzyme	Source	Type[b]	Reaction catalyzed
3,4-Dihydroxyphenylacetate-2,3-dioxygenase	*Pseudomonas ovalis*	ED	Converts 3,4-dihydroxyphenylacetate to α-hydroxy-δ-carboxymethylmuconic semialdehyde
Protocatechuate-4,5-dioxygenase	*Pseudomonas testeroni*	ED	Converts protocatechuate to α-hydroxy-γ-carboxymuconic semialdehyde
Collagen biosynthesis			
Prolyl hydroxylase	Fetal skin, embryos	D	Converts proline to hydroxyproline
Lysyl hydroxylase	Chick embryo	D	Converts lysine to hydroxylysine
γ-Butyrobetaine hydroxylase	Liver, *Pseudomonas*	D	Converts γ-butyrobetaine to carnitine
Pyrimidine oxygenases			
Thymine-7-hydroxylase	Neurospora	D	Converts thymine to 5-hydroxymethyluracil
5-Hydroxymethyluracil dioxygenase	Neurospora	D	Converts 5-hydroxymethyluracil to 5-formyluracil
Pyrimidine deoxyribonucleoside-2-hydroxylase	Neurospora	D	Converts thymidine to thymine ribonucleoside (hydroxylates deoxyribose)

[a] Compiled from Hayaishi (1974).
[b] Abbreviations: M, monoxygenase; D, dioxygenase; ID, intradiol dioxygenase, i.e., cleavage of benzene ring occurs between contiguous hydroxyl groups; ED, extradiol dioxygenase, i.e., benzene ring cleavage occurs meta to a hydroxyl group.
[c] This compound may cyclize to form a precursor of NAD.
[d] Pistorius and Axelrod (1974).

Table 9-4. Physical–Chemical Properties of Some Nonheme Iron Containing Oxygenases

Enzyme	Mol. wt.	Subunits	Iron (moles/mole)	K_m & substrate	Reference
Phenylalanine hydroxylase	110,000	2, identical	2, Fe^{3+}	4×10^{-5} *M*, phenylalanine	Kaufman and Fisher, 1970;[a] Fisher *et al.*, 1972
Tyrosine hydroxylase	200,000	Unknown	Fe^{2+}	5×10^{-5} *M*, tyrosine	Kuczenski, 1973[b]
	135,000–155,000	Unknown	Fe^{2+}	—	Kaufman and Fisher, 1974[c]
Pyrocatechase	81,000	2, identical	2, Fe^{3+}	—	Patel *et al.*, 1976[d]
4-Hydroxyphenylpyruvate hydroxylase	80,000–87,000	2, identical	0.4, Fe^{2+}	50 μ*M*, O_2; 0.05 m*M*, phenylpyruvate; 0.03 m*M*, *p*-OH-phenylpyruvate	Lindblad *et al.*, 1977[e]
Protocatechuate-3,4-dioxygenase	700,000	8, identical	7–8, Fe^{3+}	—	Fujisawa and Hayaishi, 1968; Fujisawa *et al.*, 1972[f]
	677,000	Unknown	7, Fe^{3+}	7.1×10^{-5} *M*, protocatechuate; 5.9×10^{-5} *M*, O_2	Hou *et al.*, 1976[g]
p-Hydroxyphenylpyruvate dioxygenase	166,000	4, identical	0.6–1.3, Fe^{2+}	30 μ*M*, *p*-hydroxyphenylpyruvate	Lindstedt *et al.*, 1977[h]
Proline hydroxylase	240,000	2, mw. 61,000 & 64,000	Fe^{2+}	5 μ*M*, Fe^{2+}	Kuutti *et al.*, 1975[i]

[a] From rat liver; EPR signal at $g = 4.23$.
[b] From rat brain; trypsin produces active enzyme with a molecular weight of 50,000.
[c] From bovine adrenal medulla; trypsin produces active enzyme with a molecular weight of 34,000.
[d] From *Acinetobacter calcoaceticus*; EPR signal at $g = 4.28$.
[e] From human liver; multiple forms with different pI values.
[f] From *Pseudomonas aeruginosa*; EPR signal with $g = 4.31$.
[g] From *Acinetobacter calcoaceticus*.
[h] From *Pseudomonas* sp. P.J. 874.
[i] From human fetal skin; requires α-ketoglutarate and ascorbate.

intermediate in the biosynthesis of melanins and norepinephrine. In fact, tyrosine hydroxylase is the rate-limiting step in the biosynthesis of the catecholamines and shows product-type inhibition by a number of catechols, especially norepinephrine. It is activated by heparin. The reaction catalyzed by the enzyme may be summarized by

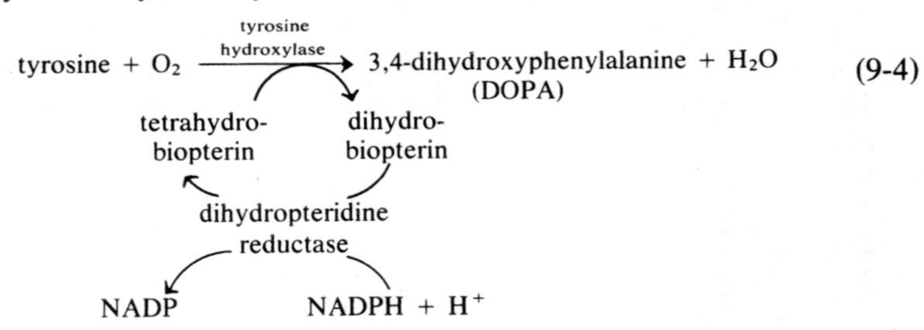

(9-4)

The enzyme is present in both the adrenal medulla and brain in the insoluble state (probably membrane-bound), and treatment of the homogenate particulate matter with proteolytic enzymes is necessary to get the enzyme into solution (Petrack *et al.*, 1968). The role of ferrous iron in the activity of the enzyme was surmised from the fact that ferrous iron served to stimulate the enzyme and because ferrous iron chelators (α,α'-dipyridyl and *o*-phenanthroline) served to inhibit its activity (Petrack *et al.*, 1972; Kaufman and Fisher, 1974).

Tryptophan hydroxylase is the third iron hydroxylase requiring tetrahydrobiopterin for its activity. It is found in the brain and catalyzes the formation of 5-hydroxytryptophan from tryptophan. Serotonin (5-hydroxytryptamine) is then formed through decarboxylation of 5-hydroxytryptophan. The tryptophan hydroxylase is believed to be the rate-limiting step in the formation of serotonin from tryptophan. It is stimulated by ferrous iron and inhibited by ferrous iron chelators (Kaufman and Fisher, 1974).

9.3.3 Pyrocatechase (Catechol-1,2-dioxygenase)

Many microorganisms have the capability of oxidatively splitting the benzene ring of catechols. The ring can be split in several places with respect to the two hydroxyl groups. Pyrocatechase is a representative of the so-called intradiol dioxygenases, where the benzene ring cleavage occurs between the two hydroxyl groups. The microorganisms from which the enzyme has been isolated are *Pseudomonas arvilla, Pseudomonas fluorescens, Brevibacterium fuscum,* and *Acinetobacter calcoaceticus*

(Nozaki, 1974; Patel *et al.*, 1976). The reaction is represented by

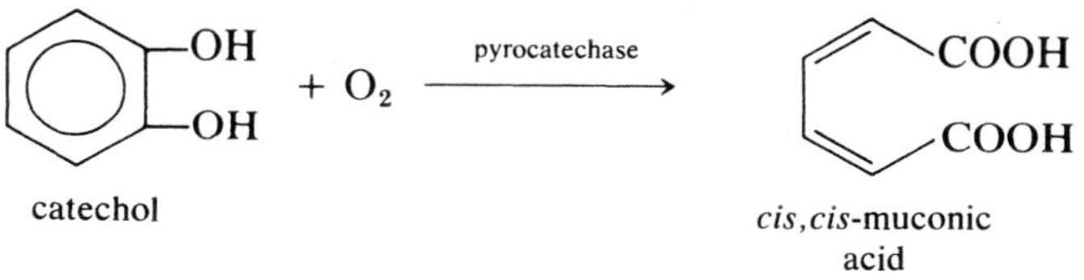

catechol (–OH, –OH) + O_2 —pyrocatechase→ *cis,cis*-muconic acid (COOH, COOH) (9-5)

Catechol may be split in other ways as well, as, for instance, by the metapyrocatechase (and also from *Pseudomonas arvilla*), as shown in

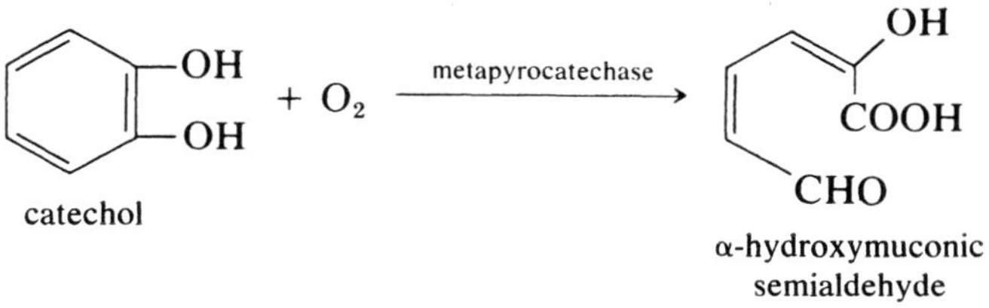

catechol (–OH, –OH) + O_2 —metapyrocatechase→ α-hydroxymuconic semialdehyde (OH, COOH, CHO) (9-6)

Metapyrocatechase is termed an extradiol dioxygenase.

Pyrocatechases have molecular weights between 60,000 and 100,000 and may contain subunits (see Table 9-4). Ferric iron is apparently required for activity, since the EPR spectrum yielded a *g* value of 4.28. In the presence of a substrate (catechol), the signal disappears, perhaps because of a reduction process. The amino acid composition of the *Acinetobacter calcoaceticus* enzyme is presented in Table 9-5.

9.3.4 Protocatechuate-3,4-dioxygenase

The title enzyme is another intradiol dioxygenase which uses protocachuate as a substrate. This enzyme is isolated from several microorganisms, most notably *Pseudomonas aeruginosa*. It catalyzes the following reaction:

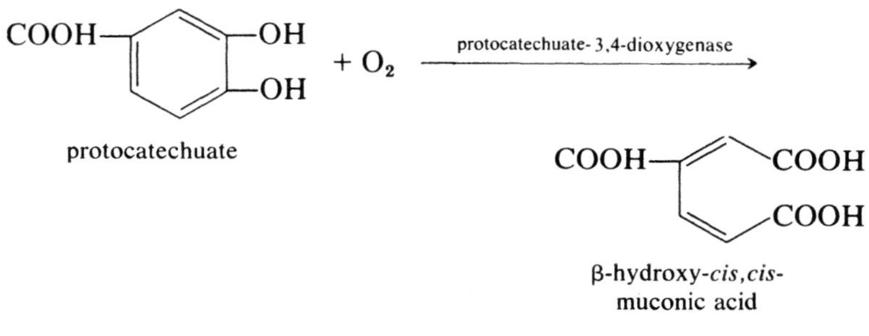

protocatechuate (COOH–, –OH, –OH) + O_2 —protocatechuate-3,4-dioxygenase→ β-hydroxy-*cis,cis*-muconic acid (COOH–, COOH, COOH) (9-7)

Table 9-5. Amino Acid Content of Iron-Containing Oxygenases (in moles/mole of protein, except for proline hydroxylase)

	Source[a]				
Amino acid	1	2	3	4	5
Lys	23	17.1	182	446	93
Arg	21	30.5	351	412	42
His	8	19.6	194	175	19
Asx	71	80.0	718	696	122
Thr	20	31.3	286	300	51
Ser	30	14.6	205	275	45
Glx	56	62.4	513	787	142
Pro	28	34.1	425	11	50
Gly	26	54.4	460	467	72
Ala	31	54.0	477	545	89
Val	21	41.1	265	294	58
Met	4	8.3	46	6	—
Ile	19	28.5	364	570	45
Leu	46	49.7	459	500	98
Tyr	14	31.4	188	130	22
Phe	21	20.3	247	318	52
½-Cys	12	3.4	95	46	—
Trp	2	9.4	158	151	—

[a] Notes and references:

1. Phenylalanine hydroxylase, molecular weight 50,000 (subunit); from Fisher *et al.* (1972); ½-Cys includes five residues of cysteine and seven residues of cysteic acid.
2. Pyrocatechase from *Acinetobacter calcoaceticus*, molecular weight 81,000; from Patel *et al.* (1976).
3. Protocatechuate-3,4-dioxygenase from *Pseudomonas aeruginosa*, molecular weight 700,000; from Fujisawa and Hayaishi (1968).
4. Protocatechuate-3,4-dioxygenase from *Acinetobacter calcoaceticus*, molecular weight 677,000; from Hou *et al.* (1976).
5. Prolyl hydroxylase from human source, molecular weight not given; in residues per 1000 residues; from Kuutti *et al.* (1975).

The molecular weight of protocatechuate-3,4-dioxygenase has been reported to be around 650,000–700,000, and alkali dissociates this into eight identical subunits with molecular weights of 90,000–120,000 each (Fujisawa and Hayaishi, 1968; Fujisawa *et al.*, 1972). Each subunit so produced is dissociable into smaller fragments by denaturants or detergents. These have molecular weights of 15,000 and 18,000. It is claimed that each "subunit" produced by alkali consists of two 15,000-dalton particles and two 18,000-dalton particles for a total molecular weight of an alkali-produced "subunit" of 66,000 (Lipscomb *et al.*, 1976). This is substantially lower than the figure reported by Fujisawa *et al.* (1972). The α-helix content of both the *Pseudomonas* and the *Acinetobacter* enzymes is only 0.85 and 7.24%, respectively (Hou *et al.*, 1976).

It is generally accepted that protocatechuate-3,4-dioxygenase binds seven to eight iron atoms per protein molecule (Hou *et al.*, 1976; Fujisawa *et al.*, 1972), though figures twice this have been reported (Lipscomb *et al.*, 1976). The iron is apparently present in a high-spin state ($S = \frac{5}{2}$), and there is an EPR signal at $g = 4.3$. This indicates a ferric oxidation state in a rhombic environment. When the substrate and/or O_2 are added to the enzyme, the EPR spectrum changes, but the iron remains in the high-spin ferric state (Que *et al.*, 1976, 1977): There appears an EPR signal with a g value of 6.0 upon substrate binding and signals with g values of 6.7 and 5.3 upon the ternary oxygen complex formation. Circular dichroism and other physical studies have suggested aromatic amino acids as iron-binding ligands in this enzyme (Hou, 1975, 1978), though because of the large negative zero-field splitting observed in the EPR spectra of the enzyme, the ligand field may be something not previously encountered in biological systems (Hou *et al.*, 1976). A mechanism for the pyrolysis of the benzene ring by the enzyme has been proposed by Que *et al.* (1977) and is shown in Figure 9-2.

9.3.5 4-Hydroxyphenylpyruvate Dioxygenase

4-Hydroxyphenylpyruvate dioxygenase catalyzes the conversion of *p*-hydroxyphenylpyruvate to homogentisic acid, a reaction of normal tyrosine metabolism:

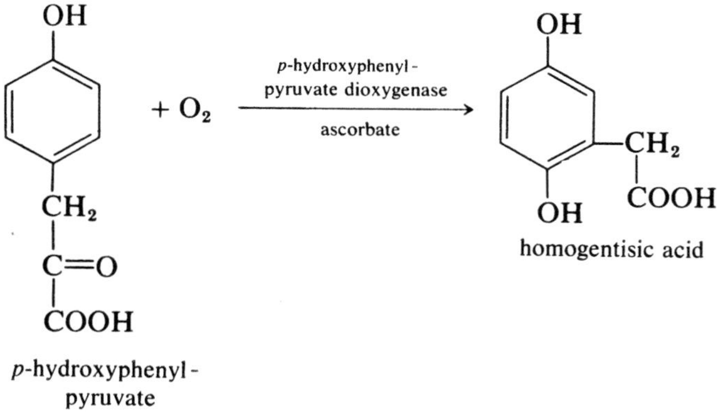

OH

$+ O_2$ —— *p*-hydroxyphenyl-pyruvate dioxygenase / ascorbate ——→

OH

CH_2

COOH

OH

CH_2

C=O

COOH

p-hydroxyphenyl-pyruvate

homogentisic acid

(9-8)

The enzyme has been isolated from mammalian as well as bacterial sources. In the human being, deficiencies of this enzyme are the cause of neonatal tyrosinemia and hereditary tyrosinemia. The former is transient and can be reversed by the administration of ascorbic acid, whereas the latter is a classical inborn error of amino acid metabolism.

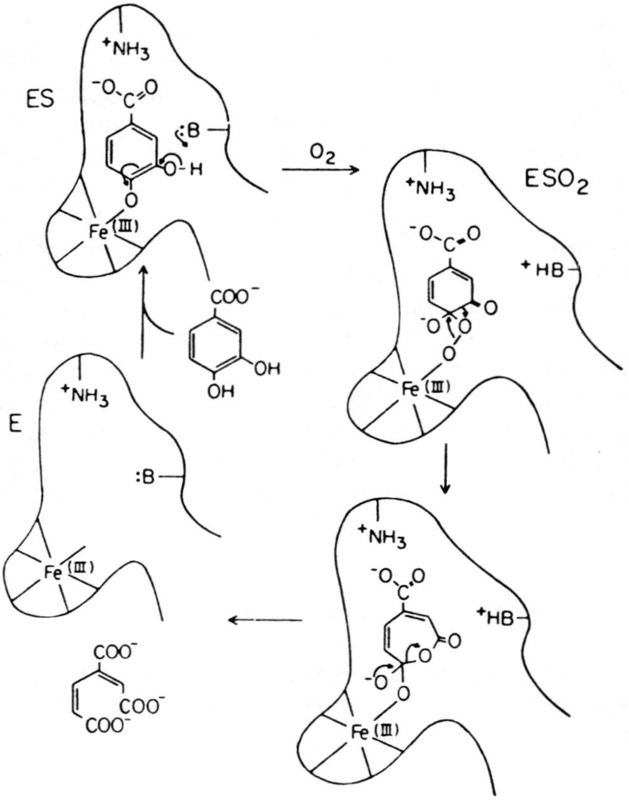

Figure 9-2. The mechanism of benzene ring cleavage by protocatechuate-3,4-dioxygenase. From Que *et al.* (1977).

Iron is present in small amounts in *p*-hydroxyphenylpyruvate dioxygenase (see Table 9-4), and there is apparently some copper present as well (Lindstedt *et al.*, 1977). Because ferrous iron chelators could inhibit the enzymatic activity, which could be restored upon ferrous iron addition, it is believed that ferrous iron is involved in the activity. However, adding ferrous iron to the native enzyme does not increase its activity; hence the iron must be rather tightly bound to the protein. The enzymatic mechanism appears to be of the ordered single-displacement (ordered bi–bi) type, where the enzyme first combines with one substrate (*p*-hydroxyphenylpyruvate) and then with the other (oxygen) to form the ternary enzyme–*p*-hydroxyphenylpyruvate–oxygen complex. The products are then released in an orderly manner, CO_2 first and then homogentisic acid (Rundgren, 1977). Hydrogen peroxide is a potent inhibitor of the enzyme, though the mechanism for this is unclear.

9.3.6 Proline Hydroxylase

Proline hydroxylase is an oxygenase whose activity is associated with the decarboxylation of α-ketoglutarate. Other enzymes of this class are lysine hydroxylase, γ-butyrobetaine hydroxylase, and pyrimidine and nucleoside hydroxylases. These enzymes are not necessarily iron–enzymes. The overall reaction of proline hydroxylation to hydroxyproline is represented by

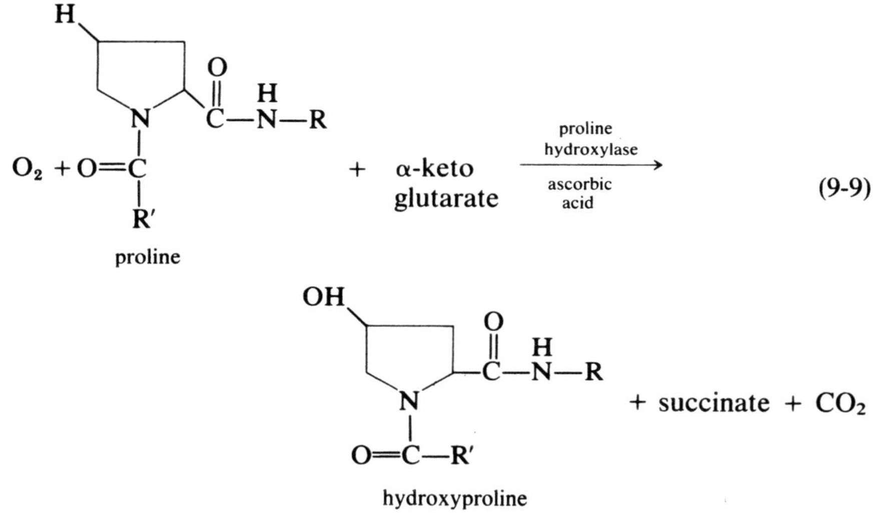

$$O_2 + \underset{\text{proline}}{\text{H-pyrrolidine-N}(\text{C(=O)}R')\text{-C(=O)-NH-R}} + \begin{matrix}\alpha\text{-keto}\\ \text{glutarate}\end{matrix} \xrightarrow[\text{ascorbic acid}]{\text{proline hydroxylase}} \qquad (9\text{-}9)$$

$$\underset{\text{hydroxyproline}}{\text{OH-pyrrolidine-N}(\text{C(=O)}R')\text{-C(=O)-NH-R}} + \text{succinate} + CO_2$$

where R and R′ are the appropriate amino acid sequences of procollagen. The hydroxylation of proline (and hydroxylysine) is thus a posttranslational event.

The enzyme has an absolute requirement for both ferrous iron and ascorbate; however, ascorbate is not stoichiometrically consumed in the reaction, and the enzyme system can operate for a few cycles without ascorbate before becoming inactive. It is suspected that the ascorbate is necessary to maintain the iron in the ferrous state when the normally ferrous iron may become oxidized to the ferric state through some side reaction (Tuderman *et al.*, 1977; Myllylä *et al.*, 1978). Ferrous iron is apparently loosely bound to the enzyme and is easily lost therefrom during purification procedures. Zinc is an inhibitor of the enzyme, probably by combination with the active site of the enzyme. Collagen and succinate show a product-type inhibition.

The mechanism of the reaction is of the ordered single-displacement type (sequential), where the enzyme first forms a complex with the ferrous iron followed by α-ketoglutarate and oxygen. At this point, a decarboxylation of the α-ketoglutarate can take place, but if the substrate peptide (procollagen) is available, it combines with the enzyme complex, and instead a hydroxylation of the prolyl residue ensues. The enzyme minus iron or α-ketoglutarate can also combine with the substrate peptide to form *dead-end* complexes. Following hydroxylation, the modified peptide separates from the enzyme complex followed by CO_2 and succinate. Iron probably is oxidized and then reduced in this process. Ascorbate apparently can combine with the enzyme–iron complex following this reaction cycle. It is believed that iron need not dissociate from the enzyme after each hydroxylation cycle (Myllylä *et al.*, 1977).

9.4 Hemerythrins

The hemerythrins are oxygen carrier proteins in the circulation of various invertebrates such as the annelids, brachiopods, priapulids, and sipunculoids. They are presumably *remote* relations of the hemoglobins from an evolutionary point of view (Llinas, 1973). Hemerythrins carry iron bound to the polypeptide chain via amino acid residues; hence they contain no heme.

A great proportion of the work on the properties of hemerythrins was performed in I. M. Klotz's laboratory using the *Golfingia gouldii* (a marine worm) material. This protein was shown to have a molecular weight of near 107,000 by a number of physical techniques. It contains eight identical subunits, each with a molecular weight of about 13,500 (Klotz and Keresztes-Nagy, 1963). Apparently the intact protein exists in an equilibrium with free subunits in the aqueous medium (Keresztes-Nagy *et al.*, 1965). Chemical modification methodology has established that the subunits interact with each other via free sulfhydryl and possibly tyrosyl and histidyl residues (York and Roberts, 1976). The subunits are therefore held together by physical interactions. They are sometimes called myohemerythrin.

There are 16 iron atoms present in each molecule of hemerythrin, i.e., two for each subunit. They are normally in the ferrous state but are easily oxidized to the ferric state either by oxygen or by ferricyanide. In the oxidized state, the protein is termed methemerythrin, and many investigators prefer to work with this species because of its greater degree of stability. The ferrous iron present in native hemerythrin (oxygen-free) is in the high-spin state ($S = 2$) with four unpaired electrons. In methemerythrin, the iron is also in the high-spin state ($S = \frac{5}{2}$), with a single unpaired electron and no paramagnetism. It is believed that the two Fe^{3+} atoms in the methemerythrin subunit are coupled antiferromagnetically

(York and Bearden, 1970; Moss *et al.*, 1971). The two iron atoms are sufficiently close to each other to be capable of such an interaction (3.44 = 0.05 Å) (Hendrickson *et al.*, 1975). Each subunit is capable of reversibly binding one oxygen molecule. Spectroscopic analyses of a number of hemerythrins have been performed, and all appear to give identical spectra, indicating indentical iron-binding ligands (Dunn *et al.*, 1977).

Amino acid sequences of a number of hemerythrins have been determined (Loehr *et al.*, 1978; Gormley *et al.*, 1978), and they are presented in Figure 9-3. As expected, all three show extensive sequence homologies. Additionally, hemerythrin crystals have been subjected to X-ray crystallography, and this has yielded information not only on the general shape of the protein molecule and the architecture of the iron sites but also the identity of amino acid side chains involved in iron-binding interactions (Hendrickson *et al.*, 1975; Stenkamp *et al.*, 1976, 1978). Such studies have revealed the overall appearance of *Themiste pyroides* myohemerythrin shown in Figure 9-4. In this particular hemerythrin, the molecular dimensions are 30 × 44 × 28 Å. The thick rods in Figure 9-4 represent α helix, estimated at 75% of all amino acid residues present. No β structure (pleated sheet) is present. This agrees with the results obtained with cir-

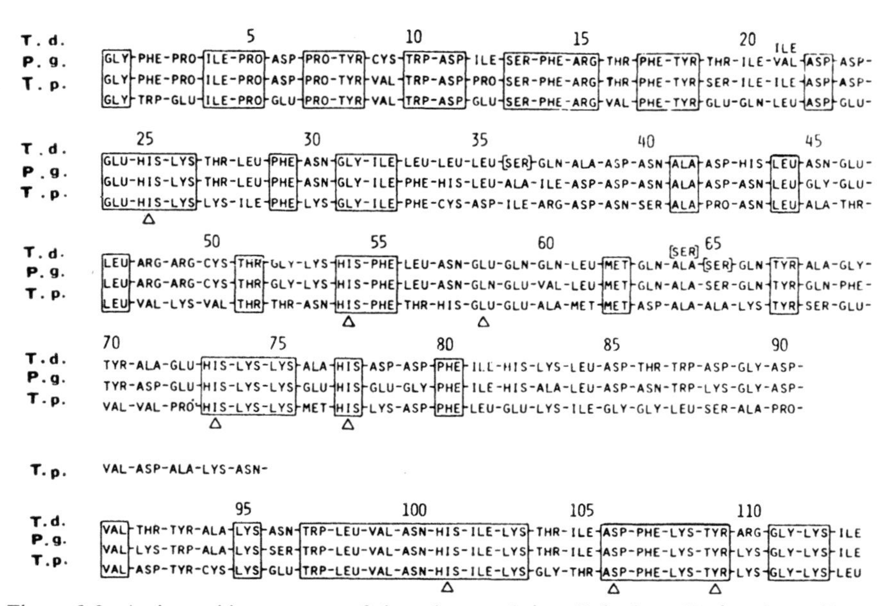

Figure 9-3. Amino acid sequences of three hemerythrins. *T.d.*, from *T. dyscritum*; *P.g.*, from *P. gouldii*; and *T.p.*, from *T. pyroides*. The boxed residues are invariant. Triangles indicate iron-binding ligands. From Loehr *et al.* (1978).

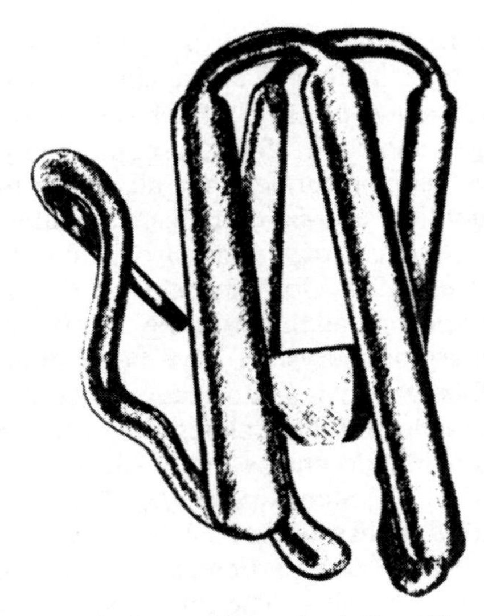

Figure 9-4. Overall appearance of a hemerythrin subunit from *Themiste pyroides*. From Hendrickson *et al*. (1975).

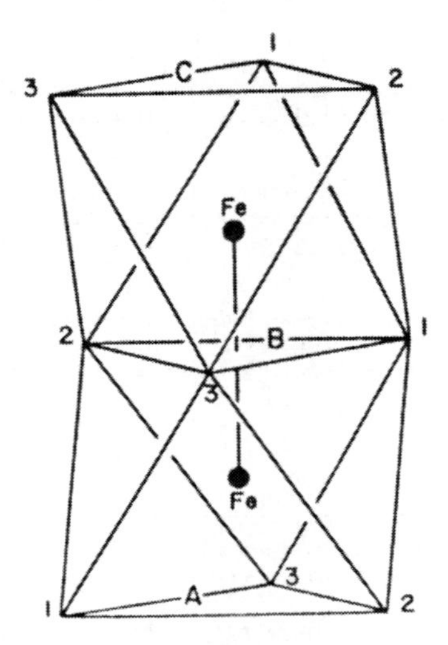

Figure 9-5. Geometry of the iron center in *Themiste dyscritum* hemerythrin. The apexes of triangles A, B, and C represent iron ligands. From Stenkamp *et al*. (1976).

cular dichroism measurements (Darnall *et al.*, 1969). Side chains implicated in the binding of iron are Tyr 67, His 73, and His 106 for one iron residue and His 25, His 54, and Tyr 114 for the other.

In the *Themiste dyscritum* hemerythrin, the iron was found to have an octahedral ligand field with the geometry shown in Figure 9-5. In both the *Themiste dyscritum* and *Golfingia gouldii* materials, the ligands are as follows: A-1, A-2, A-3: His 25, His 54, and Tyr 109; B-1, B-2: Glu 58, Asp 106; C-1, C-2, C-3: His 73, His 77, His 101 (Stenkamp *et al.*, 1976; Loehr *et al.*, 1978). Position B-3 is presumably occupied by water in deoxyhemerythrin and by oxygen in oxyhemerythrin. The exact identity of the amino acid side chains involved in subunit interactions is at the present unclear.

9.5 Iron Bacteria

Iron bacteria are defined as those microorganisms that can derive their energy requirements by oxidizing iron. Most authorities further restrict this classification to those bacteria which are iron autotrophs. Such bacteria are a part of the chemolithotrophic group of organisms which can utilize inorganic materials to derive energy and reducing equivalents for the sustenance of their growth. Other chemolithotrophic organisms are the sulfur bacteria and the nitrite reducers. These organisms do not utilize light as an energy source (Kelly, 1971).

It is said that Serge Vinogradsky in 1888 was first to recognize the fact that the iron bacteria needed little if any organic matter for growth and that energy therefore was derivable entirely from the oxidation of ferrous iron. He used the *Leptothrix* organism in his study (Pringsheim, 1949). Today, the iron bacteria are of considerable economic interest, as organisms excreting ferric iron, autotrophic or otherwise, are a major cause of water supply pollution as well as a major problem in the plumbing business. On the other hand, iron-precipitating bacteria and other microorganisms have been largely responsible for the deposition of iron sulfide and iron oxide ores (Silverman and Ehrlich, 1964).

There are numerous bacteria that can oxidize ferrous iron. Pringsheim (1949) has published an extensive list of these, but most are heterotrophs. Among the autotrophs, we find the *Gallionella, Thiobacillus ferrooxidans*, and certain strains of the *Leptothrix* and *Sideromonas* (Silverman and Ehrlich, 1964). *Thiobacillus ferrooxidans* has been utilized in research as a typical representative of the autotrophic iron bacterial group of organisms.

In the autotrophic iron bacteria, the simple process of converting ferrous iron to ferric iron must be able to generate ATP as well as to provide for the reduction of NAD and NADP, which in turn are utilized to reduce (fix) CO_2 for the biosynthesis of proteins and other organic compounds (Lundgren *et al.*, 1974). The ultimate electron acceptor in these organisms is oxygen, so that the overall reaction can be represented by (Suzuki, 1974)

$$4\ Fe^{2+} + O_2 + 4\ H^+ \rightarrow 4\ Fe^{3+} + 2\ H_2O \tag{9-10}$$

Electrons flow from ferrous iron to oxygen through the participation of the cytochrome system. Two types of cytochromes, cytochrome c and a, have been identified in this process, so that the known sequence of reactions may be represented by

$$\begin{array}{ccccccc} Fe^{3} & & \text{cyt } c_{red} & & \text{cyt } a_{ox} & & H_2O \\ & \times & & \times & & \times & \\ Fe^{2+} & & \text{cyt } c_{ox} & & \text{cyt } a_{red} & & \tfrac{1}{2}\ O_2 \\ \uparrow & & & & & & \nwarrow \\ \text{cytochrome c reductase} & & 2\ \text{ADP} & \curvearrowright & 2\text{ATP} & & \text{cytochrome oxidase} \end{array} \tag{9-11}$$

Somewhere along the pathway outlined in equation (9-11), two molecules of ADP are phosphorylated to ATP. Most likely, this step occurs at the cytochrome c level.

The enzymes cytochrome c reductase and cytochrome oxidase have been purified from *Thiobacillus ferrooxidans*. The former is apparently a nucleoprotein composed of a protein subunit (molecular weight 27 - 000–30,000) and an RNA subunit (molecular weight 315,000–330,000). The mechanism of the reaction is of the ping-pong bi-bi type (Lundgren *et al.*, 1974). There are apparently at least two cytochromes c and two cytochromes a involved in the transport of electrons in *Thiobacillus ferrooxidans*. In addition, there appears to be a blue copper-containing protein associated with the electron transport mechanism. The name rusticyanin has been suggested for it. Rusticyanin is normally associated with the cell membrane but is released therefrom into the supernatant if the cells are disrupted by sonication at pH 2 (Cobley and Haddock, 1975). It is possible that rusticyanin mediates the electron transfer from the very high-redox-potential Fe^{2+}–Fe^{3+} system ($\Delta E_0' = 0.78$ V) to the cytochromes, which have a relatively low redox potential.

The second process that the metabolic machinery of the iron bacteria must accommodate is providing reducing equivalents for the purpose of CO_2 fixation via the Calvin cycle and perhaps reverse Krebs cycle (Kelly, 1971). This process is carried out by the so-called reverse electron flow mechanism, this term being invoked because electrons must flow from the Fe^{2+}–Fe^{3+} system with a $\Delta E_0'$ of 0.78 V to the NADH–NAD system with a $\Delta E_0'$ of −0.32 V. Such a process has been demonstrated to occur in chemolithotrophs including *Tiobacillus ferrooxidans* and requires considerable amounts of ATP. Reduced cytochrome c, which can be produced from Fe^{2+} as in equation (9-11), can reduce NAD in the presence of ATP using a flavoprotein as an apparent intermediate, as indicated in

$$\text{cyt } c_{red} + \text{flavoprotein}_{ox} \rightarrow \text{cyt } c_{ox} + \text{flavoprotein}_{red}; \quad \text{flavoprotein}_{red} + \text{NAD} \rightarrow \text{flavoprotein}_{ox} + \text{NADH} + \text{H}^+ \tag{9-12}$$

Normally, the overall reaction in equation (9-12) would require about 4 moles of ATP per mole, calculated as follows: Overall $\Delta E_0' = -0.32 - (-0.254) = -0.574$ V, and $\Delta G = 2 \times 23{,}000 \times (-0.574) = 26{,}404$ cal. The number of ATPs required is then 26,404/7300 = 3.62; however, the extracts required some 10–20 moles of ATP to reduce 1 mole of NAD (Aleem, *et al.*, 1963). Many microorganisms that can derive energy from ferrous iron oxidation are heterotrophic organisms, and their machinery for generating reducing equivalents can assume numerous other forms. For a review, consult Forrest and Walker (1971).

9.6 Metalloserotransferrin as a Lymphocyte Growth Promoter

The best understood biological activity of serotransferrin is the delivery of iron to the erythropoietic system (Chapter 6). An additional biological activity may be its role of delivering iron and perhaps zinc to *transformed* lymphocytes. It is well established that phytohemagglutinin (PHA), a plant protein, can induce certain transformations in lymphocyte cultures *in vitro* which closely resemble the immunological response of these cells *in vivo*. Human serum had been required to obtain maximum effect. It has been found that the active component of serum that can replace serum in tissue culture as a growth promoter of PHA-stimulated lymphocytes is iron-laden serotransferrin (Tormey *et al.*, 1972; Phillips and Azari, 1975). Iron chloride or iron-free serotransferrin are not active; iron-laden transferrins from other sources, e.g., conalbumin and lactoferrin, are only partially active. The iron is apparently quantitatively removed from the iron-laden serotransferrin. The iron-laden protein is required in the $C_1 \rightarrow S$ phase of cell replication, which is the step where active DNA synthetic processes take place. Steps preceding this phase do not require serotransferrin. It is believed that the lymphocyte has specific serotransferrin receptors on its cell membrane surface (Tormey and Mueller, 1972).

Lymphocyte transformation apparently requires zinc as well as iron, since zinc–enzymes play a prominent role in DNA and RNA biosynthesis. Such zinc requirement is met by zinc serotransferrin, though in contrast to iron–serotransferrin, only some 10% of the zinc present in the serotransferrin is incorporated into the cells (Phillips and Azari, 1974).

Summary

In this chapter we have dealt with a variety of subjects related to iron metabolism which are not adequately handled in Chapters 1–8.

Phosvitins are phosphoglycoproteins present in the egg yolks of birds, reptiles, and amphibians. Their function is apparently to act as ferroxidases and to carry iron to the developing egg yolk. These proteins are especially rich in phosphate residues, which are esterified to seryl and threonyl side chains. The phosphate groups act as iron ligands, the iron being present in the form of polynuclear complexes.

The oxygenases are a group of enzymes many of which contain iron ligated through nonsulfur amino acid side chains. These enzymes catalyze the *fixation* of molecular oxygen unto organic molecules. It is possible to distinguish the monoxygenases, which promote the assimilation of only one of the two atoms of O_2, and the dioxygenases, which catalyze the fixation of the entire O_2 molecule. Iron-containing oxygenases participate in the hydroxylation of aromatic amino acids, ring cleavage in catechol-like substances, and the hydroxylation of proline and hydroxylysine during collagen biosynthesis.

Hemerythrins are nonheme oxygen carriers in the bloodstream of a number of nonvertebrates. Each hemerythrin molecule, molecular weight of about 107,000, consists of eight identical subunits, and each subunit carries a cluster of two ferrous iron atoms. The iron-binding ligands are the side chains of histidine, tyrosine, aspartate, and glutamate.

The iron bacteria are organisms which can derive their energy and reducing equivalents from the oxidation of ferrous iron to ferric iron. Oxygen is the electron acceptor. Ferrous iron can reduce a cytochrome c with the generation of ATP. Cytochrome c, in turn, by the ATP-dependent process of reverse electron flow, can reduce NAD to NADH.

An as yet incompletely understood biological function of serotransferrin is the delivery of iron and zinc to immunologically active lymphocytes. The iron is required during the phase of active DNA biosynthesis.

References

Aleem, M. I. H., Lees, H., and Nicholas, D. J. D., 1963. Adenosine triphosphate-dependent reduction of nicotinamide adenine dinucleotide by ferrocytochrome c in chemoautotrophic bacteria, *Nature (London)* 200:759–761.

Allerton, S. E., and Perlmann, G. E., 1965. Chemical characterization of the phosphoprotein phosvitin, *J. Biol. Chem.* 240:3892–3898.

Clark, R. C., 1970. The isolation and composition of two phosphoproteins from hen's egg, *Biochem. J.* 118:537–542.

Clark, R. C., 1972. Sephadex fractionation of phosvitins from duck, turkey and ostrich egg yolk, *Comp. Biochem. Physiol. B* 41:891–903.

Cobley, J. G., and Haddock, B. A., 1975. The respiratory chain of *Thiobacillus* ferrooxidans: The reduction of cytochromes by Fe^{2+} and the preliminary characterization of rustacyanin, a novel "blue" copper protein, *FEBS Lett.* 60:29–33.

Darnall, D. W., Garbett, K., Klotz, I. M., Aktipis, S., and Keresztes-Nagy, S., 1969. Optical rotatory properties of hemerythrin in the ultraviolet range, *Arch. Biochem. Biophys.* 133:103–107.

Dunn, J. B. R., Addison, A. W., Bruce, R. E., Loehr, J. S., and Loehr, T. M., 1977. Comparison of hemerythrins from four species of sipunculids by optical absorption, circular dichroism, fluorescence emission, and resonance Raman spectroscopy, *Biochemistry* 16:1743–1749.

Fisher, D. B., Kirkwood, R., and Kaufman, S., 1972. Rat liver phenylalanine hydroxylase, an iron enzyme, *J. Biol. Chem.* 247:5161–5167.

Forrest, W. W., and Walker, D. J., 1971. The generation and utilization of energy during growth, *Adv. Microb. Physiol.* 5:213–274.

Fujisawa, H., and Hayaishi, O., 1968. Protocatechuate 3,4-dioxygenase. I. Crystallization and characterization, *J. Biol. Chem.* 243:2673–2681.

Fujisawa, H., Uyeda, M., Kojima, Y., Nozaki, M., and Hayaishi, O., 1972. Protocatechuate 3,4-dioxygenase. II. Electron spin resonance and spectral studies on interaction of substrates and enzyme, *J. Biol. Chem.* 247:4414–4421.

Gormley, P. M., Loehr, J. S., Brimhall, B., and Hermodson, M. A., 1978. New evidence for glutamic acid as an iron ligand in hemerythrin, *Biochem. Biophys. Res. Commun.* 85:1360–1366.

Gray, H. B., 1975. Polynuclear iron(III) complexes, *in Proteins of Iron Storage and Transport in Biochemistry and Medicine*, R. R. Crichton (ed.), North-Holland, Amsterdam, pp. 3–13.

Grizzuti, K., and Perlmann, G. E., 1970. Conformation of the phosphoprotein phosvitin, *J. Biol. Chem.* 245:2573–2578.

Gunsalus, I. C., Pederson, T. C., and Sligar, S. G., 1975. Oxygenase-catalyzed biological hydroxylations, *Annu. Rev. Biochem.* 44:377–407.

Hayaishi, O. (ed.), 1974. *Molecular Mechanisms of Oxygen Activation*, Academic Press, New York.

Hegenauer, J., Saltman, P., and Nace, G., 1979. Iron III–phosphoprotein chelates. Stoichiometric equilibrium constant for interaction of iron III and phosphoserine residues of phosvitin and casein, *Biochemistry* 18:3865–3875.

Hendrickson, W. A., Klippenstein, G. L., and Ward, K. B., 1975. Tertiary structure of myohemerythrin at low resolution, *Proc. Natl. Acad. Sci. U.S.A.* 72:2160–2164.

Hou, C. T., 1975. Circular dichroism of holo- and apoprotocatechuate-3,4-dioxygenase from *Pseudomonas aeruginosa*, *Biochemistry* 14:3899–3902.

Hou, C. T., 1978. Iron-binding ligands in the catalytic site of protocatechuate-3,4-dioxygenase, *Bioinorg. Chem.* 8:237–243.

Hou, C. T., Lillard, M. O., and Schwartz, R. D., 1976. Protocatechuate-3,4-dioxygenase from *Acinetobacter calcoaceticus, Biochemistry* 15:582–588.

Kaufman, S., and Fisher, D., 1970. Purification and some physical properties of phenylalanine hydroxylase from rat liver, *J. Biol. Chem.* 245:4745–4750.

Kaufman, S., and Fisher, D., 1974. Pterin-requiring aromatic amino acid hydroxylases, *in Molecular Mechanisms of Oxygen Activation,* O. Hayaishi (ed.), Academic Press, New York, pp. 285–369.

Kelly, D. P., 1971. Autotrophy: Concepts of lithotrophic bacteria and their organic metabolism, *Annu. Rev. Microbiol.* 25:177–210.

Keresztes-Nagy, S., Lazer, L., Klapper, M. H., and Klotz, I. M., 1965. Hybridization experiments: Evidence of dissociation equilibrium in hemerythrin, *Science* 150:357–359.

Klotz, I. M., and Keresztes-Nagy, S., 1963. Hemerythrin: Molecular weight and dissociation into subunits, *Biochemistry* 2:445–452.

Kuczenski, R., 1973. Rat brain tyrosine hydroxylase, *J. Biol. Chem.* 248:2261–2265.

Kuutti, E. R., Tuderman, L., and Kivirikko, K. I., 1975. Human prolyl hydroxylase. Purification, partial characterization, and preparation of antiserium to the enzyme, *Eur. J. Biochem.* 57:181–188.

Lindblad, B., Lindstedt, G., Lindstedt, S., and Rundgren, M., 1977. Purification and some properties of human 4-hydroxyphenylpyruvate dioxygenase(I), *J. Biol. Chem.* 252:5073–5084.

Lindstedt, S., Odelhög, B., and Rundgren, M., 1977. Purification and some properties of 4-hydroxyphenylpyruvate dioxygenase from *Pseudomonas sp. P. J. 874, Biochemistry* 16:3369–3377.

Lipscomb, J., Howard, J., Lorsbach, T., and Wood, J., 1976. Protocatechuate-3,4-dioxygenase (PCA se): Subunit structure, *Fed. Proc. Fed. Am. Soc. Exp. Biol.* 35:1536.

Llinas, M., 1973. IV. Hemerythrin, *Struct. Bonding (Berlin)* 17:169–176.

Loehr, J. S., Lammers, P. J., Brimhall, B., and Hermodson, M. A., 1978. Amino acid sequence of hemerythrin from *Themiste dyscritum, J. Biol. Chem.* 253:5726–5731.

Lundgren, D. G., Vestal, J. R., and Tabita, F. R., 1974. The iron oxidizing bacteria, *in Microbial Iron Metabolism,* J. B. Neilands (ed.), Academic Press, New York, pp. 457–473.

Mackler, B., Person, R., Miller, L. R., and Finch, C. A., 1979. Iron deficiency in the rat: Effects on phenylalanine metabolism, *Pediatr. Res.* 13:1010–1011.

Mecham, D. K., and Olcott, H. S., 1949. Phosvitin, the principal phosphoprotein of egg yolk, *J. Am. Chem. Soc.* 71:3670–3679.

Moss, T. H., Moleski, C., and York, J. L., 1971. Magnetic susceptibility evidence for a binuclear iron complex in hemerythrin, *Biochemistry* 10:840–842.

Myllylä, R., Tuderman, L., and Kivirikko, K. I., 1977. Mechanism of the prolyl hydroxylase reaction. 2. Kinetic analysis of the reaction sequence, *Eur. J. Biochem.* 80:349–357.

Myllylä, R., Kuutti-Savolainen, E.-R., and Kivirikko, K. I., 1978. The role of ascorbate in the prolyl hydroxylase reaction, *Biochem. Biophys. Res. Commun.* 83:441–448.

Nozaki, M., 1974. Non-heme iron dioxygenases, *in Molecular Mechanisms of Oxygen Activation,* O. Hayaishi (ed.), Academic Press, New York, pp. 135–165.

Osaki, S., Sexton, R. C., Pascual, E., and Frieden, E., 1975. Iron oxidation and transferrin formation by phosvitin, *Biochem. J.* 151:519–525.

Patel, R. N., Hou, C.-T., Felix, A., and Lillard, M. O., 1976. Catechol 1,2-dioxygenase from *Acinetobacter calcoaceticus*: Purification and properties, *J. Bacteriol.* 127:536–544.

Perlmann, G. E., and Grizzuti, K., 1971. Conformational transition of the phosphoprotein phosvitin. Random conformation → β-structure, *Biochemistry* 10:258–264.

Petrack, B., Sheppy, F., and Fetzer, V., 1968. Studies on tyrosine hydroxylase from bovine adrenal medulla, *J. Biol. Chem.* 243:743–748.

Petrack, B., Sheppy, F., Fetzer, V., Manning, T., Chertock, H., and Ma, D., 1972. Effect of ferrous iron on tyrosine hydroxylase of bovine adrenal medulla, *J. Biol. Chem.* 247:4872–4878.

Phillips, J. L., and Azari, P., 1974. Zinc transferrin. Enhancement of nucleic acid synthesis in phytohemagglutinin-stimulated human lymphocytes, *Cell. Immunol.* 10:31–37.

Phillips, J. L., and Azari, P., 1975. Effect of iron transferrin on nucleic acid synthesis in phytohemagglutinin-stimulated human lymphocytes, *Cell. Immunol.* 15:94–99.

Pistorius, E. K., and Axelrod, B., 1974. Iron, an essential component of lipoxygenase, *J. Biol. Chem.* 249:3183–3186.

Planas, J., and Frieden, E., 1973. Serum iron and ferroxidase activity in normal, copper deficient, and estrogenized roosters, *Am. J. Physiol.* 225:423–430.

Pringsheim, E. G., 1949. Iron bacteria, *Biol. Rev.* 24:200–245.

Que, L., Lipscomb, J. D., Zimmermann, R., Münck, E., Orme-Johnson, N. R., and Orme-Johnson, W. H., 1976. Mössbauer and EPR spectroscopy on protocatechuate 3,4-dioxygenase from *Pseudomonas aeruginosa*, *Biochim. Biophys. Acta* 452:320–334.

Que, L., Lipscomb, J. D., Münck, E., and Woo, J. M., 1977. Protocatechuate 3,4-dioxygenase. Inhibitor studies and mechanistic implications, *Biochim. Biophys. Acta* 485:60–74.

Rundgren, M., 1977. Steady state kinetics of 4-hydroxyphenylpyruvate dioxygenase from human liver (III), *J. Biol. Chem.* 252:5094–5099.

Shainkin, R., and Perlmann, G. E., 1971a. Phosvitin, a phosphoglycoprotein. I. Isolation and characterization of a glycopeptide from phosvitin, *J. Biol. Chem.* 246:2278–2284.

Shainkin, R., and Perlmann, G. E., 1971b. Phosvitin, a phosphoglycoprotein: Composition and partial structure of carbohydrate moiety, *Arch. Biochem. Biophys.* 145:693–700.

Silverman, M. P., and Ehrlich, H. L., 1964. Microbial formation and degradation of minerals, *Adv. Appl. Microbiol.* 6:153–206.

Stenkamp, R. E., Sieker, L. C., and Jensen, L. H., 1976. Structure of the iron complex in methemerythrin, *Proc. Natl. Acad. Sci. U.S.A.* 73:349–351.

Stenkamp, R. E., Sieker, L. C., Jensen, L. H., and McQueen, J. E., 1978. Structure of methemerythrin at 2.8 Å resolution: Computer graphics fit of an averaged electron density map, *Biochemistry* 17:2499–2504.

Suzuki, I., 1974. Mechanisms of inorganic oxidation and energy coupling, *Annu. Rev. Microbiol.* 28:85–101.

Taborsky, G., 1963. Interaction between phosvitin and iron and its effect on a rearrangement of phosvitin structure, *Biochemistry* 2:266–270.

Taborsky, G., 1974. Phosphoproteins, *Adv. Protein. Chem.* 29:1–210.

Tormey, D. C., and Mueller, G. C., 1972. Biological effects of transferrin on human lymphocytes *in vitro*, *Exp. Cell Res.* 74:220–226.

Tormey, D. C., Imrie, R. C., and Mueller, G. C., 1972. Identification of transferrin as a lymphocyte growth promoter in human serum, *Exp. Cell Res.* 74:163–169.

Tuderman, L., Myllylä, R., and Kivirikko, K. I., 1977. Mechanism of the prolyl hydroxylase reaction. I. Role of co-substrates, *Eur. J. Biochem.* 80:341–348.

Webb, J. H., Multani, J. S., Saltman, P., Beach, N. A., and Gray, H. B., 1973. Spectroscopic and magnetic studies of iron(III) phosvitins, *Biochemistry* 12:1797–1802.

York, J. L., and Bearden, A. J., 1970. Active site of hemerythrin. Iron electronic states and the binding of oxygen, *Biochemistry* 9:4549–4554.

York, J. L., and Roberts, M. P., 1976. The iron and subunit binding sites of hemerythrin. The role of histidine, tyrosine, and tryptophan, *Biochim. Biophys. Acta* 420:265–278.

Suggested Reading

Aisen, P., and Brown, E. B., 1975. Structure and function of transferrins, *Prog. Hematol.* 9:25–56.

Brown, E. B., Aisen, P., Fielding, J., and Crichton, R. R. (eds.), 1977. *Proteins of Iron Metabolism*, Grune & Stratton, New York.

Chasteen, D., 1977. Human serotransferrins: Structure and function, *Coord. Chem. Rev.* 22:1–36.

Crichton, R. R. (ed.), 1975. *Proteins of Iron Storage and Transport in Biochemistry and Medicine*, North-Holland, Amsterdam.

Eichhorn, G. (ed.), 1973. *Inorganic Biochemistry*, Vols. I and II, Elsevier, Amsterdam.

Gould, R. F. (ed.), 1971. *Bioinorganic Chemistry*, Vol. I, American Chemical Society, Washington, D.C.

Harrison, P. M., and Hoare, R., 1978. *Metal Ions in Biology*, Chapman & Hall, London.

Jacobs, A., and Worwood, M. (eds.), 1974. *Iron in Biochemistry and Medicine*, Academic Press, New York.

Lovenberg, W. (ed.), 1974 and 1977. *Iron–Sulfur Proteins*, Vols. I, II, and III, Academic Press, New York.

Munro, H. N., and Linder, M. C., 1978. Ferritin: Structure, biosynthesis, and role in iron metabolism, *Physiol. Rev.* 58:317–396.

Neilands, J. B. (ed.), 1974. *Microbial Iron Metabolism*, Academic Press, New York.

Raymond, K. N. (ed.), 1977. *Bioinorganic Chemistry*, Vol. II, American Chemical Society, Washington, D.C.

Worwood, M., 1977. The clinical biochemistry of iron, *Semin. Hematol.* 14:3–30.

Index

Acinetobacter calcoaceticus oxygenases, 401, 404–406
Aconitase, 385, 388
Actinomyces griseus siderophore, 312
Actinomycin, 233
Adrenodoxin
  amino acid composition, 381
  amino acid sequence, 380
  EPR signal, 353
  evolution, 380
  hydroxylation reactions, 379–382
  physical properties, 381
  and putidaredoxin, 361
  steroid hormone biosynthesis, 379
Aerobacter aerogenes siderophores, 306, 308, 314
Aerobactin-type siderophores, 314, 315, 318
Aging
  and iron absorption, 88–89
  and iron deficiency, 110
  and iron metabolism, 13, 17, 38, 56
Agrobacterium tumefaciens siderophore, 310
Agrobactin, 307, 310
Albomycins, 312, 320
Alcaligenes lutropus hydrogenase, 385
Aldehyde oxidase, 385, 387
Alfalfa ferredoxin
  amino acid composition, 355
  amino acid sequence, 356
  evolution, 355, 357–358, 374
  iron, 353
  isolation, 351–352
Amidophosphoribosyl transferase, 386, 388
Anabaena cylindrica nitrogenase, 377
Andral, G., 4
Anemias: *see also* Hemolytic disorders, Iron deficiency anemia
  classification, 28, 37, 40
  hematologic parameters, 259
  hypoplastic, 37, 56
  macrocytic, 37, 40
  pernicious, 28, 56
  refractory, 37
  sickle cell, 38
  sideroblastic, 277–278
Antiferromagnetic coupling, 353, 377
Apoferritin: *see also* Ferritin, Isoferritins
  definition, 210
  oxidase activity, 236–239
Ascorbic acid
  as ferrireductase, 247
  in ferritin iron release, 239
  in iron absorption, 72
  in proline hydroxylase reaction, 409
Aspergillic acid, 317
Atomic absorption spectroscopy, 19
Azobacter vinlandii
  ferredoxin, 363
    amino acid composition, 365
    physical properties, 364
  molybdenum–iron protein, 375, 377
  siderophore, 309

Bacillus megaterium siderophore, 314
Bacillus polymyxa ferredoxin, 367, 368, 369, 377
Bacillus stearothermophilus ferredoxin, 353
Bacillus subtilis, 308, 386
Bacterium W3A1 trimethylamine dehydrogenase, 386
Band, P., 4
Barkan, G., 4

Bathophenanthroline, 18
Berzelius, J. J., 4
Bestuzhev-Ryumin, Field Marshal, 4
Biantennary structure, 176
Bicarbonate: *see* Carbonate
Bile, 72, 133
Bipyridyl: *see* 2,2′-Dipyridyl
Blood volume, 37, 57
Bone marrow
  in iron metabolism model, 38
  removal of plasma iron, 27, 32, 33
Brevibacterium fuscum pyrocatechase, 404
Brown phosvitin, 398
Bunge, Gustav, 4
Butyrobetaine hydroxylase, 402, 409

Cadmium, 145, 210
Calorimetric methods, 156–157
Calvin cycle, 373, 414
Camphor hydroxylation, 360
Candida albicans growth, 333, 334
Carbon dioxide fixation, 358–359, 372, 373
Carbonate
  binding by transferrins, 146–148, 155, 159
  removal of iron by reticulocytes, 296
Caroline, L., 5
Catalases, 3, 7, 9
Catechol oxygenases, 401-407, 408
Celiac disease, 92
Cerebrospinal fluid, 10, 12, 133
Ceruloplasmin, 1, 241–243, 247
Chang cells, 245, 250
Chlorobium limicola ferredoxins, 363
Chlorobium thiosulfatophilum ferredoxin, 363, 365
Cholestyramine, 101
Cholylhydroxamic acid, 326–328
Chromatium vinosum ferredoxins
  amino acid composition, 365, 369
  amino acid sequence, 370
  EPR signal, 353
  evolution, 367, 374
  four-iron type, 363, 365
  high-potential iron protein, 353
Chromium, 145, 151, 152, 154

Citrate
  iron chelator, 145, 149
  iron translocator in plants, 3
  in iron uptake by reticulocytes, 277, 295
  mobilization of iron from ferritin, 239
  as potentiator of microbial virulence, 331
  randomization of iron in transferrins, 154
  in siderophores, 314
Clostridium acidiurici
  ferredoxin, 361, 366, 374
  pyruvate dehydrogenase, 385
Clostridium butyricum ferredoxin, 361, 366, 374
Clostridium pasteurianum
  ferredoxin
    amino acid composition, 365
    amino acid sequence, 366
    in conversion of CO_2 to formate, 372
    EPR signal, 353
    evolution, 367, 374
    nitrogen fixation, 375, 376, 377
    physical properties, 364
  iron–sulfur enzymes, 385, 386
  rubredoxin, 348–351, 353
Clostridium tartarivorum ferredoxin, 361, 366, 367, 374
Clostridium thermosaccarolyticum ferredoxin, 361, 366, 367, 374
Clostridium welchii growth, 330, 331
Cobalt, 145, 152
Collagen biosynthesis enzymes, 402, 409–410
Colorimetric methods, 18–19
Conalbumin: *see also*: Transferrins
  amino acid composition, 169
  amino acid sequences, 179, 181, 190
  antimicrobial agent, 395
  bilobal nature, 183
  binding of metals, 144–167
  biosynthesis, 190
  carbohydrate, 13, 167–177, 187
  in circulation of laying hens, 13
  discovery, 5, 127
  EPR spectra, 153, 154
  fragmentation, CNBr, 177–178, 185

Conalbumin (*cont.*)
  fragmentation, enzymes, 180–184, 185
  interaction with immature red cells, 272
  iron-binding ligands, 160–167
  isolation, 135
  levels in egg whites, 132, 134
  lymphocyte growth promoter, 415
  messenger RNA, 190
  microheterogeneity, 186–187
  physical properties, 137–144
  secondary structure, 143–144
Copper
  binding by transferrins, 145, 146, 154, 162, 164
  effect on shape of serotransferrin, 279
  in 4-hydroxyphenylpyruvate hydroxylase, 408
  in iron bacteria, 414
  role in iron metabolism, 241–243, 247
Coprogen, 314
Coulometry, 19
Crohn's disease, 92
Cyanidium caldorium ferredoxin, 355
Cycloheximide, 105, 192
Cytochromes
  iron in, 3, 7, 9, 10
  iron absorption, 260
  hydroxylation reactions, 360, 380–382
  photosynthesis, 359, 372
  in physiology of iron bacteria, 414, 415

Desferal: *see* Desferrioxamine
Desferrioxamine, 101, 311, 324–328
Desulfovibrio gigas ferredoxin
  amino acid composition, 369
  amino acid sequence, 367
  evolution, 367, 368, 374
  function in sulfur metabolism, 371
  nitrogen fixation, 377
  physical properties, 368
Diabetes, 248
Diethylenetriaminepentaacetic acid, 327–328
Dihydroorotate dehydrogenase, 385, 388
2,3-Dihydroxybenzoic acid, 306, 308–310, 319
2,3-Dihydroxybenzoyl glycine, 307, 308
bis-(2,3-Dihydroxybenzoyl)-L-lysine, 307, 309
2,3-Dihydroxy-*N*-benzoyl-L-serine, 307, 308
$N'N^8$-bis-(2,3-Dihydroxybenzoyl) spermidine, 309
2,3-Dihydroxybenzoyl threonine, 308
Dimerum acid, 314
2,2′-Dipyridyl
  as ferrous iron chelator, 128
  inhibitor of iron uptake by reticulocytes, 277
  in iron determination, 18, 63
  removal of iron from mitochondria, 297
Diurnal fluctuation of plasma iron, 13–14, 36
Drabkin, D. L., 5

Early iron reflux, 28, 36–37
Endotoxin, 332
Enterobactin: *see* Enterochelin
Enterochelin
  association constant with iron, 307, 318
  biosynthesis, 318
  combination with serotransferrin, 332
  EPR spectrum, 306
  mechanism of action, 319, 320
  model for synthetic iron chelators, 325
  reaction with iron, 306–307
  receptors in microorganisms, 320, 322, 323
  redox potential, 319
  structure, 306–307, 325
EPR spectroscopy
  ferredoxins, 345, 352–354, 361
  iron–sulfur enzymes, 385, 386
  nitrogenase, 378
  oxidative phosphorylation components, 382–383
  phenylalanine hydroxylase, 400
  photosynthetic machinery, 359
  principle of, 151–152
  protocatechuate-3,4-dioxygenase, 407
  pyrocatechase, 405
  rubredoxins, 347–348, 350
Erythrocytes: *see* Red cells

Escherichia coli
growth inhibition by trnasferrins, 330–331
infection and total iron, 332
source of enterochelin, 306
source of various siderophores, 308
sulfite reductase, 386
uptake of iron, 320, 321
Evolution
adrenodoxin, 380
ferredoxins, 357, 358, 363–364, 367, 373–374, 380
iron-containing proteins, 3
rubredoxins, 351
transferrins, 130, 178–179, 184
Exochelins, 315, 317, 320, 322
Extrinsic label, 66–70

Fatty acids
hydroxylation, 347, 401
mycobactins, 315, 316
Fecal excretion test, 58–59
Ferredoxin–CO_2 oxidoreductase, 372, 386
Ferredoxins: *see also* High-potential iron proteins; *names of individual microorganisms or plants*
amino acid composition, 355, 365
amino acid sequences, 356, 366
Azobacter-type, 363
bacterial type, 351, 361–364
classification, 343–346
clostridial type: *see* Ferredoxins, bacterial type
evolution: *see* Evolution, ferredoxins
four-iron cluster-type, 365, 367–368, 369
functions, 369–373
isolation, 351–352, 361
in nitrogen fixation, 375–378
in photosynthesis, 356, 358–360, 372–373
photosynthetic bacterial-type, 363
physical properties, 354, 363, 364
plant-type, 7, 345, 346, 351–361
Ferrichrome siderophores, 311-312, 313, 318, 320
Ferrimycin, 311, 312
Ferrioxamine siderophores, 311, 318
Ferrireductase, 240, 241, 247
Ferritin: *see also:* Apoferritin, Isoferritins
amino acid composition, 227
amino acid sequences, 218, 226
biosynthesis in response to iron, 230–235
biosynthetic control mechanism, 233–235
carbohydrate, 228, 234
circular dichroism, 229–230
degradation, 234
denaturation, 229
dietary iron source, 101
discovery, 207–208
electrophoresis, 212, 218, 219, 220, 222
and iron absorption, 2, 76–78, 107–109
iron content, 9, 10, 212–215, 224–225
and iron incorporation into reticulocytes, 293, 294, 298–299
iron overload, 251–253
iron oxide micelles, 212, 213, 236–239
isolation, 209–210, 211
in liver, 4, 207, 230
mechanism of iron uptake and release, 236–240
messenger RNA, 233
metabolism, 230–235
molecular weights, 213–215
occurrence, 207
redox cycles in iron uptake and release, 236, 237
secondary structure, 229–230
and serotransferrin biosynthesis, 191–192
serum, 50, 77, 107, 253–258, 259
in subcellular particles, 207
subunits, 215–230
three-dimensional structure, 210, 211–215, 217, 229
topography, 218
turnover, 231, 232
Ferrochelatase, 297–298
Ferrokinetics, 1, 25–45
deterministic, 34
model, 38
stochastic, 35
Ferroxidase: *see* Ceruloplasmin
Ferrozine, 18
Fever, 332–333
Finch, C. A., 35–38

Flavoproteins: *see also* Riboflavin derivatives in iron mobilization
as complex iron–sulfur proteins, 344
in hydroxylation reactions, 380, 382
in oxidative phosphorylation, 383
as oxygenases, 399
in physiology of iron bacteria, 415
in putidaredoxin hydroxylations, 360
Fletcher–Huehns hypothesis, 245, 279–286
Fodisch, F., 4
Folic acid, 146
Fortification, iron
compounds used, 103–105, 110–112
milk, 85, 336
Fungi, inhibition by transferrins, 331, 333, 334
Fusarinine siderophores, 317, 328

Gallionella, 413
Gallium, 146, 153
Gastric achylia, 91–92
Gastroferrin, 71–72
Geophagia, 103
Gloecapsa nitrogenase, 377
Golfingia gouldii hemerythrin, 410–413
Granick, S., 209
Green phosvitin, 398
Griseins, 312

Hahn, P. F., 5, 25, 209
Halobacterium halobium ferredoxin, 356
Heilmeyer, L., 4, 5
Hematocrit, 10, 37
Hematologic disease classification, 28, 37, 40
Heme, 50–51, 277–278
Hemerythrins, 410–413
Hemochromatosis, idiopathic, 249–253
chelation therapy, 324–328
development, 251, 253
hematologic parameters, 259
iron absorption, 57, 92
iron turnover, 28, 30, 35, 40, 249
isoferritins, 255
plasma iron, 14, 17, 249–251
reticuloendothelial system lesion as cause, 248, 249
Hemoglobin
delivery of iron for the synthesis of, 292–296
Hemoglobin (*cont.*)
in estimation of body iron stores, 6, 7
iron content, 9, 10
measurement, 10
response in iron absorption, 58
Hemolytic disorders: *see also* Anemias
chelation therapy in, 324–328
hematologic parameters, 259
iron abosrption, 90–91
iron turnover, 17, 28, 31–32, 38, 40
Hemosiderin
discovery, 207–208
ferritin as precursor, 207–208, 235
iron content, 9, 10
in liver, 4, 207
in reticuloendothelial system, 247, 251
High-potential iron proteins *see also:* Ferredoxins; *names of specific organisms*, 346, 351, 365, 367–368
amino acid composition, 369
amino acid sequences, 367, 370
function, 371
in oxidative phosphorylation, 384
physical properties, 368
History of iron biology, 3–5
Hodgkin's disease, 37, 56
Holmberg, C. G., 5
Homogentisic acid oxygenase, 401
Hydrogenase, 371, 384, 385
Hydroxamic acids as siderophores, 311–317
3-Hydroxyanthranilate oxygenase, 401
2-Hydroxybenzoyl-*N*-L-threonyl-(*N'N*[8]-bis-(dihydroxybenzoyl)) spermidine, 309
Hydroxylases: *see* Oxygenases; *names of specific hydroxylases*
Hydroxylation
camphor, 360–361
cholesterol, 379, 382
fatty acids, 346–347, 401
proline, 409–410
role of adrenodoxin, 378–392
in vitamin D biosynthesis, 382
5-Hydroxymethyluracil dioxygenase, 402
4-Hydroxypyruvate hydroxylase, 401, 403, 407–408

Infection, systemic
iron metabolism, 332–335

Infection systemic, (*cont.*)
resistance to, in breast-fed infants, 336–337
Inflammation and iron metabolism, 332–333
Interlocking sites model, 148, 149
Intestinal mucosal cells
biosynthesis of ferritin, 259
iron metabolism, 258—261
iron pools, 258
role in iron absorption, 1, 2, 73–78, 105–106
uptake of iron, 260–261, 279
Intrinsic label, 64–66
Immature red cell interaction with serotransferrin and iron uptake
abstraction of iron by cell machinery, 292–299
bovine serotransferrin, 183
effect of divalent metals, 278–279
effect of iron saturation levels of serotransferrin, 276, 278–279, 283–286
effect of metabolic inhibitors, 273, 275, 277, 296
energy requirement, 273, 275
inhibitors, 277, 294
internalization of serotransferrin, 293–296, 298
iron-binding sites in serotransferrin, 279–286
mechanism of reticulocyte–serotransferrin interaction, 273–278
operation of the Fletcher–Huehns hypothesis, 279–286
pathway of iron incorporation into protoporphyrin, 292–299
rat isotransferrins, 283–284
redox cycles of iron, 298–299
role of carbohydrate moiety of serotransferrin, 167, 272–273
role of serotransferrin-bound carbonate, 296, 298
role of ferritin, 293, 294
role of iron receptors, 292, 294
role of microtubules, 294, 295
role of mitochondria, 293–298
serotransferrin binding sites, 273–276
Immature red cell interaction (*cont.*)
and serotransferrin shape, 159, 278–279
thermodynamics, 274
tissue receptors: *see* Receptors
Iron
absorption: *see* Iron absorption
analysis in blood, 10, 17–19, 57
in biological fluids, 7, 10, 12
in blood: *see* Total iron, serum
in cell fractions, 75–76, 241–242, 246–247
cell metabolism, 240–261
chelators used in clinical practice, 324–328
clearance: *see* Iron, turnover in plasma
compartments, 27–35
concentration required for bacterial growth, 305
daily intake, 10, 48–49
deficiency: *see* Iron deficiency anemia
delivery to lymphocytes, 415–416
delivery to microorganisms, 319–323
distribution in serotransferrin *in vivo*, 158–159
excretion: *see* Iron excretion
food fortification with: *see* Fortification, iron
half-life, plasma, 26, 28, 35
in hemerythrins, 410, 412
in intestinal mucosal cells, 258–261
labile pool, 27–35, 248
liver, 207, 244–247, 249–253
in milk, 14, 17, 18
in nutrition, 47–113
in oxygenases, 400–410
in phosvitin, 398
pools, 241–242, 248
potentiation of infection, 326–338
reaction with siderophores, 306–337
receptors, 292, 294
redox cycles, 236–237, 241–243, 298
refluxes, 28, 36–41
return from reticuloendothelial system, 248
storage, 8, 48, 207–261
tissue content, 6, 8, 326
in transferrins, 144–145, 148–167
turnover in plasma, 26–45, 38

Iron (*cont.*)
turnover, tissue, 28, 37, 39
turnover in various diseases, 28, 37–38, 40
uptake into various tissues from serotransferrin, 281–286
utilization by iron bacteria, 414, 415
in various body compounds, 7, 9, 10
whole body stores, 6–7, 9, 10, 48–50, 253–259
Iron absorption
adults, 88–89
availability, 103
in children, 87–88
daily, 1, 10, 258
in disease, 90–93
effect of alcohol, 106
effect of carbohydrate, 94–95
effect of oral contraceptives, 105–106
effect of lipids, 95–96
effect of minerals, 99–101
effect of proteins, 96–98
effect of sequestering agents, 101–103
effect of vitamins, 98–99
in the elderly: *see* Aging
factors affecting, 93–106
from fortified foods, 103–105
in infancy, 82–87
involvement of ferritin, 2, 258–259
involvement of mucosal transferrin, 2, 259–260
in iron deficiency anemia, 57, 70
mechanism of, 70–78
oxidation state of iron, 71, 104
regulation, 106–110, 260–261
relation to serum ferritin, 256–258
role of bile, 72
role of human milk, 336
role of the intestine, 72–78
role of the pancreas, 72
role of serotransferrin, mucosal transferrin, and ferritin, 76–78
role of the stomach, 70–72
sexes, 78
techniques for measurement, 51–63
from various foods, 63–70
Iron bacteria, 413–415
Iron deficiency anemia
by age, 110
Iron deficiency anemia (*cont.*)
classification, 57
diagnosis via serum ferritin, 50, 254–255
iron turnover, 28, 40, 56
and nutrition, 49–50
and susceptibility to infection, 333–335
various hematologic parameters, 259
in various populations, 110
Iron excretion
with chelators, 324–328
children, 48–49
feces, 58–59, 326–327, 328
in iron-retention time studies, 61–62
in males and females, 48–49
urine, 326–327
Iron overload, 248–253; *see also* Hemochromatosis, idiopathic
artificially produced, 207
diagnosis via serum ferritin, 253–259
reticuloendothelial system, 247
susceptibility to infection, 333–335
tissue iron content, 326
treatment with iron chelators, 3, 324–328
Iron protein of nitrogenase, 375–378
Iron–sulfur proteins: *see also* Ferredoxins, Rubredoxins
classification, 344, 384
clusters, 345, 368, 385–386
definitions, 3, 343–391
enzymes, 384–388
evolution, 3
iron-binding ligands, 345–346
mammalian, 378–384
in nitrogen fixation, 375–378
oxidative phosphorylation, 382–384
oxygenases, 399
photosynthesis, 359
removal of clusters from, 368
Isoferritins
amino acid composition, 228
biosynthesis, 234
immunochemical relationships, 220
intestinal mucosa, 77
iron absorption, 107–108
iron content, 224–225
iron overload, 250, 255
isoelectric points, 219, 225, 255
subunits, 219–223

Isonicotinic acid hydrazide, 277, 297
Itoic acid: *see* 2,3-Dihydroxybenzoyl glycine

Jandl, J. H., 5

Klebsiella oxytoca siderophore, 308
Klebsiella pneumoniae nitrogenase, 375–376, 377
Koa ferredoxin, 356, 357–358, 374
Kupffer cells
  hemosiderin, 207
  iron overload, 251, 252
  iron release, 247
  iron uptake, 208, 244

Lactoferrin: *see also* Transferrins
  amino acid compositions, 172–173
  amino acid sequences, 179
  bacterial growth inhibitor, 131, 329–332, 336–337
  binding of metals, 129, 144–167
  carbohydrate, 167–176
  chemical modification, 166
  circular dichroism, 143, 145, 162
  clearance from bloodstream, 191
  definition, 2
  discovery, 128–129
  effect of pH on iron binding, 149, 184
  fragmentation with enzymes, 180–184, 185
  interaction with immature red cells, 272
  interaction with intestinal mucosal cells, 261
  internal homology, 180
  iron-binding ligands, 160–167
  isolation, 135–137
  levels in biological fluids, 129–133
  lymphocyte growth promoter, 415
  in mastitis, 337
  in neutrophils, 131, 132, 333
  physical properties, 137–144
  secondary structure, 143–144
  synovial fluid, 132
Late iron reflux, 28, 36–37
Laufberger, V., 4, 209
Laurell, C.-B., 5
Lemery, N., 4
Leptothrix, 413
Lipoxygenase, 401
Liver
  aldehyde oxidase, 385
  iron excretion, 326–320
  iron histology, 4
  iron metabolism, 244–247
  iron overload, 249–253
  iron storage, 2, 8, 244
  iron turnover, 27, 29, 30, 31, 244
  iron uptake from serotransferrin, 244–247, 279
  iron uptake in infection, 332
  source of oxygenases, 401–402
  total iron content, 207
  transferrin biosynthesis, 191–192
Lymphocytes, 415–416
Lysine hydroxylase, 402, 409

Magendie, F., 4
Magnesium carbonate, 18, 19
Manganese, 145
Mastitis, 337
Menardes, N., 4
Menghini, V., 4
Metapyrocatechase, 401, 405
6-Methylsalicylic acid, 315
Micrococcus aerogenes
  ferredoxin, 374
  rubredoxin, 349
Micrococcus denitrificans
  nitrate reductase, 386
  siderophores, 309
Microorganisms: *see also individual organism names*
  ferredoxins and rubredoxins, 343–390
  iron metabolism, 305–338
  source of oxygenases, 401–402
  virulence, 335
Milk
  bacterial growth inhibition, 329, 336–337
  effect on iron absorption, 84–85, 96
  iron absorption from, 10, 84–85
  iron content, 17, 18
  lactoferrin content, 131–133
  rabbit, 176
  serotransferrin content, 130–131
  xanthine oxidase, 385
Molybdenum
  in iron–sulfur proteins, 344

Molybdenum (*cont.*)
nitrogenase, 375–378
various enzymes, 385–386
xanthine oxidase, 384, 387
Molybdenum–iron protein of nitrogenase, 375–378
Mucosal block theory, 258
Mucosal transferrin, 76, 107, 259–260
Murchison meteorite, 374
Muscle, iron content, 6, 7, 8
Mycobacterium flavum ferredoxin, 363, 365, 377
Mycobacterium paratuberculosis siderophore, 314
Mycobacterium phlei siderophore, 314
Mycobacterium smegmatis
ferredoxin, 363, 365
siderophores, 315, 321–322
Mycobactins, 314–315, 316, 320–322
Myoglobin, 7, 9, 10

NADH-dehydrogenase, 383
Neumann, E., 208
Neurospora crassa
oxygenases, 402
siderophores, 314, 320
Neutrophils, 131, 132, 133
Nickel, 146
Nisseria gonorrheae, 330
Nitrate reductase, 372, 386
Nitriloacetate
effect on iron uptake by reticulocytes, 277
iron chelator, 145, 148, 155
mobilization of iron from ferritin, 239
Nitrite reductase, 371–372, 386, 413
Nitrogen fixation, 375–378
Nitrogenase, 375–378
Nitroso-R salt, 19
Nostoc ferredoxins, 354, 355, 356, 357–358
Nutritional immunity, 329–337

Osborne, T. B., 5
Oxidative phosphorylation, 383–384
Oxygenases, 399–410
amino acid composition, 406
classification, 343, 399, 402
hydroxylation of cholesterol, 379
physical properties, 403

Oxygenases (*cont.*)
putidaredoxin, 360
sources, 401–402

Pancreas, 72
Paracoccus denitrificans ferredoxin, 365, 369
Pasteurella multocida, 332
Penicillamine, 277
Peptococcus aerogenes ferredoxin, 361–367
Peptostreptococcus elsdenii ferredoxin, 361
Perchloric acid, 152
Perls' solution, 207
Peroxidases, 3
1,10-Phenanthroline, 18, 236, 277
Phenylalanine hydroxylase, 400, 401, 403, 406
3-(4-Phenyl-2-pyridyl)-5,6-diphenyl-1, 2, 4-triazine, 19
Phosphoroclastic reaction, 371
Phosvitins, 395–399
Photosynthesis
bacterial ferredoxins in, 372–373
plant-type ferredoxins in, 356, 358–360
Phytate, 102–103
Phytohemagglutinin, 415
Picolinic acid, 324
Plants
ferredoxins, 351–360
ferritin, 207, 212, 227, 228
iron in, 3, 7, 11
source of iron, 51
Platinum, 146
Plötner, K., 4, 5
Pollycove–Mortimer iron compartment model, 27–35
Polycythemia vera, 26, 28
Porphyria umbilicalis ferredoxin, 355, 357
Probability theory in ferrokinetics, 35–41
Proline hydroxylase, 402, 403, 409–410
Protocatechuate-3,4-dioxygenase, 401, 403, 405–408
Protocatechuate-4,5-dioxygenase, 402
Prussian blue, 4, 207, 208
Pseudomonas aeruginosa
oxygenase, 401, 405, 406
siderophore, 321

Pseudomonas arvilla oxygenases, 401, 404, 405
Pseudomonas elsdenii rubredoxin, 349
Pseudomonas fluorescens oxygenases, 401, 404
Pseudomonas oleovorans rubredoxin, 347–348, 349, 350, 351
Pseudomonas ovalis oxygenase, 402
Pseudomonas putida ferredoxin, 354, 355, 360–361
Pseudomonas testeroni oxygenase, 402
Purines, 384, 388
Putidaredoxin: *see Pseudomonas putida* ferredoxin
Pyochelins, 321
Pyrexia, 90
Pyrimidine deoxyribonucleotide-2-hydroxylase, 402
Pyrocatechase, 401, 403, 404–406
Pyruvate dehydrogenase, 371, 384, 385

Ramsay procedure, 18
Receptors, serotransferrin
  amino acid and carbohydrate composition, 291
  destruction by trypsin, 272–273
  energetics of interaction, 291
  initial reaction, 295
  isolation, 286–290
  loss of, upon reticulocyte maturation, 288–290
  in lymphocytes, 416
  physical–chemical properties, 289
  Scatchard plots, 291, 292
Red cells: *see also* Immature red cell interaction with serotransferrin and iron uptake
  ineffective turnover, 39
  in iron metabolism model, 38
  iron turnover (EIT), 28, 33, 35, 37, 39
  mean hemoglobinization time (MEEHT), 34, 35
  mean life span (MELS), 34, 35
  uptake of radioactive iron, 27, 32, 56
  utilization time (RBCU), 37
Reticuloendothelial system: *see also* Kupffer cells
  in iron metabolism, 38, 247–248
  iron overload, 249, 250, 324–328
  iron removal from, 326–328
Reticuloendothelial system (*cont.*)
  lactoferrin uptake, 131
  response to systemic infection, 332–335
  source of serum ferritin, 256
  source of serum iron, 1, 36, 158, 247, 271
Reverse Krebs cycle, 373, 413
Riboflavin derivatives in iron mobilization, 2, 239–240
Rieske's protein, 381, 384
Rivanol, 134
Rhodopseudomonas gelatinosa ferredoxin, 365, 370
Rhodopseudomonas palustrus ferredoxin, 365
Rhodospirillum capulatus ferredoxin, 365
Rhodospirillum rubrum
  ferredoxin, 363
  nitrogenase, 377
Rhodospirillum spheroides ferredoxin, 365
Rhodospirillum tenue ferredoxin, 365, 369, 370
Rhodotorula pilimanae siderophores, 312, 321
Rhodotorulic acid, 312, 314, 318, 320, 321, 326, 328
Rubredoxins, 343–351; *see also individual microorganism names*
  amino acid composition, 349
  amino acid sequence, 347, 350
  classification, 343–344
  EPR signals, 348, 353
  evolution, 351
  function, 347
Rusticyanin, 414

Salmonella typhimurium siderophore, 306, 308, 332
Salicylate, 322
Scandium, 146
Schade, A. L., 5
Schizokinen, 314, 315, 328
Schmiedeberg, O., 209
Serosal iron release, 78, 109
Serotonin, 404
Serotransferrin: *see also* Transferrins
  amino acid compositions, 168, 170–171
  amino acid sequences, 179–182

Serotransferrin (*cont.*)
association constants for iron, 158
bilobal nature, 138, 184
binding of metals, 144–167
carbohydrate sequences, 167–176, 186–187
change in properties upon binding of iron, 159–160, 278–279
chemical modification, 142, 160, 165–166
circular dichroism, 143, 145, 162
control of biosynthesis, 191–192
denaturation, 141–142
discovery, 127–128
electrophoretic mobilities of polymorphs, 188
EPR spectra, 151–154, 161–165
exchange with extravascular serotransferrin, 36
fragmentation with CNBr, 177–180
gene frequencies, 187–190
genetic polymorphism, 184, 187–190
half-life, 191, 192, 285
interaction with immature red cells: *see* Immature red cell interaction with serotransferrin and iron uptake
interaction with intestinal mucosal cells, 261, 279, 283
interaction with liver cells, 244–247, 279
interaction with phosvitin, 399
interaction with reticuloendothelial system cells, 247–248
internal homology, 180, 182
in intestinal mucosal cells, 76–77
iron-binding fragments, 180–184, 185
iron-binding ligands, 160–167
iron-binding sites, 2, 149–159, 279–286
iron content, 9, 130
iron distribution *in vivo*, 158–159
isolation, 132–135, 136
levels in biological fluids, 130–132, 133
as lymphocyte growth promoter, 415–416
measurement from TIBC, 10, 130
metabolism, 190–192, 285
microbial growth inhibitor, 329
microheterogeneity, 184, 186–190

Serotransferrin (*cont.*)
nonidentity of iron-binding sites, 151–160, 279–280
nonspecific binding of iron, 147, 148
pH effect upon the binding of iron, 149, 153–155, 297
physical properties, 137–144, 159–160
rat isotransferrins, 184, 186, 283–284
receptors in tissues, 286–292
removal of iron from by chelators, 149, 297–298
in RNA and DNA biosynthesis, 416
role in iron absorption, 260–261
secondary structure, 143–144
sialic acid role in clearance, 191
in tissue iron metabolism, 241
titration with iron, 145, 146
in women taking oral contraceptives, 105–106
Shen Nung, Emperor of China, 3
Siderophores: *see also names of individual siderophores*
association constants with iron, 318
biosynthesis, 318, 319, 322–323
catechol-like, 306–311, 315
combination with transferrins, 331
definition, 3, 305
hydroxamate-like, 311–317
inhibition by antibodies, 330
mammalian, 323–324
as models for clinical application, 324–328
physical–chemical properties, 317–318
potentiation of infections, 329, 331
redox potentials, 319, 320
receptors for in microorganisms, 320, 322, 323
role in iron delivery to microorganisms, 319–323
structures, 306–317
yields, 308
Spermidine, 307, 309–310
Spinach
ferredoxin, 353–358, 374
nitrite reductase, 386
Spirulina platensis ferredoxin, 352, 354
Spleen, 8, 255
Staphylococcus aureus growth, 329
Steroid hormone biosynthesis, 378–380
Stomach and iron absorption, 70–72, 91

Streptomyces pylosus siderophore, 311
Succinic dehydrogenase, 383
Sulfate reductase, 371
Sulfite reductase, 371, 386
Superoxide dismutases, 3
Swirski, G., 4
Sydenham, T., 4
Synergistic anions, 146–148, 149

Tannins, 101–102
Taro ferredoxin, 356–358, 374
Terosite, 18
2,3,7,8-Tetrachlorodibenzo-*p*-dioxin (TCDD), 105
Tetrahydrobiopterin, 400, 404
Themiste dyscritum hemerythrin, 411, 412, 413
Themiste pyroides hemerythrin, 411, 412, 413
Thiobacillus ferroxidans physiology, 413–415
Thiocyanate, 19
Thioglycollic acid, 19
Thymine-7-hydroxylase, 402
TI: *see* Total iron, serum
TIBC: *see* Total iron-binding capacity
Total iron, serum
  definition, 10
  determination, 17–19
  in ferrokinetic experiments, 26–45
  in infections, 332–333
  normal values, 13–16
  variation with age, 13, 17
  variation in disease, 17
Total iron-binding capacity, serum
  definitions, 10, 13
  determination, 17–19
  normal values, 13–15
  serotransferrin values from, 130
  in systemic infection, 332–333
  variation with age, 13, 16
  variation with disease, 17
Transferrins: *see also* Serotransferrin, Lactoferrin, Conalbumin
  association constants with iron, 305
  bacterial growth inhibitors, 5, 127–128, 131, 329–337
  binding of synergistic anions, 146–148
  carbohydrate, 167–176
  chemical modification, 165–166
Transferrins (*cont.*)
  combination with siderophores, 331–332
  EPR spectra, 153, 156, 161–165
  evolution: *see* Evolution
  fragmentation with CNBr, 177–180
  fragmentation with enzymes, 180–184, 185
  immunochemical relationships, 129
  iron-binding halves, 180–184, 185
  iron-binding ligands, 160–167
  iron-binding sites, 149–167
  isolation, 132–137
  levels in biological fluids, 129–132, 133
  metabolism, 190–192
  metal-binding properties, 144–167
  molecular weights, 137, 139, 140
  nomenclature and definitions, 129–130
  nonspecific binding of iron, 147, 148
  removal of iron from, 149, 297–298
  secondary structure, 143–144
  separation of different iron forms, 155–157
  single-chain nature, 142-143
Trimethylamine dehydrogenase, 386, 388
Tryptophan hydroxylase, 400, 401, 403
2,2′,2″-Tripyridine, 18
Tripyridyl-*s*-triazine, 19, 63
Tyrosine hydroxylase, 400, 401, 403, 404

UIBC: *see* Unbound iron-binding capacity
Unbound iron-binding capacity
  definition, 13
  determination, 17–19
  of plasma and serum, 15–16
  in systemic infection, 332–333

Vanadium, 146, 154
Vinogradsky, Serge, 413
Vitamin D biosynthesis, 383

Wastage erythropoiesis, 1, 36, 38
Wheat ferredoxin, 355
Whole body retention, 59–62
Widdowson, E. M., 5

Xanthine oxidase, 384, 385

Yersinia pestis siderophores, 321

Zinc
  binding by transferrins, 145, 146
  delivery to lymphocytes, 415
  inhibition of iron uptake by ferritin, 239
Zinc (*cont.*)
  inhibition of proline hydroxylase, 409
  RNA and DNA biosynthesis, 416
  serotransferrin as carrier, 146
Zymobacterium oroticum enzymes, 385

Lightning Source UK Ltd.
Milton Keynes UK
UKOW05f1818210214
226931UK00009B/179/P